WITHDRAWN
FROM
UNIVERSITIES
AT
MEDWAY
LIBRARY

Microbial Life in Extreme Environments

Extremely halophilic bacteria growing in salterns in San Francisco Bay have turned the saturated salt water bright red. (See Chapter 8.) (Reproduced with permission from J. Lanyi.)

Microbial Life in Extreme Environments

Edited by
D. J. KUSHNER
University of Ottawa

1978

ACADEMIC PRESS

LONDON NEW YORK SAN FRANCISCO
A Subsidiary of Harcourt Brace Jovanovich, Publishers

ACADEMIC PRESS INC. (LONDON) LTD.
24–28 Oval Road
London NW1

US edition published by
ACADEMIC PRESS INC.
111 Fifth Avenue
New York, New York 10003

Library of Congress Catalog Card Number 77–85106
ISBN: 0–12–430250–5

Text set in 10/12 pt Monotype Times New Roman, printed by letterpress,
and bound in Great Britain at The Pitman Press, Bath

Contributors

R. E. AMELUNXEN *Department of Microbiology, University of Kansas Medical Centre, Kansas City, KS 66103, USA*
J. A. BAROSS *Department of Microbiology, Oregon State University, Corvallis, OR 97331, USA*
T. D. BROCK *Department of Bacteriology, University of Wisconsin, Madison, WI 53706, USA*
H. L. EHRLICH *Department of Biology, Rensselaer Polytechnic Institute, Troy, NY 12181, USA*
J. L. INGRAHAM *Department of Bacteriology, University of California, Davis, CA 95616, USA*
W. E. INNISS *Department of Biology, Faculty of Science, Waterloo, Ontario, Canada N2L 3GI*
A. P. JAMES *Division of Biological Sciences, National Research Council of Canada, Ottawa, Ontario, Canada K1A 0R6*
D. J. KUSHNER *Department of Biology, Faculty of Science and Engineering, University of Ottawa, Ottawa, Ontario, Canada K1N 6N5*
T. A. LANGWORTHY *Department of Microbiology, School of Medicine, University of South Dakota, Vermillion, SD 57069, USA*
R. E. MARQUIS *Department of Microbiology, University of Rochester School of Medicine and Dentistry, Rochester, NY 14642, USA*
P. MATSUMURA *Department of Microbiology, University of Rochester School of Medicine and Dentistry, Rochester, NY 14642, USA*
R. Y. MORITA *Department of Microbiology, Oregon State University, Corvallis, OR 97331, USA*
A. L. MURDOCK *Department of Microbiology, University of Kansas Medical Centre, Kansas City, KS 66103, USA*
A. NASIM *Division of Biological Sciences, National Research Council of Canada, Ottawa, Ontario, Canada K1A 0R6*
D. W. SMITH *Department of Biological Sciences, University of Delaware, Newark, DE 19711, USA*
M. R. TANSEY *Department of Biology, Indiana University, Bloomington, IN 47401, USA*

To my father and mother,
Sam and Lily Kushner

Preface

It has long been known that certain microorganisms can live under conditions of high and low temperatures, great pressure, high concentrations of solutes or normally toxic substances, acid or alkaline pH values and intense irradiation. These microorganisms were at one time studied as biological curiosities and also because some of them were important in food preservation and spoilage. In the past few years a wider and more fundamental interest in these microorganisms has developed. It has become apparent that they may provide fascinating objects for studying the deepest aspects of cellular and molecular biology, they reveal the scope of biochemical and physiological mechanisms that living things can use and they also help us to understand modes of life in the deep sea. Their study is essential for considerations of the origin of life and of the possibilities of life in outer space.

The chapters in this book, written by a number of specialists, deal with the ecological distribution of microorganisms that live in extreme environments, their physiology and modes of adaptation. The book is designed to be read by specialized workers, advanced students of microbiology and biochemistry and by others interested in the scope and diversity of the microbial world.

D. J. KUSHNER
FEBRUARY, 1978

Contents

Contributors v

Preface vii

1. Introduction: A Brief Overview 1
D. J. KUSHNER

2. Microbial Life at Low Temperatures: Ecological Aspects
J. A. BAROSS and R. Y. MORITA

I. Introduction 9
II. Definition of psychrophiles, psychrotrophs and psychrophilic environments 10
III. Atmospheric environment 13
IV. Caves 16
V. Freshwater environments 18
VI. Rivers and streams 27
VII. Soils 31
VIII. Marine environments 42
IX. Snow and ice 49
X. Discussion 51
XI. Concluding remarks 56
Acknowledgements 57
References 57

3. Microbial Life at Low Temperatures: Mechanisms and Molecular Aspects
W. E. INNISS and J. L. INGRAHAM

I. Kinetics of growth at low temperature 73
II. Determinants of low maximum temperature of growth of psychrophilic and psychrotrophic microorganisms 77
A. Protein synthesis 78
B. Cell structural modification 83
C. Enzyme inactivation 87
III. Biochemical factors affecting the minimum temperature of growth 90
A. Cold sensitive mutants as a model 90
B. Lipid composition and membrane permeability at low temperature 95
C. Ribosome function at low temperature 98
D. The minimum temperature of growth: conclusions . . . 98
References 99

4. Microbial Life Under Pressure
R. E. MARQUIS and P. MATSUMURA

I. Introduction 105
II. General mechanisms underlying biological pressure responses . 110
A. Effects of pressure on biological equilibria 110
B. Effects of pressure on rates of biochemical reactions . . 119
C. Conclusions 121
III. A few historical notes on microbial life under pressure . . . 122
IV. Microbial growth under pressure 126
V. Killing of microorganisms by pressure 135
VI. Pressure, pollution, petroleum microbiology and deep-sea mining . 138
VII. Biological effects of compressed gases 140
VIII. Epilogue 143
Addendum 144
Acknowledgements 147
References 147

5. Microbial Life at High Temperatures: Ecological Aspects
M. R. TANSEY and T. D. BROCK

I. Introduction 159
II. Prokaryotes 161
A. Photosynthetic prokaryotes 164
B. Chemolithotrophic bacteria 166
C. Methanogenic bacteria 167
D. Nonphotosynthetic bacteria 167
III. Eukaryotes 168
A. Fungi 168
B. Algae 186
C. Lichens 187
D. Protozoa 187
Addendum 189
Acknowledgements 193
References 194

6. Microbial Life at High Temperatures: Mechanisms and Molecular Aspects
R. E. AMELUNXEN and A. L. MURDOCK

I. Introduction 217
II. Classification and general characteristics 218
III. Proposed mechanisms of thermophily and related aspects: general considerations 220
A. Lipids and membrane involvements 220
B. Enzymes from cell free extracts, transferable stabilizing factors, and rapid resynthesis 223
C. Macromolecular thermostability 226
D. Highly charged macromolecular environment 235
E. Allosterism, thermoadaptation and transformation 238
IV. Detailed consideration of selected proteins from obligate and caldo-active thermophiles 243

A. Glyceraldehyde-3-phosphate dehydrogenase 243
B. Ferredoxin 252
C. Thermolysin 258
D. Formyltetrahydrofolate synthetase 262
V. Conclusions 264
Acknowledgements 267
References 267

7. Microbial Life in Extreme pH Values
T. A. LANGWORTHY

I. Introduction 279
II. Occurrence of life at extremes of pH 280
A. Life at low pH 280
B. Life at high pH 285
III. Nature of the extreme pH life forms 287
A. The external environment 287
B. Internal pH 289
C. Cell surfaces and membranes 294
IV. Conclusion 305
References 306

8. Life in High Salt and Solute Concentrations: Halophilic Bacteria
D. J. KUSHNER

I. Introduction 318
II. Natural zones of high salt and solute concentrations . . . 319
III. Biological effects of high solute concentrations 320
IV. Taxonomic distribution of solute-tolerant and halophilic micro-organisms 323
A. Isolation of more and less solute-requiring mutants . . 325
V. Effects of extracellular solutes on internal solute concentrations . 325
A. "Compatible solutes" 329
VI. Physiology of extreme halophiles 330
A. Nutrition and metabolism 330
B. Enzyme activity 331
C. *In vitro* protein synthesis 334
D. Mechanisms of the effects of salts on enzymes and other salt-dependent proteins of extremely halophilic bacteria . . 336
E. Peculiarities of the DNA and bacteriophages of extreme halophiles 340
F. External layers of extremely halophilic bacteria 340
VII. Special problems posed by moderately halophilic bacteria and salt-tolerant microorganisms 347
A. *In vitro* protein synthesis 351
VIII. Concluding remarks 354
A. Water and solutes 354
B. What next with the halophiles? 355
Addendum 356
Acknowledgements 357
References 357

9. Water Relations of Microorganisms in Nature
D. W. SMITH

I. Introduction 369
II. Fungi 370
III. Lichens 371
IV. Bacteria 373
V. Algae 374
VI. Mosses 376
VII. Mechanisms of damage by water stress 376
References 377

10. How Microbes Cope with Heavy Metals, Arsenic and Antimony in their Environment
H. L. EHRLICH

I. Introduction 381
II. The sources of heavy metals and of arsenic and antimony in the environment 382
III. Microbial responses to heavy metals, arsenic and antimony in the environment 385
IV. The bacteriology of acid mine-drainage 389
V. Microbial interaction with manganese and iron compounds 392
VI. Microbial interactions with mercury 395
VII. Conclusion 397
Addendum 398
References 398

11. Life Under Conditions of High Irradiation
A. NASIM and A. P. JAMES

I. Introduction 409
II. Radiation in the environment 410
A. Non-ionizing radiation 411
B. Ionizing radiation 411
III. Variation in radiation sensitivity 412
IV. Protection mechanisms 416
V. Nature of radiation damage 419
VI. DNA repair mechanisms 421
A. Dark repair systems 424
VII. *Micrococcus radiodurans* 426
VIII. Radiation resistant strains 429
IX. Concluding remarks 430
Acknowledgements 431
References 431

Subject index 441

Chapter 1

Introduction: A Brief Overview

D. J. KUSHNER

University of Ottawa

Those of us whose biological training and experience is mainly concerned with mammalian biochemistry, or the ways of such "standard" microorganisms as *Escherichia coli* or *Bacillus subtilis*, tend to think of life as a rather cosy affair, taking place at a pressure of one atmosphere, in well-aerated sera or growth media at approximately neutral pH and temperatures near that of our own bodies. Biologists more concerned with life in the outside world realize that "Nature" is much more hostile; that living creatures must contend with temperatures close to freezing (as do most of the forms of life in the oceans), or much colder. It has long been known that life can exist in such harsh environments as hot springs, some of which may also be very acid; salty natural lakes and salterns (ponds from which salt is prepared by evaporating sea water); very acid streams, such as mine effluents, which may also contain high concentrations of toxic heavy metals; in acid laboratory reagents with even higher concentrations of toxic metals; on dry rock surfaces; in deserts and in depths of the sea where the pressure may reach over 1000 atm.

Interest in such forms of life is not confined to microorganisms. Observations on animals, plants and microbes living in cold, hot, salty and deep places of the earth go back for well over one hundred years (some of these are cited in Heilbrunn, 1943; and many other reviews in this volume). When we consider that the brine shrimp, *Artemia salina*, can grow in saturated salt lakes, that invertebrates thrive in the deepest seas, that some fish do very well in sea water at −2°C, that insects and trees survive cooling to −40°C, we might ask, 'What is so special about the ability of microorganisms to come to terms with extreme environments?" In fact, microorganisms are not qualitatively different from other forms of life in this respect. However, they are able to withstand somewhat more stringent conditions. In the hottest parts of hot springs and the saltiest of lakes (the Dead Sea), only microorganisms are found. Only bacteria can survive the very highest temperatures (Brock, 1969). Then, too, with microorganisms, at least the outside of the whole creature is exposed to the environment. We do not have to consider the possibility that special tissues may have been evolved to deal with the external conditions, such as the salt glands of sea birds that permit them to drink sea-water, or the specialized channels in halophytic plants that grow in salty soil. It might be easier to understand the mechanism of adaptation of a microorganism than of a whole animal—though, in fact, a good deal is understood of ways in which several animals adapt to cold, salty or other unusual conditions (Hochachka and Somero, 1973).

A great deal of interest in microbial adaptation to extreme environments has been aroused (and supported) by the search for life on other planets. Even the most likely of these, Mars, has a harsh environment from the terrestrial point of view: low temperatures, which periodically rise above freezing, and very dry conditions. The only comparable earthly habitats are found in the dry valleys of the Antarctic, whose soil is sterile in places (Heinrich, 1976, and Chapter 2 by Baross and Morita). While this is being written, the search for life on Mars by the Viking Lander is going on, and a definite answer for the particular site studied may be obtained by the time this book appears.

Several symposia on microbial life in extreme environments have appeared in recent years (Heinen, 1974; Heinrich, 1976). The last is the report of the 1974 NASA conference on life in extreme environments, which has been held biannually since 1970. The 1976 volume is dedicated to Dr Wolf Vishniac, a pioneer student of such environments, who died in the Antarctic in 1973 while studying the dry valleys. Among the shorter general review articles written on the theme of this book are those by Brock (1969), Alexander (1976) and Kushner (1964, 1966, 1971). The last was part of a symposium dealing with the origins of life, a subject of obvious interest for the search for life on other plants. It is clear that knowing the limits of life on earth helps us to gauge the physical and chemical limits under which life may have arisen on earth and, for that matter, on other planets. The two studies go together and attract similar audiences.

The last few years have seen a gratifying growth of research on life in extreme environments. This is evidenced by the proliferation of reviews on specialized aspects of this subject, especially reviews on life in hot, cold and salty conditions. This book is an attempt to present a more integrated treatise on the subject, bringing together detailed essays on its different facets. For the most-studied subjects; life in hot, cold and salty conditions, separate chapters were obtained for the physiological and environmental aspects. There is necessarily some overlap between such chapters, as well as between chapters on different major topics. Microorganisms living in acid hot springs are considered in chapters on high temperature and pH extremes. Those that live in acid mine waters are also exposed to high concentrations of normally toxic metal ions. Effects of temperature and pressure are interrelated. Both Chapter 2 and Chapter 4 by Marquis and Matsumura discusses the ecology of microorganisms in the deep sea, take note of the fact that high pressures can have a much greater effect on microbial growth at low temperatures. Some of the coldest places on earth, the dry valleys of the Antarctic, are, as it happens, also some of the dryest.

Neither here nor in the succeeding chapters is an attempt made to define "extreme" precisely, except to point out that many of the conditions discussed would seem quite extreme to man. Rather, we are dealing with the ranges of environmental conditions that microorganisms can withstand and with the mechanisms by which they do so. These ranges are limited by the nature of matter and by the physical nature of the Earth. Liquid water is needed for all forms of life. Of all stresses to which living things can be exposed, they are probably most sensitive to drying. It is very doubtful that any organism can grow if its internal water activity (a_w) is less than 0·6, that is, less than 60% of the activity of pure water, and this limit is reached by only a few fungi. Most microorganisms need considerably higher values. The actual temperature range in which living organisms can grow is that in which liquid water can exist, approximately 273–373 K. (I speculated on conditions that might permit liquid water to exist at much lower temperatures, such as those prevailing on some of the moons of Jupiter (Kushner, 1976) but these were admittedly extremely fanciful.) In absolute terms, the most heat-resistant microorganisms can live at a temperature only about 37% greater than the lowest temperature of the most cold-resistant ones—which still does not hinder us from regarding boiling water as an extreme condition.

Much wider environmental variations are possible. Life is possible over concentrations of H^+ ions, varying by several orders of magnitude, and some individual microorganisms can grow over a range of 10 pH units or more. Many microorganisms can easily stand hundred-fold or greater variations in pressure. Indeed, the highest pressure possible in the depths of the sea is only moderately inhibitory to growth of many microorganisms. Heavy metal concentrations as low as 10^{-8}M can inhibit the growth of some microorganisms, while others may be resistant to concentrations a million-fold greater. Different microbial species show thousand-fold differences in susceptibility to irradiation.

With such a diverse subject, any order of presentation is somewhat arbitrary. To emphasize that extreme environments are not small, specialized niches, I have chosen to begin with the chapter by Baross and Morita, which deals with the microbial ecology of cold environments. These include soils, the atmosphere and the oceans, indeed the great bulk of our biosphere. Most of the seas' waters are near zero degrees. Many of the microorganisms in the atmosphere are also exposed to very low temperatures, though it is uncertain if any of them grow there. This subject is of special interest in considering the forms of life that might exist on the giant plants which have only gaseous portions in which conditions at all compatible with life exist (Ponnamperuma, 1976). Other cold environments are considered in detail, especially the Antarctic, which has been a region of increasing study in the last few years.

In Chapter 3, Inniss and Ingraham examine reasons why microorganisms are able to grow in cold conditions or require such conditions. The latter may be explained in terms of lesions in protein synthesis, in cell envelopes or in other vital parts and processes that occur at moderate temperatures. As for the former, there is no obvious reason that the growth of many mesophiles stops, instead of just slowing down, at low temperatures above zero. This has been investigated most precisely and elegantly with cold-sensitive mutants of *E. coli* and other mesophiles. This work points out that, below about 10°C, misfunction in regulatory enzymes or in ribosome formation may prevent growth. Modification of membrane lipids, and hence of membrane function, is also a very important aspect of temperature adaptation, though it is not always clear that a failure to make such modifications can set the lowest temperature at which growth is possible.

Many microorganisms that live in the sea are subjected not only to low temperatures but to pressures that, in the extreme deeps, can go to over 1100 atm. The average pressure is almost 400 atm. It has been known for some years, nevertheless, that many living organisms are found in sediment from all these depths. However, combinations of low temperature and high pressure are so highly inhibitory to microbial growth that the role of microorganisms found in the sea bottom in biodegradation is still mysterious. Recently, the intriguing possibility has arisen that microorganisms in the guts of deep sea amphipods, and other marine animals, are especially active in biodegradation. These matters, as well as the history of barobiology, and the effects of pressures—including those of several thousand atmospheres—are reviewed in Chapter 4. Pressure effects at the subcellular level have been studied in detail. The well-known ability of high pressures to dissociate hydrophobic bonds and thus disassemble cell structures poses the problem, still little explored at the cellular level, of the ways in which organisms that live in great pressures, or over a wide range of pressures, can adapt.

The fact that certain microorganisms can live at high temperatures has long caught the imagination of biologists and, for some time, this was the most studied extreme environment. Ecological relations were examined, especially in hot springs and other volcanic effluents. Work on these is reviewed briefly in Chapter 5 by Tansey and Brock, but the most detailed survey is made of the ecological distribution of the thermophilic fungi. The chapter shows that there is no lack of additional hot places in the world, including such diverse ones as the upper layers of the soil, compost heaps, hay, wood chip piles, coal dumps, slag heaps, alligator dung and cooling towers from nuclear reactors. Except for the first and the last examples, the high temperature in such sites, which may reach 60°C or more, is caused by microbial action. This is the first instance cited of an extreme condition caused by microorganisms themselves; others are cited in later chapters.

Mechanisms of resistance to high temperatures have been studied in more detail than of resistance to any other extreme condition. In Chapter 6 by Amelunxen and Murdoch, changes in metabolic rates and in the structure of membranes, ribosomes and individual proteins are considered as causes of temperature adaptation. Though all are important, it seems generally agreed that the key to high temperature adaptation lies in changes in protein structures. Many enzymes of thermophiles maintain both activity and regulation at the high temperatures at which these organisms grow. Several such enzymes have been highly purified, and they would seem to offer an excellent opportunity to correlate chemical properties with temperature adaptation. Despite this, and despite a great deal of work, the structural attributes that determine the ability of proteins to function at high temperatures are still uncertain. Thermophily in enzymes does not seem due to a major change in proportions of nonpolar amino acids or any other class of amino acids. The key seems to lie in more subtle structural differences, which may become evident only through three-dimensional analysis of these proteins. Studies of such proteins emphasize dramatically the great dependence of protein function on subtle changes in structure.

Many hot springs are also acidic, and a number of other acidic regions exist. In some of these, such as bogs and acid mine waters, the low pH is caused by microbial action. Though less studied, zones of high pH exist, and some very curious microorganisms may be found in them. One cause of high pH, urea decomposition to ammonia, is also due to microbial action. Chapter 7 by Langworthy points out that it has long been known that many microorganisms can grow over a range of pH in which their internal enzymes cannot function. The inference has been that they maintain their internal pH constant in face of a varying external pH. In recent years, this has been confirmed for a number of organisms, especially those living in acid conditions. Extracellular structures of such organisms must adapt directly to pH extremes, and a few have been shown to do so. The external layers of microorganisms that live in hot, acid conditions seem to have a distinctive chemical composition which may be needed for them to withstand such conditions. Some have rigid membranes with very low lipid content; in this, and in the properties of their lipids, they bear some relation to the membranes of extremely halophilic bacteria.

The latter organisms, together with a number of others that live in very salty or solute-rich conditions, are discussed in Chapter 8 by Kushner. The extreme halophiles have a special place in the ranks of microorganisms that live in extreme environments because they represent a complete adaptation, internally and externally, to very salty conditions and because of their unique biochemical properties. More recently, it has been realized that organisms which live in other high solute concentrations, or that can grow over a wide range of concentrations, pose quite fascinating problems in their own right.

Anything living in high solute concentrations is also subject to low water activity (a_w). Often, it is not possible to distinguish one effect from another. Studies of the effects of water availability on microorganisms growing in natural conditions are discussed in Chapter 9 by Smith. One can consider two kinds of stress, osmotic for microorganisms in solution and matric for those growing on surfaces such as soil particles. Microorganisms seem more sensitive to the latter stress. Though a certain water level is needed for growth, it is certainly not needed for survival. As Smith points out, many microorganisms can survive for long periods under dry conditions, and grow when water becomes available again.

The last two chapters in this book are partly concerned with extreme conditions that have come about through human activity. Heavy metal toxicity may be more of a problem to man than to microorganisms; the latter have many ways of adapting to such substances. Indeed, as Chapter 10 by Ehrlich points out, microorganisms are themselves responsible for many of the transformations that heavy metals undergo in the environment. These include the release of metals from ores in acid mine effluents, their changes in valence as in the transformation of mercury to more and less toxic forms, as well as in the production of such specialized metal accretions as the manganese nodules.

The earth and the first forms of life were exposed to high levels of irradiation before the production of oxygen led to the formation of a protective screen against the sun's short ultraviolet rays. Through our own efforts, we are again threatened by the possibility of exposure to very high levels of irradiation, especially ionizing radiations. As Chapter 11 by Nasim and James indicates, microorganisms exhibit wide variation in radiation sensitivity. Many can survive doses quite lethal to other forms of life. Though several factors contribute to this resistance, the ability to repair radiation-damaged DNA seems the most important one. The versatility of microorganisms in dealing with the challenges of radiation, combined with their high rates of growth and their ability to live in sheltered nooks and crannies may give them the distinction of being the last survivors or the first new colonists of a war-ravaged earth.

One condition which may reasonably be considered extreme, that is, life under very low nutrient supply (Hanson, 1976), has not been dealt with. Many microorganisms can grow on traces of nutrients such as one finds in natural clean waters, or even distilled water in the laboratory. Such organisms, which may make up a very important part of the biota, have been neglected in favour of those that need more food but grow more quickly. Indeed, many organisms that live in extreme environments have been unfairly neglected, partly because of the difficulty in studying them and obtaining publishable results. Admittedly, it is trying to study microorganisms whose growth media fills the laboratory with steam, on the centrifuge heads with salt, or which grow so slowly that weeks, instead of hours, may be required for experiments and whose genetics are unknown or almost impossible to study. Those who have persisted have found their rewards, both in the satisfaction and leisure for contemplation available to the student of an out-of-the-way field, and in the fascination afforded by the microorganisms themselves and the very clever ways they have found to adapt to such a wide range of environmental conditions. It is my hope that some of this fascination will be transmitted to the readers of this book.

References

Alexander, M., (1976). *In* "Extreme Environments: Mechanisms of Microbial Adaptation" (Ed. M. W. Heinrich). Academic Press, New York and London.

Brock, T. D. (1969). Microbial growth under extreme conditions. *Symp. Soc. Gen. Microbiol.* **19,** 15–42.

Hanson, R. S. (1976). Dormant and resistant stages of procaryotic cells. *In* "Chemical Evolution of the Giant Planets" (Ed. C. Ponamperuma), pp. 107–120. Academic Press, New York and London.

Heilbrunn, L. V. (1943). "Outline of General Physiology" (2nd revised ed.) W. B. Saunders, London and Philadelphia.

Heinen, W. (1974). Proceedings of the First European Workshop on Microbial Adaptation to Extreme Environments. *Biosystems* **6,** 57–80.

Heinrich, M. W. (Ed.) (1976). "Extreme Environments: Mechanisms of Microbial Adaptation." Academic Press, New York and London.

Hochachka, P. W. and Somero, G. N. (1973). "Strategies of Biochemical Adaptation." W. B. Saunders, London and Philadelphia.

Kushner, D. J. (1964). Microbial resistance to harsh and destructive environmental conditions. *Exp. Chemother.* **2,** 113–168.

Kushner, D. J. (1966). Microbial resistance to harsh and destructive environment conditions. *Exp. Chemother.* **4,** 512–514.

Kushner, D. J. (1971), Life in Extreme Environments. *In* "Chemical Evolution and the Origin of Life" (Eds R. Buvet and C. Ponnamperuma), pp. 485–491. North-Holland Publishing Co., Amsterdam.

Kushner, D. J. (1976). Microbial life in the cold. *In* "Chemical Evolution of the Giant Plants" (Ed. C. Ponamperuma). Academic Press, New York and London.

Ponamperuma, C. (Ed.) (1976). "Chemical Evolution of the Giant Planets." Academic Press, New York and London.

Chapter 2

Microbial Life at Low Temperatures: Ecological Aspects

J. A. BAROSS and R. Y. MORITA

Oregon State University

I. Introduction 9
II. Definition of psychrophiles, psychrotrophs and psychrophilic environments 10
III. Atmospheric environment 13
IV. Caves 16
V. Freshwater environments 18
VI. Rivers and streams 27
VII. Soils 31
VIII. Marine environments 42
IX. Snow and ice 49
X. Discussion 51
XI. Concluding remarks 56
Acknowledgements 57
References 57

The intellect is characterized by a natural inability
to comprehend life

Henri Bergson, *Creative Evolution*, 1911

I. Introduction

The temperature variations encountered in various aquatic and terrestrial environments are extreme enough to include at least three distinct thermal groups of organisms. At the very low and high extremes of this temperature

Published as Technical Report No. 4750, Oregon Agricultural Experiment Station.

range, microorganisms and primitive plants are the predominant life forms. Moreover, in some permanently cold soil, lake and air samples from polar regions, only bacteria can be found.

A vast proportion of the earth's surface is cold (less than 5°C). The world's oceans occupy 71% of the earth's surface of which approximately 90% is colder than 5°C. The polar regions, including the continent of Antarctica and the permanently cold areas surrounding the Arctic Circle represent approximately 14% of the earth's surface. The actual area which is colder than 5°C increases by several orders of magnitude if the volume of the oceans is taken into account. In spite of the fact that over 80% of the earth's biosphere is permanently cold, our knowledge of the microorganisms and their ecological significance in these regions is quite lacking.

Generally, cold environments can be divided into categories based on whether or not the characteristic low temperature is stable, such as deep oceans, or unstable, such as alpine lakes or most terrestrial environments in temperate regions. Even within environments which are permanently cold, the temperature characteristics can be quite markedly unstable, such as some polar air and soils which exhibit temperature variations from −88·3 to greater than +5°C (Weyant, 1966). These unstable polar regions can be contrasted with most temperate terrestrial environments which exhibit temperature fluctuations from less than 0 to greater than 25°C. In both instances, at the low temperature extreme, it is quite probable that there is little or no microbial growth or activity, and what is important, therefore, is the ability of microorganisms to survive at temperatures outside their growth range.

In this chapter, the incidence of bacteria capable of growth and/or metabolic activity in permanently or seasonally cold (less than 5°C) environments will be discussed without particular regard for the organisms maximum or optimum growth temperature. All marine and terrestrial environments, including lakes and streams, soils, arid caves and ice, will be discussed. The bacteriology associated with man-made cold environments, generally including frozen and iced foods, will not be discussed in this review except where food items might have originated from low-temperature environments, such as various seafoods.

II. Definition of Psychrophiles, Psychrotrophs and Psychrophilic Environments

Ever since Forster (1887) first isolated bacteria capable of growing at 0°C, similar low temperature bacteria have been isolated from various natural and man-made environments (Ingram, 1965; Inniss, 1975; Morita, 1975). Various descriptive terms have been proposed to describe this seemingly distinct physiological group of microorganisms (Morita, 1975). The word "psychrophile" with and without various restrictive adjectives is perhaps the most common term employed. However, very few words in microbiology have been subjected to as many different definitions as "psychrophile". Not even the characteristic ability to grow at 0°C has been a universal criterion for psychrophiles, and frequently the definition might stress the optimum or maximum growth temperatures, or, in some cases, an arbitrary growth rate at 0°C. In defining a word, what is important is what is right, which unfortunately sometimes implies whoever's definition is currently accepted. It is essential, however, to establish a workable definition that is free of the many ambiguities regarding temporal factors, degree of growth, etc. The definition recently proposed by Morita (1975) restricts the temperature growth range of psychrophiles from 0°C, or less, to 20°C, or less, with an optimum growth temperature of 15°C, or less. This definition is based on the repeated isolations of microorganisms from various environments which fulfil these criteria. Moreover, some of the recent psychrophiles isolated from Antarctic waters have a maximum growth temperature of 10°C, or less (Christian and Wiebe, 1974; Morita, 1975). Thus it is apparent that there exists a thermal group of microorganisms that is clearly as unique physiologically as are thermophiles, for example.

The real problem, therefore, is not the definition of psychrophiles. The more difficult issue to resolve, if in fact there is a need for resolution, is to account for those organisms that grow at 0°C, but also at temperatures above 20°C, as well as other organisms that grow below the minimum temperature of mesophiles such as 5 to 10°C. There is no doubt that there is a continuum of upper and lower growth temperatures within the microbial world, and the categorization of these microorganisms according to their temperature limits for growth would be at best quite tenuous. It is not unusual, for example, to isolate bacteria from the marine environment which gave growth temperature ranges from less than 0°C to 35°C, or higher. Furthermore, many soil organisms considered to be mesophilic, such as *Bacillus megatherium* and *B. subtilis*, and some *Arthrobacter* and *Corynebacterium* can grow at temperatures below 5°C, and *Yersinia pestis* has been reported to grow at −2°C as well as at 40°C (Buchanan and Gibbons, 1974). It is possible that many more microorganisms than reported could grow at temperatures below 5°C and, perhaps, below 0°C if their temperature range of growth were tested.

The two most common terms used to describe low temperature growing microorganisms not fitting Morita's definition are facultative psychrophile* and psychrotroph (Hucker, 1954; Eddy, 1960). The use of the term facultative would imply that these organisms could be (occasionally) capable of

* This term is used by Inniss and Ingraham, Chapter 3. They call "psychrophiles", as defined in the present chapter, "obligate psychrophiles". (Editor)

psychrophilic growth according to the definition of Morita (1975). The term psychrotroph, as defined by Eddy (1960), refers to any organism capable of growing at 5°C, or below, regardless of its upper or optimum growth temperatures. This latter definition is quite practical since most low temperature growing microorganisms, particularly those originating from seasonally variable environments and chilled and frozen foods, fit this definition. In this chapter only microorganisms fitting the definition of Morita will be called psychrophilic, and all other low temperature growing microorganisms will be referred to as psychrotrophic.

Most terrestrial, aquatic and atmospheric environments in temperate regions are seasonally subjected to freezing and subfreezing temperatures as well as temperatures definitely within the mesophilic range. This is particularly true for soils, some surface waters of lakes and oceans, and shallow rivers and streams. These environments could best be described as psychrotrophic and it is assumed that if any microbial activity occurs at temperatures below 5°C *in situ*, it is likely to result from psychrotrophs. On the other hand, ocean and lake sediments and waters, below the thermocline, are generally regarded as thermostable environments having temperatures rarely exceeding 5°C. These environments are psychrophilic, and indeed most of the authentic psychrophiles which have been described have originated from seasonally stable cold environments. It is an oversimplification, however, to say that only psychrophiles occur in stable cold environments, because most of the microorganisms isolated from permanently cold areas are psychrotrophic. No doubt, psychrophilic microorganisms, particularly those which have a very narrow temperature range for growth, evolved in stable cold environments since all known psychrophiles tested, with the exception of some sporeformers, are quite sensitive to temperatures above their maximum for growth. On the other hand, most microorganisms show moderate to good survival at temperatures below their minimum growth temperature (Baross *et al.*, 1975). The adherence to a classification of environments as psychrophilic or psychrotrophic based on whether or not they are permanently or seasonally cold is due to convenience and not to a naivety of the many problems inherent in such an oversimplification.

The final question concerns whether or not microorganisms capable of growing at low temperatures on rich culture media are also capable of growth at similar temperatures *in situ*. This question is important since some psychrotrophs isolated from temperate regions, which might have a minimum growth temperature of 2 to 5°C in the laboratory, could thus be isolated and counted as psychrotrophic or in some cases as a psychrophiles [*sic*]. It has been reported, for example, that although *Vibrio parahaemolyticus* has a minimum growth temperature between 7 and 10°C, this organism rarely occurs *in situ* at temperatures below 15°C (Baross, 1972). At the present time sufficient information is not available to ascertain if most of the reports on the incidence of low temperature growing microorganisms from environmental samples represent those that are active *in situ*. In this chapter, again for convenience sake, it will be assumed that microorganisms isolated at low temperatures will be active at similar temperatures *in situ* and, furthermore, that they are representative of the active microbial flora of these environments.

III. Atmospheric Environment

Bacteria, fungi, pollen and other microscopic particles have been collected from air samples near the earth's surface up into the stratosphere at altitudes exceeding 27,000 m (Burch, 1967). In general, a relatively high incidence of viable bacteria have been reported within the troposphere which extends from the ground to a height of approximately 10,000 m (Gregory, 1961). A general feature of the troposphere is the continuous decrease in temperature with increasing altitude to the extent that in some air masses, the temperature is lower than −40°C (Proctor, 1935). In most of the lower atmosphere at altitudes greater than 1000 m, the temperature is lower than 10°C and definitely within the temperature range for survival and possibly growth and activity of psychrophiles. Surprisingly, the highest incidence of bacteria observed by Fulton (1966a, b) and Gregory (1961) was at altitudes exceeding 3000 m where the temperature was consistently lower than 5°C. At the higher altitudes, particularly over polar regions where the temperature is consistently below freezing, viable microorganisms have been isolated apparently capable of surviving long periods of exposure to temperatures below −20°C.

It is obvious that the air masses harbour high number of microorganisms which can exceed 500 m^{-3} air space volume (Fulton, 1966a). Moreover, Fischer *et al.* (1969) observed over 100 particles per cm^3 air within the size range of 0·2 to 2 μm, some of which were presumably bacteria. Parker and Wachtel (1971) observed that rain water contained relatively high levels of the vitamins cobalamin, biotin and niacin, again presumed to be bacterial in origin. These relatively high concentrations of viable bacteria and associated organic material quite definitely originate from marine and terrestrial environments. Air masses which pass over land and ocean surfaces become contaminated with terrestrial and marine microorganisms. Generally, large particles (greater than 100 μm) "fall out" while smaller particles can be carried to higher altitudes (Gregory, 1961). Judging from the high incidence of bacteria in the upper atmosphere, this process must be quite efficient.

There are many reports of microorganisms in surface air masses (Gregory, 1961, 1973); however, there has been very little work done on the incidence of psychrophiles in the atmosphere. Early biological investigators of polar regions frequently exposed nutrient agar plates to the frigid air. Ekelöf

(1908), for example, isolated bacteria from over 50% of the plates exposed to Antarctic air, and it was estimated that there was a settling rate of one bacterium per 2 h. Some of these organisms which Ekelöf presumed to originate from Antarctic soils were recognized as psychrotrophic. McLean (1918, 1919) microscopically observed the presence of an active microflora associated with falling snow and glacier ice. Based on these observations, McLean proposed that the autochthonous microflora capable of dividing upon thawing of ice, originates from the air-borne microorganisms which are frozen to snow flakes. Furthermore, there is evidence that perhaps bacteria are important in the production of ice-nucleated particles, particularly those associated with decaying tree leaves (Schnell and Vali, 1972, 1973). Specifically, *Pseudomonas syringae*, isolated from alder leaves, plays a significant role in the production of ice nuclei particles in temperate regions (Maki *et al.*, 1974). The ice-nucleation activity of this species occurs at the relatively warm temperature of −2°C and apparently is associated with intact cells. Most of the other early investigators who sampled air for microorganisms used incubation temperatures exceeding 20°C. These investigations are discussed by Sieburth (1965).

In some recent investigations, Lacy *et al.* (1970) could only isolate a single mesophilic bacillus from air sampled near the Antarctic interior. The organism was presumed to be a contaminant. Cameron *et al.* (1972a) isolated several organisms from Antarctic air sampled near a saline pond. Only *Corynebacterium sepedonicum* and *Micrococcus* were observed to have optimum growth temperatures near 25°C, whereas all of the other isolates tested grew best at 37 to 45°C. Cameron *et al.* (1972b) and Cameron *et al.* (1973) reported the presence of 0 to 160 bacteria m^{-3} air in Antarctic dry valleys. Most of the bacterial isolates were *Arthrobacter* and *Brevibacterium* and were typical of the organisms most frequently isolated from Antarctic dry valley soils. No *Bacillus* were isolated. The temperature ranges for growth of these isolates were not determined.

One of the primary sources of microorganisms in the atmosphere is surface oceanic waters. ZoBell and Matthews (1936) isolated seawater-requiring bacteria 30 miles from the coast. It was observed that the ratio of seawater- to freshwater-requiring bacteria decreased with increasing distance from shore. ZoBell (1942) considered that marine bacteria became air-borne in droplets of water, particularly from rugged coastal regions. Other investigators have substantiated the work of ZoBell and have further observed that bacteria readily become associated with bubbles which are then ejected into the atmosphere (Blanchard and Syzkek, 1970). There is also some evidence that these bubbles can concentrate organic matter (Carlucci and Williams, 1965; Barber, 1966) thus increasing the possibility that marine bacteria, including psychrophiles, could be metabolically active in the atmosphere, utilizing the organic material concentrated by the bubbles. It is quite possible that relatively high concentrations of marine bacteria could be ejected into the atmosphere since Sieburth (1971) and others have observed bacterial populations in excess of 10^5 ml^{-1} in the marine neuston layer.

Obviously, the air bacteria originating from marine waters would not necessarily be seawater-requiring or rigidly psychrophilic according to the definition of Morita (1975). However, considering that most of the atmosphere is cold, psychrophiles would definitely be the predominant organism capable of growth and activity if, in fact, these processes occur in the atmosphere.

There has been considerable interest in the presence of microorganisms and other terrestrial-originating particles in the upper atmosphere, ever since it became possible for man to fly. Apparently, Pasteur proposed to use a hot air balloon to sample the upper atmosphere (Proctor and Parker, 1942). The early efforts to show the presence of microorganisms in the upper atmosphere were always successful, whether they were observed directly on air exposed petrolatum-coated slides (Meier and Lindbergh, 1935) or by direct plate counts (Proctor, 1935). Proctor (1935) sampled the upper atmosphere (greater than 7000 m) over both the polar Pacific and Atlantic Oceans by aeroplane. The air temperatures at these altitudes were generally below −20°C and occasionally as low as −40°C. The predominant bacterial types isolated from 20°C plate counts were typical soil organisms. Proctor (1935b) and Proctor and Parker (1942) reported that 36 of 105 pure cultures of bacteria isolated from 3000 m grew on nutrient agar plates at 0°C. Of the organisms which grew at 0°C, 31 of the 36 could survive 48 h exposure to temperatures ranging from −26 to −39°C. These authors considered most of these organisms to be *Bacillus*. The work of Proctor is the only report of upper atmosphere bacteria capable of growing at psychrotrophic temperatures. Burch (1973) reported the presence of 2 to 3 bacteria/28·3 m^3 at 9144 to 18,288 m and 5 bacteria/28·3 m^3 at 18,388 and 27,432 m. None of these bacteria were characterized as to their cardinal temperatures, and it is presumed that most of the bacteria isolated from upper atmosphere air represent terrestrial forms capable of surviving below-freezing temperatures. The survival at temperatures below the minimum for growth does not appear to be an esoteric property of few organisms since many other terrestrial bacteria, including *E. coli*, *Pasteurella tularensis*, *Flavobacterium* and *Bacillus subtilis* spores were shown to survive equally well at temperatures between −40 and +24°C, whereas at temperatures above 24 to 49°C, there was an accelerated death rate for all of the isolates (Ehrlich, 1974).

Quite obviously, the question remains as to the presence of metabolically active microorganisms in the atmosphere. All of the prerequisites necessary for microbial growth and activity occur in the atmosphere, including a population of viable bacteria, sufficient moisture, and ample quantities of both organic and inorganic nutrients. The low temperatures observed in the troposphere would be favourable for the growth of psychrophiles, particularly bacteria originating from the marine environment (ZoBell and Conn, 1940). Gregory (1961) also indicated that there could be a "biological zone" in the atmosphere where microorganisms could grow in organic enriched droplets and possibly fix nitrogen. Parker and Wachtel (1971) indicated that microorganisms could grow in clouds, and the resulting organic metabolites could in part account for the high concentrations of organics (8 mg litre^{-1}) found in rain water. When these organisms are exposed to freezing, which could occur in high altitude snow clouds, soluble organics including vitamins could be "squeezed out" (perhaps by reverse osmosis) from cells. These organics are then washed down into surface waters and soils in rain. The role of these rain-associated organics on the ecology of terrestrial environments or the origin of these substances is not precisely known; however Collins (1960) and others have indicated that the nutrient levels in rain would definitely affect the bacterial population and thus the overall productivity of lakes.

Unfortunately, there are no reports on the incidence of seawater-requiring psychrophilic bacteria from air samples, or whether marine bacteria are active when associated with air-borne oceanic bubbles.

It can be concluded with some certainty that if there is a "biological zone" in the atmosphere, the organisms which are actively metabolizing and growing must be psychrophilic or psychrotrophic and would probably have originated from the marine environment.

IV. Caves

There are many subterranean and glaciated caves, both in temperate and arctic regions, that have permanently cold temperatures which range from about 10°C to below freezing. Some of the general features of these caves, in addition to low temperature, are absence of light, low levels of organic material, and relatively high moisture. Furthermore, in Karst caves, very high levels of calcareous substances are present. The stability of these properties of caves, particularly low temperature and the absence of light, has had a selective effect on the permanent flora and fauna, such as the absence of functional eyes in fish and bats and the general lack of pigment in most indigenous organisms. The presence of permanently adapted indigenous cave microflora, however, has not been unequivocally demonstrated.

There are only few reports on the microflora of caves, and particularly, on the incidence of psychrotrophs in permanently cold caves. Gounot

(1968b, 1969, 1973a) investigated the incidence of different thermal groups of bacteria in permanently cold caves in the Arctic, Lapland, the Pyrenees, the Alps, and in Romania. The temperatures in these caves ranged from −0·8 to 5°C and were generally quite stable throughout the year. Invariably, the plate counts at 20°C were higher than the 2°C and 28°C counts and ranged from 1–11·3 × 10^6 bacteria g^{-1} soil (dry wt) at 20°C to 0·02–4·3 × 10^6 bacteria g^{-1} at 2°C. Most of the organisms isolated were of the genera *Arthrobacter*, *Pseudomonas* and *Flavobacterium*, and in one cave all of the 140 isolates obtained from plates incubated at 2, 20 and 28°C were *Arthrobacter* (Gounot, 1967; Moiroud and Gounot, 1969). None of the arthrobacters isolated at 2°C could grow at 28°C. Many of the psychrophilic species isolated from cave soils resembled *A. glaciales* which was isolated from frozen soils by Moiroud and Gounot (1969). This psychrophile grew between −5 and 18°C with an optimum of approximately 13°C (Gounot, 1973b). Respiration, however, increased with increasing temperature exceeding the maximum temperature for growth.

Gounot (1973a) considered the psychrotrophic *Arthrobacter* isolated from caves to be part of the permanent indigenous flora. She reasoned that the ancient permanently dark caves definitely selected out for non-pigmented bacteria since less than 1% of 515 cultures examined produced pigment. In contrast, approximately 20% of the *Arthrobacter* from soil and an even greater percentage of the total microflora of glaciers formed pigment (Flint and Stout, 1960). Moreover, in contrast to soil arthrobacter types, Gounot (1973a) reported that none of the cave psychrotrophs could produce pigment when cultured in continuous light.

Psychrophilic bacteria were not detected in cold spring-water and sediments from caves in the Karst region of southern Indiana (Brock *et al.*, 1973). It was anticipated *a priori* that a permanent microbial population could have developed which would have optimum temperatures, for growth and activity coincident with the *in situ* cave temperatures which were constant at 10 to 12°C (temperature range in cave spring-water was 5·5 to 15°C). This was not the case, and only a psychrotrophic microbial flora, displaying optimum growth at 25 to 30°C, was isolated. The uptake of ^{14}C-acetate by the cave bacteria, which had developed on immersed microscopic coverslips, was also optimal at temperatures between 25 and 30°C. These authors speculated that the psychrotrophs in these cave springs probably originated from outside sources such as other terrestrial and aquatic environments. It was further pointed out that the caves were formed during the retreat of the Wisconsin ice from Indiana (20,000 years ago) and that, perhaps, this time period would be too short for psychrophilic bacteria to evolve from other thermal groups, particularly when the generation time of non-psychrophilic bacteria may be in the order of days at these low temperatures. However, 10 to 12°C (cave temperature) is 5 to 10°C warmer than the permanently cold environments where true psychrophiles are usually isolated, and it is generally expected that most organisms will grow at temperatures at last 10°C higher than *in situ* temperatures, particularly in environments having stable temperatures (Braarud, 1961; ZoBell, 1962). Therefore, true psychrophiles would not be expected to exist in environments warmer than 5°C.

Bacteria definitely exist in frozen caves; however, their rates of activity and their specific *in situ* physiological processes, including source of energy, substrate transformations, etc., are not known. It has been suggested that perhaps caves would be a suitable environment for chemoautotrophs (Caumartin, 1963; Gounot, 1973a), however, there are no reports indicating the presence of organisms capable of autotrophic growth. It is possible, for example, that silicate rock-decomposing bacteria capable of fixing molecular nitrogen, such as the *Arthrobacter* described by Smyk and Ettlinger (1963), could exist in caves. Moreover, Sweeting (1973) suggested that the formation of "rock or moon milk" found on the floors, walls and fissures of frozen and nonfrozen Karst caves, may be the result of microbial decomposition of limestone, resulting in the formation of microscopic calcite granules. Whatever physiological groups of microorganisms exist in frozen caves, they must either be capable of surviving on very low concentrations of organic nutrients or have quite esoteric enzymes and/or metabolites in order to carry out autotrophic processes in the presence of low levels of some inorganic nutrients, which are frequently present only in the solid state.

V. Freshwater Environments

Fresh water environments include lakes, ponds, rivers, streams and creeks. It is, however, quite misleading to call all such environments freshwater since this is often not the case. There are many landlocked saline lakes and ponds as well as areas of brackish to definitely saline water associated with rivers and streams which flow into marine and estuarine environments. For practical purposes, any body of water not directly derived from, or associated with, marine environments will be discussed in this section and will include permanently or seasonally cold saline ponds and lakes. The emphasis, moreover, will be on lakes since there is very little data on the incidence, activity and taxonomy of psychrophilic and psychrotrophic bacteria in flowing water environments.

Solar radiation is the primary source of heat in lakes, and depending on regional climatic variations, the lake water temperatures will vary accordingly. Lakes in temperate regions will generally have a definite thermocline during the summer with highest temperatures at the surface. The depth of the thermocline, or epilimnon layer, is determined by atmospheric conditions and the wind distribution of heat. Below the thermocline is the hypolimnon which rarely gets warmer than 4°C at any time of the year. The thermocline acts as the barrier between the epilimnon and hypolimnon. Frequently, in early spring, the lake water temperature is uniform throughout depth at approximately 4°C.

Lakes located in tropical regions, in contrast to temperate lakes, frequently will show no differences between surface and bottom water temperatures. The lake water temperatures are thus influenced by the high atmospheric temperatures which are relatively constant throughout the year. Polar and alpine lakes are characterized as having temperatures generally below 4°C throughout the year and having either a seasonal or a permanent ice cover. Highest temperatures and maximum photosynthetic activity occurs just below the ice when the summer water temperatures can approach 4°C.

Obviously, the heterotrophic microbial activities in polar and alpine lakes is carried out by psychrophilic bacteria. In temperate lakes, in the water column below the thermocline or in the sediments, the microbial mineralization processes occur at temperatures approximating 4°C. Microbial activity in the epilimnon during the summer would likely involve psychrotrophs or mesophiles since the surface water temperatures frequently get higher than 25°C.

It is well established that the primary roles of bacteria in freshwater environments are the degradation of particulate organics including planktonic organisms, the recycling of nutrients, the synthesis of vitamins and other growth factors, and as a food source for other trophic level organisms. In general, the highest population of bacteria and the highest heterotrophic activities occur soon after algal peaks. The incidence of bacteria in lakes is not necessarily directly due to algal blooms since other factors, such as temperature, may be more important, particularly in lakes located in polar regions. Jones (1971), for example, showed that temperature, pH and dissolved oxygen were the most important parameters affecting bacterial populations in a high nutrient lake; whereas, in a nutrient poor lake, the level of particulate matter, pH and rainfall were the most important factors. In each case, the bacterial activity continued after the algae bloom into the fall months with much of the degradation of particulate matter occurring either in the hypolimnion regions or sediments at temperatures below 5°C.

The principal generic types of bacteria in lakes are similar to their salt-requiring counterparts in marine waters. Gram negative motile and non-motile rods of the genera *Pseudomonas*, *Vibrio*, *Flavobacterium*, *Acinetobacter*, *Moraxella*, and various myxobacteria are most frequently isolated, although Gram positive typical soil organism can also generally be isolated (Klein, 1962; Jones, 1971; Tilzer, 1972). Christensen (1974) also reported that the bacteria from Arctic lake sediments were psychrotrophic or mesophilic;

65% of the isolates grew better at 30°C than at 5°C, and only 1% showed no growth at 30°C. Most of the bacteria from these lake sediments were Gram negative rods, and 40 to 50% were *Cytophaga*. Many of these organisms are well adapted for existence in lakes having low levels of nutrients, temperatures approaching freezing, and quite high salinities. Rao and Dutka (1974), for example, found that a *Flavobacterium*, which was the most abundant organism isolated from Lake Ontario and Lake Superior, showed greater uptake of O_2 at 0°C than at 20°C (isolation temperature). Moreover, Jones (1971) showed a seasonal fluctuation in separate populations of protease-, amylase- and lipase-producing bacteria in lakes. These population responded primarily to the availability of specific substrates rather than to temperature.

The incidence and activity of microorganisms in lakes, rivers and streams have received rather intense interest in the past few years primarily as a result of the development of methods for directly measuring the numbers of bacteria (Zimmerman and Meyer-Reil, 1974; Daley and Hobbie, 1975) and microbial heterotrophic activity (Wright and Hobbie, 1966; Hobbie and Crawford, 1969). Prior to the acceptance of these procedures, plate counting techniques were used exclusively for enumeration of bacterial biomass; however, these were performed without any attempt to standardize the media and the procedures or to use *in situ* temperatures for incubation, pH, salinity, etc. Consequently, plate counts were just "count data" which were surrounded by a familiar aura of functional mystery. In general, there was no correlation between the numbers of bacteria from plate counts with the *in situ* microbial activity although the more productive lakes also have the highest populations of bacteria and the highest rates of microbial activity. Moreover, Morgan and Kalff (1972) showed quite good correlation between the numbers of bacteria determined by direct counts and the maximum velocity of uptake of glucose (V_{max}) in Char Lake. Obviously, many of the bacteria responsible for heterotrophic activity in lakes and other bodies of water are not enumerated using plate counting techniques. These organisms, particularly in the sediments and in water below the thermocline, are capable of activity at temperatures below 4°C.

Lakes are generally classified into various trophic types depending on the level of primary productivity. Oligotrophic lakes have between 30 and 100 mg C $m^{-2}day^{-1}$ photosynthetic activity and mesotrophic lakes between 300 and 1000 mg C $m^{-2}day^{-1}$, whereas in eutrophic lakes, there can be planktonic productivity from 1500 to 3000 mg C $m^{-2}day^{-1}$ (latter value for polluted lakes) (Rodhe, 1969). The maximum velocity of glucose uptake by heterotrophic bacteria is also a sensitive index in determining the trophic characteristics of lakes. Generally, the V_{max} in oligotrophic lakes is 10^{-2} to greater than 10^{-3} μg glucose $litre^{-1}$ h^{-1}; whereas in eutrophic lakes, the

V_{max} is frequently from 10 to greater than 100 μg glucose litre^{-1} h^{-1} (Morgan and Kalff, 1972; Burnison and Morita, 1974). Based on this classification, most Arctic and Antarctic lakes, particularly in dry regions, are oligotrophic (Kalff, 1970; Goldman, 1970). Most Arctic lakes are considered to be "ultra oligotrophic" since they are less productive than most oligotrophic lakes in the north temperate zone (Kalff, 1970). The low productivity in polar lakes is presumably the result of nutrient deficiency and is not due to the consistently low water temperature or the short phytoplankton growing season characteristic of these environments. This has been substantiated by several investigators who have shown that the highest photosynthetic activity and the shortest turnover time for planktonic organisms have been at water temperatures approximating 0°C in some lakes (Hobbie, 1964; Kalff, 1970). Obviously, temperature would be the most important factor in some seasonally frozen Antarctic lakes where the depth of the ice might exceed the depth for light penetration.

Table 1 shows some of the published information on the numbers of bacteria and their activity (heterotrophic potential) in various Arctic, alpine and seasonally ice-covered lakes. In all cases, the V_{max} varied from very active indicative of eutrophic lakes such as Meretta, Lotsjon, Marion and Upper Klamath Lakes, to the definitely oligotrophic Char and Lapland Lakes. Again, it must be stressed that the role of temperature in differentiating between lakes showing markedly different heterotrophic activity is frequently not as important as other physical or chemical factors (Jones, 1971). However, there are some instances where temperature effects are quite pronounced, particularly in shallow, eutrophic lakes in temperate regions. Allen (1969), for example, reported a seasonal variation in the bacterial uptake of glucose in a eutrophic pond to be 0·1 to 15 μg glucose h^{-1}. Thompson and Hamilton (1973), moreover, showed that the turnover time for sucrose in a lake enriched with 5·54 g carbon (C) m^{-2} year^{-1} of sucrose, was 2 to 15 h from July to October, but 55 h from October to January when the temperature was below 2°C. In some of the eutrophic lakes, it is possible that only one thermal class of bacteria are present such as reported by Boylen and Brock (1973) in Lake Wingra.

One of the obvious problems with stressing heterotrophic activity in the water column as an index of lake microbial activity (particularly turnover rates) is that, more than likely, much of the secondary microbial productivity and mineralization occurs in the sediments. Harrison *et al.* (1971) for example, reported 10^7 to 10^8 bacteria g^{-1} dry wt sediment, and that glucose was mineralized 24 times faster in the top square centimetre of sediment than in a square centimetre of the overlying water. Moreover, the turnover time of glucose in the sediment was 100 times shorter than in the overlying water. It seems likely that these results would be typical for other lakes.

Table 1

Incidence and activity of bacteria in some arctic, alpine and seasonally ice covered lakes

Lake	Number of bacteria	V_{max} μglitre^{-1} h^{-1}	Turnover time (h)	Phytoplankton productivity (mg C m^{-2} day^{-1})	References
Char, arctic	0·1–2 × 10^{8} litre^{-1}	1–8 × 10^{-3} (glucose)	43–1700	25	Morgan and Kalff (1972)
Meretta, arctic	2–80 × 10^{8} litre^{-1}	0·1–7·5 × 10^{-1} (glucose)	5–175	125	Morgan and Kalff (1972)
Vorderer Finstertaler. See (alpine) (8 months ice and snow cover)	2–8 × 10^{8} litre^{-1}	ratio of bacteria to phytoplankton winter: 3:1 summer: 0·03:1		1·8	Tilzer (1972)
Lake Lotsjon, Sweden	winter temperature 0·4–0·6	(acetate) winter: 3·3–8·7 × 10^{-1} summer: 80 × 10	4–22 0·5		Allen (1968)
Lapland lakes (6), ice- and snow-covered most of year		1–2 × 10^{-3} (glucose)	—	50	Rodhe *et al.* (1966), Hobbie and Wright (1968)
Erken (ice-covered)	estimate to be 50 × 10^{6} litre^{-1} during winter (less than 4°C)	(water beneath the ice) winter samples 2·3–4·5 × 10^{-2} (glucose) 1·9–5·4 × 10^{-2} (acetate)	60–100 200–430	500	Wright and Hobbie (1965, 1966), Hobbie and Wright (1968), Rodhe *et al.* (1966)
Wingra, Wisconsin (winter ice cover)	bacterial no./g sediment 4°C counts: 0·7–44 × 10^{5} 20°C counts: 0·7–70 × 10^{5}		optimum uptake of ^{14}C-glucose by winter population is 25°C		Boylen and Brock (1973)
Marion Lake, Canada	bacteria cm^{-3} sediment	V_{max} μgC g^{-1} h^{-1} sediment: glycine / acetate / glucose		31	Hall *et al.* (1972), Hall and Hyatt (1974)
winter (less than 3°C)	5 × 10^{4}	0·1–0·5 / 3–5 / 1–3			
summer (greater than 10°C)	2 × 10^{6}	2–3 / 40–50 / 10–20			

West Blue Lake, Manitoba (ice surface during winter)	organisms ml^{-1} on different substrates[a]				18–67 (summer only)	Robinson *et al.* (1973), Ward and Robinson (1974)
	a. succinic:	24·6	$2·8 \times 10^{-2}$	33		
	b. pyruvic:	21·7	$7·7 \times 10^{-2}$	300		
	c. fumaric:	13·1	$4·8 \times 10^{-2}$	50		
	d. malic:	12·0	$3·6 \times 10^{-2}$	8		
	e. lactic:	6·0	$1·2 \times 10^{-1}$	303		
	f. acetic:	5·8	$3·1 \times 10^{-2}$	416		
	g. citric:	2·4	$8·0 \times 10^{-3}$	100		
	h. glycollic:	2·1	$9·0 \times 10^{-3}$	950		
	l. complete medium	92·4				
Upper Klamath Lake (naturally eutrophic; ice surface during winter)	February samples only, 5°C					
	a. glutamate	4–6		10–15	no data	Burnison and Morita (1974)
	b. aspartate	4–5		5–15		
	c. asparagine	5		20–40		
	d. lysine	2		10–15		
	e. proline	>5		25–135		
	f. alanine	2–3		20–60		
	Bacterial counts ml^{-1} at 2°C					
Dolomite Lake	65					Boyd and Boyd (1967)
Shell Lake	290					
Gravel Pit Lake	85					
Hospital Lake	110					
Duck Lake	83					
Twin Lake	93					
Boot Lake	60					
Hidden Lake	240					

[a] Average of 4 different sample times; complete medium is *Cytophaga* medium of Anderson and Ordall (1961)

There is very little liquid water in the Antarctic continent. Some lakes located near the sea can undergo either partial or complete thawing during the short Antarctic summer. In contrast, most inland Antarctic lakes, particularly in dry valleys, are permanently frozen to the bottom. Some of the frozen lakes retain a bottom layer of water which is generally quite brackish, such as Lakes Vanda and Bonney (Goldman *et al.*, 1972). Microbiological and photosynthetic activity has been observed to occur just below the ice layer in these permanently frozen lakes (Goldman, 1970). Some of the unique features of permanently frozen lakes are the absence of fish and zooplankton and the lack of circulation (no wind activity); the latter creates an inverse thermal stratification such as occurs in Lake Vanda where the temperature in the bottom layers approaches 25°C (Goldman *et al.*, 1967). Table 2 shows some of the physical, chemical and bacteriological information for three Antarctic lakes, two of which are permanently frozen. Lakes Bonney and Vanda are extremely clear lakes, with extinction coefficients as low as 0·031 for blue light in Lake Vanda and 0·069 for green light in Lake Bonney (Goldman *et al.*, 1967). Light penetrates through the ice surface to the bottom of both lakes. In both of these lakes, the temperature and salinity increase with increasing depth. Photosynthetic activity occurs primarily just below the ice. The primary productivity is quite low, however, and in the order of 30 mg C m^{-1} day^{-1}. This is in contrast to some other Antarctic lakes located near Cape Evens, such as Skua, Alga and Coast Lakes, which have photosynthetic activities as high as 1900 mg C m^{-2} day^{-1} (Goldman, 1970; Goldman *et al.*, 1972). Unfortunately, there have been no microbial studies performed on these relatively productive lakes.

Different investigators have reported somewhat contradictory information on the numbers and activity of bacteria in Lake Bonney. Koob and Leister (1972) showed the presence of distinct vertical populations of bacteria, with highest densities (6×10^5 $litre^{-1}$) at 8 to 9 m, and 12 to 15 m. Within these depths, the water temperature ranged from −2 (beneath the ice) to 7°C. Maximum photosynthetic activity occurred in the 5 to 10 m range which was also the same depth noted for maximum uptake of ^{14}C acetate by heterotrophic bacteria. Benoit *et al.* (1971) noted that the highest 0°C counts were located in the 5 m layer under the ice and were generally greater than 400 ml^{-1}. Slightly higher counts were obtained using a soil extract medium without additional salts than in a medium containing the levels and kinds of salts found in the bottom waters of Lake Bonney. There was no growth observed on a sea-water agar medium. Significantly higher numbers were reported by Goldman *et al.* (1972) using direct count methods which would measure both viable and dead cells; however, as Benoit *et al.* (1971) pointed out, an unusual microflora might exist at the greater depths which are not readily cultivatable and may account for the significantly high fixation of

Table 2

Incidence and activity of microorganisms in three Antarctic lakes

Lake and properties	Primary productivity (mg C m^{-2} day^{-1})	Number of bacteria	Isolation temperature °C	Microbial activity	References
Bonney depth: 32 m temperature: perennially ice covered, −2 to 7°C, high temperature at 15 m	31	high concentration just below zone of photosynthetic activity (20 m), $1{\cdot}2 \times 10^5$ ml^{-1} to less than 10^3 ml^{-1} at 30 m	direct counts		Goldman *et al.* (1967): Goldman (1970)
salinity: 0·25% at 11 m; 21·7% at 30 m; mostly NaCl and $MgCl_2$		spring: 0–400 ml^{-1} summer: 3–460 ml^{-1} decrease with depth	0°C plate counts		Benoit *et al.* (1971)
		distinct vertical populations at 4, 6, 8–9, 12, and 15–16 m; 10^2–10^3 ml^{-1}	4°C plate counts	^{14}C acetate max. activity 15×10^3 cpm at 5 m, decrease with depth	Koob and Leister (1972)
Vanda depth: 60 m temperature: perennially ice covered; 0°C at surface to 24·2°C at 60 m (solar trap)	29	no zone of high bacterial densities noted, summer $0{\cdot}4$–1×10^3 ml^{-1} blue-green algae		^{14}C glucose surface: no counts; 10 m 180 cpm	Goldman *et al.* (1967)
salinity: 12%, 10% $CaCl_2$		summer surface water, 210 ml^{-1}; 30 m is 5 ml^{-1}, 40–50 m is 175–180 ml, bottom 0 ml^{-1}	0°C plate counts		Benoit *et al.* (1971)
Don Juan depth: 11 cm temperature: no ice cover −24 to −3°C salinity: 47·4% solutes most of which is $CaCl_2$	no photosynthesis	*Bacillus*, *Micrococcus Corynebacterium* and a yeast species	representative isolates grew at both 0 and 25°C; grew well in nutrient medium made with Don Juan water, particularly at low temperatures		Meyer *et al.* (1962)
		apparent toxicity of pond water; 10^{-2} dilution gave 3 organisms, whereas 10^{-3} dilution gave 90 Only microorganism recovered was *A. parvulus*	plate counts at 20°C		Cameron *et al.* (1972a)

carbon (up to 0·7 mg C m^{-1} h^{-1}) observed by Koob and Leister (1972) at 15 m depth. There is no information on the kinds of bacteria present in Lake Bonney other than the observation made by Benoit *et al.* (1971) that most of the organisms isolated at 10 to 15 m depth were yeasts of the genera *Candida* and *Cryptococcus*. The temperature range for growth and activity of bacteria from Lake Bonney has not been investigated.

In contrast to Lake Bonney, Lake Vanda is deeper (60 m) exhibiting a more exaggerated increase in temperature with depth to approximately 25°C. Wilson *et al.* (1974) consider Lake Vanda a solar trap since there is no evidence of geothermal activity. Benoit *et al.* (1971) observed over 300 bacteria ml^{-1} in water just below the ice using 0°C plate counts. The numbers decreased with depth to approximately 5 ml^{-1} at 30 m and increased again at 40 to 50 m depth to 175 to 180 ml^{-1}. Even at 40 to 50 m depth, the water temperature did not exceed 10°C. Below 50 m the water temperature rapidly increases to 25°C and the levels of oxygen decrease. Benoit *et al.* (1971), observed that the water mass below 50 m contained concentrations of sulphate as high as 7000 mg $litre^{-1}$; however, no attempt was made to isolate sulphate reducing microorganisms from deep water or sediments. Benoit and Hall (1970), however, reported the presence of sulphate reducing bacteria in some other Antarctic pools, and Barghoorn and Nichols (1961) isolated *Desulfovibrio* from anaerobic sediments of Antarctic kettle holes which were characteristically quite saline (13%) and consistently cold (−51 to 4·5°C). The sulphate reducing organisms grew both at 5 and 25°C. It is quite likely that anaerobic sulphate reducers and possibly methane bacteria are important ecological components in permanently ice covered lakes which are anaerobic at depths below the inverse thermocline. This could be particularly interesting since gas bubbles from these anaerobic processes might be the only method by which nutrients can be transported from the bottom to the surface photosynthetic layer. On the other hand, the nutrients from bacterial decomposition might remain stratified on the bottom layer, at least in Lake Vanda, and thus could help explain the observation made by Goldman *et al.* (1967) that significant photosynthetic activity occurs in deep water at 50 to 60 m. The plankton at these depths were reported to be different from the surface population and appeared to be coccoid blue-green algae. These authors speculated that the more favourable warm temperatures might have accounted for the photosynthetic activity at these depths.

It is assumed that nitrogen is the limiting nutrient in permanently ice covered lakes, and in Lake Vanda it was found that there was very little heterotrophic uptake of acetate unless a source of nitrogen was added to the water (Goldman *et al.*, 1967). Bacteria have been isolated from Lake Vanda, and it might be that during the summer when the acetate uptake experiments were performed, nitrogen was limiting due to the relatively intense photosynthetic activity. During the winter months, however, nitrogen might be recycled by bacteria through algal decomposition. There still would be a shortage of nitrogen, and nitrogen fixation must occur either by blue green algae such as described by Bunt (1971), or by bacteria. The source of molecular nitrogen and whether or not gas exchange occurs through the ice cover are presently not known.

There are several known high saline ponds and lakes in the Antarctic. Don Juan Lake is particularly unusual in that the extremely high salinity (approximating 45% $CaCl_2$) prevents freezing unless the water temperature falls below −48°C (Meyer *et al.*, 1962). There is no photosynthetic activity in the lake, and bacteria and yeast are the only living organisms detected. All of the microorganisms isolated from Don Juan water have been found to be facultatively halophilic and capable of growing in media made with both Don Juan water and distilled water (Meyer *et al.*, 1962). The organisms isolated included *Micrococcus, Bacillus* and *Corynebacterium* all of which grew at 0°C as well as 25°C; however, it was noted that all of these isolates grew more rapidly at 0°C if the growth medium was prepared with Don Juan water. The organisms from Don Juan were found to be different from Lake Bonney and Lake Vanda isolates, and in fact, Meyer *et al.* (1962) could not recover any microorganisms in Lake Vanda water below 60 m. Cameron *et al.* (1972a), moreover, recovered only one bacterial species from Don Juan water, *Achromobacter parvalus*, and further indicated that the saline water was toxic to microorganisms since higher plate counts were obtained from the 10^{-3} dilution than lower dilutions. These authors indicated that *A. parvalus* is probably not active in the pond water, but no attempt was made to isolate organisms using medium prepared with pond water and incubated under *in situ* temperatures (0 to −3°C) and pH (8 to 5·4). Based on the report of Meyer *et al.* (1962), there are facultatively halophilic psychrotrophs capable of growing in Don Juan water with added nutrients. It is not known whether these organisms are capable of growth *in situ*, particularly in view of the low concentrations of organic carbon (0·03%) found in Don Juan waters (Cameron *et al.*, 1972). This pond is also discussed by Kushner in Chapter 8.

VI. Rivers and Streams

There are very few reports on the incidence and activities of microorganisms in rivers, streams and creeks at temperatures below 5°C. No doubt, psychrophilic microorganisms usually found in soils from both temperate and polar regions could be isolated from rivers and streams. There is, however, no reports on the incidence and activity of a true psychrophilic microbial flora in rivers and streams. Some investigators have isolated bacteria and fungi at temperatures from 0 to 3°C in rivers and streams, and these data are summarized in Table 3. It is obvious that bacteria capable of growing at temperatures below 3°C can probably be found from any body of water from temperate, alpine and polar regions; however, the temperature ranges for growth have not been determined from any of these studies, and thus it is not known whether the organisms isolated at 0°C are psychrophiles according to the definition of Morita (1975). Stokes and Redmond (1966) indicated that 16 to 47% of the bacteria associated with rivers and streams in Washington state were psychrotrophic based on their growth at 0°C. Bott (1975) did not detect the presence of true psychrophiles in a small stream, but did show that the bacterial population during the winter had a lower optimum temperature for uptake of ^{14}C glucose than did the summer population. This is in contrast to information presented by Ziekus and Brock (1972) indicating the presence of a stable, well adapted population in river water capable of reproducing at essentially the same rate over the temperature range of 8 to 26°C. Bacteria isolated from some temperate lakes have also been shown to be well adapted to grow and utilize dissolved nutrients over the seasonal temperature range (Boylen and Brock, 1973; Rao and Dutka, 1974).

In general, highest numbers of bacteria occur in lakes during the warm season and usually coincide with the algal blooms. No apparent seasonal changes in numbers of psychrophiles or psychrotrophs occurred in White Clay creek water or silt, and the bacterial counts were relatively stable throughout the year (Bott, 1975). Larkin (1970), on the other hand, found significantly higher numbers of 0°C and 30°C growing organisms in Mississippi River water during the winter than in the summer. No 0°C growing organisms were isolated from river mud at any time of the year, even though there was consistently greater than 10^6 bacteria g^{-1} mud when cultured at 30°C.

The only report of *in situ* bacterial growth rates in a stream during the winter (0 to 5°C) showed a marked reduction in the generation times of both unicells and filamentous cells with decreasing seasonal temperature (Bott, 1975). The generation times were determined using submerged microscope coverslips and corrected for rate of microbial attachment by having one set of the coverslips periodically exposed to ultraviolet light to prevent division by attached microorganisms. Table 4 summarizes these data and clearly shows the effect of temperature on microbial growth rates in a stream. In all cases, the generation times of both unicellular and filamentous forms were 8 to 20 times greater during the summer (16 to 21°C) than during the winter. This marked decrease in the generation times of river organisms with decreasing temperature is somewhat surprising since the bacteria isolated during the winter showed a lower optimum temperature for ^{14}C-glucose uptake than the summer population. In all cases, the stream organisms were

Table 3

Incidence of psychrophiles and psychrotrophs in rivers, streams and springs

Water type and geographical area	Counts and isolation temperature		References
	0°C bacteria ml^{-1}	20°C bacteria ml^{-1}	Stokes and Redmond (1966)
River, Washington	4–12 × 10^2	1–3·8 × 10^2	
Stream, Washington	7–87 × 10^2	3·6–24 × 10^3	
Mississippi River	0°C counts	20°C counts	Larkin (1970)
mud a. winter	10 g^{-1}	1·2 × 10^6 g^{-1}	
b. summer	10 g^{-1}	6·9 × 10^4 g^{-1}	
water a. winter	0 ml^{-1}	9 × 10^4 ml^{-1}	
b. summer	1·3 × 10^3 ml^{-1}	1·5 × 10^6 ml^{-1}	
White Clay Creek, Pennsylvania	Winter (0 to 10°C)	Summer (14 to 19°C)	Bott (1975)
a. silt	17·5 ± 9·9 × 10^6 g^{-1}	15·7 ± 5·8 × 10^6 g^{-1}	
b. rock and gravel	50·2 ± 23·5 × 10^6 cm^{-2}	11·9 ± 4·8 × 10^4 g^{-1}	
c. water	6·8 ± 2·7 × 10^3 ml^{-1}	13·9 ± 4 × 10^3 ml^{-1}	
Farm water supply (wells and springs)	3°C counts	30°C counts	Druce and Thomas (1970)
a. nonchlorinated	360	690	
b. chlorinated	21	510	
	4°C counts		
River water (France) three rivers	0·1 to 2·6 × 10^6 ml^{-1}		Breuil and Gounot (1972)
River sediments, Ottawa, Canada	0·1 to 3·5 × 10^7 g^{-1}		

Table 4

Generation times for unicellular and filamentous bacteria in situ *in White Clay Creek as influenced by seasonal temperature changes*[a]

Season and temperature range	Generation time (h)	
	unicells	filaments
Winter 0–5°C		
riffle	48	42
pool	58	45
Spring 8–20°C		
riffle	11·5	9·6
pool	10·8	8·4
Summer 16·5–21°C		
riffle	4·8	4·0
pool	7·0	2·8

[a] Taken from Bott (1975).

not found to be optimally adapted to environmental temperatures, but rather to temperatures 5 to 20°C higher than *in situ* temperatures (Bott, 1975).

Schallock *et al.* (1970) reported that during the winter, ice-covered Arctic and subarctic rivers become depleted of dissolved oxygen as a result of bacterial activity occurring at near-zero temperatures. Plate counts in Alaskan rivers frequently yield 10^4 to 10^6 bacterial ml^{-1} when incubated at low temperatures (Gordon, 1970). Furthermore, Gordon (1970) reported 550 to 9000 bacterial ml^{-1} from the Chena River in Alaska. All of the cultures studies grew at 0, 5 and 10°C, but not at higher temperatures. The indigenous microflora was quite active in assimilating organic substrates at temperatures below 10°C.

Even though there is little information on psychrophilic and psychrotrophic microbial activities in rivers and streams there is little doubt as to the presence of an active microbial flora. The degradation of organic matter originating from plants and soil, particularly in mountainous rivers and streams, is the result of microorganisms active at temperatures frequently below 5°C (Kaushik and Hynes, 1968, 1971). It is well established, for example, that an important, and perhaps the most significant source of energy in streams, is the allochthonous detritus particularly from deciduous leaves (Teal, 1957; Peterson and Cummins, 1974). There is considerable evidence that bacteria and fungi are responsible for leaf processing in streams and that frequently animals derive their nutritional needs from the leaf decomposing microorganisms and not from the leaf itself (Mackay and

Kalff, 1973; Triska, 1970; Cummins *et al.*, 1972, 1973). The rate of decomposition of leaves in a woodland stream having fall and winter water temperatures ranging from 0·1°C to 11°C was from 0·5% day^{-1} to a high of 2% day^{-1} (Peterson and Cummins, 1974). The rate of microbial decomposition of leaves was dependent on the type of leaf and not on the seasonal temperature variations. Moreover, microbial respiration rates on leaf surfaces ranged from 0·02 to 7·0 $\mu l\ O_2\ mg^{-1}$ ash free dry wt h^{-1}. The highest activity (7·0 $\mu l\ O_2$) occurred on hickory leaves at 5°C. Considerable microbial activity occurred even at temperatures approaching freezing. Triska (1970), however, observed that the greatest bacterial respiration occurred on leaves during the summer, while fungi were the principal leaf decomposing microorganisms during the winter in a woodland stream having a seasonal temperature range from freezing to 22°C. Egglishaw (1972) observed that the *in situ* microbial decomposition of cellulose in fast flowing streams in the Scottish highlands required from 12 to 40 weeks. Even though no temperature data was given, the geographical location of these streams indicates that the temperature was below 5°C.

The incidence of psychrophilic and psychrotrophic bacteria associated with springs and wells has received particular attention by individuals investigating the source of psychrotrophic contaminants in dairy products (Dempster, 1968) and other foods. Druce and Thomas (1970) showed that farm, well and spring water generally harboured greater than 10^2 psychrotrophs ml^{-1} (3°C incubation). Over 80% of the psychrotrophs isolated were found to be Gram negative non-fermenting rods including *Achromobacter* and fluorescent and nonfluorescent pseudomonads. In contrast, over 30% of the 30°C isolates from the same water samples were coliforms. These results are consistent with others, and it is generally accepted that up to $10^3\ ml^{-1}$ psychrotrophs are associated with farm water. Some of these organisms, including *Pseudomonas fluorescens* and a *Flavobacterium* are frequently involved in the spoilage of dairy products (Rhodes, 1959; Druce and Thomas, 1970). Both of these organisms have been shown to grow quite well at temperatures below 5°C (Stanier *et al.*, 1966).

VII. Soils

Soils in most regions are among the most unstable environments with regard to temperature. Most terrestrial environments are exposed to sub-freezing conditions during the winter and temperatures sometimes exceeding 30°C during the summer. However, in many temperate regions, the soil temperatures may never reach 20°C (Gray and Williams, 1971; Okafor, 1966). In polar regions, the soil temperatures may be considerably below freezing throughout the year; whereas, in some Antarctic dry valleys, the surfaces of rocks, which are frequently colonized by bacteria and primitive plants, have been reported to undergo temperature variations from below freezing to greater than 32°C (Rudolph, 1966). The instability in the temperature of most terrestrial environments is reflected in the temperature growth range of most of the resident microorganisms. The commonly encountered soil bacteria in temperate regions are psychrotrophic and can usually grow over a wide temperature range (Druce and Thomas, 1970; Ingram, 1965). Many "typically" mesophilic soil bacteria, such as various *Bacillus* and *Clostridium*, *Pseudomonas* and other Gram negative bacilli, can grow at temperatures from below 0°C as well as to 40°C (Buchanan and Gibbons, 1974; Druce and Thomas, 1970). *Yersinia pestis* has been reported to grow at −2°C as well as at 40°C, and in soils (Domardski *et al.*, 1968). Stokes and Redmond (1966), moreover, reported that from less than 1 to 86% of the bacteria found in temperature soils were capable of growing at 0°C. Other investigators have reported that approximately 5 to 15% (Biederbeck and Campbell, 1971; Campbell *et al.*, 1973; Lockhead, 1924) and 75% (Druce and Thomas, 1970) of the microflora of temperate soils are capable of growing at temperatures below 5°C. Even in permanently cold polar soils, however, the predominant microorganisms isolated have optimum growth temperatures consistently higher than the highest *in situ* temperatures encountered (Flint and Stout, 1960; Straka and Stokes, 1960). It seems obvious that, in temperate environments, most of the microbial activity in soils occurs during the warmest season, and in some instances, no psychrophilic bacteria were found in temperate soils, even during the winter (Larkin, 1970).

The early reports on the microbial flora of frozen soils indicated that the numbers of bacteria increased or decreased in direct proportion to the moisture content of the soil, except that a significant increase in the bacterial population was observed in soils after freezing (Conn, 1910). This was later confirmed by Conn (1914a, b) who further suggested, without any confirming data, that there possibly existed a separate summer and winter microbial population. Conn (1911) and Vanderleek (1917, 1918) reported that slightly frozen soils promote microbial growth; whereas there was a decrease in the incidence of bacteria after thawing. Brown and Smith (1912) confirmed these reports and noted that ammonification increased in soils after freezing. Contrarily, Lochhead (1924; 1926) could not confirm these earlier investigations and indicated that the availability of moisture was the single most important factor in determining whether or not microbial growth could occur in frozen soils. Even the microorganisms capable of growth at 0°C were inactive in frozen soils. Many of Lochhead's conclusions, particularly his recognition of the important correlation between the availability of moisture and microbial activity, have since been substantiated by others.

One of the characteristics of soils in temperate regions, particularly during the autumn and spring, is the diurnal temperature fluctuations which can vary from near 0°C to greater than 15°C (Biederbeck and Campbell, 1971; Campbell *et al.*, 1973). This is significant since much of the initial processing of plant material occurs during the fall and continues through the spring. A psychrophilic or psychrotrophic bacterial population would be considerably more efficient in carrying out these mineralization processes than most mesophiles. There is, however, the possibility that as a result of intense microbial activity in soils, particularly associated with the rhizosphere, that soil temperatures could increase by several degrees (Clark *et al.*, 1962; Bartholomew and Norman, 1944).

There are surprisingly few reports on the quantitative incidence of psychrophilic and psychrotrophic bacteria in temperate soils. In general, however, bacteria, fungi and yeasts, capable of growing at 0°C or below, are almost invariably isolated from temperate soils. Table 5 summarizes some of the

Table 5

Incidence of psychrophilic and psychrotrophic bacteria in non-polar soil samples

Location	Soil types	Incubation temperature °C	Bacterial counts g^{-1} dry wt soil	References
Temperate soil		4·5	7×10^5	Biederbeck and Campbell (1971)
Temperate (France)	garden	4	$4{\cdot}3 \times 10^7$ g^{-1} wet wt soil	Breuil and Gounot (1972)
French Alps	glaciers	4	$0{\cdot}6$–$3{\cdot}6 \times 10^6$ g^{-1} wet wt soil	
1. vegetable gardens (13)[a] 2. salt marsh pastures (4) 3. sand dune pastures (7) 4. lowland pastures (23) 5. lapland pastures (17) 6. rhizosphere (10) 7. lowland raised bogs (3) 8. heathmoors (5) 9. arable soils (8)		3·5	$2{\cdot}4 \times 10^5$ $1{\cdot}0 \times 10^5$ $6{\cdot}0 \times 10^5$ $39{\cdot}8 \times 10^5$ $107{\cdot}0 \times 10^5$ $230{\cdot}0 \times 10^5$ $16{\cdot}0 \times 10^5$ $13{\cdot}0 \times 10^5$ $5{\cdot}9 \times 10^5$	Druce and Thomas (1970)
Lapland (near glacier)	forest soil	2	below ice: $2{\cdot}8$–$3{\cdot}9 \times 10^6$ no ice : $1{\cdot}5$–2×10^6	Gounot (1973)
Louisiana	garden, footpath, lawn	0	summer <10 winter <100	Larkin (1970)
Eastern Canada	frozen soil	3	surface to 2 cm: $6{\cdot}1 \times 10^6$ 6 cm: $6{\cdot}5 \times 10^6$ 10 cm: $1{\cdot}3 \times 10^6$	Lochhead (1926)
Washington State	garden soil cultivated soil uncultivated soil	0	$0{\cdot}92$–1700×10^3 23– 810×10^3 420–3100×10^3	Stokes and Redmond (1966)
Temperate soil (near area receiving cannary waste water)		2	$1{\cdot}5$–$7{\cdot}4 \times 10^5$ g^{-1} wet wt soil	Vela (1974)

[a] Total number of samples.

reports on the quantitative incidence of bacteria in soils cultured at temperatures below 5°C. It is obvious that bacteria capable of growing at temperatures below 5°C are quite abundant in soils, and it is not uncommon to enumerate 10^7 psychrotrophs g^{-1} soil. Druce and Thomas (1970) found that a high proportion of the total microflora in rhizosphere soil, moor, pastures and uncultivated salt marsh was psychrotrophic. Surprisingly, the lowest numbers of psychrotrophs seem to occur in cultivated soils (Druce and Thomas, 1970; Stokes and Redmond, 1966), while, the highest counts are reported from garden soils (Ingraham, 1958; Stokes and Redmond, 1966).

It is generally presumed that very little if any microbial activity occurs in temperate soils at temperatures below 5°C. No apparent activity occurs in frozen soils, but at temperatures above 0°C, microbial activity has been indicated by the evolution of CO_2 (Soulides and Allison, 1961), O_2 uptake (Ross, 1965) and nitrification (Sabey *et al.*, 1956; Stanford *et al.*, 1973; Seifert, 1961). Moreover, Seifert (1961) indicated that quite high nitrification occurs in soils during the winter (2°C) due to the availability of ammonium; this activity was frequently equal to summer rates. Okafor (1966) found that chitin was mineralized in English soils throughout the year, but that bacteria and fungi were the dominant organisms on buried chitin during the winter; whereas, actinomycetes, protozoans and nematodes were most abundant at 20°C.

The predominant types of bacteria occurring in temperate soils vary with the soil types, moisture, location, etc. In general both sporeforming and nonsporeforming Gram positive rods, actinomycetes and *Pseudomonas* probably comprise over 90% of the microbial flora (Conn, 1948; Alexander, 1961). Gram negative bacteria generally comprise a low percentage of the soil population and usually vary from 7% in soil samples to 20% in the rhizosphere (Holding, 1960, 1973). Similar kinds of microorganisms are found in temperate soils when isolated at temperatures below 5°C. Frequently, however, very few sporeformers and actinomycetes are isolated at low temperatures from cold soils (Lochhead, 1926; Mishoustine, 1964; Druce and Thomas, 1970). The predominant bacterial types isolated at low temperatures are arthrobacter-like organisms including *Corynebacterium*, *Cellulomonas*, *Microbacterium*, *Arthrobacter*, *Brevibacterium*, and several genera of Gram negative bacilli although Sinclair and Stokes (1964) and Beerens *et al.* (1965) have isolated psychrotrophic clostridia from temperate soils. Druce and Thomas (1970) showed that over 60% of the psychrotrophic bacteria isolated from both lowland and upland pastures were arthrobacter types. On the other hand, *Pseudomonas* species were found to be the most dominant bacterial type isolated at low temperatures from low moor peat (Janota-Bassalik, 1963a) and from various garden and glacial soils in France (Breuil and Gounot, 1972). Other Gram negative organisms and some Gram positives including *Brevibacterium*, *Kurthia* and *Corynebacterium*, were also isolated. Most of these organisms were capable of growing at 0°C or below and at 25°C, but not at 37°C (Janota-Bassalik, 1963b).

Although the existence of a large population of psychrotrophic bacteria is well established, it cannot be presumed that these organisms are significantly active in soils during the winter. The optimum growth temperature of almost all of the reported psychrotrophic bacteria isolated from cold temperate soils is generally above 20°C. In the few instances when some physiological index of microbial activity was measured in winter soils, activity was less than in spring or summer. The proportion of the whole microbial population, and the specific bacterial types, involved in winter activity in soils has not been unequivocally determined. It is not known, for example, what extent of organic matter processing, N_2 fixation and other known microbial transformation processes occur in cold soils, or in fact, whether or not there is measurable activity.

Polar environments are characterized by long periods of subfreezing temperatures. This characteristic, however, is perhaps the only common property shared by both the Arctic and Antarctic. The Arctic terrestrial environments are generally less harsh than the Antarctic, and in areas of the subarctic, complex populations of vascular plants, mosses and lichens, in addition to various animals, can be found (Bliss *et al.*, 1973; Wrigley, 1974). Most of the Arctic is treeless tundra, underlain with permafrost. The air and soil temperatures within the Arctic vary with season and location, and generally during the summer months the surface ground temperatures can reach 20 to 40°C (Boyd and Boyd, 1971; Bliss, 1962). In several areas of interior Alaska located near the intersection of the Yukon and Tanana Rivers, winter temperatures down to −30°C and summer temperatures below 10 to 15°C are common (Streten, 1974). The ground temperature will vary depending on the extent of precipitation and the presence of a snow cover; the latter is known to have an insulating effect on underlying soil and permafrost layers (MacKay and MacKay, 1974). The precipitation in the Arctic interior is generally low, ranging from 160 to 300 cm; however, flood rains have been known to occur if sufficient moisture, originating in lower latitudes of the North Pacific, enters the interior (Streten, 1974).

The Antarctic terrestrial environment is frequently divided into the maritime and continental zones. The maritime Antarctic is characterized as having less intensely cold temperatures, higher moisture content and a greater variety of soil types. In some maritime regions there has been an accumulation of organic soils resulting primarily from penguins and seals. Also, various forms of vegetation consisting chiefly of lichens and mosses, and a number of flowering plants have been observed (Llano, 1961; Boyd *et al.*, 1966). Seal and penguin rookeries are found in most Antarctic coastal areas and thus guano contributes significant quantities of organic material, nitrogen and phosphorus to the soils. These soils, named ornithogenic, support some plant life, but perhaps more important, may contribute nitrogen and phosphorus to nutrient deficient areas through leaching and wind transport (Bunt, 1971; Ugolini, 1972). The mean annual temperature of various coastal Antarctic areas vary between −10 and −20°C with summer temperatures ranging from −1 to −6°C; however, temperatures exceeding 8°C have been recorded (Ugolini, 1970). In general, however, liquid water, and not temperature, is the single most important factor which determines if various soils harbour microorganisms. Any area, therefore, having a permafrost layer shallow enough to undergo some thawing will support life since the average annual water equivalent from precipitation, which is 15 cm in the maritime Antarctic and probably less than 10 cm in the dry valleys is too low to support an active microbial population (Boyd *et al.*, 1966; Cameron, 1972). The soils of the continental Antarctic zones are primarily composed of sand or gravel and the old soils of the dry valleys could be classified as desert (Tedrow and Ugolini, 1966; Cameron, 1972). The mean annual air temperatures in the dry valleys vary between −20 and −25°C, and in general the summer air temperature does not rise above 0°C; however, diurnal cycles of freezing and thawing can occur. At such times the ground surface can heat up to 15°C (Horowitz *et al.*, 1972; Ugolini, 1970; Weyant, 1966). Moreover, during the summer months in the Antarctic interior, rock surface temperatures of 32°C have been reported (Rudolph, 1966). Microorganisms represent the exclusive flora of these dry soils, and a significant portion of samples tested (26%) were reported to be sterile (Horowitz *et al.*, 1972; Boyd *et al.*, 1966).

The existence of an extensive microbial flora in Arctic and Antarctic soils is quite well established. Frequently, the numbers of bacteria which are associated with the "active layer" of the soil above the permafrost are similar to counts obtained from temperate soils (Boyd, 1958; Boyd and Boyd, 1971). Table 6 summarizes most of the reports on the incidence of bacteria in Arctic and other cold northern soil samples enumerated at low temperatures. In general, bacterial counts ranged from 10^4 to 10^6 g^{-1} of soil at the surface and decreased markedly with depth (Boyd and Boyd, 1971; Ivarson, 1965). As in temperate regions, the highest bacterial populations were detected in uncultivated rather than cultivated soils (Boyd and Boyd, 1971; Ivarson, 1965). Similar psychrotrophic bacterial counts have been reported in Antarctic soil samples (Table 7). In the Antarctic, however, it is quite obvious that the incidence of bacteria is dependent on the availability of moisture, and in dry soils, less than 10 bacteria g^{-1} are generally reported (Benoit and Hall, 1970; Boyd and Boyd, 1962a; Cameron *et al.*, 1970a; Cameron *et al.*, 1970b).

The incidence of bacteria in soils, as measured by viable plate counts, definitely underestimate the actual numbers. Parinkina (1974) compared direct counts with viable plate counts using various tundra soils. The direct counts in various Antarctic soils were 1600 to 7000 times greater than the

Table 6

Incidence of psychrophilic and psychrotrophic bacteria in Arctic and other cold northern soil samples

Location	Soil type	Incubation temperature °C	Bacterial counts (g^{-1} dry soil)		References
Arctic coast, Alaska					
Point Lay	peat	2	51×10^4		Boyd (1958)
	loam	2	30×10^4		
Wainwright	peat	2	38×10^4		
	loam	2	4×10^4		
Barrow	peat	2	10×10^4		
	loam	2	7×10^4		
Cape Simpson	peat	2	14×10^4		
	loam	2	5×10^4		
Pitt Point	peat	2	17×10^4		
	clay	2	9×10^4		
Barter Island	loam	2	$13–26 \times 10^4$		
Alaskan Arctic	peat to permafrost layer	2	surface: $9·6 \times 10^3$		Boyd and Boyd (1964)
			15 cm: $9·3 \times 10^3$		
			30 cm: $5·4 \times 10^2$		
			>30 cm: <1		
Inuvik, North West Territories	uncultivated	2	surface: $0·7–1500 \times 10^4$		Boyd and Boyd (1971)
	cultivated	2	surface: $4·9–280 \times 10^4$		
Alaska tundra, Napaskiak	boggy soil	3–5	Sept. range: $1·2–150 \times 10^3$		Fournelle (1967)
			average: 35×10^3		
			June range: $0·1–1300 \times 10^3$		
			average: 150×10^3		
MacKenzie Valley, North West Territories		4	actinomycetes	other bacteria	Ivarson (1965)
	organic soils		2·5–30 cm: 71×10^4	$<10^3$	
			30–35 cm: 33	$<10^3$	
	subarctic gleyed acid, brown wooded		0–23 cm: $5–10 \times 10^5$	17×10^3	
			23–70 cm: 3×10^4	2×10^2	
	subarctic brown wooded (cultivated)		0–10 cm: $5·6 \times 10^6$	$<10^3$	
			10–70 cm: $4–50 \times 10^4$	$<10^3$	
	subarctic brown wooded (uncultivated)		0–40 cm: $80–212 \times 10^4$	$<10^3–60 \times 10^3$	
			>40 cm: 1×10^4	$<10^3$	
Baffin Island, North West Territories		10	counts g^{-1} dry wt soil: actinomycetes	other bacteria	Strzelczyk *et al.* (1969)
	sandy soil		$14–17·8 \times 10^3$	$8–159 \times 10^3$	
	gravel		$1–11·7 \times 10^3$	$10–25 \times 10^3$	
	wet soil or rock edges		$0·8 \times 10^3$	$11·2 \times 10^3$	

Table 7

Incidence of psychrophilic and psychrotrophic bacteria in Antarctic soil samples

Location	Soil type	Incubation temperature °C	Bacterial counts g^{-1} dry wt soil	References
Signy Island	peat	10	surface 1–2 cm: 486×10^3 6–7 cm: 1160×10^3 11–12 cm: 2200×10^3	Baker (1970a, b)
Victoria Land (dry valley)	dry soil	2	0–2 cm: <10 15 cm: 8×10^3	Benoit and Hall (1970)
	permafrost		25 cm: $1{\cdot}8 \times 10^4$	
	moist soil (near pond)	2	$1{\cdot}2$–14×10^5 (range of three samples)	
Ross Island				
Cape Royal	dump (human contamination)	2	$2{\cdot}8 \times 10^5$	Boyd and Boyd (1962)
	soil 1		0	
	soil 2		$5{\cdot}2 \times 10^3$	
McMurdo Sound				
	areas inhabited by man	2	$0{\cdot}18$–63×10^6	
	soil 3		$1{\cdot}1 \times 10^3$	
	soil 4		73×10^3	
	soil 5		38	
Dry Valley			1	
Cape Evans (inhabited by man)			$6{\cdot}8 \times 10^5$	
Wright Valley	mummified seal carcasses	2	0–$1{\cdot}1 \times 10^3$	Boyd *et al.* (1966)
Dry Valleys	1. Taylor Valley soil	2	0–$2{\cdot}6 \times 10^3$	
	2. Wright Valley soil		0–$3{\cdot}9 \times 10^3$	
	3. Marble Point soil		$0{\cdot}46$–25×10^3	
	4. Strand Moraines		$0{\cdot}053$–27×10^3	
Macquarie Island	1. soils	10		Bunt and Rovira (1955)
	a. gravel		$4{\cdot}3 \times 10^5$	
	b. basalt		$2{\cdot}9 \times 10^6$	
	2. rhizosphere			
	a. soil		$1{\cdot}67 \times 10^6$ $14{\cdot}2 \times 10^6$	
	b. root		$4{\cdot}81 \times 10^6$ g^{-1} root 2870×10^8 g^{-1} root	
	algal soil crests	2	$1{\cdot}6$ –$2{\cdot}6 \times 10^5$ $0{\cdot}96$–$1{\cdot}2 \times 10^5$	Cameron and Devaney (1970)
Wheeler Dry Valley	sand	2	surface to 2 cm: $0{\cdot}2$–150×10^3 2–15 cm: 1–100×10^3 18–33 cm: $0{\cdot}5$–$4{\cdot}8 \times 10^3$ 60 cm: 100	Cameron *et al.* (1970a)
Matterhorn Valley	sand	2	surface to 2 cm: <10–370 2–10 cm: <10–20	
	loam or sandy loam	2	surface to 2cm: 10–180 2–10 cm: $3{\cdot}5 \times 10^4$	
Coalsack Bluff Antarctic interior	arid desert soil	2	surface: <10–13,000 2–10 cm: 0–2700	Cameron *et al.* (1970b)
farthest south soil sample	desert soil	2	5–25	Cameron *et al.* (1971)
Victoria Valley	sand	2	0– 2 cm: $2{\cdot}8 \times 10^4$ 2–15 cm: $2{\cdot}6 \times 10^3$ 15–25 cm: $5{\cdot}6 \times 10^4$ 25–30 cm: $2{\cdot}7 \times 10^7$	Cameron (1972)
	sandy loam	2	0– 2 cm: <10 2–15 cm: 30 15–25 cm: 0 25–30 cm: 5	
	sand (miscellaneous samples)		0–25 cm: <10–730	
Dry Valley		5	surface to 2 cm: <10–$2{\cdot}5 \times 10^5$	Cameron *et al.* (1973)
South Victoria Land dry valleys		2	surface: 10–10^4 subsurface: 10–$1{\cdot}7 \times 10^5$	Horowitz *et al.* (1972)
Taylor Dry Valley	saline soil	2	0% NaCl[a]: $5{\cdot}1 \times 10^3$ 5% NaCl: $1{\cdot}2 \times 10^3$ 15% NaCl: <40	
	nine soil samples	0	<100–20,000	Straka and Stokes (1960)
	animal faeces (four samples)	0	$0{\cdot}32$–6000×10^4	

[a] Per cent NaCl concentration in the isolation medium.

viable counts. In various other subarctic and tundra soils, the ratio frequently exceeded 20,000 to 1. The viability of most of the bacteria observed by direct methods is uncertain. Hubbard *et al.* (1968) and Horowitz *et al.* (1972) observed that bacteria present in subsurface soil samples from Antarctic dry valleys were enzymatically active but not viable. In contrast, Vishniac and Mainzer (1972) and Vishniac (1973) demonstrated the presence of a viable population of bacteria 40 cm within the ice-cemented permafrost in Antarctica. These bacteria showed a high portion of their cell volume to be DNA. Viable bacteria were also demonstrated in soils taken from Antarctic dry valleys (Vishniac and Mainzer, 1973). The evolution of ^{14}C-CO_2 was insignificant, but these authors observed microorganisms with an electron microscope and detected bacterial growth using a light scattering device.

There are a number of reports on the rate or extent of microbial activity in polar soils. In general, the extent of *in situ* microbial activity is measured both directly and indirectly. The direct procedures involve the actual evolution of CO_2 or uptake of O_2 resulting from the addition of some low molecular weight organic substrate to soil samples. In the indirect methods, the extent of microbial activity is shown by actual decomposition rates of naturally occurring particulate organic material, such as leaf litter, cellulose, chitin, etc. In some of the early investigations, the extent of CO_2 evolution was shown to be primarily dependent on the organic content of the soil (Bunt and Rovira, 1955), whereas Tedrow and Douglas (1959) reported that temperature and the degree of moisture were the principal factors influencing microbial activity in Arctic soils. The rate of organic matter decomposition was very low at 3°C (*in situ* soil temperature in June) and increased significantly with increasing temperatures. Cameron (1972) also reported marked decreases in microbial activity with depth in Antarctic soils. At depths from 2 cm to 30 cm, no microbial activity could be detected. The fact remains, however, that bacterial decomposition of organic material does occur in polar soils at temperatures below 5°C, and Sieburth (1963) indicated that the microbial decomposition of penguin guano occurs at temperatures below 5°C. The resulting "humus" is most important in supporting a bryophytic flora which is apparently indigenous to penguin rookeries.

The primary factors influencing the decay rates of naturally occurring organic material are temperature, moisture and substrate quality (Flanagan and Veum, 1974). In polar soils, the decomposition of leaf litter is slow compared to rates in temperate soils. The *in situ* rate of litter processing in Antarctic and Arctic soils is between 1 and 24% per year (Heal and French, 1974). Moreover, in some instances, plant material was not fully decomposed after five years in tundra soils (Heal and French, 1974). It is apparent that during most months of the year, the *in situ* soil temperatures in most polar regions are too low for significant microbial activity to occur. Flanagan and

Veum (1974) reported that microbial respiration from leaf litter decomposition occurred at $-7{\cdot}5$°C in soils from Barrow, Alaska. The rate of activity increased with temperature to 25°C and then rapidly fell off at 30°C (soil temperature at Barrow rarely exceeds 17°C). An estimate of the respiration rate for Barrow soils is in the order of 75 to 120 ml CO_2 m^{-2} h^{-1}, which is approximately 1/5 the rate for temperate forest soils. The overall decomposition of organic matter in Alaskan tundra is estimated to be about 200 g m^{-2} $year^{-1}$ in the top 2 cm of soil. Flanagan and Veum (1974) speculated that the ability of microorganisms to respire at $-7{\cdot}5$°C might allow active decomposition of plant material to continue for an additional month both in the autumn and spring. There is no doubt, however, that most of the microbial activity in polar soils occurs during short periods when there is a sharp increase in temperature to 10°C or above. Carbon loss from leaf litter at these higher temperatures is 1·5 to 2·5 times greater than at 0°C (Flanagan and Veum, 1974).

The decomposition of added organic substrates, such as cellulose, to soil is frequently used as an index of *in situ* microbial activity and organic carbon turnover times. Heal and French (1974) and Rosswall (1974) estimated that only about 4% of added cellulose or cotton strips were decomposed after one year in soils from Barrow, Alaska. Similar low decomposition rates were also reported from tundra soils of Finland and Norway. In all cases, the mean annual soil temperatures in these regions were between -1°C and -14°C. Temperature is again the most important factor influencing microbial activity in soils having a permafrost layer, whereas nutrient limitations and water logging are not limiting in bog tundra (Heal and French, 1974).

The overall efficiency of microbial decomposition of leaf and other plant material into CO_2, and microbial biomass is dependent on the availability of various inorganic compounds including nitrate and ammonium. The fact that most higher plants have a high carbon to nitrogen (C:N) ratio underscores the importance of an alternate source of organic nitrogen for efficient microbial mineralization. In the absence of significant rainfall, soils in many polar regions would show little microbial activity unless active N_2-fixing microorganisms were present. This would be particularly true in some inland Arctic and Antarctic regions which are removed from penguin rookeries and receive very little precipitation. In some Antarctic soils, N_2-fixing organisms were not detected (Flint and Stout, 1960); whereas in most other reports, N_2-fixing bacteria and blue-green algae were found to be most abundant. Boyd and Boyd (1962) isolated *Azotobacter chroococcum* and *A. indicus* from frozen Antarctic soils, and Arctic peat and clay soils. These authors pointed out that the *in situ* temperatures at the time of sampling were between 0·9 and $-5{\cdot}5$°C. Boyd and Boyd (1971) also detected *Azotobacter* in all soil samples from Inuvik, Northwest Territories. Moreover, these authors reported the isolation of an N_2-fixing spirillum from river mud. Bunt (1971) reported the isolation of nitrogen fixing blue-green algae, *Azotobacter* and *Clostridium* from polar terrestrial environments. *Rhizobium* nodulated legumes were also found in the Arctic (Bunt, 1971). Also, blue-green algae which are capable of fixing N_2 at 1°C, such as *Nostoc commun*, have been detected in the Arctic (Fogg and Stewart, 1960). The extent of N_2-fixation by blue-green algae in Barrow, Alaska, tundra soils is from less than 1 to 228 $\mu g\ N\ g^{-1}$(dry wt) h^{-1} (Alexander and Schell, 1973). Soils from the Arctic and subarctic had an annual nitrogen input from N_2-fixation ranging from 23 to 380 mg N m^{-3}, which represents most of the nitrogen available to microorganisms since rainfall at these locations was usually insignificant (Alexander, 1974). In polar regions, the primary factors, such as the availability of water and the temperature, affecting N_2 fixation are the same as for other physiological groups of microorganisms. In some polar regions the most efficient N_2-fixers are the species that can withstand desiccation such as blue-green algae (Horne, 1972).

The extent of microbial species diversity in polar soils varies with the location and is most influenced by the availability of moisture, nutrients and temperature. Early investigations on the microflora of Antarctic soils showed a predominance of *Bacillus* and *Micrococcus* (Darling and Siple, 1941). Straka and Stokes (1960) found that most of the psychrotrophic and psychrophilic isolates were Gram negative cocci or coccobacilli. Some of these isolates grew at −7°C as well as at 30°C, or above. Cameron *et al.* (1970a) and Cameron (1972) reported that the dominant bacterial types isolated from dry valleys of Antarctica were diphtheroid types including *Arthrobacter*, *Brevibacterium*, *Cellulomonas*, *Corynebacterium*, *Kurthia*, etc. Most of these isolates grew at 2°C as well as at 20°C. *Bacillus* were not found to be common in Antarctic dry valley soils (Cameron, 1972; Cameron *et al.*, 1973), and obligate anaerobes were not detected (Cameron *et al.*, 1970a). Other groups of bacteria isolated from Antarctic dry soils included *Pseudomonas*, actinomycetes, sulphate reducers, ammonia oxidizers and lactate fermenters (Boyd *et al.*, 1966; Cameron *et al.*, 1970a). The psycrhrotrophic microflora found in antarctic peat is composed of remarkably similar genera to those found in most other Antarctic soil samples (Baker and Smith, 1972). This is surprising since the numbers of microorganisms and the levels of nutrients in peat are considerably higher than found in dry soils (Baker, 1970). *Brevibacterium*, *Arthrobacter*, *Corynebacterium* and *Cellulomonas* were the most dominant genera isolated. *Brevibacterium* comprised over 50% of the 119 strains isolated (Baker and Smith, 1972).

In general, the numbers and generic composition of microorganisms in Arctic soils are more typical of temperate regions than Antarctic soils. High numbers of Gram negative rods, including *Pseudomonas* and *Achromobacter*

(*Moraxella/Acinetobacter* group), are frequently isolated (Fournelle, 1967). Brockman and Boyd (1963) also isolated the cellulolytic myxobacterial species *Myxococcus fulvus* and *Sorangium sorediatum*; these isolates, however, were incapable of growing below 6 to 8°C. Strzelczyk *et al.* (1969) and Ivarson (1965) reported a high incidence of actinomycetes which frequently approached half the numbers of bacteria from the same samples. Boyd (1958) did not detect any autotrophic nitrifers; however, in a later report Boyd and Boyd (1971) found that ammonia oxidizers, nitrogen fixers and sulphate reducers were commonly present in Arctic soils. Dunican and Rosswall (1974) also found a predominance of *Pseudomonas* and *Achromobacter* in tundra soils from Barrow, Alaska. These authors further reported that more than 2×10^5 obligate anaerobes g^{-1} of soil were enumerated in tundra soil. *Actinomyces* were the dominant anaerobic types; however, *Clostridium* were also isolated. Rosswall and Clarholm (1974) indicated that over 90% of the bacterial isolates from some tundra soils could grow at 2°C but not at 37°C. There did not appear to be any significant phsysiological differences between tundra and temperate soil bacterial isolates.

VIII. Marine Environments

The many factors, physical, chemical and biological, that affect the distribution, numbers, types and activity of microorganisms in marine environments are not well understood. Taken singularly, it is known that variations in temperature, salinity, pressure, oxygen, pH, etc., influence all organisms to some extent, both on a population-ecology and a biochemical level. However, the specific effects these various oceanic variables have on microbial populations, and their activities *in situ* are not known. In a discussion of temperature, as it relates to microbial activity in the sea, many factors must be considered. These include thermal stability, synergistic effects of temperature on other physical and chemical properties, such as hydrostatic pressure and salinity, and levels of organic and inorganic nutrients. What is understood is that most life in the oceans occurs at temperatures below 5°C and in 35 to 40‰ salinity.

The difficulty with discussing the ecology of psychrophiles in marine environments is that there is no single marine environment. Generalizations are difficult to make and are always open to criticism. In high and low latitudes, there is very little measurable fluctuation in water temperature (usually below 5°C); whereas in middle latitudes, surface water temperatures vary with season usually between 2 and 4°C to 16°C. In contrast to surface waters, the open ocean below the thermocline is generally below 5°C, and in the Atlantic, Pacific and Indian Oceans, bottom water temperatures are between 2 and 3°C at low latitudes and near 0°C in the Antarctic. Temperatures as low as −1·9°C exist in the deep waters of the Arctic.

It is obvious, then, that the oceans may have markedly different temperature ranges and stability depending on the season and the latitude. The characteristic temperatures of specific areas would influence the types and numbers of microorganisms and the extent of their activity. Indeed, the little information we do have indicates that psychrophilic and psychrotrophic "marine" bacteria occur offshore below the thermocline in most oceans. Mesophilic bacteria, on the other hand, are frequently isolated from inshore and estuarine environments, particularly during the summer months. Psychrotrophic to mesophilic pigmented bacteria are most abundant in the neuston layers.

There are many reports on the incidence of bacteria in marine waters and sediments and associated with animals, plants, and detritus. Most of these are based on plate counts, frequently using incubation temperatures considerably higher than *in situ* temperatures. Moreover, plate counts are known to underestimate the actual direct counts by as much as 10,000 to 1 (Jannasch and Jones, 1959).

The distribution of psychrophilic and psychrotrophic bacteria in marine waters, based on viable counts, varies considerably with the sample location, depth, isolation medium, etc. In near shore waters and sediments, seasonal temperature variations are accompanied by changes in the numbers of psychrophilic and psychrotrophic bacteria (Sieburth, 1967; Nedwell and Floodgate, 1971). It is also generally accepted that the incidence of bacteria decreases significantly with depth in the water column. Sieburth (1971), however, observed patchiness in the microbial population in some water masses. Frequently, 10^3 to 10^5 bacterial ml^{-1} were observed from the surface through the water column to greater than 1000 m. In most instances, however, the counts were less than 10 ml^{-1} below 2000 m. Wiebe and Hendricks (1974) also observed low densities of psychrophiles in Antarctic waters. Counts usually ranged from less than one to a few hundred per ml. In all cases, the number of colony forming units appeared to be stable from day to day at any location. These authors also noted that the bacterial types seen throughout the water column appeared to be surface types. Morita *et al.* (1977) also observed low counts in Antarctic waters and further noted that the incidence of bacteria did not correlate with heterotrophic activity. Similarly, Morita and Burton (1970) reported less than 10 to greater than 600 ml^{-1} of 0°C growing bacteria in various Alaskan waters.

In the water column, an important consideration is that perhaps most bacteria do not exist freely suspended, but attached to particles of detritus, or as bacterial clumps. The ability of marine bacteria to adhere to various surfaces is now well established (Marshall *et al.*, 1971; Corpe and Winter, 1972; Floodgate, 1972). Attachment is usually preceded by bacterial synthesis of a polysaccharide adhesive pad. The obvious importance of bacterial adhesion and colonization on particles of detritus is that bacteria can then effectively utilize refractile organic material, a process that requires extracellular enzymes.

The actual incidence of detritus in the water column has been determined, at least to the extent that its presence in the open ocean is unequivocal. In the deep sea, approximately 1×10^6 particles litre^{-1} have been detected (Lenz, 1972). In the Arctic Ocean, Holm-Hansen (1972) reported particulate organic carbon concentrations to range from 7 μg C litre^{-1} in the euphotic zone to less than 1 μg C litre^{-1} at depths exceeding 1000 m. Concentrations of organic detritus throughout the water column in the North Pacific was 3·0 μg C litre^{-1}. The levels of organic nitrogen and organic phosphorus decreased significantly with depth, strongly indicating bacterial activity (Holm-Hansen, 1972). In general, the size of the particles found in deep waters rarely exceeded 30 μ (Lenz, 1972; Gordon, 1970). The question is, in deep waters do bacteria exist primarily attached to detrital particles? Kriss (1963) believed bacteria exist primarily as free floating single cells, and Wiebe and Pomeroy (1972) indicated that most detrital particles, observed microscopically, were free of bacteria. However, Meyer-Reil (personal communication), using the more sensitive epi-fluorescent microscopic techniques, observed that detritus from deep Pacific Ocean waters (700 m) was heavily colonized by bacteria, many of which were less than 0·2 μ in diameter. Moreover, it is now well established that bacterial clumps are quite readily formed in surface waters both *in situ* and in the laboratory (Sheldon *et al.*, 1967; Seki, 1971). It has been suggested that many of the small particles in ocean waters are aggregates of bacteria and that these aggregates might serve as one of the most important sources of food for the benthos (Sheldon *et al.*, 1967). In summary, then, many organic aggregates in oceanic waters may be the result of bacteria. Initially, these microorganisms are very actively metabolizing both the utilizable particulate and surrounding soluble organic material. If the deep ocean, the bacteria associated with detritus have probably utilized all of the available organic nitrogen and phosphorus, and thus either are endogenously metabolizing or are dormant. What is important is that this whole process from the initial attachment to the subsequent utilization of refractile organic material by bacteria takes place at low temperatures.

Bacterial biomass in the water column, as determined by various methods, usually substantiates the fact that more bacteria exist than are shown by plate counts. Kriss (1963) reported the biomass of bacteria as determined by slide cultures, to be from 7·8 mg m^{-3} at the surface to 0·007 mg m^{-3} at 3400 m in the ocean near the North Pole. Average daily increases in microbial biomass was from 2·5 mg m^{-3} to 0·0008 mg m^{-3}. Sorokin (1973) also indicated that bacteria can serve as a significant source of food for zooplankton, particularly in some tropical oceans where the bacterial productivity of 300 to 500 mg C m^{-2} would be twice the phytoplankton productivity. The validity of these data, however, has been justifiably criticized (Sieburth, 1971). The use of other indirect methods for measuring bacterial biomass in the water column, such as ATP and DNA determinations, might eventually lead to information on the standing stock of bacteria (Holm-Hansen, 1973; Holm-Hansen *et al.*, 1968). These methods, however, would not give answers to the important question as to the average daily increase in bacterial biomass (growth rates *in situ*) which would be a more realistic indication of microbial productivity.

The *in situ* activity of heterotrophic bacteria in the water column has only recently been studied using the uptake of ^{14}C-organic compounds as an index. Table 8 summarizes some of the data obtained from heterotrophic potential studies of Arctic and Antarctic surface water, ice and sediment samples. These data clearly show that the heterotrophic potential (V_{max}) in these polar waters is similar to the productivity in some oligotrophic and mesotrophic lakes and in surface waters of the Eastern Tropical Pacific (Hamilton and Preslan, 1970). The relatively long turnover times for glutamic acid (20 to 12,300 h) would seem to preclude any significant high bacterial population in these waters. However, as Banoub and Williams (1973) indicated, the turnover rate of nutrients is high in areas of low levels of nutrients and low phytoplankton activity, such as the Mediterranean Sea and Gulf Stream.

It is becoming increasingly apparent that microbial activity in most deep waters is negligible. This suppression of microbial activity in the deep ocean might be due primarily to either the presence of a very small microbial population as suggested by Hobbie *et al.* (1972) or due to the co-effect of temperature and pressure (Jannasch *et al.*, 1971; Jannasch and Wirsen, 1973; Wirsen and Jannasch, 1974), or both. In all instances, microbial activity at depths from 1500 to 5300 m was only a fraction of the 1 atm activity. Morita (1976), moreover, reasoned that low metabolic activity by deep sea bacteria may have survival value since the rate of energy supply in deep waters is low. There are, however, some reports of the presence of a significant population of pigmented cells, 1 to 4 μ in diameter, that exist throughout the water column to depths exceeding 3000 m (Fournier, 1966; Hamilton *et al.*, 1968). The frequency of occurrence and the significance of these organisms is presently not known. (See also Chapter 4.)

Although the incidence of bacteria and bacterial activity in deep waters may be negligible, there is some evidence that a large, active bacterial population exists in deep sediments and in the sediment-water interface. Even in shallow benthic regions, the incidence of bacteria in the sediment is one to several orders of magnitude higher than the counts in the water column. The existence of a large population of low temperature growing bacteria in

Table 8

Heterotrophic activity of natural bacterial populations in Arctic and Antarctic marine water, sediment and ice

Location	Sample	Substrate	*In situ* temperature °C	Total V_{max} (μg litre^{-1} h^{-1})10^{-2}	V_{max} CO_2 (μg litre^{-1} h^{-1})10^{-2}	Turnover time (h × 10^2)	% Respired	References
Arctic	ice	glutamic acid	0	0·1–2·2		1·4–13·3	60–86	Morita *et al.* (1977)
	seawater	glutamic acid	0	0·1–17	0·1–7·1	0·2–68	44–76	
	sediment	glutamic acid	0	2·0–170	1·0–59	—	32–65	
Antarctic	seawater	glutamic acid	0	0·11	—	—	—	Gillespie *et al.* (1976)
Antarctic	seawater	glutamic acid	0	0·2–8·7	—	1·1–123	51–77	Morita (1975)
		glucose	0	0·05–1·8	—	5·7–48	21–36	
		tyrosine	0	0·4	—	50	9	
		lysine	0	0·07	—	37	4	

sediments was discovered by Certes (1884). Since that time many reports have established that the incidence of bacteria in the sediments is from less than 100 to greater than 10^8 bacteria g^{-1} of sediment (Waksman and Cary, 1933; ZoBell and Anderson, 1936; Rittenberg, 1941; Morita and ZoBell, 1955).

Sediments from the Arctic were shown to contain low levels of viable bacteria ranging from 10 to 10^3 g^{-1} wet wt of sediment (Rozenberg, 1954; Boyd and Boyd, 1963, Kriss, 1963). These low bacterial counts are difficult to understand in view of the very high mineralization rates in Arctic sediments as noted in Table 8. Wiebe and Liston (1972) and Liston (1968) reported approximately 10^4 bacteria ml^{-1} mud slurry in sediments off the Oregon and Washington coasts and in Puget Sound using 5°C incubation. Most of the organisms isolated from the Columbia River were psychrotrophic and seawater–requiring; whereas, approximately 75% of the isolates from the Pacific Ocean sediment samples were eurythermic, and 95% were euryhaline. Sediments from inshore regions show seasonal variations similar to the fluctuations reported in near shore waters. Nedwell and Floodgate (1971) substantiated these findings both *in situ* and as a result of incubating sediment in the laboratory at different temperatures. These studies showed that the optimum isolation temperature for bacterial populations corresponded to the temperature of incubation of the sediments. Similar population shifts were also observed *in situ*.

The viable counts from intertidal sediments apparently grossly underestimate the actual numbers of bacteria. Dale (1974) reported that the direct counts from intertidal sediments (using acridine orange and fluorescent microscopy) ranged from $1{\cdot}17 \times 10^8$ to $9{\cdot}97 \times 10^9$ bacteria g^{-1} dry wt sediment. These counts represent a biomass ranging to 30 g m^{-2} dry wt bacteria which is at least equal to the faunal standing crop. Moreover, since the metabolic and division rates for bacteria are significantly greater than the benthic invertebrate population, the actual bacterial biomass might be substantially higher. The role of bacteria as food in the benthos appears to be substantiated.

The other marine environments that harbour high numbers of both psychrophilic and psychrotrophic bacteria are the guts and outer surfaces of marine animals and micro- and macrophytic plant populations. In the benthos, the resident invertebrate population usually represents 10 g dry wt m^{-2}, but can, in exceptionally productive areas, be as high as 100 to 150 g m^{-2} (Tait and DeSanto, 1972). If bacteria are a main source of food for these animals, then bacterial production must be at least 10 times the productivity of the invertebrate fauna.

Bacterial counts in marine animals range from less than 10^3 in non-feeding animals to greater than 10^7 g^{-1} gut in feeding animals (Chan, 1970;

Baross, 1972). The incidence of psychrophilic and mesophilic bacteria in near shore invertebrate populations is dependent on seasonal variations which is a reflection of available nutrients and animal feeding rates. Seasonal variations in bacteria associated with finfish have also been observed (Liston, 1957). Total viable counts in skate and sole gut samples ranged from less than 100 to greater than 5×10^6 g^{-1}. Moreover, populations exceeding 3000 luminous bacteria g^{-1} on fish gills and fins were also noted.

Little is known about the activity of psychrophilic and psychrotrophic bacteria in sediments or associated with benthic animals. Goodrich (1976) showed that much of the breakdown of chitin occurred in the guts of marine animals at temperatures below 5°C as a result of a commensal chitinoclastic bacterial flora. There is also some evidence that significant levels of nitrogen are fixed by bacteria in some productive sediments (Maruyama *et al.*, 1974). Even in hadal sediments there appears to be a substantially active microbial flora. Wada *et al.* (1975), for example, observed bacterial nitrate reduction at rates up to 0·42 μg-at N $litre^{-1}$ day^{-1} in the sediment-water interface at

Table 9

Reported isolations and temperature characteristics of psychrophilic marine bacteria

Organism	Origin	Optimum growth temperature °C	Maximum growth temperature °C	References
Vibrio	fish excrement	16	20	Tsiklinsky (1908)
Cytophaga psychrophila	silver salmon	15–20	20	Borg (1960)
Vibrio marinus	water, Oregon coast	15	20	Morita and Haight (1964)
Arthrobacter sp	water, Narragansett Bay			Sieburth (1964)
Vibrio psychroerythrus	flounder eggs	?	19	D'Aoust and Kushner (1972)
Clostridium sp 1	sediment-Washington coast and Puget sound	10·4[a]	17·5	Matches and Liston (1973) Finne and Matches (1974)
2		10·7	18·3	
5		9·8	17·0	
19		8·3	16·3	
40		9·4	17·5	
41		10·0	16·4	
54		8·2	16·3	
Vibrio 18–300	water, Antarctic	7	13	Baross *et al.* (1974)
Vibrio 18–500		10	16	
E5–4–4	water, Antarctic	6–7	10	Christian and Wiebe (1974)
E5–22–8		?	<15	
Vibrio AP–2–24	water, Antarctic	4	9	Morita (1975)
Pseudomonas, *Vibrio* and *Spirillum*	water, North Sea		<20	Harder and Veldkamp (1967)

[a] Minimum log time.

depths to 5845 m and at temperatures less than 2°C. Seki *et al.* (1974) reported $1{\cdot}5 \times 10^6$ bacteria litre $^{-1}$ in the sediment-water interface at depths of 5207 m using direct microscopic counts. The activity of some of these microbial populations was found to be similar to the activity observed in some productive surface waters. These authors further reported that rod-shaped barophilic bacteria had generation times of about 10 h *in situ*. Similar high rates of activity were also noted by Schwartz *et al.* (1976) from the gut microflora of an invertebrate obtained from the Aleutian Trench (7050 m). Growth and mineralization rates were similar at both 1 and 750 atm and at 3°C.

Most of the bacteria isolated from marine waters and sediments and associated with plants and animals are psychrotrophic. Psychrophilic bacteria occur, but at a significantly reduced level. The most common occurring marine bacteria are Gram negative, non sporeforming, aerobic rods. Table 9 summarizes all of the characterized marine psychrophiles. Most of these psychrophiles are Gram negative *Vibrio* although psychrophilic clostridia and myxobacteria have also been isolated. It is particularly noteworthy that the psychrophiles with extremely narrow temperature ranges for growth originated from Antarctic waters. Out of 150 Antarctic isolates, 44 would not grow above 15°C and 52 would not grow at 20°C (Morita, 1975). Morita (1966) presents a detailed description of most of the psychrotrophic marine bacteria and their temperature characteristics.

IX. Snow and Ice

There are many reports on the incidence of bacteria and other microorganisms occurring in snow and in fresh and seawater ice. In most of these reports no attempt was made to isolate and characterize psychrophilic bacteria, and usually just the fact of the existence of viable microorganisms was related. The results of many of these studies are discussed in appropriate sections of this review. There are, however, some reports which indicate the possibility that a unique microflora might exist in ice, particularly from high mountain glaciers and sea ice.

The early evidence dates back to observations made by Aristotle that occasionally snow would be stained a bright red. This was subsequently shown to be the result of several species of "snow algae". The principal distinguishing characteristic of snow algae is that they do not grow at temperatures above 10°C and thus fit Morita's definition of psychrophiles (Morita, 1975). Eight species of snow algae have been described of which three have been isolated and purified as axenic cultures (Hoham, 1975). The optimum growth temperatures for these isolates are 5°C for *Raphidonema nivale*, 1°C for *Chloromonas pichinchae*, and 10°C for *Cylindrocystis brebissonii*. Snow algae have also been reported in Signy Island, Antarctica, and Fogg (1976) estimated their productivity to be 10 mg C^{-2} snow surface day^{-1}. There are no data on the possibility of an active microbial flora associated with snow algae; however, in the few instances that an attempt was made to isolate psychrophilic bacteria from glacier ice, their presence was unequivocally confirmed (Moiroud and Gounot, 1969). There is little doubt in our minds that in all probability a large population of psychrophilic bacteria exist in association with snow algae. In contrast, no microorganisms were detected in snow samples from the interior of the Antarctic (Lacy *et al.*, 1970).

There have been few reports on the incidence of an active microbial flora in the surface ice of lakes and streams. In an interesting note, Barsdate and Alexander (1970) observed large bubbles on the surface-ice of No-name Lake located on the Tanana Valley, Alaska. These ice bubbles contained a mixed population of predominantly purple, pink and green photosynthetic bacteria and very small numbers of algae. All microbial activity was confined to the ice layer since the underlying water was anoxic.

The occurrence of a highly productive population of algae and bacteria in polar sea ice has been well documented (Meguro *et al.*, 1966, 1967; Bunt, 1971). The predominant photosynthetic organisms are diatoms which apparently are active in the loosely aggregated ice crystals associated with the undersurface and the snow-ice interface (Bunt, 1963; Bunt and Wood, 1963). At McMurdo Sound, Antarctica, Bunt and Lee (1970) reported that the standing stock of ice algae was 0·5 g C m^{-3} in an area with surface snow and 10·4 g C m^{-3} at a snow-free site. These snow algae were shown to be active at *in situ* temperatures of −2°C. Sorokin and Konovalova (1973) found a very high biomass of diatoms in a bay of the Japan Sea during the winter. This bay, which was covered with a 60 cm layer of ice at a temperature of −1·8°C, harboured a population of diatoms equivalent to 10 to 20 g m^{-3}. The incident of bacteria associated with sea ice has been documented to the extent that the populations are known to be quite extensive. Cameron and Benoit (1970), for example, reported that 10^7 bacteria ml^{-1} are associated with sea ice which harbours a diatom population. Bacteria have also been observed throughout the 15 cm ice layer in the Arctic Ocean (Horner and Alexander, 1972). The V_{max} using ^{14}C-glycine and ^{14}C-glucose was $1{\cdot}3 \times 10^{-4}$ mg $litre^{-1}h^{-1}$ (respiration not measured). The bacteria in the ice were observed to be motile, and these authors speculated that microorganisms are possibly active in brine pockets. Sorokin and Konovalova (1973) found 0·5 to $0{\cdot}8 \times 10^6$ bacterial ml^{-1} in sea surface ice from the Bay of Japan. These investigators estimated that the rate of productivity per day for bacteria in ice was 2·4 to 3·6 mg C m^{-3} which is only 5 to 15 times lower than the phytoplankton productivity in this same area. Iizuka *et al.* (1966)

observed $7{\cdot}4 \times 10^4$ bacteria g^{-1} plankton ice from the Antarctic Ocean. These counts may be significantly low since these investigators incubated their samples at 25°C. The types of bacteria found in sea ice are apparently quite different from the dominant generic types found in surface waters. Iizuka *et al.* (1966) reported that approximately 70% of the sea ice bacterial isolates were *Brevibacterium minutiferula*.

What is important from all of these studies is that both the ice algae and the bacteria are apparently active at temperatures approaching −2°C, and in some polar regions this ice or snow is a permanent characteristic feature.

X. Discussion

It is quite apparent from the information discussed in this review that undue attention has been given in the past to the fitting of low temperature growing organisms into a rigidly defined set of boundaries that can be used as a guide for a definition. What is frequently neglected is whether or not these low temperature organisms are active under *in situ* conditions at low temperature and to what extent; instead, exclusive emphasis is placed on their maximum or optimum rate of growth under ideal laboratory conditions. Even growth rates *in situ* may not indicate the extent of mineralization activity or production of vital ectocrine compounds, etc., which may be most important ecologically. From the point of view of the microbe residing in cold environments, the primary concern is survival during periods when both physical and chemical environmental factors prevent growth and metabolism. Obviously, organisms would not evolve physiological systems that function optimally only under extremely narrow environmental conditions unless that environment was completely stable. Microorganisms which grow only within a narrow temperature range would not be expected to survive in environments that have very unstable temperature characteristics unless they possessed a special mechanism for survival during periods of extreme temperatures, such as spore or cyst formation, for example. Even with higher life forms, survival in unstable harsh environments, such as found in the dry valleys of Antarctica, usually requires the ability to withstand desiccation or temperature variations extending above and below their growth range. Blue-green algae have been reported to withstand desiccation in polar soils (Horne, 1972), and some lichens, particularly those residing on rock surfaces in the Antarctic, have been reported to have a maximum growth temperature below 20°C even though they are sometimes exposed to temperatures above 30°C (Rudolph, 1966).

It can be said, therefore, that there are two major classes of microorganisms capable of growth and activity at temperatures below 5°C. One class, which is adapted to a stable cold environment, is typified by the stenothermal bacteria isolated from marine environments (Morita and Haight, 1964; D'Aoust and Kushner, 1972; Christian and Wiebe, 1974) and some ice caves (Gounot, 1973b). In general, these organisms have a maximum growth temperature below 20°C and are quite susceptible to temperatures exceeding their maximum for growth (Haight and Morita, 1966; D'Aoust and Kushner, 1971; Geesey and Morita, 1975). In some cases these psychrophiles have also adapted quite rigid salinity requirements, the extent of which is dependent on the temperature (Stanley and Morita, 1968; Morita, 1975). In contrast, organisms isolated from unstable cold environments display a wider temperature range for growth and activity, which is frequently 20 to 30°C above the maximum environmental temperature. However, these organisms can usually survive long-term exposure to temperatures considerably above their maximum (Evison and Rose, 1965) and, based on their repeated isolations from the upper atmosphere and from various Antarctic terrestrial environments, can withstand temperatures down to 50°C below their minimum for growth. It is also possible that this second class of low temperature-growing organisms can tolerate extremes in other physical and chemical parameters, such as salinity, desiccation, hydrostatic pressure and pH.

How can we reconcile the disparity that psychrophiles are not usually found in unstable cold environments even though the maximum temperature *in situ* may never exceed 15°C? The organisms frequently isolated are said to grow better at 20 to 30°C and thus would only show slight activity even at the highest *in situ* temperatures. It appears likely from the data presented that there probably exists in nature two types of microorganisms, categorized primarily on whether or not they are continuously metabolizing *in situ*. This is somewhat analogous to the autochthonous and zymogenous bacterial populations described by Winogradsky (1925)*, except that factors other than nutrients can determine whether or not microorganisms continue to metabolize and grow. In stable cold environments, for example, there exists both a autochthonous and a zymogenous population; whereas, in unstable cold environments, such as the upper atmosphere or in some polar regions where the temperature variation encountered within a season might range from −50 to 5°C, only a zymogenous population would exist. Obviously, in these unstable harsh environments, microbial activity will occur only during the short "thaw" period of the summer (depending on the temperature and availability of water). As the temperature decreases, the organisms must transform into a physiological state that will allow them to remain viable but dormant throughout the winter. In contrast, the zymogenous bacterial populations in a stable cold environment, such as most marine

* The autochthonous populations have a low but steady rate of activity and live on native soil organic matter; the zymogenous populations arise to utilize fresh materials reaching the soil. (Editor)

waters and sediments below the thermocline, when not active, are still within the temperature range for respiration to continue. The survival of these organisms will be dependent on a different set of factors than that of the active microbial population found in harsh environments.

It is well established that when microorganisms undergo a burst of growth in nature, it is in response to some physical factors, such as temperature and moisture in unstable environments and probably nutritional factors in stable environments. This burst of growth is followed by a lag period in which microorganisms presumably begin their secondary metabolic functions. During secondary metabolism, the organism stops dividing and produces synthetases that convert primary metabolites to secondary metabolites. In general, these secondary metabolites are biologically inert, but some, like antibiotics and toxins, can have a secondary effect on other organisms but usually not on the producer cells. Long-term survival of bacterial cells is correlated with unimpaired progress of secondary metabolism (Smith *et al.*, 1974). Furthermore, the formation of various morphological structures, such as spore formation in *Bacillus* and stalk elongation in *Caulobacter crescentus*, are considered secondary metabolic processes (Weinberg, 1974). Presumably, the formation of coccoid cells by *Arthrobacter* and related organisms and cyst formation by *Azotobacter* are also the result of secondary metabolism.

It is well established that greatest survival of microorganisms occurs when secondary metabolism is allowed to proceed under optimum conditions. Invariably, the optimum conditions for secondary metabolism are not the same as the optimum conditions for growth or primary metabolism. In particular, it has been shown that the optimum temperature for secondary metabolism to occur in organisms is frequently 20°C below the optimum for growth and usually can occur only within a narrow temperature range such as 5 to 10°C (Weinberg, 1974). If this is the case, then microorganisms should have two distinct optimum temperatures, one for growth and the other for secondary metabolism, which would then be an indication of survival. A true autochthonous population of bacteria would not be expected to have developed a secondary metabolic system since these organisms, by definition, would never shut off primary metabolic functions. In the absence of external metabolites, these autochthonous bacteria would have little chance for long-term survival. The true marine autochthonous microorganisms would be expected to adapt to the point where they can become true psychrophiles as defined by Morita (1975). This would be particularly true for the recent Antarctic isolates which have maximum growth temperatures around 10°C. In contrast, the zymogenous population in the marine environment would have to retain the capacity to grow at temperatures considerably higher than *in situ* temperatures so as to carry out secondary metabolic functions at *in situ* temperatures, which should be 20°C below the optimum growth temperature. Thus, it is conceivable, that the psychrotrophs, which are most abundant in cold marine environments, might be the zymogenous population; whereas, the true psychrophiles might be the autochthonous bacteria.

In very unstable cold regions, such as Antarctic soils and some lakes, the microorganisms isolated were not true psychrophiles but psychrotrophs. These isolates almost invariably grew at temperatures over 20°C higher than the warmest *in situ* temperature. Survival of these organisms at temperatures below their minimum for growth might involve secondary metabolic processes which would occur just a few degrees below the temperature in which these organisms grow *in situ* since the difference between the temperature for growth and freezing might be just 3 to 5°C. Secondary metabolism would occur just prior to freezing and the resulting dormancy of the cells. This transition would be necessary since it is obvious that bacteria which are in the log phase are more susceptible to abrupt changes in environmental conditions than are cells in the lag phase.

It can be concluded, therefore, that it might be possible to differentiate autochthonous from zymogenous bacterial populations in stable cold environments by their thermal range for growth. In unstable cold environments, only a zymogenous population exists. These zymogenous bacteria generally have a wide temperature range of growth since the most important factor for their survival is efficient secondary metabolism which must occur at temperatures below the optimum for growth.

One of the obvious advantages that psychrotrophic bacteria would have in both stable and unstable cold environments is that their rate of primary metabolism would be very low at the *in situ* suboptimum (for psychrotrophs) temperatures and, thus, would be more efficient. This is particularly important when the levels of nutrients in most cold environments are considered. In marine waters below the photic zone and in most Antarctic terrestrial environments, for example, the levels of organic metabolites are very low and extremely rapid metabolism would quickly exhaust the available supply, thus forcing the cells into starvation conditions, perhaps irreversibly. A corollary to this hypothesis is that the zymogenous bacteria in marine environments might not ever be active in the water column or in sediments because they normally reside in marine sub-environments which consistently have high levels of organic nutrients. Examples of these sub-environments would include the guts of fish and invertebrates, the surfaces of micro- and macrophytic plants and animals, surface waters within the photic zone and some shallow sediments. It is interesting that the majority of psychrophiles isolated from marine samples have been *Vibrio* or related organisms. *Vibrio* have been repeatedly isolated from the guts of both marine fishes (Liston, 1957) and invertebrates (Liston and Colwell, 1963; Baross and Liston, 1970), and it has been suggested that these organisms might be part of the commensal flora of these animals. Goodrich (1976) and Chan (1970) reported that most of the *in situ* decomposition of chitin takes place in the guts of marine animals as a result of chitinoclasts which are usually *Vibrio*. Goodrich (1976) also indicated that bacterial chitinase activity was more efficient in animal guts than in sea-water chitin medium. It is possible that the psychrophilic vibrios isolated from marine waters might actually represent excreted bacteria and thus would be the indigenous marine "coliforms". The fate of these bacteria after excretion would not be as mineralizers but as food for the benthos.

As previously mentioned, the first class of low temperature organisms, which fit the definition of Morita (1975) for psychrophiles, is most frequently isolated from marine environments. Several investigators have considered that an exclusive property of psychrophiles is that they were Gram negative rods (Ingraham and Stokes, 1959; Farrell and Rose, 1967) and that this could be interpreted as a characteristic of Gram negative membranes. This seems to be particularly true of some psychrophiles that have maximum temperatures for growth at 10 to 15°C. Many Gram positive bacteria, however, have been isolated that fit Morita's definition or have a maximum growth temperature just slightly above 20°C (Larkin and Stokes, 1966; Marshall and Ohye, 1966; Matches and Liston, 1973). What is important to emphasize, however, is that in unstable cold environments, such as found in the Antarctic, Gram negative bacteria and usually Gram positive aerobic and anaerobic sporeformers are rarely isolated (Flint and Stout, 1960; Cameron *et al.*, 1973). In stable cold environments, most of the bacterial types isolated are Gram negative rods, usually *Pseudomonas*, *Vibrio* and other related oxidative genera. Gram positive organisms, usually *Arthrobacter*, *Corynebacterium*, *Brevibacterium* and other related types and various sporeformers have been isolated much less often than Gram negatives.

The second class of cold temperature bacteria, which are found predominently in unstable cold environments, are Gram positive cocci and rods, predominantly of the genera *Arthrobacter*, *Corynebacterium*, *Brevibacterium*, *Kurthia*, *Cellulomonas* and related types. These organisms are the most common types found in Antarctic and Arctic soils, lakes, glaciers, ice and snow, in ice caves and in the upper atmosphere. A particular property of these genera is their apparent ability to survive under very harsh environmental conditions. It is interesting that when raw sewage samples were exposed to the extreme conditions of low temperature (4°C) and high hydrostatic pressure (500 to 1000 atm) found in some ocean trenches, the only genera of bacteria which could be recovered after 500 h exposure were arthrobacter types (Baross *et al.*, 1975). Moreover, Ensign (1970) and Boylen and Ensign (1970) reported that *Arthrobacter crystallopoietes* could survive longer periods of time under starvation conditions than Gram negatives. This long-term survival was due to the efficiency of this species to endogenously metabolize in the absence of an external source of nutrients. Evison and Rose (1965) also reported that *Corynebacterium erythrogenes*, in contrast to marine psychrophiles, could survive without loss of viability at temperatures 3 to 5°C above the maximum growth temperature. Sieburth (1964) showed that a marine *Arthrobacter* showed multiple temperature optima over its growth range from approximately 0 to 40°C. At each temperature optimum this species showed morphological variations. A characteristic of *Arthrobacter* and related organisms is that they form coccoid cells during stationary phase or under starvation conditions (Mulder and Antheunisse, 1963; Cure and Keddie, 1973). There is also evidence that soil arthrobacters can survive long periods of starvation *in situ*, thus accounting for their widespread occurrence in unstable terrestrial environments (Zevenhuizen, 1966; Ensign, 1970). The relative sparsity of various *Bacillus* in Antarctic dry soils and other polar environments is probably due to the inability of these species to germinate after spore formation. The temperature in some of the more harsh cold environments might not get high enough to allow spores to germinate. The *Bacillus* that have been isolated from polar regions were usually isolated from subantarctic or subarctic regions. Most of these species grew at 25 to 30°C but were capable of sporulation at 0°C (Marshall and Ohye, 1966).

XI. Concluding Remarks

It can be concluded that psychrophilic bacteria and *in situ* microbial activity at psychrophilic temperatures are not necessarily mutually inclusive. In psychrophilic environments, the stability of the temperature characteristic is the most important factor determining the types of microorganisms that occur and the extent of their activity. Psychrophiles that fit the definition of Morita (1975) are almost exclusively isolated from stable cold environments. Presumably, these psychrophiles have evolved enzymes and synthetic systems that function optimally at temperatures near *in situ*. Rapid mineralization activity and short generation times *in situ* might be advantageous to these organisms, particularly if they reside in environments that either continuously or seasonally contain high levels of organic nutrients. In most cold environments, however, the nutrient levels are low and frequently temperatures which are in the range that will allow microbial activity and growth may only occur for one or two months of the year. In these environments, psychrotrophs, all of which apparently can survive extended periods of extremes in temperature and dryness, predominate.

It is obvious, therefore, that an overwhelming portion of the biologically active areas on earth are either continuously or seasonally cold. Some of the most biologically productive oceanic and terrestrial environments are continuously at psychrophilic temperatures, yet, our knowledge of the types of microorganisms and the kinds and extent of their activity *in situ* is embarrassingly small. The precipitous increase in man's influence on and utilization of resources from the worlds' oceans and from polar terrestrial environments underlines the importance of our need for greater understanding of microbial activities at low temperatures.

Acknowledgement

This paper was supported by NSF grant DES73–06611–A02.

References

Alexander, M. (1961). *In* "Introduction to Soil Microbiology." Wiley, New York.

Alexander, V. (1974). A synthesis of the IBP tundra biome circumpolar study of nitrogen fixation. *In* "Soil Organisms and Decomposition in Tundra" (Eds A. J. Holding, O. M. Heal, S. F. Maclean, Jr. and P. W. Flanagan), pp. 109–121. Swedish IBP Committee, Stockholm.

Alexander, V. and Schell, D. M. (1973). Seasonal and spatial variation of nitrogen fixation in the Barrow, Alaska tundra. *Artic Alpine Res.* **5,** 77–88.

Allen, H. L. (1968). Acetate in fresh water. Natural substrate concentrations determined by dilution bioassay. *Ecology* **49,** 346–349.

Allen, H. L. (1969). Chemo-organotrophic utilization of dissolved organic compounds by planktonic algae bacteria in a pond. *Int. Rev. Ges. Hydrobial.* **54,** 1–33.

Anderson, R. L. and Ordal, E. J. (1961). *Cytophaga succinicans* sp. n., a facultatively anaerobic aquatic myxobacterium. *J. Bacteriol.* **81,** 130–146.

Baker, J. H. (1970a). Yeast, molds, and bacteria from an acid peat on Signy Island. *In* "Antarctic Ecology" (Ed. M. W. Holdgate), Vol. 2, pp. 717–722. Academic Press, London and New York.

Baker, J. H. (1970b). Quantitative study of yeasts and bacteria in a Signy Island peat. *Br. Antarctic Surv. Bull.* **23,** 51–55.

Baker, J. H. and Smith, D. G. (1972). The bacteria in an antarctic peat. *J. Appl. Bacteriol.* **35,** 589–596.

Bonoub, M. W. and Williams, P. J. LeB. (1972). Measurement of microbial activity and organic material in the western Mediterranean Sea. *Deep-Sea Res.* **19,** 433–443.

Barber, R. T. (1966). Interaction of bubbles and bacteria in the formation of organic aggregates in sea water. *Nature, Lond.* **211,** 257–258.

Barghoorn, E. S. and Nichols, R. L. (1961). Sulfate-reducing bacteria and pyritic sediments in Antarctica. *Science* **134,** 190.

Baross, J. A. (1972). Some influences of temperature, bacteriophage, and other ecological parameters on the distribution and taxonomy of marine vibrios. Ph.D. thesis, University of Washington, Seattle.

Baross, J. A. and Liston, J. (1970). Occurrence of *Vibrio parahaemolyticus* and related hemolytic vibrios in marine environments of Washington State. *Appl. Microbiol.* **20,** 179–187.

Baross, J. A., Hanus, F. J. and Morita, R. Y. (1974). The effects of hydrostatic pressure on uracil uptake, ribonucleic acid synthesis, and growth of three obligately psychrophilic marine vibrios, *V. alginolyticus* and *Escherichia coli. In* "Effect of the Ocean Environment on Microbial Activities" (Eds R. R. Colwell and R. Y. Morita), pp. 180–202. University Park Press, Baltimore.

Baross, J. A., Hanus, F. J. and Morita, R. Y. (1975). The survival of human enteric and other sewage microorganisms under simulated deep sea conditions. *Appl. Microbiol.* **21,** 309–318.

Barsdate, R. J. and Alexander, V. (1970). Photosynthetic organisms in subarctic lake ice. *Arctic* **23,** 201.

Bartholomew, W. V. and Norman, A. G. (1944). Microbial thermogenesis in the decomposition of plant materials. III. *J. Bacteriol.* **47,** 499–504.

Beerens, H., Sugama, S. and Tahon-Castel, M. (1965). Psychrotrophic clostridia. *J. Appl. Bacteriol.* **28,** 36–48.

Benoit, R. E. and Hall, C. L. Jr. (1970). The microbiology of some dry valley soils of Victoria Land, Antarctica. *In* "Antarctic Ecology" (Ed. M. W. Holdgate), Vol. 2, pp. 697–701. Academic Press, New York and London.

Benoit, R., Hatcher, R. and Green, W. (1971). Bacteriological profiles and some chemical characteristics of two permanently frozen Antarctic lakes. *In* "The Structure and Function of Freshwater Microbial Communities" (Ed. J. Cairns, Jr.), pp. 281–293. Virginia Polytechnic Institute and State University, Blacksburg, Virginia.

Biederbeck, V. O. and Campbell, C. A. (1971). Influence of simulated fall and spring conditions on the soil system. I. Effect of soil microflora. *Soil Sci. Soc. Am. Proc.* **35,** 471–479.

Blanchard, D. C. and Syzdek, L. D. (1970). Mechanism for the water to air transfer and concentration of bacteria. *Science* **170,** 626–628.

Bliss, L. C. (1962). Adaptations of arctic and alpine plants to environmental conditions. *Arctic* **15,** 117–134.

Bliss, L. C., Courtin, G, M., Pattie, D. L., Riewe, R. R., Whitfield, D. W. A. and Widden, P. (1973). Arctic tundra ecosystems. *Ann. Rev. Ecol. Syst.* **4,** 359–399.

Borg, A. F. (1960). Studies on myxobacteria associated with diseases in salmonid fishes. *Wildlife Dis.* **8,** 1–85.

Bott, T. L. (1975). Bacterial growth rates and temperature optima in a stream with a fluctuating thermal regime. *Limnol. Oceanogr.* **20,** 191–197.

Boyd, W. L. (1958). Microbiological studies of Arctic soils. *Ecology* **39,** 332–336.

Boyd, W. L. and Boyd, J. W. (1962a). Soil microorganisms of the McMurdo Sound area, Antarctica. *Appl. Microbiol.* **11,** 116–121.

Boyd, W. L. and Boyd, J. W. (1962b). Presence of *Azotobacter* species in polar regions. *J. Bacteriol.* **83,** 429–430.

Boyd, W. L. and Boyd, J. W. (1963). Enumeration of marine bacteria of the Chukchi Sea. *Limnol. Oceanogr.* **8,** 343–348.

Boyd, W. L. and Boyd, J. W. (1964). The presence of bacteria in permafrost of the Alaskan Arctic. *Can. J. Microbiol.* **10,** 917–919.

Boyd, W. L. and Boyd, J. W. (1967). Microbial studies of aquatic habitats of the area of Inuvik, Northwest territories. *Arctic* **20,** 27–41.
Boyd, W. L. and Boyd, J. W. (1971). Studies of soil microorganisms. Inuvik, Northwest territories. *Arctic* **24,** 162–176.
Boyd, W. L., Staley, J. T. and Boyd, J. W. (1966). Ecology of soil microorganisms of Antarctica. *Antarctic Res. Ser.* **8,** 125–159.
Boylen, C. W. and Brock, T. D. (1973). Bacterial decomposition processes in Lake Wingra sediments during winter. *Limnol. Oceanogr.* **12,** 628–634.
Boylen, C. W. and Ensign, J. C. (1970). Intracellular substrates for endogenous metabolism during long-term starvation of rod and spherical cells of *Arthrobacter crystallopoietes*. *J. Bacteriol.* **103,** 578–587.
Braarud, T. (1961). Cultivation of marine organisms as means of understanding environmental influences on populations. *In* "Oceanography" (Ed. M. Sears), pp. 271–298. Am. Assoc. Adv. Sci., Publ. No. 67, Washington, D.C.
Breuil, C. and Gounot, A.-M. (1972). Recherches préliminaries sur les bactéries lipolytiques psychrophiles des sols et des eaux. *Can. J. Microbiol.* **18,** 1445–1451.
Brock, T. D., Passman, F. and Yoder, I. (1973). Absence of obligately psychrophilic bacteria in constantly cold springs associated with caves in southern Indiana. *Am. Midland Nat.* **90,** 240–246.
Brockman, E. R. and Boyd, W. L. (1963). Myxobacteria from soils of the Alaskan and Canadian Arctic. *J. Bacteriol.* **86,** 605–606.
Brown, P. E. and Smith, R. E. (1912). Bacterial activities in frozen soils. *Iowa Agric. Exp. Stn. Res. Bull.* **4,** 155–184.
Buchanan, R. E. and Gibbons, N. E. (1974). *In* "Bergey's Manual of Determinative Bacteriology" (8th ed.) Williams and Wilkins, Baltimore.
Bunt, J. S. (1963). Diatoms of antarctic sea ice as agents of primary production. *Nature, Lond.* **199,** 1255–1257.
Bunt, J. S. (1971). Microbial productivity in polar regions. *In* "Microbes and Biological Productivity" (Ed. D. E. Hughes and A. H. Rose) pp. 333–354. Twenty-first Symposium of the Society of General Microbiology, Cambridge University Press, London.
Bunt, J. S. and Lee, C. C. (1970). Seasonal primary production in antarctic sea ice at McMurdo Sound in 1967. *J. Mar. Res.* **28,** 304–320.
Bunt, J. S. and Rovira, A. D. (1955). The effect of temperature and heat treatment on soil metabolism. *J. Soil Sci.* **6,** 129–136.
Bunt, J. S. and Wood, E. J. F. (1963). Microalgae and Antarctic sea ice. *Nature, Lond.* **199,** 1254–1255.
Burch, C. W. (1967). Microbes in the upper atmosphere and beyond. *In* "Airborne Microbes" (Eds P. H. Gregory and J. L. Monteith), pp. 354–374. Seventeenth Symposium of the Society for General Microbiology, Cambridge University Press, London.
Burnison, B. K. and Morita, R. Y. (1974). Heterotrophic potential for amino acid uptake in a naturally eutrophic lake. *Appl. Microbiol.* **27,** 488–495.
Cameron, R. E. (1972). Microbial and ecologic investigations in Victoria Valley, Southern Victoria Land, Antarctica. *In* "Antarctic Terrestrial Biology", Vol. 20 (Ed. G. A. Llano), pp. 195–260. Antarctic Research Series American Geophysical Union, Washington, D.C.
Cameron, R. E. and Benoit, R. E. (1970). Microbial and ecological investigations of recent cinder cones, Deception Island, Antarctica. A preliminary report. *Ecology* **21,** 802–807.

Cameron, R. E. and Devaney, J. R. (1970). Antarctic soil algal crests: scanning electron and optical microscope study. *Trans. Am. Microsc. Soc.* **89,** 264–273.

Cameron, R. E., King, J. and David, C. N. (1970a). Microbiology, ecology and microclimatology of soil sites in dry valley of Southern Victoria Land, Antarctica. *In* "Antarctic Ecology" (Ed. M. W. Holdgate), Vol. 2, pp. 702–716. Academic Press, London and New York.

Cameron, R. E., Hanson, R. B., Lacy, G. H. and Morelli, F. A. (1970b). Soil microbial and ecological investigations in the Antarctic interior. *Antarctic J. U.S.* **5,** 87–88.

Cameron, R. E., Lacy, G. H. and Morelli, F. A. (1971). Farthest south soil microbial and ecological investigations. *Antarctic J. U.S.* **6,** 105–106.

Cameron, R. E., Morelli, F. A. and Randall, L. P. (1972a). Aerial, aquatic, and soil microbiology of Don Juan Pond, Antarctica. *Antarctic J. U.S.* **7,** 254–258.

Cameron, R. E., Morelli, F. A. and Johnson, R. M. (1972b). Bacterial species in soil and air of the Antarctic Continent. *Antarctic J. U.S.* **7,** 187–189.

Cameron, R. E., Morelli, F. A. and Honour, R. C. (1973). Aerobiological monitoring of dry valley drilling sites. *Antarctic J. U.S.* **8,** 211–214.

Campbell, C. A., Biederbeck, V. O. and Warder, F. G. (1973). Influence of simulated fall and spring conditions on the soil system. III. Effect of method of simulating spring temperatures on ammonification, nitrification, and microbial populations. *Soil Sci. Soc. Am. Proc.* **37,** 382–386.

Carlucci, A. F. and Williams, P. N. (1965). Concentration of bacteria from sea water by bubble scavenging. *J. Con. Perm. Int. Explor. Mer.* **30,** 28–33.

Certes, A. (1884). Sur la culture, à l'abrides germes atmospherique des caux et des sediments rapportes par les expeditions du travailleur et du talisman: 1882–1883, *Compt. Rend. Acad. Bulg. Sci.* **90,** 690–693.

Chan, J. G. (1970). The occurrence, taxonomy and activity of chitinoclastic bacteria from sediment, water and fauna of Puget Sound. Ph.D. Thesis, University of Washington, Seattle.

Christensen, P. J. (1974). A microbiological study of some lake water and sediments from the Mackenzie Valley with special reference to cytophagas. *Arctic* **27,** 390–311.

Christian, R. R. and Wiebe, W. J. (1974). The effects of temperature upon the reproduction and respiration of a marine obligate psychrophile. *Can. J. Microbiol.* **20,** 1341–1345.

Clark, F. E., Jackson, R. D. and Gardner, H. R. (1962). Measurement of microbial thermogenesis in soil. *Soil Sci. Am. Proc.* **26,** 155–160.

Collins, V. G. (1960). The distribution and ecology of gram-negative organisms other than *Enterobacteriaceae* in lakes. *J. Appl. Bacteriol.* **23,** 510–514.

Conn, H. J. (1910). Bacteria of frozen soil. *Zent. Bakt.* Abt. II, **28,** 422–434.

Conn, H. J. (1911). Bacteria in frozen soils. *Zent. Bakt.* Abt. II, **32,** 70–97.

Conn, H. J. (1914a). Bacteria of frozen soil. III. *Zent. Bakt.* Abt. II, **42,** 510–519.

Conn, H. J. (1914b). Bacteria in frozen soils. N.Y. Agric. Stn. (Geneva) Tech. Bull. No. 35, 20 pp.

Conn, H. J. (1948). The most abundant groups of bacteria in soil. *Bacteriol. Rev.* **12,** 257–273.

Corpe, W. A. and Winters, H. (1972). Hydrolytic enzymes of some periphytic marine bacteria. *Can. J. Microbiol.* **18,** 1483–1490.

Coumartin, V. (1963). Review of the microbiology of underground environments. *Nat. Speleol. Soc. Bull.* **25,** 1–14.

Cummins, K. W., Klug, M. J., Wetzel, R. G., Petersen, R. C., Suberkropp, K. F., Manny B. A., Wuycheck, J. C. and Howard, F. P. (1972). Organic enrichment with leaf leachate in experimental lotic ecosystems. *Bioscience* **22**, 719–722.

Cummins, K. W., Petersen, R. C., Howard, F. O., Wuycheck, J. C. and Holt, V. I. (1973). The utilization of leaf litter by stream detritorvores. *Ecology* **54**, 336–345.

Cure, G. L. and Keddie, R. M. (1973). Methods for the morphological examination of aerobic coryneform bacteria. *In* "Sampling–Microbiological Monitoring of Environments" (Eds R. G. Board and D. W. Lovelock), pp. 123–135. Academic Press, London and New York.

Dale, N. G. (1974). Bacteria in intertidal sediments: factors related to their distributions. *Limnol. Oceanogr.* **19**, 509–518.

Daley, R. J. and Hobbie, J. E. (1975). Direct counts of aquatic bacteria by a modified epifluorescence technique. *Limnol. Oceanogr.* **20**, 875–882.

Darling, C. A. and Siple, P. A. (1941). Bacteria of Antarctica. *J. Bacteriol.* **42**, 83–98.

D'Aoust, J. Y. and Kushner, D. J. (1971). Structural changes during lysis of a psychrophilic marine bacterium. *J. Bacteriol.* **108**, 916–927.

D'Aoust, J. Y. and Kushner, D. J. (1972). *Vibro psychoerythrus* sp. n.: classification of the psychrophilic marine bacterium, NRC 1004, *J. Bacteriol.* **111**, 340–342.

Dempster, J. F. (1968). Distribution of psychrophilic microorganisms in different dairy environments. *J. Appl. Bacteriol.* **31**, 290–301.

Domaradski, I. V., Grigoryan, E. G., Borzenkova, V. I. and Val'kov, B. G. (1968). Multiplication of plague causative agents in sterile and non-sterile soil. *J. Microbiol. Epidemiol. Immunobiol.* **45**, 104–108.

Druce, R. G. and Thomas, S. B. (1970). An ecological study of psychrotrophic bacteria of soil, water, grass, and hay. *J. Appl. Bacteriol.* **33**, 420–435.

Dunican, L. K. and Rosswall, T. (1974). Taxonomy and physiology of tundra bacteria in relation to site characteristics. *In* "Soil Organisms and Decomposition in Tundra" (Eds A. J. Holding, O. M. Heal, S. F. Maclean, Jr. and P. W. Flanagan), pp. 79–92. Swedish IBP Committee, Stockholm.

Eddy, B. P. (1960). The use and meaning of the term 'psychrophilic'. *J. Appl. Bacteriol.* **23**, 189–190.

Egglishaw, H. J. (1972). An experimental study of the breakdown of cellulose in fast-flowing streams. *In* "Detritus and Its Role in Aquatic Ecosystems" (Eds U. Melchiorri-Santolini and J. W. Hopton), pp. 405–428. Mem. Int. Ital. Idrobial., 29, Suppl. Pallanza, Italy.

Ehrlich, R. (1974). Survival of airborne microorganisms at different environmental temperatures. *Dev. Ind. Microbiol.* **15**, 28–32.

Ekelof, E. (1908). "Wissenschaftliche ergebnisse der Schwedischen Sudpolar-Expedition, 1901–1903". Lithographisches Institut des Generalstabs, Stockholm.

Ensign, J. C. (1970). Long-term starvation survival of rod and spherical cells of *Arthrobacter crystallopoietes*. *J. Bacteriol.* **103**, 569–577.

Evison, J. M. and Rose, A. H. (1965). A comparative study on the biochemical basis of the maximum temperatures for growth of three psychrophilic microorganisms. *J. Gen. Microbiol.* **40**, 349–364.

Farrell, J. and Rose, A. (1967). Temperature effects on microorganisms. *Ann. Rev. Microbiol.* **21**, 101–120.

Finne, G. and Matches, J. R. (1974). Low temperature-growing clostridia from marine sediments. *Can. J. Microbiol.* **20**, 1639–1645.

Fischer, W. H., Lodge, J. P., Jr., Pate, J. B. and Cadle, R. D. (1969). Antarctic atmosphere chemistry: Preliminary exploration. *Science* **164**, 66–67.

Flanagan, P. W. and Veum, A. K. (1974). Relationships between respiration, weight loss, temperature and moisture in organic residues on tundra. *In* "Soil Organisms and Decomposition in Tundra" (Eds A. J. Holding, O. M. Heal, S. F. Maclean, Jr. and P. W. Flanagan), pp. 249–277. Swedish IBP Committee, Stockholm.

Flint, A. E. and Stout, J. D. (1960). Microbiology of some soils from Antarctica. *Nature, Lond.* **188,** 767–768.

Floodgate, G. D. (1972). The mechanism of bacterial attachment to detritus in aquatic systems. *In* "Detritus and its Role in Aquatic Ecosystems" (Eds U. Melchiorri-Santolini and J. W. Hopton), pp. 309–323. Mem. Ist. Ital. Idrobiol 29, Suppl. Pallanza, Italy.

Fogg, G. E. (1967). Observations on the snow algae of the South Orkney Islands. *Phil. Trans. R. Soc. Lond., B.* **252,** 279–287.

Fogg, G. E. and Stewart, W. D. P. (1968). *In situ* determinations of biological nitrogen fixation in Antarctica. *Br. Antarctic Surv. Bull.* **15,** 39–46.

Forster, J. (1887). Ueber einige Eigenschaften Leuchtender Bakterien. *Cent. Bacteriol. Parasitenk.* **2,** 337–340.

Fournelle, H. J. (1967). Soil and water bacteria in the Alaskan subarctic tundra. *Arctic* **20,** 104–113.

Fournier, R. O. (1966). North Atlantic deep-sea fertility. *Science* **153,** 1250–1252.

Fulton, J. D. (1966a). Microorganisms of the upper atmosphere. III. Relationship between altitude and micropopulation. *Appl. Microbiol.* **14,** 233–340.

Fulton, J. D. (1966b). Microorganisms of the upper atmosphere. IV. Microorganisms of a land air mass as it traverses an ocean. *Appl. Microbiol.* **14,** 241–244.

Geesey, G. G. and Morita, R. Y. (1975). Some physiological effects of near maximum growth temperatures on an obligately psychrophilic marine bacterium. *Can. J. Microbiol.* **21,** 811–818.

Gillespie, P. A., Morita, R. Y. and Jones, L. P. (1976). The heterotrophic activity for amino acids, glucose and acetate in Antarctic water. *J. Oceanogr. Soc. Japan.* **32,** 74–82.

Goldman, C. R. (1970). Antarctic freshwater ecosystems, *In* "Antarctic Ecology" (Ed. M. W. Holdgate), pp. 609–627. Academic Press, New York and London.

Goldman, C. R., Mason, D. T. and Hobbie, J. E. (1967). Two Antarctic desert lakes. *Limnol. Oceanogr.* **12,** 295–310.

Goldman, C. R., Mason, D. T. and Wood, B. J. B. (1972). Comparative study of the limnology of two small lakes on Ross Island, Antarctica. *In* "Antarctic Terrestrial Biology" (Ed. B. A. Llano), pp. 1–50. Antarctic Res. Ser., Vol. 20, American Geophysical Union, Washington, D.C.

Goodrich, T. D. (1976). Incidence and significance of bacterial chitinase in the marine environment. Ph.D. thesis, Oregon State University, Corvallis.

Gordon, D. C., Jr. (1970). A microscopic study of organic particles in the north atlantic ocean. *Deep-Sea Res.* **17,** 175–186.

Gordon, R. C. (1970). Depletion of oxygen by microorganisms in Alaskan rivers at low temperatures. *In* "International Symposium on Water Pollution Control in Cold Climates" (Eds R. S. Murphy and D. Nyquist), pp. 71–95. University of Alaska, Institute of Water Resources and Federal Water Quality Administration.

Gounot, A.-M. (1967). Role biologique des arthrobacter dans les limone souterraine. *Ann. Ist. Pasteur.* **113,** 923–943.

Gounot, A.-M. (1968a). Étude microbiologique de boues glaciares arctiques. *Comp. Rendus Acad. Sci. Paris.* **D.266,** 1437–1438.

Gounot, A.-M., (1968b). Étude microbiologique des limons de deux grottes antique. *Comp. Rendus Acad. Sci. Paris.* **D.266,** 1619–1620.

Gounot, A.-M. (1969). Contribution a l'etudes des bacteries des grottes froides. *V. Int. Kongr. Speläologie, Stuttgart* **4,** 1–6.

Gounot, A.-M. (1973a). Bacteries des glaciers et des grottes froides. *Abst. Int. Cong. Bacteriol., Jerusalem* **1973,** 235.

Gounot, A.-M. (1973b). Importance of temperature factor in the study of cold soils microbiology. *Bull. Ecol. Res.Comm.* (*Stockholm*) **17,** 172–173.

Gounot, A.-M., Breuil, C., Borgere, P. and Simeon, D. (1970). Action selective de la temperature sur le micropeuplement des grottes froides. *Compt. Rend. 9th Congr. Nat. Spéléol. Dijon. Spelunca Mem.* **7,** 123–126.

Gray, T. R. G. and Williams, S. T. (1971). "Soil Microorganisms." Oliver and Boyd, Edinburgh.

Gregory, P. H. (1961). "The Microbiology of the Atmosphere," 257 pp. Leonard Hill (Books) Ltd. London.

Gregory, P. H. (1973). "The Microbiology of the Atmosphere," 377 pp. Halsted Press, Wiley, New York.

Haight, R, D. and Morita, R. Y. (1966). Thermally induced leakage from *Vibrio marinus*, an obligately psychrophilic bacterium. *J. Bacteriol.* **92,** 1388–1393.

Hall, K. J. and Hyatt, K. O. (1974). Marion Lake (IBP)—from bacteria to fish. *J. Fish. Res. Bd. Can.* **31,** 893–911.

Hall, K. J., Kleiber, P. M. and Yesaki, I. (1972). Heterotrophic uptake of organic solutes by microorganisms in the sediment. *In* "Detritus and Its Role in Aquatic Ecosystems" (Eds U. Melchiorri-Santolini and J. W. Hopton), pp. 441–471. Mem. Ist. Ital. Idrobiol., 29 Suppl. Pallanza, Italy.

Hamilton, R. D. and Preslan, J. E. (1970). Observation on heterotrophic activity in the eastern tropical pacific. *Limnol. Oceanogr.* **15,** 395–401.

Hamilton, R. D., Holm-Hansen, O. and Strickland, J. D. H. (1968). Notes on the occurrence of living microscopic organisms in deep water. *Deep-Sea Res.* **15,** 651–656.

Harder, W. and Veldkamp, H. (1967). A continuous culture study of an obligately psychrophilic *Pseudomonas* species. *Arch. Microbiol.* **59,** 123–130.

Harder, W. and Veldkamp, H. (1968). Physiology of an obligately psychrophilic marine *Pseudomonas* species. *J. Appl. Bacteriol.* **31,** 12–23.

Harrison, M. J., Wright, R. T. and Morita, R. Y. (1971). Method for measuring mineralization in lake sediments. *Appl. Microbiol.* **21,** 698–702.

Heal, O. W. and French, D. D. (1974). Decomposition of organic matter in tundra. *In* "Soil Organisms and Decomposition in Tundra" (Eds A. J. Holding, O. M. Heal, S. F. Maclean, Jr. and P. W. Flanagan), pp. 279–309. Swedish IBP Committee, Stockholm.

Hobbie, J. E. (1964). Carbon-14 measurements of primary production in two Arctic Alaskan lakes. *Verh. Int. Verein. Theor. Angew. Limnol.* **1,** 360–364.

Hobbie, J. E. and Crawford, C. C. (1969). Respiration corrections for bacterial uptake of dissolved organic compounds in natural waters. *Limnol. Oceanogr.* **14,** 528–532.

Hobbie, J. E. and Wright, R. T. (1968). A new method for the study of bacteria in lakes: description and results. *Mitt. Int. Verein. Limnol.* **14,** 64–71.

Hobbie, J. E., Holm-Hansen, O., Packard, T. T., Pomeroy, L. R., Sheldon, R. W., Thomas, J. P. and Wiebe, W. J. (1972). A study of the distribution and activity of microorganisms in ocean water. *Limnol. Oceanogr.* **17,** 544–555.

Holding, A. J. (1960). The properties and classification of the predominant Gram negative bacteria occurring in soil. *J. Appl. Bacteriol.* **23**, 515–525.

Holding, A. J. (1973). The isolation and identification of certain soil Gram negative bacteria. *In* "Sampling—Microbiological Monitoring of Environments" (Eds R. G. Board and D. W. Lovelock), pp. 137–141. Academic Press, London and New York.

Hoham, R. W. (1975). Optimum temperatures and temperature ranges for growth of snow algae. *Arctic Alpine Res.* **7**, 13–24.

Holm-Hansen, O. (1972). The distribution and chemical composition of particulate material in marine and fresh waters. *In* "Detritus and Its Role in Aquatic Ecosystems" (Eds U. Melchiorri-Santolini and J. W. Hopton), pp. 37–51. Mem. Ist. Ital. Idrobiol. 29 Suppl., Pallanza, Italy.

Holm-Hansen, O. (1973). Determination of total microbial biomass by measurement of adenosine triphosphate. *In* "Estuarine Microbial Ecology" (Eds L.H. Stevenson and R. R. Colwell), pp. 73–89. University South Carolina Press, Columbia.

Holm-Hansen, O., Sutcliff, H. and Sharp, J. (1968). Measurement of deoxyribonucleic acid in the ocean and its ecological significance. *Limnol. Oceangr.* **13**, 507–514.

Horne, A. J. (1972). The ecology of nitrogen fixation of Signy Island, South Orkney Islands. *Br. Antarctic Surv. Bull.* **27**, 1–18.

Horner, R. and Alexander, V. (1972). Algae populations in arctic sea ice: an investigation of heterotrophy. *Limnol. Oceanogr.* **17**, 454–458.

Horowitz, N. H., Cameron, R. E. and Hubbard, J. S. (1972). Microbiology of the dry valleys of Antarctica. *Science* **176**, 242–245.

Hubbard, J. S., Cameron, R. E. and Miller, A. B. (1968). Soil studies—desert microflora. XV. Analysis of Antarctic dry valley soils by cultural and radiorespirometric methods. Space Prog. Summary No. 37–52, **3**, pp. 172–175. Jet Propulsion Laboratory, California Technical Institute, Pasadena.

Hucker, G. J. (1954). Low temperature organisms in frozen vegetables. *Food. Technol.* **8**, 79–108.

Iizuka, H., Ianabe, I. and Meguro, H. (1966). Microorganisms in planton-ice of the Antarctic ocean. *J. Gen. Appl. Microbiol.* **12**, 101–102.

Ingraham, J. L. (1958). Growth of psychrophilic bacteria. *J. Bacteriol.* **75**, 75–80.

Ingraham, J. L. and Stokes, J. L. (1959). Psychrophilic bacteria. *Bacteriol. Rev.* **23**, 97–108.

Ingram, M. (1965). Psychrophilic and psychrotrophic microorganisms. *Annals. Inst. Pasteur, Lille.* **16**, 111–118.

Innis, W. E. (1975). Interaction of temperature and psychrophilic microorganisms. *Ann. Rev. Microbiol.* **29**, 445–465.

Ivarson, K. C. (1965). The microbiology of some permafrost soils in the MacKenzie Valley, N. W. T. *Arctic* **18**, 256–260.

Jannasch, H. W. and Jones, G. E. (1959), Bacterial populations in sea water as determined by different methods of enumeration. *Limnol. Oceanogr.* **4**, 128–139.

Jannasch, H. W. and Wirsen, C. O. (1973). Deep-sea microorganisms: *In situ* response to nutrient enrichment. *Science* **180**, 641–643.

Jannasch, H. W., Eimhjellen, K., Wirsen, C. O. and Farmanfarmaian, A. (1971). Microbial degradation of organic matter in the deep sea. *Science* **171**, 672–675.

Janota-Bassalik, L. (1963a). Psychrophiles in low-moor peat. *Acta Microb. Pol.* **12**, 25–40.

Janota-Bassalik, L. (1963b). Growth of psychrophilic and mesophilic strains of peat bacteria. *Acta Microb. Pol.* **12,** 41–54.

Jones, J. G. (1971). Studies on freshwater bacteria: factors which influence the population and its activity. *J. Ecol.* **59,** 593–613.

Kalff, J. (1970). Arctic lake ecosystems. *In* "Antarctic Ecology (Ed. M. W. Holdgate), Vol. 2, pp. 651–663. Academic Press, New York and London.

Kaushik, N. K. and Hynes, H. B. N. (1968). Experimental study on the role of autumn-shed leaves in aquatic environments. *J. Ecol.* **56,** 229–243.

Kaushik, N. K. and Hynes, H. B. N. (1971). The fate of the dead leaves that fall into streams. *Arch. Hydrobiol.* **68,** 229–243.

Klein, L. (1962). "River Pollution. II. Causes and Effects." Butterworths, London.

Koob, D. D. and Leister, G. L. (1972). Primary productivity and associated physical, chemical, and biological characteristics of Lake Bonney: A perennially ice-covered lake in Antarctica. *In* "Antarctic Terrestial Biology" (Ed. G. A. Llano), Vol. 20, pp. 51–68. Antarctic Research Series, American Geophysical Union, Washington, D.C.

Kriss, A. E. (1963). "Marine Microbiology: Deep Sea." Oliver and Boyd, Edinburgh and London.

Lacy, G. H., Cameron, R. E., Hanson, R. B. and Morelli, F. A. (1970). Microbiological analysis of snow and air from the Antarctic interior. *Antarctic J. U.S.* **5,** 88–89.

Larkin, J. M. (1970). Seasonal incidence of bacterial temperature types in Louisiana soil and water. *Appl. Microbiol.* **20,** 286–288.

Larkin, J. M. and Stokes, J. L. (1966). Isolation of psychrophilic species of *Bacillus. J. Bacteriol.* **91,** 1667–1671.

Lenz, J. (1972). The size distribution of particles in marine detritus. *In* "Detritus and Its Role in Aquatic Ecosystems" (Eds U. Melchiorri-Santolini and J. W. Hopton), pp. 17–35. Mem. Ist. Ital. Idrobiol. 29 Suppl., Pallanza, Italy.

Liston, J. (1957). The occurrence and distribution of bacterial types on flatfish. *J. Gen. Microbiol.* **16,** 205–216.

Liston, J. (1968). Distribution, taxonomy and function of heterotrophic bacteria on the sea floor. *Bull. Misaki Marine Biol. Inst. Kyoto Univ.* **12,** 97–104.

Liston, J. and Colwell, R. R. (1963). Host and habitat relationships of marine commensal bacteria. *In* "Symposium on Marine Microbiology" (Ed. C. H. Oppenheimer), pp. 611–624. Charles C. Thomas, Springfield, Illinois.

Llano, G. A. (1961). Status of lichenology in Antarctica. *In* "Science in Antarctica", Natl. Acad. Sci., Natl. Res. Council Publ. **839,** 13–19.

Lochhead, A. G. (1924). Microbiological studies of frozen soil. *Trans. R. Soc. Can.*, Ser. III **18,** 75–96.

Lochhead, A. G. (1926). The bacterial types occurring in frozen soils. *Soil Sci.* **21,** 225–231.

Mackay, R. J. and Kalff, J. (1973). Ecology of two related species of caddisfly larvae in the organic substrates of a woodland stream. *Ecology* **54,** 499–511.

Mackay, J. R. and MacKay, D. K. (1974). Snow cover and ground temperatures, Garry Island, N. W. T. *Arctic* **27,** 287–296.

Maki, L. R., Galyan, E. L., Chang-Chien, M. and Caldwell, D. R. (1974). Ice nucleation induced by *Pseudomonas syringae. Appl. Microbiol.* **28,** 456–459.

Marshall, B. J. and Ohye, D. F. (1966). *Bacillus macquariensis* n. sp. a psychrotrophic bacterium from sub-Antarctic soil. *J. Gen. Microbiol.* **44,** 41–46.

Marshall, K. C., Stout, R. and Mitchell, R. (1971). Mechanism of the initial events in the sorption of marine bacteria to surfaces. *J. Gen. Microbiol.* **68,** 337–348.

Maruyama, Y. Suzuki, T. and Otobe, K. (1974). Nitrogen fixation in the marine environment: the effect of organic substrates on acetylene reduction. *In* "Effect of the Ocean Environment on Microbial Activities" (Eds R. R. Colwell and R. Y. Morita), pp. 341–353. University Park Press, Baltimore.

Matches, J. R. and Liston, J. (1973). Methods and techniques for the isolation and testing of clostridia from estuarine environment, *In* "Estuarine Microbial Ecology" (Eds L. H. Stevenson and R. R. Colwell) pp. 345–361. University of South Carolina Press, Columbia.

McLean, A. L. (1918). Bacteria of ice and snow in Antarctica. *Nature, Lond.* **102,** 35–39.

McLean, A. L. (1919). Bacteriological and other researches, Australian Antarctic Expedition, 1911–1914. *Sci. Rep. C* 7, No. 4, 13–19.

Meguro, H., Ito, K. and Fukushima, H. (1966). Diatoms and the ecological conditions of their growth in sea ice in the Arctic Ocean. *Science* **152,** 1089–1090.

Meguro, H., Ito, K. and Fukushima, H. (1967). Ice flora (bottom type): a mechanism of primary production in polar seas and the growth of diatoms in sea ice. *Arctic* **20,** 114–133.

Meier, F. C. and Lindbergh, C. A. (1935). Collecting microorganisms from Arctic atmosphere. *Sci. Month.* **40,** 5–20.

Meyer, G. M., Morrow, M. B., Wyss, O., Berg., T. E. and Littlepage J. Q. (1962). Antarctica: the microbiology of an unfrozen saline pond. *Science* **138,** 1103–1104.

Mishoustine, E. N. (1964). Les differents types de sol et la specificite de leur micropopulation. *Ann. Ist. Pasteur* **107,** 63–77.

Moiroud, A. and Gounot, A.-M. (1969). Sur une bacterie pschrophilile obligatoire isolee de limons glaciaries. *Comp. Rend. Acad. Sci. Paris.* **D.269,** 2150–2152.

Morgan, K. G. and Kalff, J. (1972). Bacterial dynamics in two high Arctic lakes. *Freshwater Biol.* **2,** 217–228.

Morita, R. Y. (1966). Marine psychrophilic bacteria. *Oceanogr. Mar. Biol. Ann. Rev.* **4,** 105–121.

Morita, R. Y. (1975). Psychrophilic bacteria. *Bacteriol. Rev.* **39,** 144–167.

Morita, R. Y. (1976). Survival of bacteria in cold and moderate hydrostatic pressure environments with special reference to psychrophilic and barophilic bacteria. *In* "The Survival of Vegatative Microbes" (Eds T. B. Gray and J. R. Postgate), pp. 279–298. Twenty-sixth Symposium of the Society for General Microbiology, Cambridge University Press, Cambridge.

Morita, R. Y. and Burton, S. D. (1970). Occurrence, possible significance, and metabolism of obligate psychrophiles in marine waters. *In* "Organic Matter in Natural Waters" (Ed. D. W. Hood), pp. 275–285. Publ. No. 1, Institute of Marine Science, University of Alaska, Fairbanks.

Morita, R. Y. and Haight, R. D. (1964). Temperature effects on the growth of an obligate psychrophilic marine bacterium. *Limnol. Oceanogr.* **9,** 102–106.

Morita, R. Y. and ZoBell, C. E. (1955). Occurrence of bacteria in pelagic sediments collected during the mid-pacific expedition. *Deep-Sea Res.* **3,** 66–73.

Morita, R. Y., Griffiths, R. P. and Hayasaka, S. S. (1977). Heterotrophic potential of microorganisms in Antarctic waters. Third SCAR/IUBS Symposium on Antarctic Biology N. A. S.

Mulder, E. G. and Antheunisse, J. (1963). Morphologie, physiologie et écologie des *Arthrobacter. Ann. Inst. Pasteur, Paris.* **105,** 46–74.

Nedwell, D. B. and Floodgate, G. D. (1971). The seasonal selection by temperature of heterotrophic bacteria in an intertidal sediment. *Marine Biol.* **11,** 306–310.

Okafor, N. C. (1966). Ecology of microorganisms on chitin buried in soil. *J. Gen. Microbiol.* **44,** 311–327.

Parinkina, O. M. (1974). Bacterial production in tundra soils. *In* "Soil Organisms and Decomposition in Tundra" (Eds A. J. Holding, O. M. Heal, S. F. Maclean, Jr. and P. W. Flanagan), pp. 65–77. Swedish IBP Committee, Stockholm.

Parker, B. C. and Wachtel, M. A. (1971). Seasonal distribution of cobalamin, biotin and niacin in rainwater. *In* "The Structure and Function of Fresh-Water Microbial Communities" (Ed. J. Cairns, Jr.), pp. 195–207. Res. Div. Mon. 3, Virginia Polytechnic Institute and State University, Blacksburg, Virginia.

Petersen, R. C. and Commins, K. W. (1974). Leaf processing in a woodland stream. *Freshwater Biol.* **4,** 343–368.

Procter, B. E. (1935a). The microbiology of the upper air. I. *Proc. Am. Acad. Arts Sci.* **69,** 315–340.

Procter, B. E. (1935b). The microbiology of the upper air. *J. Bacteriol.* **30,** 363–375.

Procter, B. E. and Parker, B. W. (1942). Microorganisms in the upper air. *In* "Aerobiology" (Ed. S. Moulton), pp. 48–54. AAAS Publ. No. 17. Washington, D.C.

Rao, S. S. and Dutka, B. J. (1974). Influence of temperature on lake bacterial activities. *Water Res.* **8,** 525–538.

Rhodes, M. E. (1959). The characterization of *Pseudomonas fluorescens. J. Gen. Microbiol.* **21,** 221–263.

Rittenberg, S. C. (1941). Studies on the marine sulfate reducing bacteria, Ph.D. thesis, University of California.

Robinson, G. G. C., Hendzel, L. L. and Gillespie, D. C. (1973). A relationship between heterotrophic utilization of organic acids and bacterial populations in West Blue Lake, Manitoba. *Limnol. Oceanogr.* **18,** 264–269.

Rodhe, W. (1969). Crystallization of eutrophication concepts in Northern Europe. *In* "Symposium on Eutrophication: Causes, Consequences, Correctives", pp. 50–64. National Academy of Sciences, Washington, D.C.

Rodhe, W., Hobbie, J. E. and Wright, R. T. (1966). Phototrophy and heterotrophy in high mountain lakes. *Mitt. Int. Verein. Theor. Angew. Limnol.* **16,** 302–313.

Ross, D. J. (1965). A seasonal study of oxygen uptake of some pasture soils and activities of enzymes hydrolyzing sucrose and starch. *J. Soil Sci.* **16,** 73–85.

Rosswall, T. (1974). Cellulose decomposition studies on the tundra. *In* "Soil Organisms and Decomposition in the Tundra" (Eds A. J. Holding, O. M. Heal, S. F. Maclean, Jr. and P. W. Flanagan), pp. 325–340. Swedish IBP Committee, Stockholm.

Rosswall, T. and Clarholm, M. (1974). Characteristics of tundra bacterial populations and a comparison with populations from forest and grassland soils. *In* "Soil Organisms and Decomposition in Tundra" (Eds A. J. Holding, O. M. Heal, S. F. Maclean, Jr. and P. W. Flanagan), pp. 93–108. Swedish IBP Committee, Stockholm.

Rozenberg, L. A. (1954), The quality of bacteria of Bering Sea bottoms. A methodical study on quantitative estimation of bacteria. *Trans. Inst. Oceanogr.* **11,** 264–270.

Rudolph, E. D. (1966). Terrestrial vegetation of Antarctica: Past and present studies. *In* "Antarctic Soils and Soil Forming Processes" (Ed. J. C. F. Tedrow), Vol. 8, pp. 109–124, Antarctic Research Series, American Geophysical Union, Washington, D.C.

Sabey, B. R., Bartholomew, W. V., Shaw, R. and Pesek, J. (1956). Influence of temperature on nitrification in soils. *Soil Sci. Soc. Am. Proc.* **20**, 357–360.

Schallock, E. W., Mueller, E. W. and Gordon, R. C. (1970). "Assimilative Capacity of Arctic Rivers" Alaska Scientific Conference, Federal Water Quality Administration, Department of the Interior, Alaska Water Laboratory, College, Alaska. Paper No. 7, 13 pp.

Schnell, R. C. and Vali, G. (1972). Atmospheric ice nuclei from decomposing vegetation. *Nature, Lond.* **236**, 163–165.

Schnell, R. C. and Vali, G. (1973). World-wide source of leaf-derived freezing nuclei. *Nature, Lond.* **246**, 212–213.

Schwarz, J. R., Yayanos, A. A. and Colwell, R. R. (1976). Metabolic activities of the intestinal microflora of a deep-sea invertebrate. *Appl. Environ. Microbiol.* **31**, 46–48.

Seifert, J. (1961). The effect of low temperature on the intensity of nitrification. *Folia. Microbiol., Praha.* **6**, 350–353.

Seki, H. (1971). Microbial clumps in seawater in the euphotic zone of Saanich Inlet (British Columbia). *Marine Biol.* **9**, 4–8.

Seki, H., Wada, H. and Hattori, H. (1974). Evidence of high organotrophic potentiality of bacteria in the deep sea. *Marine Biol.* **26**, 1–4.

Sheldon, R. W., Evelyn, T. P. T. and Parsons, T. R. (1967). On the occurrence and formation of small particles in seawater. *Limnol. Oceanogr.* **12**, 367–375.

Sieburth, J. McN. (1963). Bacterial habitats in the Antarctic environment. *In* "Symposium on Marine Microbiology" (Ed. C. H. Oppenheimer), pp. 533–548. Charles C. Thomas, Springfield, Illinois.

Sieburth, J. McN. (1964). Polymorphism of a marine bacterium (*Arthrobacter*) as a function of multiple temperature optima and nutrition. *In* "Symposium on Experimental Marine Ecology," pp. 11–16, University of Rhode Island Occasional Publication No. 2, University of Rhode Island, Kingston.

Sieburth, J. McN. (1965). Microbiology of Antarctica. *In* "Biogeography and Ecology in Antarctica" (Eds P. Van Oye and J. Van Mieghem), pp. 267–295. Monographiae Biologicae, Vol. 15. W. Junk, The Hague.

Sieburth, J. McN. (1967). Seasonal selection of estuarine bacteria by water temperature. *J. Exp. Marine Biol. Ecol.* **1**, 98–121.

Sieburth, J. McN. (1971). Distribution and activity of oceanic bacteria. *Deep-Sea Res.* **18**, 1111–1121.

Sinclair, N. A. and Stokes, J. L. (1964). Isolation of obligately anaerobic psychrophilic bacteria. *J. Bacteriol.* **87**, 562–565.

Smith, D. K., Benedict, C. D. and Weinberg, E. D. (1974). Bacterial culture longevity: Control by inorganic phosphate and temperature. *Appl. Microbiol.* **27**, 292–293.

Smyk, B. and Ettinger, L. (1963). Recherches sur quelques espéces d'arthrobacter fixatrices d'azote isolées des roches karstiques alpines. *Ann. Inst. Pasteur, Paris.* **105**, 341–348.

Sorokin, Yu. I. (1973). Productivity of bacterioplankton in the western pacific. *Oceanology* **13**, 70–80.

Sorokin, Yu. I. and Konovalova, I. W. (1973). Production and decomposition of organic matter in a bay of the Japan Sea during the winter diatom blooms. *Limnol. Oceanogr.* **18**, 962–967.

Soulides, D. A. and Allison, R. E. (1961). Effect of drying and freezing soils on carbon dioxide production, available mineral nutrients, aggregation, and bacterial populations. *Soil Sci.* **91,** 291–298.

Stanford, G., Frere, M. H. and Schwaninger, D. H. (1973). Temperature coefficient of soil nitrogen mineralization. *Soil Sci.* **115,** 321–323.

Stanier, R. Y., Palleroni, N. J. and Doudoroff, M. (1966). The aerobic pseudomonads: a taxonomic study. *J. Gen. Microbiol.* **43,** 159–271.

Stanley, S. O. and Morita, R. Y. (1968). Salinity effect on the maximum growth temperature of some bacteria isolated from marine environments. *J. Bacteriol.* **95,** 169–173.

Stokes, J. L. and Redmond, M. L. (1966). Quantitative ecology of psychrophilic microorganisms. *Appl. Microbiol.* **14,** 74–78.

Straka, R. P. and Stokes, J. L. (1959). Metabolic injury to bacteria at low temperatures. *J. Bacteriol.* **78,** 181–185.

Straka, R. P. and Stokes, J. L. (1960). Psychrophilic bacteria from Antarctica. *J. Bacteriol.* **80,** 622–625.

Streten, N. A. (1974). Some features of the summer climate of interior Alaska. *Arctic* **27,** 273–286.

Strzelczyk, E., Rouatt, J. W. and Peterson, E. A. (1969). Studies on actinomycetes from soils of Baffin Island. *Arctic* **22,** 130–139.

Sweeting, M. M. (1973). "Karst Landforms." Columbia University Press, New York.

Tait, R. V. and DeSanto, R. S. (1972). "Elements of Marine Ecology." Springer Verlag, New York.

Teal, J. M. (1957). Community metabolism in a temperate cold spring. *Ecol. Monog.* **27,** 283–302.

Tedrow, J. C. F. and Douglas, L. A. (1959). Organic matter decomposition rates in arctic soils. *Soil Sci.* **88,** 305–312.

Tedrow, J. C. F. and Ugolini, F. C. (1966). Antarctic soils. *In* "Antarctic Soils and Soil Forming Processes" (Ed. J. C. F. Tedrow), pp. 161–177. Antarctic Research Series, Vol. 8. American Geophysical Union, Washington, D.C.

Thompson, B. M. and Hamilton, R. D. (1973). Heterotrophic utilization of sucrose in an artificially enriched lake. *J. Fish Res. Bd. Can.* **30,** 1547–1552.

Tilzer, M. (1972). Dynamics and productivity of phytoplankton and pelagic bacteria in high mountain lakes. *Arch. Hydrobiol.* **40,** 201–273.

Triska, F. J. (1970). Seasonal distribution of aquatic hyphomycetes in relation to the disappearance of leaf litter from a woodland stream. Ph.D. Thesis, Univ. Pittsburgh, Pittsburgh.

Tsiklinsky, M. (1908). "La flore microbienne dans les regions du Pôle Sud. Expedition Antarctique francaise, 1903–1905." Masson et Cie, Paris.

Ugolini, F. C. (1970). Antarctic soils and their ecology. *In* "Antarctic Ecology" (Ed. M. W. Holdgate), Vol. 2, pp. 673–692. Academic Press, New York and London.

Ugolini, F. C. (1972). Ornithogenic soils of Antarctica. *In* "Antarctic Terrestrial Biology" (Ed. G. A. Llano), pp. 181–193. Antarctic Research Series, Vol. 20. American Geophysical Union, Washington, D.C.

Vanderleek, J. (1917). Bacteria of frozen soils in Quebec. *Trans. R. Soc.* (*Canada*), Ser. III, Sect. IV, **11,** 15–37.

Vanderleek, J. (1918). Bacteria of frozen soils in Quebec, II. *Trans. R. Soc.* (*Canada*) Ser. III, Sect. IV, **12,** 15–39.

Vela, G. R. (1974). Effect of temperature on cannery waste oxidation. *J. Water Pollut. Control. Fed.* **46,** 198–202.
Vishniac, W. V. (1973). Analysis and related work on Antarctic soil microbiology. *Antarctic J. U.S.* **8,** 303.
Vishniac, W. V. and Mainzer, S. E. (1972). Soil microbiology studied *in situ* in the dry valleys of Antarctica. *Antarctic J. U.S.* **7,** 88–89.
Vishniac, W. V. and Mainzer, S. E. (1973). Antarctica as a martian model. *In* "Cospar Life Science and Space Research, XI" (Ed. P. H. A. Sneath), pp. 3–31. Akademic Verlag, Berlin.
Wada, E., Koike, I. and Hattori, A. (1975). Nitrate metabolism in abyssal waters. *Marine Biol.* **29,** 119–124.
Waksman, S. A. and Carey, C. L. (1933). Role of bacteria in decomposition of plant and animal residues in the ocean. *Proc. Soc. Exp. Biol. Med.* **30,** 526–527.
Ward, F. J. and Robinson, G. G. C. (1974). A review of research on the limnology of West Blue Lake, Manitoba. *J. Fish. Res. Bd. Can.* **31,** 977–1005.
Weinberg, E. D. (1974). Secondary metabolism: Control by temperature and inorganic phosphate. *Dev. Ind. Microbiol.* **15,** 70–81.
Weyant, W. S. (1966). The antarctic climate. *In* "Antarctic Soils and Soil Forming Processes" (Ed. J. C. F. Tedrow), pp. 47–49. Antarctic Research Series Vol. 8. Geography Union Publ., Washington, D.C.
Wiebe, W. J. and Hendricks, C. W. (1974). Distribution of heterotrophic bacteria in a transect of the Antarctic Ocean. *In* "Effect of the Ocean Environment on Microbial Activities" (Eds R. R. Colwell and R. Y. Morita), pp. 524–535. University Park Press, Baltimore.
Wiebe, W. J. and Liston, J. (1972). Studies of the aerobic, nonexacting heterotrophic bacteria of the benthos. *In* "The Columbia River Estuary and Adjacent Ocean Waters" (Eds A. T. Pruter and D. L. Alverson), pp. 281–312. University of Washington Press, Seattle.
Wiebe, W. J. and Pomeroy, L. R. (1972). Microorganisms and their association with aggregates and detritus in the sea: a microscopic study. *In* "Detritus and Its Role in Aquatic Ecosystems" (Eds U. Melchiorri-Santolini and J. W. Hopton), pp. 325–352. Mem. Ist. Ital. Idrobiol. 29 Suppl. Pallanza, Italy.
Wilson, A. T., Holdsworth, R. and Hendy, C. H. (1974). Lake Wanda: Source of heating. *Antarctic J. U.S.* **9,** 137–138.
Winogradsky, S. (1925). Etudes sur la microbiologie du sol I. Sur la methode. *Ann. Inst. Pasteur, Paris.* **39,** 299–354.
Wirsen, C. O. and Jannasch, H. W. (1974). Microbial transformations of some ^{14}C-labeled substrates in coastal water and sediment. *Microb. Ecol.* **1,** 25–37.
Wright, R. T. and Hobbie, J. E. (1965). The uptake of organic solutes in lake water. *Limnol. Oceanogr.* **10,** 22–28.
Wright, R. T. and Hobbie, J. E. (1966). Use of glucose and acetate by bacteria and algae in aquatic ecosystems. *Ecology* **47,** 447–464.
Wrigley, R. E. (1974). Ecological notes on animals of the Churchill Region of Hudson Bay. *Arctic* **27,** 201–214.
Zeikus, J. G. and Brock, T. D. (1972). Effects of thermal additions from the Yellowstone geyser basins on the bacteriology of the Firehole River. *Ecology* **53,** 282–290.
Zevenhuizen, L. P. T. M. (1966). Formation and function of the glycogen-like polysaccharide of *Arthrobacter*. *Antonie van Leeuwenhoek* **32,** 356–372.

Zimmerman, R. and Meyer-Reil, L.-A. (1974). A new method for fluorescence staining of bacterial populations on membrane filters. *Kieler Meersforsch.* **30,** 24–27.

ZoBell, C. E. (1942). Microorganisms in marine air. *In* "Aerobiology" (Ed. S. Moulton), pp. 55–68. Am. Assoc. Adv. Sci. Publ. No. 17., Washington, D.C.

ZoBell, C. E. (1962). Importance of microorganisms in the sea. *In* "Proceedings Low Temperature Microbiology Symposium" pp. 107–132. Campbell Soup Co., Camden, New Jersey.

ZoBell, C. E. and Anderson, D. O. (1936). Vertical distribution of bacteria in marine sediments. *Am. Ass. Petrol. Geol. Bull.* **20,** 258–269.

ZoBell, C. E. and Conn, J. E. (1940). Studies on the thermosensitivity of marine bacteria. *J. Bacteriol.* **40,** 223–238.

ZoBell, C. E. and Mathews, H. M. (1936). A qualitative study of the bacterial flora of land and sea breezes. *Proc. Natl. Acad. Sci. USA* **22,** 567–572.

Chapter 3

Microbial Life at Low Temperatures: Mechanisms and Molecular Aspects

W. E. INNISS and J. L. INGRAHAM

University of Waterloo and University of California

I. Kinetics of growth at low temperature 73
II. Determinants of low maximum temperature of growth of psychrophilic and psychrotrophic microorganisms 77
A. Protein synthesis 78
B. Cell structural modification 83
C. Enzyme inactivation 87
III. Biochemical factors affecting the minimum temperature of growth . 90
A. Cold-sensitive mutants as a model 90
B. Lipid composition and membrane permeability at low temperature . 95
C. Ribosome function at low temperature 98
D. The minimum temperature of growth: conclusions . . . 98
References 99

I. Kinetics of Growth at Low Temperature

Microbial growth involves an interrelated sequence of chemical reactions, which are influenced by temperature. The basis for the relationship between temperature and growth rate is expressed by the Arrhenius equation

$$\log_{10} k = \frac{-\Delta H_a}{2{\cdot}303RT} + C$$

where k = reaction rate constant, ΔH_a = heat of activation, R = gas constant, T = absolute temperature, and C = constant. Sometimes ΔE, the energy of activation, is substituted for ΔH_a. Application of the equation to the growth of microorganisms results in k being considered as the growth rate and ΔH_a or ΔE being referred to as the temperature characteristic (μ). Consequently, if a plot of $\log_{10} k$ versus $1/T$ is made, then the slope of the linear portion of the curve equals $-\mu/2{\cdot}303R$ and the temperature characteristic value can be calculated. The linear portion of the curve indicates the obedience of the growth rate to temperature on the Arrhenius principle regarding a chemical reaction (Fig. 1). However, at temperatures below this certain range, a deviation from linearity occurs and rather than just a continual decrease in growth rate as would follow from the Arrhenius relationship, the curve finally becomes vertical and growth ceases (minimum

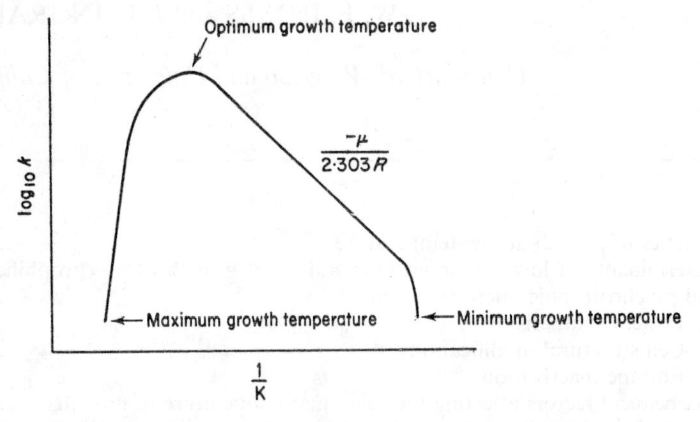

Fig. 1. Effect of temperature on microbial growth rate.

growth temperature). Similarly, the growth rate decreases at temperatures above the optimum until, again, growth finally terminates (maximum growth temperature).

Using an Arrhenius type plot, the relationship between temperature and the rate of growth of microorganisms capable of growth at relatively low temperatures, has been examined. When the appearance of the curve from such a plot for a psychrophilic (the terms psychrophile, obligate and facultative psychrophile, and psychrotroph are the ones used in the papers cited), *Pseudomonas* is compared with that for the mesophile, *Escherichia coli*, a marked difference is readily apparent (Ingraham, 1958). The slopes of the linear descending sections of the curves were different by 1·6-fold; the psychrophilic slope being the lesser one. Calculation of the temperature characteristics for two other psychrophilic strains revealed essentially the same comparative values. Similarly, Arrhenius plots of growth rate and temperature data for *Micrococcus cryophilus*, an obligate psychrophile, and its two mesophilic mutants, T8 and M19, resulted in a temperature characteristic of 10,000 calories for the psychrophilic parent and values of 16,000 and 15,000 calories, respectively, for the two mesophilic mutants (Tai and Jackson, 1969b). In this latter investigation, any species differences have been removed by comparing the wild type with mutants developed from it. A similar relationship has been found for 11 psychrophiles isolated from natural water sources. Calculation of their temperature characteristics for growth over the range of 5 to 30°C indicated that the microbes with the fastest rates of growth at 5 and 10°C had the lowest temperature characteristics (Baig and Hopton, 1969).

However, such a relationship has not always been found. The temperature characteristics for three species of vibrios, namely, *Vibrio marinus* MP-1 (obligate psychrophile), *V. marinus* PS 207 (facultative pyschrophile) and *Vibrio metschnekovis* (mesophile) have been reported to be similar (Hanus and Morita, 1968). Also, temperature characteristics for seven psychrophilic yeasts were similar to those of two mesophilic yeasts (Shaw, 1967). The explanation for the difference in such results is unclear.

Other aspects of the effect of relatively low temperatures, both above and below 0°C, on the growth of a variety of psychrophilic microbes have been reported (Adams and Stokes, 1968; Albright and Morita, 1968; Dejardin and Ward, 1971a; Frank *et al.*, 1972; Geesey and Morita, 1975; Gray and Jackson, 1973; Griffiths and Haight, 1973; Kimmel and Maier, 1969; Larkin and Stokes, 1968; Ledebo and Ljunger, 1973; Mosser *et al.*, 1976; Ward, 1966). For example, the kinetics of growth of various psychrophilic, yeasts is readily affected by abrupt temperature shifts. Various types of "step-up" and "step-down" experiments have been performed (Shaw, 1967), based on the range of temperature from which or to which the shift is made. The results for psychrophiles and mesophiles have been compared to elucidate any fundamental differences in the way the rate of growth of such different microbes react to such temperature shifts. For the psychrophilic microorganisms, the specific growth rate converted immediately to a new higher rate (the same rate as for the microorganisms continuously growing at that temperature) when the temperature increase was within the relatively "moderate" temperature range, i.e. the range in which the temperature characteristic changed little. Within this same range, when the temperature shift was downward, again no transient state occurred and the new lower growth rate was established immediately. On the other hand, when the temperature transfer was either upward or downward from a temperature which was above or below the "moderate" temperature range, then a temporary period consisting of a few generations existed. During these temporary periods the growth rates were intermediate between the final steady-state rates of growth typically associated with microorganisms growing at the starting and finishing temperatures involved in the shift. Comparison of the data obtained for both the psychrophilic and the mesophilic microbes indicates that when the shift is to a lower temperature, transient rates of growth are obtained at much lower temperatures for psychrophiles than for mesophiles. Similarly, upward temperature shifts result in transient growth rates occurring at much lower temperatures for psychrophiles as compared to mesophiles. In order for no transient growth rates to be obtained, the minimal transfer temperature for mesophiles must be much higher than for psychrophiles. The evidence indicates that when the temperature is shifted,

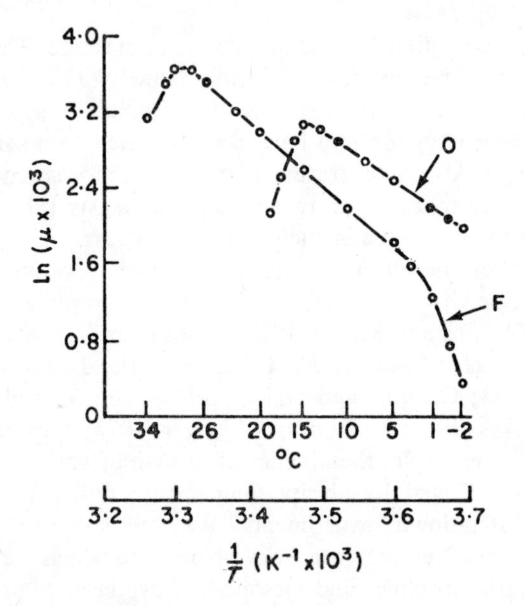

Fig. 2. Arrhenius plot of the maximum specific growth rate of an obligately (O) and a facultatively (F) psychrophilic *Pseudomonas* species at different temperatures (Harder and Veldkamp, 1971).

the growth reaction obeys the Arrhenius function and the change of the transient growth rate to the rate of growth normal for that particular temperature involves a corrective mechanism (Shaw, 1967).

The kinetics of the growth of both obligately and facultatively psychrophilic bacteria have been measured and compared. Representative results (Fig. 2) show that the difference between the two types of psychrophiles is in the range of temperature in which their Arrhenius curves are linear, rather than in the slopes of the linear portion of each curve (Harder and Veldkamp, 1971). The curve for the obligate psychrophile does not become non-linear till at least −2°C whereas the curve for the facultative psychrophile deflects in a downward fashion from linearity at about 5°C. Thus, the obligate psychrophiles have a greater rate of growth at the lower temperatures, suggesting that they might possess a competitive advantage at such temperatures.

A very useful approach to examine the kinetics of growth of microorganisms at relatively low temperatures has been the use of continuous culture conditions in a chemostat. With such a system, different growth rates can be established and readily held constant and the effect of temperature on such rates examined. For example, with an obligately psychrophilic species of *Pseudomonas*, the molar growth yields decreased above and below the optimum temperature of 14°C, but concomitantly, the cell protein content and the "actual-QO_2" values increased (Harder and Veldkamp, 1967). Consequently, the energy requirements for a certain rate of growth becomes greater relative to the degree of deviation from the optimum growth temperature. Within certain limits, the cells appear to be able to compensate for the effect of the temperature on growth by increasing the production of enzymes of the energy-generating systems.

Using the chemostat again to good advantage, the kinetics of the growth at low temperatures of both obligate and facultative psychrophiles have been compared to determine if the obligate psychrophile's faster rate of growth actually resulted in a competitive advantage under the usually natural nutritional condition of carbon- and energy-limitation (Harder and Veldkamp, 1971). Equal numbers of an obligately psychrophilic *Pseudomonas* and a facultatively psychrophilic *Spirillum* were placed in the chemostat and their growth rates compared at different dilution rates (dilution rate being defined as the number of culture volumes of medium passing through the growth vessel per unit time) and temperatures. The facultative psychrophile grew less quickly at −2°C at all the dilution rates used than did the obligate psychrophile. At either 4 or 10°C, the obligate psychrophile grew faster at high dilution rates whereas, conversely, at low dilution rates the facultative psychrophile exhibited faster rates of growth. The obligate psychrophile was discriminated against at 16°C and the facultative psychrophile predominated at all the dilution rates used. In sum, it is evident that the faster growing psychrophilic microorganism (1) has a competitive advantage at low temperature regardless of the substrate concentration, (2) may or may not have the advantage depending on the substrate concentration at relatively intermediate temperatures and (3) is definitely at a disadvantage at higher temperatures.

II. Determinants of Low Maximum Temperature of Growth of Psychrophilic and Psychrotrophic Microorganisms

The molecular basis for the relatively low maximum temperature of growth for microorganisms which normally grow at relatively low temperatures is not completely understood and, indeed, probably varies depending on the microbe. In general terms, the upper temperature limit for microorganisms is often considered to be due to the thermolability of one or more essential chemical or structural components. Clearly, when a cellular component (or activity) is sensitive to temperatures approximating the maximum growth temperature, it is suggestive that a causal relationship exists between the sensitivity of the component and the upper limit for growth. In addition, sometimes a positive correlation is found between a decrease in the thermolability of one or more cellular characteristics and the increasing maximum growth temperatures of psychrophilic, mesophilic and thermophilic microbes. Such comparative results may also indicate the basis(es) for the upper limits for growth of these microorganisms. With other experimental approaches, temperatures both above and below the actual maximum temperature of growth have been used. All such evidence can contribute to the understanding of the upper limit for growth of psychrophilic and psychrotrophic microorganisms.

A. Protein Synthesis

1. Cellular Synthesis of Protein

Relatively moderate temperatures are capable of directly preventing cellular protein synthesis by intact cells of certain psychrophilic microorganisms. For example, protein synthesis was terminated at 22·5°C in an obligately psychrophilic *Pseudomonas* (Harder and Veldkamp, 1968). Similarly a fast decrease in protein synthesis occurred at about 20°C with the same *Pseudomonas*, when it was continuously cultured in a chemostat (Harder and Veldkamp, 1967). Prevention of protein synthesis, as measured by radioactive amino acid incorporation into protein, and loss of viability has been found to be related to *M. cryophilus* (Malcolm, 1968b). In this case, alteration of the thermal death rate by variation of temperature corresponded to the amount of protein synthesis inhibition observed. The beginning of viability loss in *V. marinus* occurred concomitantly with the commencement of protein synthesis inhibition at 22°C (Cooper and Morita, 1972). In *Bacillus psychrophilus* and *Bacillus insolitus*, *in vivo* protein synthesis and growth followed each other, with both activities being lower at 30°C than at 20°C (Bobier *et al.*, 1972).

As might be expected, synthesis of specific cellular proteins can be temperature-sensitive. For example, the synthesis of various induced enzyme systems in psychrophilic microorganisms are affected by the relatively moderate temperature of 25°C. The synthesis of the luminescent system of a psychrophilic photobacterium, tentatively identified as a strain of *Photobacterium fischeri*, was prevented by 25°C (Makemson, 1973). Also, no induction of formic hydrogenlyase (Quist and Stokes, 1969) or of formic hydrogenase

(Quist and Stokes, 1972) occurred at 25°C in psychrophilic bacterium strain 82, originally isolated by Upadhyay and Stokes (1962). A temperature of 35°C prevents the formation of protease in a psychrophilic *Pseudomonas*, number 548 (Kato *et al.*, 1972) and the synthesis of a benzoate oxidation system by a psychrophilic strain of *Pseudomonas fluorescens* (Quist and Stokes, 1972).

2. *Sub-cellular Synthesis of Protein*

a. Ribosomes. It is clear that relatively moderate temperatures can adversely affect microbial protein synthesis, and various investigations have been conducted using sub-cellular fractions to determine the sensitive steps. For example the physical stability of ribosomes from psychrophiles, mesophiles and thermophiles to heat was compared by measuring their hyperchromicity at 260 nm at different temperatures. A positive correlation has been found between the ribosomal T_m (temperature at which absorbance at 260 nm has increased to half its maximum value) values of 19 such microorganisms and their maximum temperature for growth (Pace and Campbell, 1967). Also, as the maximum growth temperature increased, the guanine plus cytosine content of rRNA, isolated from each microorganism, increased. Since the temperatures at which the denaturation of the isolated rRNA commenced were less than those at which hyperchromicity of whole ribosomes was observed, it was concluded that the basis for the ribosomal heat stability is the fine structure of the interaction between RNA and protein.

Using a similar comparative approach, the ribosomes from one psychrophilic, one mesophilic, and two thermophilic species of clostridia have been measured for their stability to heat (Irwin *et al.*, 1973). The T_m values for the psychrophile and the mesophile were about the same, wheras the thermophilic T_m values were higher. When rRNA was isolated from the microbes and examined, the T_m values were similar, although some degree of denaturation was observed at lower temperatures for the psychrophile and mesophile than for the thermophiles. Denaturation profiles of ribosomes from the psychrophile, *Candida gelida*, and the mesophile, *Candida utilis* (Nash and Grant, 1969) showed that the ribosomes of the former were more susceptible to heat than those of the latter (T_m values of 49 and 52, respectively).

It will be noted that the temperatures at which ribosomes from psychrophilic microorganisms denature, as measured by T_m values, are considerably greater than the upper limit for growth for the same microbes. On the other hand, a positive correlation between the ribosomal or rRNA denaturation temperature and the maximum growth temperature exists for various psychrophiles relative to both mesophiles and thermophiles. Consequently, it is not clear whether such relationships have a direct implication regarding the upper temperature limit for psychrophilic growth.

Another procedure which has indicated the physical breakdown of psychrophilic ribosomes by temperature is sucrose gradient centrifugation. Sedimentation profiles of unheated ribosomes from the psychrophile, *C. gelida*, demonstrated the presence of two distinct components (Nash and Grant, 1969). Upon subjection of the ribosomes to 45°C for 5 min, the amount of lighter component increased. A 5°C raise in the temperature for the same period of time caused the complete eradication of the heavier component, thus indicating the thermal degradation of the ribosomes. Of particular interest regarding the possible relationship between such ribosomal degradation and the maximum growth temperature is that under identical experimental conditions, ribosomes from a mesophilic species of the same genus, namely *C. utilis*, were relatively unchanged with no lighter component developing. In addition, the amount of heavier component from both species was essentially the same. On the other hand, sucrose gradient analysis of the ribosomes from a psychrophilic and mesophilic species of *Clostridium* showed that, although structural alterations at 55°C occurred, the psychrophilic ribosomes aggregated whereas the mesophilic ones dissociated (Irwin *et al.*, 1973). Consequently, the type of thermally-caused physical change in ribosomes does not always bear a direct relationship to the maximum growth temperature of psychrophiles.

Ribosomal function, as opposed to ribosomal structure, in psychrophilic microorganisms is affected by temperatures much closer to those of the upper limit of growth. Psychrophilic ribosomes may lose activity at a temperature very near the maximum growth temperature but at which growth can still occur (Bobier *et al.*, 1972). The effect of 30°C on washed ribosomes (W-RIB), as well as on the supernatant fraction (IS-100), isolated from *B. psychrophilus* and *B. insolitus* is indicated by Table 1. The ribosomes and the soluble fraction were each subjected separately to 30°C for 10 min and then combined and their capacity to perform protein synthesis, as measured by poly-U-directed ^{14}C-polyphenylalanine synthesis at 15°C determined. Such results were compared to identical ribosomal and supernatant fractions subjected to 15°C for 10 min prior to measurement of activity. The relatively moderate temperature of 30°C caused an 82% and 95% decrease in polypeptide synthesis by the ribosomes of *B. psychrophilus* and *B. insolitus*, respectively.

The protein biosynthetic activity of ribosomes from *C. gelida* is also thermally inhibited. Temperatures of 30, 35 and 40°C for 5 min were sufficient to cause a decrease of 70, 90 and 100%, respectively, in the ability of the ribosomes to carry out ^{14}C-polyphenylalanine synthesis (Nash and Grant, 1969). Additionally, measurement of the ability of such ribosomes to bind ^{14}C-phenylalanyl-tRNA also shows ribosomal temperature sensitivity. Decreases of 30, 50 and 60% were found in the binding of charged-tRNA to ribosomes preheated for 5 min at 30, 35, and 40°C, respectively. By

Table 1
Effect of temperature on W-RIB and IS-100 fractions from B. psychrophilus *and* B. insolitus

Fraction(s) present	pmol ^{14}C-L phenylalanine incorporated *B. psychrophilus*	*B. insolitus*
W-RIB (15°C) and IS-100 (15°C)	14·44	23·3
W-RIB (15°C) and IS-100 (30°C)	13·95	19·7
W-RIB (30°C) and IS-100 (15°C)	2·64	0·98
W-RIB (15°C)	0·30	1·41
IS-100 (15°C)	0·46	1·94
W-RIB (30°C)	0·59	1·74
IS-100 (30°C)	0·32	1·34

Value within parentheses indicates temperature to which the fraction was exposed for 10 min before mixture and the measurement of ^{14}C-polyphenylalanine-synthesizing ability at 15°C (Bobier *et al.*, 1972)

comparison, the ribosomal activity of the mesophilic species, *C. utilis* was unaffected by such temperatures and even after 5 min at 55°C, only a 15% loss in protein synthesis occurred. However, such a distinct difference is not always found. For example, raising the temperature from 37°C to 55°C reduced protein synthesis by polyribosomes from both a psychrophilic and a mesophilic *Clostridium* as measured by endogenous valine incorporation, by approximately the same extent, i.e. 60% (Irwin *et al.*, 1973). Ribosomes from psychrophilic, mesophilic, and thermophilic clostridia lost no more than 20% of their *in vitro* polypeptide-synthesizing ability, after subjection to 37°C for 60 min. At 55°C, the activity of the ribosomes from the psychrophile was completely eliminated after 10 min of exposure, but 60 min were required to decrease the mesophilic ribosomal activity by 90% and the thermophilic ribosomes lost no activity.

b. Soluble components. Relatively moderate temperatures are capable of exerting considerable inhibitory effect on the ability of soluble sub-cellular fractions from certain psychrophilic microorganisms to synthesize protein. For example, a thorough examination of the effect of 30°C on the S-100 fraction of *M. cryophilus* has been conducted (Malcolm, 1968a). In such fractions the location of the thermosensitivity must occur before or at the stage of attachment of activated amino acids to tRNA, since *in vitro* polyphenylalanine synthesis from ^{14}C-phenylalanine-tRNA was uninhibited by

30°C. Examination of the attachment activities of the aminoacyl-tRNA synthetases in the S-100 fraction demonstrated that glutamyl-, histidyl-, and prolyl-tRNA synthetases were inhibited by 98, 87 and 86%, respectively. The capacity of these three enzymes to carry out their charging function was also distinctly reduced. Additional investigation revealed that treatment with 30°C for 10 min caused a 50% loss of the tRNA charging activity of glutamyl- and prolyl-tRNA synthetase (Malcolm, 1969b). Exposure of the cognate tRNA species itself to 30°C also caused the same degree of reduction in charging activity with glutamic acid and proline. However, purified glutamyl- and prolyl-tRNA were not notably sensitive to 30°C, indicating that the moderate temperature-sensitive reaction involves the acceptance of activated amino acid by tRNA; after combination occurs the complex is stable.

Additional studies showed that the most important temperature-sensitive components were the enzymes (glutamyl- and prolyl-tRNA synthetase) themselves. The charging ability of mixtures of sub-cellular components from *M. cryophilus* with those from a mesophile and a thermophile were measured and the effect of 30°C on such reactions was determined (Malcolm, 1969b).

Table 2

Glutamic acid-tRNA charging in heterologous systems

Enzyme	tRNA	Activity	% Activity
P	P	28·2	100
M	P	26·1	100
T	P	18·0	100
P	(P)	14·7	52
M	(P)	30·0	115
T	(P)	19·4	108
(P)	P	14·4	51
(P)	M	14·7	52
(P)	T	11·7	42

The enzyme and tRNA fractions derive from: P, the psychrophile *M. cryophilus*; M, the mesophile *E. coli*; and T, the thermophile *B. stearothermophilus*. An enclosed symbol, e.g. (P), indicates that the component was preincubated in buffer plus ions for 10 min at 30°C before assay at 18°C. Activity is expressed either as μmol of amino acid charged/h/mg of protein or as a percentage of the activity noted in the unheated homologous or heterologous system (Malcolm, 1969b)

As seen in Table 2, the psychrophilic glutamyl-tRNA synthetase is readily inactivated by 30°C for 10 min. Subjection of the psychrophilic tRNA to 30°C for 10 min also caused a decrease in the degree of glutamic acid charged.

However, the same tRNA is still capable of being totally charged in the presence of either the appropriate mesophilic or thermophilic synthetase. Consequently, the tRNA is considered to be unaffected with regard to recognition by glutamyl-tRNA synthetase. Similar results were observed for prolyl-tRNA synthetase.

Examination of the subunit structure of the temperature-sensitive glutamyl-tRNA synthetase showed the occurrence of thermally-caused changes in structure (Malcolm, 1969a). When subjected to 30°C for 20 min, the enzyme permanently dissociates into subunits which possess no catalytic activity. By contrast the same enzyme from a relatively heat-resistant strain of *M. cryophilus* was much more resistant to thermal dissociation.

The soluble sub-cellular fraction from the psychrophile *C. gelida* also appears to contain temperature-sensitive components as indicated by the fact that the poly-U-directed ^{14}C-polyphenylalanine-synthesizing activity of S-100 fractions was eliminated after 35°C for 30 min (Nash *et al.*, 1969). Further investigation revealed that, again, aminoacyl-tRNA synthetases were the thermolabile components. Treatment with 35°C for 30 min was capable of completely eradicating the activity of glycyl-, isoleucyl-, and threonyl-tRNA synthetase, and histidyl-, methionyl-, and phenylalanyl-tRNA synthetase lost at least 60% of their activity. Six other synthetases were slightly or not at all inactivated, but at least two of them were 70–100% inactivated by the same temperature when they were partially purified.

The difference in effect that a relatively moderate temperature can exert on different microorganisms which can grow at low temperatures is readily seen by comparing *M. cryophilus* with *B. psychrophilus* and *B. insolitus*. The first microorganism possessed temperature-sensitive synthetases, the latter two did not (Bobier *et al.*, 1972). In fact, when the activity of their aminoacyl-tRNA synthetases were measured, it was found that the majority of them were at least three times as active at 30°C than at 5°C. All but two of the remaining enzymes showed a minimum increase of 2·5-fold, and even they (leucyl-tRNA synthetase from *B. psychrophilus* and seryl-tRNA synthetase from *B. insolitus*) showed increases in activity of 1·6-fold and 1·8-fold, respectively. In addition, protein biosynthesis by the soluble sub-cellular components of various other psychrophiles is not prevented. For example, subjection to 37°C did not cause an activity loss in the 150,000 × *g* supernatant fraction from *Pseudomonas* 412 (Szer, 1970).

B. Cell Structural Modification

Relatively moderate temperatures can alter structural constituents of microorganisms that grow at low temperatures. In certain instances an actual change in the fine structure of the microorganisms can be seen. In other situations some physiological manifestations of a change in structure is observed. Examination of the psychrophile *Vibrio psychroerythrus* (D'Aoust and Kushner, 1972) by electron microscopy after subjection to 37°C for 2 h (D'Aoust and Kushner, 1971) showed that cell wall substances are broken down and probably liberated as vesicles, demonstrating the temperature susceptibility of the outer layers. Even when no extensive degradation of the cells were visible, the usual trilaminar type of structure of the membrane was generally missing.

Temperature can also cause ultrastructural alterations in *B. psychrophilus*. The cell wall of this psychrophile was rapidly degraded at 40°C, concomitant with clumping and cell death (Alsobrook *et al.*, 1972). Destruction of the outer layer of the cell wall took place within 15 min. Analysis of the released products suggested that a lack of protein cross-linking might account for some of its thermal sensitivity. Extensive damage to the middle wall layer took longer and was only completely removed after 2 h. Even so, cells so affected maintained their shape with the inner wall layer intact. As opposed to the results for the Gram negative psychrophilic vibrio (D'Aoust and Kushner, 1971), no alteration of the cellular membrane was seen.

Thermal modification of the structure of psychrophilic cells may also be reflected by changes in overall cellular appearance. For example, the psychrophile *B. insolitus* forms filamentous cells at 30°C but only normal sized cells at 20°C (Ferroni and Inniss, 1973). After 4 h at 20°C, 97% of the cells were less than 4·4 μm long with 60% being less than 2·2 μm in length; whereas at 30°C, 70% of the cells were greater than 4·4 μm in length. Incubation for a further 8 h at 30°C resulted in 83% of the cells being at least 7·4 μm long, as compared to 87% of the cells at 20°C being only 2·2 μm or less in length (a minimum of 3·4-fold difference). Electron microscopic studies confirmed the complete absence of septation, including septum initiation, in the elongated cells. Such results indicate interference in the cell division process by near-maximum growth temperatures and suggest the possibility that the septum-forming capability of the microorganism is reduced. This thermally-caused filament formation was reversed by a return to lower temperature (Ferroni and Inniss, 1974). Examination by both light and electron microscopy showed that incubation of elongated cells at 20°C resulted in the reoccurrence of normal septation. In addition, multiple septation periodically occurred with some irregularity in the division-site location, resulting in the partitioning off of some very short cells from the filaments.

Cell morphology (and thus some structural component) in the fungal psychrophile, *Sclerotinia borealis*, is altered by relatively low temperatures. After 25°C for 24 h, the hyphae are often disarranged and swollen considerably and have lost their hyalinity whereas, normally, as at 0°C, they are straight with tapering towards the tip, and are hyaline (Ward, 1968).

The temperature modification of an isolated cellular component from a psychrophilic microorganism, the isolated cell walls of *B. psychrophilus*, has been found. Such cell walls are broken down by temperatures of 20°C and greater (Mattingly and Best, 1971). At 20°C a linear rate of cell wall degradation occurred, whereas when the temperature was raised to 37°C, the breakdown reaction was linear only for the initial 10 min period. At 45°C and above, the degradation was totally non-linear and, as the temperature increased, the rate of dissolution increased. The data suggested that a non-enzymatic breakdown of cell wall occurred at temperatures above the upper limits for growth but that below such a temperature, the degradation reaction was enzymatic. Further investigations (Mattingly and Best, 1972) supply additional evidence for such an interpretation. Treatment of isolated cell walls with cold 10 M lithium chloride eliminated their enzymic degradation at 20°C, a result similar to that obtained for other microorganisms (Fan, 1970; Pooley *et al.*, 1970). However, such treatment had no effect on the breakdown of cell wall at 37°C and 45°C. Also, calcium ions do not affect the cell wall degradation at 20°C but prevent such activity at temperatures above the maximum growth temperature. Thus, the cell walls of this psychrophilic bacterium are quite susceptible to temperature-mediated modification.

The alteration of cell surface structure by relatively moderate temperature of the obligate psychrophile, *V. psychroerythrus*, is probably reflected by the observed change in its surface charge. At temperatures greater than 21°C, this microbe lyses and micro-electrophoresis has revealed that, after such thermal treatment, washed cells exhibit a much greater electrophoretic mobility over a pH range of 3 to 9 than non-treated cells (Madeley *et al.*, 1967). Chemical changes were also evident. For example, more than half of the lipid phosphorus of such cells was liberated and broken down (Korngold and Kushner, 1968). Thus, the change of the surface charge could be due to an alteration resulting in the loss of cell surface material, with the resultant exposure of additional charged groups (Madeley *et al.*, 1967).

In addition to the directly observed or measured changes in cell structure caused by relatively moderate temperatures, alteration of such structures as the cell permeability layer has been demonstrated by the leakage of intracellular substances or by actual complete lysis. For example, as shown in Table 3, various soluble cellular materials are released to the extracellular environment by the obligate psychrophile, *Candida nivalis* (Nash and Sinclair, 1968). The liberated substances included amino acids or short polypeptides, inorganic phosphate and nucleotide monophosphates; none of which were released because of lysis of the cells since no reduction in total cell counts or turbidity occurred. Also, since no extracellular liberation of protein or nucleic acids were observed, the possibility was suggested that the cell membrane was specifically altered. This leakage correlated with the viability

Table 3
Release of various cellular components from C. nivalis *cells during heating at 15°C and 35°C*

Temperature (°C)	Time (h)	Phosphate, μmol ml^{-1}	Ninhydrin positive, μmol ml^{-1}	Total Carbo-hydrate, μg ml^{-1}	Pentose, μg ml^{-1}	NH_3-N, μmol ml^{-1}
15	0	0	0·15	8	9	0·05
	3	1	0·19	24	17	0·10
	6	0	0·20	24	19	0·10
	12	0·044	0·28	32	23	0·14
35	0	0	0·15	8	9	0·05
	3	0·133	0·50	44	34	0·10
	6	0·233	0·73	72	56	0·11
	12	0·340	0·98	88	87	0·12

(Nash and Sinclair, 1968)

loss at 25°C, indicating that the low maximum growth temperature for this microorganism may be due, at least in part, to a temperature-sensitive cell membrane.

Thermally caused liberation of intracellular constituents can also occur from the mycelia of a psychrophilic fungus. Above optimum temperature brought about the leakage of nucleotides from *Merulius lacrymans* as measured by ultraviolet-absorption analysis and radioactive-labelling procedures (Langvad and Goksoyr, 1967).

Actual lysis, rather than just leakage, can occur when phychrophilic cells are subjected to relatively moderate temperatures. Whole cells of *B. psychrophilus* have been found to undergo lysis in potassium phosphate buffer, pH 6·5, at temperatures above 28°C (Mattingly and Best, 1971). Other psychrophilic bacilli are also susceptible to lysis (Stokes and Larkin, 1968).

Thermal alteration of the cell permeability layer such that leakage of intracellular constituents or even lysis occurs is known for other microorganisms capable of growth at low temperature but without correlation with loss of viability. Even in such instances, however, the possibility of a relatively small amount of leakage, perhaps of a specific essential compound(s), contributing to viability loss may be a consideration. For example, the amount of liberation of intracellular substances from the obligate psychrophile, *V. marinus* MP-1, was related to temperature; no release was evident at 15°C but occurred at about 22°C (Haight and Morita, 1966). The rate of protein liberation was the fastest but DNA, RNA, and amino acids (at decreasing rates) also leaked from the cells. Such release of intracellular material was suggested to be one of the reasons for the temperature-caused viability loss for this microbe. However, later work by the same group showed that in a growth medium, the leakage occurred after death of the cells (Kenis and Morita, 1968). Only after about 95% of the cells had lost their viability at 20 or 25°C did cells in the logarithmic phase of growth commence to liberate intracellular constituents. No such release of material was observed with maximum stationary phase cells even though only 0·1% of the cells remained viable after 120 min at 25°C. Temperature-caused leakage of intracellular material from *B. psychrophilus* appears to be unrelated to cell death in either distilled water, phosphate buffer or a growth medium, i.e. a large amount of leakage occurred without any noticeable effects on cell viability (Alsobrook *et al.*, 1972). Various compounds including RNA, protein, carbohydrate, and inorganic phosphate leaked from the cells with increasing temperature. However, calculation of the concentrations of such substances released extracellularly showed that at least 150% to 450% greater amounts of each cellular compound was released at 25°C than at 40°C, whereas, at 25°C death was very slow but at 40°C, death was rapid with a 90% loss of viability within 15 min.

C. Enzyme Inactivation

The susceptibility to relatively moderate temperature of enzymes of certain psychrophilic microorganisms may be responsible, partly or completely, for the low maximum growth temperature exhibited by such microbes. However, it should be kept in mind that the appropriate assignment of responsibility can be influenced by whether the enzyme is temperature-sensitive both in an isolated condition and in the whole cell. Sometimes, cellular integrity affords total or partial protection to thermolabile enzymatic activity. For example, cell free extracts of *V. marinus* MP-1, an obligate psychrophile, possess a malic dehydrogenase which is readily heat-sensitive (Langridge and Morita, 1966). The enzyme exhibited a decreased activity even at 0°C, and after 10 min at 30°C, i.e. a temperature above that for maximum growth, had lost almost all activity. A 20-fold purified preparation was similarly inactivated. On the other hand, the malic dehydrogenase activity of intact cells was unaffected by temperatures less than that optimum for growth, thus indicating that in this temperature range, protection was provided by cellular integrity. However, even whole cells of this psychrophile show a pronounced decrease in malic dehydrogenase activity at superoptimum temperatures, i.e. above 15°C, and thus the relatively low-temperature sensitivity of the enzyme may account for the low upper limit for bacterial growth.

With other psychrophilic microorganisms, no protection is supplied to thermal-sensitive enzymes by the intact cell. The influence of increasing temperature from 5 to 45°C on the ability of disrupted cells of *B. psychrophilus* to carry out glucose and endogenous oxidation was the same as that of whole cells (Stokes and Larkin, 1968). Also, comparison between the psychrophile and a mesophilic *Bacillus* indicated that, with both cell free preparations and intact cells, a larger amount of glucose oxidative activity was lost at a lower temperature by the psychrophile.

Similarly, the temperature-sensitive enzymes of the psychrophilic bacterium designated as strain 82 were not protected by the cellular integrity of the microorganism. Both the oxidative and fermentative activities of whole cells and cell free preparations of this psychrophile were equally affected by temperature and were much more thermosensitive than the same activities of the mesophile, *Esherichia coli* (Purohit and Stokes, 1967). A marked decrease in glucose oxidation occurred after 10 min at 46°C with total inactivation occurring within 60 min, whereas the mesophilic activity was undiminished even after 120 min. Essentially the same results were obtained when pyruvate, glycerol, or glucosamine were used as substrates. Fermentative activity in the psychrophile exhibited even greater temperature-sensitivity being eliminated by only 15 min at 46°C, but again, the mesophilic activity was not reduced. Individual enzymic examination showed that different enzymes of the psychrophile, such as reduced nicotinamide adenine dinucleotide oxidase, lactic and glycerol dehydrogenases, and cytochrome c reductase, possessed varying degrees of thermal sensitivity. Cytochrome c reductase was especially vulnerable, losing 70% of its activity at 40°C and 100% at 45°C and, consequently, was suggested to be an important factor in determining the microbe's maximum growth temperature of about 35°C.

Psychrophilic fungi also have temperature-sensitive enzymes which might be a determinant in their upper limit of growth. The snow mold fungus, *Typhula idahoensis* exhibited reduced endogenous metabolic activity at 20°C, the temperature at which growth is restricted (Dejardin and Ward, 1971b). A psychrophilic basidiomycete possesses a β-glucosidase which loses 44% of its activity within 20 min at 45°C (Stevens and Strobel, 1968).

With certain psychrophilic microorganisms the loss of overall activity and growth at relatively low temperatures results from the presence of only one certain sensitive enzyme. In the obligate psychrophile, *C. gelida*, the fermentation of glucose by intact cells and cell free preparations was inhibited 80% and 100%, respectively, by 35°C for 30 min (Grant *et al.*, 1968). As opposed to ten other enzymes which were essentially unaffected, pyruvate decarboxylase was very thermosensitive losing about 75% of its activity within 30 min at 35°C, whereas the same enzyme from the mesophile, *C. utilis*, was unaffected (Fig. 3). Further, the glucose fermentative activity of cell free extracts of the psychrophile was re-established by the addition of purified pyruvate decarboxylase. Thus, it appears that the temperature-sensitivity of this specific enzyme is at least one determinant of its relatively low upper growth limit. The psychrophile, *Clostridium* strain 69, also appears to possess a single marked temperature-sensitive enzyme. Out of the ten glycolytic enzymes investigated, only triose phosphate isomerase exhibited a pronounced thermosensitivity being 80% inactivated by 32°C for 30 min when in cell free preparations (Shing *et al.*, 1972). Up to 40°C all the other

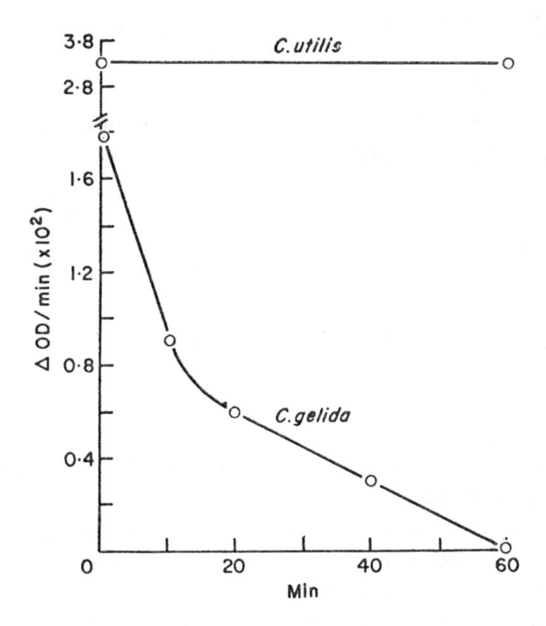

Fig. 3. Effect of heating at 35°C for indicated time intervals on pyruvate decarboxylase from *C. gelida* and *C. utilis* (Grant *et al.*, 1968).

enzymes had essentially the same degree of activity as in the mesophile, *Clostridium pasteurianum*. Similar results were obtained in intact cells. Consequently, the interruption of glycolysis in this psychrophile by the susceptibility of one enzyme was suggested to be at least partly responsible for its low maximum growth temperature. Additional work with purified triosephosphate isomerase from psychrophilic, mesophilic and thermophilic clostridia also indicated that the enzyme possessed different heat stabilities, losing 50% of initial activity after 10 min at 28, 45 and 64°C when from the psychrophile, mesophile and thermophile, respectively (Shing *et al.*, 1975). However, on the other hand, heat-labile triosephosphate isomerase has been reported to occur in mesophilic as well as psychrophilic clostridia (Finne *et al.*, 1975). Therefore, these authors have proposed that if the heat lability of this enzyme is to be used as an explanation for clostridial psychrophilic nature, it should be considered to influence the maximum growth temperatures of both types of microorganism.

III. Biochemical Factors Affecting the Minimum Temperature of Growth

As discussed, those factors that determine the upper temperature limit for the growth of microorganisms have a common denominator: the thermostability of one or more cellular components. However, those that determine the lower limit cannot be so succinctly defined. Foter and Rahn (1936) described the problem quite clearly when they wrote,

> The object of this paper was to study the minimum temperature of fermentation separately from that for growth, and to explain, if possible, the cause for a minimum temperature. Since both of these life functions are chemical reactions they should continue, though at greatly reduced speed, until the medium freezes solid. This is certainly not always the case with the growth of bacteria; most of them cease completely to grow at temperatures 5 to 10 or more degrees above the freezing point.

Although the focus of this chapter is the mechanism of how growth occurs at low temperature, Foter and Rahn (1936) point to the lack of growth at low temperature as being the phenomenon that requires explanation. Accordingly we will discuss the factors which preclude the growth of mesophiles at low temperature, assuming that it is the absence of these factors that permits the growth of psychrophiles. In this respect two different phenomena are important: the dramatic increase in the temperature characteristic of growth rate that occurs in the lower portion of the temperature range, and the minimum growth temperature *per se*. The minimum temperature of growth of a particular microorganism is a biological parameter which can be determined with a precision of less than one degree (Shaw *et al.*, 1971; Hoffmann, 1967); growth does not occur below this temperature regardless of the period of incubation.

A. Cold Sensitive Mutants as a Model

A practical difficulty associated with determining the cause of the minimal growth temperature of bacteria is the probability that a number of vital functions cease simultaneously, a probability which is supported by the familiar observations that, with most bacteria, mutants with decreased minimum temperatures of growth cannot be isolated. Although the selective pressure is intense, and serious efforts have been made, mutants of enteric bacteria with decreased minimum growth-temperature have not been isolated. We can only conclude that a number of mutations in genes encoding a variety of functions are required to extend significantly the lower growth-temperature range of these organisms. However, such mutants have been isolated in other groups of bacteria.

Aerobic sporeforming bacteria frequently undergo changes with respect to minimum growth-temperature during laboratory cultivation. Allen (1953) compared the minimum growth-temperature of 21 strains of bacilli when freshly isolated and after four transfers in the laboratory at 55°C. When freshly isolated, only four strains were able to grow at a temperature as low as 30°C. Following laboratory culture, eight of the strains were capable of growth at 30°C.

Mutants with extended low growth-temperature range have been sought and found among the pseudomonads. Azuma *et al.* (1962) isolated a psychrophilic mutant of *Pseudomonas aeruginosa*; their results were confirmed by Olsen and Metcalf (1968) who isolated similar mutants of the same species. The latter group found that both the upper and lower growth-temperature were changed by the mutation. Whereas the growth-temperature range of the parent was 44 to 11°C, that of the mutant was 32 to 0°C. Quite significantly, they found that not all strains of *P. aeruginosa* could serve as a source of psychrophilic mutants. Whereas they were repeatedly able to isolate psychrophilic mutants in certain strains of *P. aeruginosa*, they were repeatedly unable to isolate comparable mutants from other strains of the same species. *P. aeruginosa* would appear to be a particularly wise choice of organism in which to seek psychrophilic mutants because it is almost unique among the various species of *Pseudomonas* in not itself being a psychrophile. Being closely related to psychrophiles, the number of mutational changes necessary to gain psychrophily might be expected to be small. Olsen and Metcalf (1968) found one psychrophilic mutant per 10^8 cells following ultraviolet irradiation sufficient to kill 99% of the population, a result compatable with only one or two mutations being required for the change to psychrophily. In support of this contention they demonstrated that psychrophily could be transduced from *P. fluorescense* to *P. aeruginosa.* The limited genetic barrier to psychrophily encountered in *P. aeruginosa* might well be atypical of mesophiles.

Mutants with increased minimum growth temperature, called cold-sensitive mutants, have proved to be useful tools with which to study the biochemical basis of the minimum temperature of growth. Most frequently they differ from their parents only in the lower range of growth-temperature. A comparison of the effect of temperature on the growth rate of cold-sensitive mutants of *E. coli* and on that of its parent is shown in Fig. 4. As a consequence of a single mutation, the minimum temperature of growth is increased by about 12°C, and the apparent temperature characteristic of growth rate in the mid range of temperature is also increased. However, the maximum temperature of growth is unaffected. Knowing the biochemical consequence of the mutation establishes the factor which determines the minimum temperature of growth of the mutant strain. Mutations conferring cold-sensitivity are reasonably frequent, about as frequent as mutations conferring heat-sensitivity. Knute Rasmussen (personal communication) in a limited but convincing experiment mutagenized a culture of *E. coli*, and without any enrichment examined the survivors for heat- and cold-sensitivity.

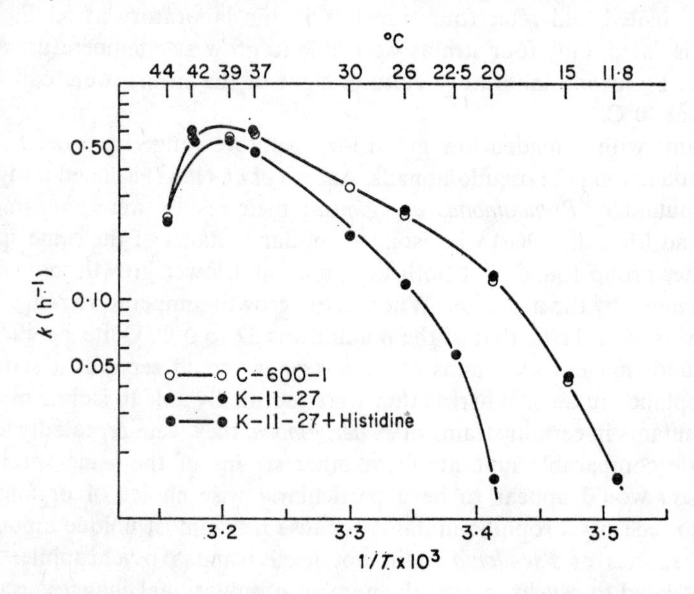

Fig. 4. Arrhenius plot of the growth rates (*k*) of a mutant (K-11-27) which is cold-sensitive for the biosynthesis of histidine and of its parent (C-600-1). In the presence of histidine, the mutant grows at parental rate at all temperatures (O'Donovan and Ingraham, 1965).

Of several hundred mutants he found the cold-sensitive class to be slightly more frequent than the heat-sensitive class. However, the genetic distribution of the two classes of temperature-sensitive mutants is distinct. Studies on temperature-sensitive mutants of the bacteriophage T4D illustrate this point. Scotti (1968) mapped 75 cold-sensitive mutants of T4D within only ten genes. The more extensive studies of Edgar and his colleagues (Epstein *et al.*, 1963; Edgar *et al.*, 1964; Edgar and Lielausis, 1964) show that mutations conferring heat-sensitivity are distributed almost randomly over the T4D chromosome, having been associated with 38 separate genes. Similar limited genetic distribution of mutations conferring cold-sensitivity to bacteria is seen in bacteria. O'Donovan and Ingraham (1965) isolated seven independent strains of

E. coli which were cold-sensitive for the biosynthesis of histidine, being auxotrophic at 20°C and prototrophic at 37°C. Although there are nine genes which encode the enzymes of the histidine biosynthetic pathway, all of the mutations conferring cold-sensitivity mapped within a single gene, that encoding the first enzyme of the pathway, phosphoribosyl-ATP pyrophosphorylase. Similarly, cold-sensitive mutants which are conditionally lethal result from changes in only a limited number of genes. It has been estimated (Tai *et al.*, 1969) that 40% of such mutations lie in genes which are co-transducible with the streptomycin-resistance locus of *Salmonella typhimurium*, i.e. in the region encoding ribosomal proteins. Guthrie *et al.* (1969) drew similar conclusions from their studies on conditional lethal cold-sensitive mutants of *E. coli*, estimating that about half of them affected the assembly of ribosomes, probably as a consequence of alterations of ribosomal proteins. Thus the target genes of mutations conferring cold-sensitivity seem fewer in number than those which confer heat-sensitivity. Studies on the mutant proteins of cold-sensitive strains have revealed certain common features, suggesting reasons for the genetic implication that a limited class of proteins are susceptible to change causing loss of function only at low temperature: proteins which are subunits of enzymes or subcellular structures the conformation of which is critical to their function. For example, O'Donovan and Ingraham (1965) found that in a class of mutants which had lost the ability to synthesize histidine at low temperature, all produced altered forms of the first enzyme of the histidine pathway, i.e. phosphoribosyl-ATP pyrophosphorylase. Although the mutant enzymes retained catalytic activity (and, indeed, functioned nearly normally at 37°C) their regulatory properties were markedly altered, being from 100- to almost 1000-fold more sensitive to feedback inhibition by the end product of the pathway, i.e. histidine. Moreover, both wild type and mutant enzyme have increased sensitivity to feedback inhibition as temperature is decreased, being about ten-fold more sensitive at 20°C than at 37°C. The loss of function at the low temperature was attributed to the combined effects of temperature on feedback inhibition and the effect of the mutation itself; i.e. at low temperature the functioning of the enzyme was prevented by a intracellular concentration of histidine which was too low to allow protein synthesis. This conclusion was supported by studies on revertants. Revertants selected for their ability to grow at 20°C in the absence of exogenous histidine were found to produce an enzyme with wild type sensitivity to feedback inhibition. Mutants selected for resistance to feedback inhibition at 37°C (by selecting for resistance to the histidine analogue, thiazole alanine) were found to be able to grow at 20°C in the absence of exogenous histidine.

The fact that mutants which are cold-sensitive for the biosynthesis of histidine can be isolated depends, therefore, on the sensitivity of the enzyme to feedback, being greater at low temperature. If such a basis for loss of function at low temperature is a general one, changes in sensitivity of enzymes to feedback inhibition must also be a general phenomenon. This is probably true. For example, Takita and Pogell (1965) have reported that fructose-1,6-diphosphatase from rat liver which is subject to feedback inhibition by AMP is half maximally inhibited by an AMP concentration of 8×10^{-4} M at 46°C whereas at 2°C a concentration of only 8×10^{-6} M gives the same level of inhibition. Similarly aspartic transcarbamylase from *Saccharomyces cereviseae* (Kaplan *et al.*, 1967) is more sensitive to feedback inhibition by UTP at low temperature (3°C) than at higher temperature, and carbamyl-phosphate synthetase from *Salmonella typhimurium* is more sensitive to activation by ornithine at 20°C than it is at 37°C (Abd-El-Al and Ingraham, 1969). Although it seems a general rule that the sensitivity of enzymes to their effectors changes with temperature, the sense of this change is unpredictable. As mentioned, aspartic transcarbamylase from *S. cereviseae* is more sensitive to inhibition at low temperature but the comparable enzymes from *E. coli* and *S. typhimurium* become increasingly less sensitive to their inhibitory effector (CTP) as temperature is decreased. Eventually, at very low temperatures (4°C), the velocity-substrate plots of the reaction catalysed by these enzymes are hyperbolic (the response typical of an unregulated enzyme) rather than sigmoid (the typical response of a regulated enzyme seen at higher temperatures). Changes in sensitivity of regulated enzymes with temperature mean that regulation can only be optimal at a single temperature and that it is more or less distorted at other temperatures. As illustrated by the example of mutants which are cold-sensitive for the biosynthesis of histidine, extreme distortions of regulation can prevent function at low temperature.

Mutations in genes encoding ribosomal proteins also frequently confer a cold-sensitive phenotype (Guthrie *et al.*, 1969; Tai *et al.*, 1969; Bayliss and Ingraham, 1974). Such mutant strains are unable to assemble ribosomal subunits at low temperatures; rather, incomplete, non-functional subunits accumulate within the cell. Shift of such a culture to low temperatures results in only a slow cessation of growth because the ribosomes previously synthesized at high temperature continue to function at a low temperature.

It seems probable that the common denominator of cold-sensitivity resulting from mutations in genes encoding ribosomal proteins and from those in genes encoding regulatory enzymes is the requirement of both classes of proteins for precise conformation. The weakening of hydrophobic bonds which occurs at low temperatures (Brandts, 1967) probably results in such changes.

Cold-sensitive mutants are increasingly being used, as heat-sensitive mutants have been for some time, to study cellular processes, the products of which cannot be supplied from the medium (e.g. Ingraham and Neuhard, 1972; Ginther and Ingraham, 1974; Reid, 1971; Wehr *et al.*, 1975; Waskell and Glaser, 1974). The biochemical basis for the cold-sensitivity of these mutant proteins is still unknown.

The cold-sensitivity of a phage lysozyme has been shown to be a consequence of a change in the activation energy of enzyme-catalysed reaction (Liebscher *et al.*, 1974; Dauberschmidt *et al.*, 1974).

B. Lipid Composition and Membrane Permeability at Low Temperatures

Temperature effects on the composition of the lipids of poikilothermic animals have been known for years. In general, as growth temperature is decreased, the lipids are composed of an increasing proportion of unsaturated fatty acids. The same phenomenon applies to bacteria (Table 4), where the

Table 4

Effect of growth temperature on fatty acid composition of Escherichia coli *ML 30 grown in glucose-minimal medium and harvested during exponential growth*

Fatty acid	Temperature (°C) 10 %	15 %	20 %	25 %	30 %	35 %	40 %	43 %
Saturated								
myristic	3·9	3·8	4·1	3·8	4·1	4·7	6·1	7·7
palmitic	18·2	21·9	25·4	27·6	28·9	31·7	37·1	48·0
methylene hexadecanoic	1·3	1·1	1·5	3·1	3·4	4·8	3·2	11·6
methylene octadecanoic	0	0	0	0	0	0	0	3·7
Unsaturated								
hexadecenoic	26·0	25·3	24·4	23·2	23·3	23·3	28·0	9·2
octadecenoic	37·9	35·4	34·2	35·5	30·3	24·6	20·8	12·2

(Marr and Ingraham, 1962)

fatty acids are largely found in the phospholipids of membranes. This general phenomenon has been interpreted as a homeostatic mechanism. The liquid to crystalline transition occurs at a lower temperature for phospholipids with a higher proportion of unsaturated fatty acids (Byrne and Chapman, 1964), and increasing unsaturation of fatty acids with decreasing growth-temperature could serve to maintain an optimal degree of fluidity in the lipid of the cell membrane (Farrell and Rose, 1967; Chapman, 1967).

There is direct evidence that membranes with greater proportions of unsaturated fatty acids are better able to function at low temperature (Wilson *et al.*, 1970). An unsaturated fatty acid auxotroph of *E. coli* was grown in a medium with oleic acid (monounsaturated) and on a medium with linoleic acid (doubly unsaturated) as the source of unsaturated fatty acid. Arrhenius plots of the uptake of β-galactosides and β-glucosides by these cells were biphasic. Oleic acid-grown cells showed an abrupt increase in slope below 13°C; the linoleic acid-grown cells showed the increase in slope below 7°C. Although linoleic acid is not a normal component of bacterial lipids, these results do show that an increase in the proportion of unsaturated fatty acids in the phospholipids of membranes, allow the membrane to remain functional at lower temperature. In a similar series of experiments, Wilson and Fox (1971) used five supplements of fatty acids. Again they found that the inflection point of Arrhenius plot of uptake decreases with the degree of unsaturation of the supplement. Further, they showed that incorporation of unsaturated fatty acids is stimulated by a lower temperature of growth. Schairer and Overath (1969) showed that the temperature characteristic of uptake is related to the fluidity of the membrane as modulated by the fatty acid supplement. Thus it seems established that the degree of fluidity of the membrane is related to its ability to function, and that the observed change in fatty acid composition of bacterial lipids with growth-temperature is, indeed, a homeostatic mechanism. Vinopal (1972) in a very interesting series of experiments showed that mutants of *S. typhimurium* which grew faster at 15°C, did so by altering their lipid composition.

By extension of the thesis that increasing unsaturation of fatty acid composition allows better membrane function at low temperature, one could conclude that the minimum growth-temperature occurs when the cell's capacity to make more highly unsaturated fatty acids is exceeded. Other experiments argue against this hypothesis. By manipulating the composition of the medium, Shaw and Ingraham (1965) showed that *E. coli* could grow (at least temporarily) at 12°C with a lipid composition typical of cells grown at 37°C. Similarly Gelman and Cronan (1972) showed that mutants of *E. coli* grown under conditions precluding their ability to increase their content of the unsaturated fatty acid, *cis*-vaccinic acid, grew quite normally at the moderately low temperature (for *E. coli*) of 15°C.

The biochemical mechanisms by which bacteria adjust their fatty acid composition in response to changes in temperature has been investigated in *E. coli* and in bacilli. Basically there are two mechanisms by which temperature could regulate the proportions of fatty acids into membranes. Either temperature could affect the incorporation of fatty acids into the membrane or affect the relative amounts of saturated and unsaturated fatty acids which are synthesized. The first of these mechanisms occurs in *E. coli* (Simensky,

1971). The ratio of oleate to stearate incorporated from the medium into the phospholipids of *E. coli* increases with decreasing growth temperature. These effects were reproduced *in vitro* (Table 5); the relative rates of transacylation

Table 5

Effect of temperature on transacetylase activity of cell free extracts of E. coli

Incubation temperature (°C)	Palmitate/oleate in phosphatidic acid	
	cells grown at 20°C	cells grown at 37°C
10	0·382	0·310
30	0·567	0·582

(Simensky, 1971)

of palmityl CoA and oleyl CoA increasingly favour the unsaturated fatty acid as temperature is decreased. This effect depends exclusively on the temperature at which the enzyme acts, not on the temperature at which it is synthesized. Thus, in the case of *E. coli*, temperature modulates the specificity of enzyme activity rather than altering the enzymatic composition of the cells.

In bacilli however (Fulco, 1969, 1970), temperature controls the composition of fatty acids in the cell membrane by modulating the synthesis of the relative quantities of saturated and unsaturated fatty acids, this modulation, in turn, being effected by a regulation of the quantities of various biosynthetic enzymes. In *Bacillus megaterium*, which does not synthesize unsaturated fatty acids when grown at 30°C or above, but does desaturate palmitic acid near 20°C, it has been shown that the desaturating enzyme is absent from cells grown at 30°C but is rapidly induced upon shift of a culture to 20°C to levels well above those of cultures growing in the steady state at the lower temperature (Fulco, 1970). The induced enzyme is rapidly inactivated *in vivo* when 20°C cultures are returned to 30°C.

Thus, three control mechanisms have been shown to regulate the level of the desaturating enzyme in *B. megaterium*; *de novo* enzyme synthesis is induced at low temperature; the enzyme is irreversibly inactivated at high temperature; and the ability to synthesize the enzyme is lost at high temperature (Fulco, 1970).

It appears, therefore, that modulation of the fatty acid composition of membranes, which is mediated by a variety of control mechanisms, is an important component of a cell's ability to grow at low temperature, but the minimum growth temperature is probably not a reflection of the limit of this modulation.

C. Ribosome Function at Low Temperature

As mentioned previously, a frequent class of cold-sensitive mutants are those which are unable to synthesize ribosomes at low temperature, but there is no evidence that any wild type organisms have similar defects growing near their minimum growth-temperature, nor when transferred to temperatures below the minimum for supporting continued growth. Polysome formation, however, does seem to be affected. Goldstein *et al.* (1964) noted in connection with their studies on the rate of peptide chain elongation that protein synthesis does take place at a constant rate in cultures of *E. coli* held below their minimum growth-temperature of a period of hours; then synthesis slows progressively and eventually stops. Later, Das and Goldstein (1968) proposed that the cold-labile step in protein synthesis was the reattachment of ribosomes to mRNA after they had "run-off" the 3′ end at the completion of a cycle of transcription. They observed that cells of *E. coli* which had been held at 0°C for 4 h were completely devoid of polysomes and unable to synthesize protein. However, the ability to synthesize more protein was almost instantly restored after only a brief exposure to 37°C. They suggested that loss of ability to form polysomes was the cause of the cessation of growth at low temperature.

Similarly, Oppenheim *et al.* (1968) observed that the polysomes of *Azotobacter vinelandii* disappear after a period of holding at low temperature, and Algranati *et al.* (1969) showed that the same applies to the polysomes of *Bacillus stearothermophilus*. Unfortunately in none of these cases has a close correspondence been established between the temperature causing disappearance of ribosomes and the minimum temperatures of growth.

Recently, Tai *et al.* (1973) showed with isolated polysomes from *E. coli* that peptide chain elongation is much less sensitive to low temperature than is initiation (polysome formation). At 14°C, elongation was observed to be one-third as rapid as at 34°C, but initiation to be negligible. They propose that the limiting reaction at low temperature may be the dissociation of run-off ribosomes which is catalysed by IF_3 (initiation factor).

D. The Minimum Temperature of Growth: Conclusions

Studies with cold-sensitive mutants have produced information about the kinds of lesions that result in loss of function at low temperature. Certainly in the case of the mutant strains they are the cause of the minimum temperature of growth. It seems only reasonable that these are analogous to the cause of the minimal temperature of growth of wild type strains.

Modulation of lipid composition with temperature is clearly an important adaptation for maximum growth rate at low temperature but deficiencies in this modulation are probably not the cause of the minimum growth-temperature.

The particular sensitivity of polysome formation to low temperature in several organisms strongly implies that this step might well be the cause of growth cessation in some strains. The reasons why this reaction is stopped at low temperature remains to be explained, as do the mechanisms of how organisms that can grow at low temperature carry out the reaction.

Progress has been made in our understanding of the growth of bacteria at low temperatures. The weight of evidence suggests that a productive question still is, why do certain bacteria cease growing at temperatures well above the freezing point of water whereas certain other bacteria do not? A full answer promises a better understanding of the entire bacterial growth process.

References*

Abd-El-Al, A. and Ingraham, J. L. (1969). Cold-sensitivity and other phenotypes resulting from mutation in *pyrA* gene. *J. Biol. Chem.* **244,** 4039–4045.

Adams, J. C. and Stokes, J. L. (1968). Vitamin requirements of psychrophilic species of *Bacillus. J. Bacteriol.* **95,** 239–240.

Albright, L. J. and Morita, R. Y. (1968). Effect of hydrostatic pressure on synthesis of protein, ribonucleic acid, and deoxyribonucleic acid by the psychrophilic marine bacterium, *Vibrio marinus. Limnol. Oceanogr.* **13,** 637–643.

Algranati, I. D., Gonzalez, N. S. and Bade, E. G. (1969). Physiological role of 70S ribosomes in bacteria. *Proc. Natl. Acad. Sci. USA* **62,** 574–580.

Allen, M. B. (1953). The thermophilic aerobic sporeforming bacteria. *Bacteriol. Rev.* **17,** 125–175.

Alsobrook, D., Larkin, J. M. and Sega, M. W. (1972). Effect of temperature on the cellular integrity of *Bacillus psychrophilus. Can. J. Microbiol.* **18,** 1671–1678.

Azuma, Y., Newton, S. B. and Witter, L. D. (1962). Production of psychrophilic mutants from mesophilic bacteria by ultraviolet irradiation. *J. Dairy Sci.* **45,** 1529–1530.

Baig, I. A. and Hopton, J. W. (1969). Psychrophilic properties and the temperature characteristic of growth of bacteria. *J. Bacteriol.* **100,** 552–553.

Bayliss, F. A. and Ingraham, J. L. (1974). Mutation in *Saccharomyces cerevisiae* conferring streptomycin and cold sensitivity by affecting ribosome formation and function. *J. Bacteriol.* **118,** 319–337.

Bobier, S. R., Ferroni, G. D., Inniss, W. E. (1972). Protein synthesis by the psychrophiles *Bacillus psychrophilus* and *Bacillus insolitus. Can. J. Microbiol.* **18,** 1837–1843.

Brandts, J. F. (1967). Heat effects on proteins and enzymes. *In* "Thermobiology" (Ed. A. H. Rose). Academic Press, New York and London.

* References marked with an asterisk show additional recent work not dealt with in the text

Byrne, P. and Chapman, D. (1964). Liquid crystalline nature of phospholipids. *Nature, Lond.* **202,** 987–988.

*Cameron, R. E., Honour, R. C. and Morelli, F. A. (1976). Antarctic microbiology—preparation for Mars life detection, quarantine, and back contamination. *In* "Extreme Environments, Mechanisms of Microbial Adaptation" (Ed. M. R. Heinrich). Academic Press, New York and London.

Chapman, D. (1967). The effect of heat on membranes and membrane constituents. *In* "Thermobiology" (Ed. A. H. Rose). Academic Press, New York and London.

Cooper, M. F. and Morita, R. Y. (1972). Interaction of salinity and temperature on net protein synthesis and viability of *Vibrio marinus*. *Limnol. Oceanogr.* **17,** 556–565.

D'Aoust, J. Y. and Kushner, D. J. (1971). Structural changes during lysis of a psychrophilic marine bacterium. *J. Bacteriol.* **108,** 916–927.

D'Aoust, J. Y. and Kushner, D. J. (1972). *Vibrio psychroerythrus* sp. n.: Classification of the psychrophilic marine bacterium. NRC 1004. *J. Bacteriol.* **111,** 340–342.

Das, H. K. and Goldstein, A. (1968). Limited capacity for protein synthesis at zero degrees centigrade in *Escherichia coli*. *J. Mol. Biol.* **31,** 209–226.

Dauberschmidt, R., Liebscher, D. H., Thiele, B. J., Rappaport, S. M., Heitmann, P. and Rosenthal, H. A. (1974). Untersuchungen ueber Phagelysozym III. Reinigung und vergleichende Charakterisierung des Lysozyms des T_4D-Wildtyps und der Kaeltempfindlichen Mutante cseBU56. *Acta Biol. Med. Germ.* **32,** 315–329.

Dejardin, R. A. and Ward, E. W. B. (1971a). Growth and respiration of psychrophilic species of the genus *Typhula*. *Can. J. Bot.* **49,** 339–347.

Dejardin, R. A. and Ward, E. W. B. (1971b). Studies on the endogenous respiration of the psychrophilic fungus *Typhula idahoensis*. *Can. J. Bot.* **49,** 2081–2087.

Edgar, R. S., Denhardt, G. H. and Epstein, R. H. (1964). A comparative study of conditional lethal mutations of bacteriophage T4D. *Genetics*, **59,** 635–648.

Edgar, R. S., Lielausis, A. (1964). Temperature sensitive mutants of bacteriophage T4D: their isolation and genetic characterization. *Genetics* **49,** 649–662.

Epstein, R. H., Bolle, A., Steinberg, C. M., Kellenberger, E., Boy De La Tour, E., Chevalley, R., Edgar, R. S., Susman, S., Denhardt, G. H. and Lielausis, A. (1963). Physiological studies on conditional lethal mutants of bacteriophage T4D. *Cold Spring Harbor Symp. Quant. Biol.* **28,** 375–394.

Fan, D. P. (1970). Cell wall binding properties of the *Bacillus subtilis* autolysin(s). *J. Bacteriol.* **103,** 488–493.

Farrell, J. and Rose, A. H. (1967). Temperature effects on microorganisms. *In* "Thermobiology" (Ed. A. H. Rose). Academic Press, New York and London.

Ferroni, G. D. and Inniss, W. E. (1973). Thermally caused filament formation in the psychrophile *Bacillus insolitus*. *Can. J. Microbiol.* **19,** 581–584.

Ferroni, G. D. and Inniss, W. E. (1974). Reversal of filament formation in the psychrophile *Bacillus insolitus*. *Can. J. Microbiol.* **20,** 1281–1283.

*Finne, G. and Matches, J. R. (1976). Spin-labeling studies on the lipids of psychrophilic, psychrotrophic, and mesophilic clostridia. *J. Bacteriol.* **125,** 211–219.

Finne, G., Matches, J. R. and Liston, J. (1975). A comparative study on the heat stability of triosephosphate isomerase in psychrophilic, psychrotrophic, and mesophilic clostridia. *Can. J. Microbiol.* **21,** 1719–1723.

Foter, M. J. and Rahn, O. (1936). Growth and fermentation of bacteria near their minimum temperature. *J. Bacteriol.* **32,** 485–499.

Frank, H. A., Reid, A., Santo, L. M., Lum, N. A. and Sandler, S. T. (1972).

Similarity in several properties of psychrophilic bacteria grown at low and moderate temperatures. *Appl. Microbiol.* **24,** 571–574.

Fulco, A. J. (1969). Biosynthesis of unsaturated fatty acids in bacilli. *J. Biol. Chem.* **244,** 889–896.

Fulco, A. J. (1970). Biosynthesis of unsaturated fatty acids in bacilli II. Temperature dependent biosynthesis of polyunsaturated acids. *J. Biol. Chem.* **245,** 2985–2990.

Geesey, G. G. and Morita, R. Y. (1975). Some physiological effects of near-maximum growth temperatures on an obligately psychrophilic marine bacterium. *Can. J. Microbiol.* **21,** 811–818.

Gelman, E. P. and Cronan, J. E. (1972). Mutant of *Escherichia coli* deficient in the synthesis of *cis*-vaccenic acid. *J. Bacteriol.* **112,** 381–387.

Ginther, C. and Ingraham, J. L. (1974). Cold-sensitive mutant of *Salmonella typhimurium* defective in nucleosidediphosphokinase. *J. Bacteriol.* **118,** 1020–1026.

Goldstein, A., Goldstein, D. P. and Lowney, L. I. (1964). Protein synthesis at 0°C in *Escherichia coli. J. Mol. Biol.* **9,** 213–235.

*Gounot, A. M. (1976). Effects of temperature on the growth of psychrophilic bacteria from glaciers. *Can. J. Microbiol.* **22,** 839–846.

*Gounot, A. M., Novitsky, T. J. and Kushner, D. J. (1977). Effects of temperature on the macromolecular composition and fine structure of psychrophilic *Arthrobacter* species. *Can. J. Microbiol.* **23,** 357–362.

Grant, D. W., Sinclair, N. A. and Nash, C. H. (1968). Temperature-sensitive glucose fermentation in the obligately psychrophilic yeast *Candida gelida. Can. J. Microbiol.* **14,** 1105–1110.

Gray, R. J. H. and Jackson, H. (1973). Growth and macromolecular composition of a psychrophile, *Micrococcus cryophilus*, at elevated temperatures. *Antonie van Leeuwenhoek*, **39,** 497–504.

Griffiths, R. P. and Haight, R. D. (1973). Reversible heat injury in the marine psychrophilic bacterium *Vibrio marinus* MP-1. *Can. J. Microbiol.* **19,** 557–561.

Guthrie, C., Nashimoto, H. and Nomura, M. (1969). Structure and function of *E. coli* ribosomes. VIII. Cold-sensitive mutants defective in ribosome assembly. *Proc. Nat. Acad. Sci. USA* **63,** 384–392.

Haight, R. D. and Morita, R. Y. (1966). Thermally induced leakage from *Vibrio marinus*, an obligately psychrophilic marine bacterium. *J. Bacteriol.* **92,** 1388–1393.

Hanus, F. J. and Morita, R.Y. (1968). Significance of the temperature characteristic of growth. *J. Bacteriol.* **95,** 736–737.

Harder, W. and Veldkamp, H. (1967). A continuous culture study of an obligately psychrophilic *Pseudomonas* species. *Arch. Mikrobiol.* **59,** 123–130.

Harder, W. and Veldkamp, H. (1968). Physiology of an obligately psychrophilic marine *Pseudomonas* species. *J. Appl. Bacteriol.* **31,** 12–23.

Harder, W. and Veldkamp, H. (1971). Competition of marine psychrophilic bacteria at low temperatures. Antonie van Leeuwenhoek, **37,** 51–63.

Hoffman, B. (1967). Wachstum and Vermehrung von *Escherichia coli* bei niedern Temperaturen. *Arch. Microbiol.* **58,** 302–304.

Ingraham, J. L. (1958). Growth of psychrophilic bacteria. *J. Bacteriol.* **76,** 75–80.

Ingraham, J. L. and Neuhard, J. (1972). Cold-sensitive mutants of *Salmonella typhimurium* defective in uridine monophosphate kinase (pyrH). *J. Biol. Chem.* **247,** 6259–6265.

Irwin, C. C., Akagi, J. M. and Himes, R. H. (1973). Ribosomes, polyribosomes, and deoxyribonucleic acid from thermophilic, mesophilic, and psychrophilic clostridia. *J. Bacteriol.* **113,** 252–262.

*Joakim, A. and Inniss, W. E. (1976). Reversible filamentous growth in the psychrophile *Bacillus psychrophilus. Cryobiology* **13,** 563–571.

Kaplan, J. G., Duphil, M. and Lacroute, F. (1967). A study of the aspartate transcarbamylase activity of yeast. *Arch. Biochem. Biophys.* **119,** 541–551.

Kato, N., Nagasawa, T., Tani, Y. and Ogata, K. (1972). Protease formation by a marine psychrophilic bacterium. *Agr. Biol. Chem.* **36,** 1177–1184.

Kenis, P. R. and Morita, R. Y. (1968). Thermally induced leakage of cellular material and viability in *Vibrio marinus,* a psychrophilic marine bacterium. *Can. J. Microbiol.* **14,** 1239–1244.

Kimmel, K. E. and Maier, S. (1969). Effect of cultural conditions on the synthesis of violacein in mesophilic and psychrophilic strains of *Chromobacterium. Can. J. Microbiol.* **15,** 111–116.

Korngold, R. R. and Kushner, D. J. (1968). Responses of a psychrophilic marine bacterium to changes in its ionic environment. *Can. J. Microbiol.* **14,** 253–263.

Langridge, P. and Morita, R. Y. (1966). Thermolability of malic dehydrogenase from the obligate psychrophile *Vibrio marinus. J. Bacteriol.* **92,** 418–423.

Langvad, F. and Goksoyr, J. (1967). Effects of supraoptimal temperatures on *Merulius lacrymans. Physiol. Plant.* **20,** 702–712.

Larkin, J. M. and Stokes, J. L. (1968). Growth of psychrophilic microorganisms at subzero temperatures. *Can. J. Microbiol.* **14,** 97–101.

Ledebo, I. and Ljunger, C. (1973). Permeability of bacteria to inorganic cations. A comparison between a psychrophilic *Achromobacter* strain and *Escherichia coli. Physiol. Plant.* **28,** 535–540.

Liebscher, D. H., Dauberschmidt, R., Rappaport, S. M., Heitmann, P. and Rosenthal, H. A. (1974). Untersuchungen ueber Phagelysozym IV. Die molekulare Ursache des kaeltempfindlichen Phaenotyps der Mutant cseBU56 des Phagen T4D: Der aminosauereaustausch Tryosin 88 → Histidin. *Mol. Gen. Genet.* **132,** 321–333.

Madeley, J. R., Korngold, R. R., Kushner, D. J. and Gibbons, N. E. (1967). The lysis of a psychrophilic marine bacterium as studied by microelectrophoresis. *Can. J. Microbiol.* **13,** 45–55.

Makemson, J. C. (1973). Control of *in vivo* luminescence in psychrophilic marine photobacterium. *Arch. Mikrobiol.* **93,** 347–358.

Malcolm, N. L. (1968a). A temperature-induced lesion in amino acid-transfer ribonucleic acid attachment in a psychrophile. *Biochim. Biophys. Acta.* **157,** 493–503.

Malcolm, N. L. (1968b). Synthesis of protein and ribonucleic acid in a psychrophile at normal and restrictive growth temperatures. *J. Bacteriol.* **95,** 1388–1399.

Malcolm, N. L. (1969a). Subunit structure and function of *Micrococcus cryophilus* glutamyl transfer RNA synthetase. *Biochim. Biophys. Acta.* **190,** 347–357.

Malcolm, N. L. (1969b). Molecular determinants of obligate psychrophily. *Nature, Lond.* **221,** 1031–1033.

Marr, A. G. and Ingraham, J. L. (1962). Effect in temperature on the composition of fatty acids in *Escherichia coli. J. Bacteriol.* **84,** 1260–1267.

Mattingly, S. J. and Best, G. K. (1971). The effect of temperature on lysis of cells and cell walls of *Bacillus psychrophilus. Can. J. Microbiol.* **17,** 1161–1168.

Mattingly, S. J. and Best, G. K. (1972). Effect of temperature on the integrity of *Bacillus psychrophilus* cell walls. *J. Bacteriol.* **109,** 645–651.

Mosser, J. L., Herdrich, G. M. and Brock, T. D. (1976). Temperature optima for bacteria and yeasts from cold-mountain habitats. *Can. J. Microbiol.* **22,** 324–325.

Nash, C. H. and Grant, D. W. (1969). Thermal stability of ribosomes from a psychrophilic and a mesophilic yeast. *Can. J. Microbiol.* **15,** 1116–1118.

Nash, C. H. and Sinclair, N. A. (1968). Thermal injury and death in an obligately psychrophilic yeast, *Candida nivalis. Can. J. Microbiol.* **14,** 691–697.

Nash, C. H., Grant, D. W. and Sinclair, N. A. (1969). Thermolability of protein synthesis in a cell-free system from the obligately psychrophilic yeast *Candida gelida. Can. J. Microbiol.* **15,** 339–343.

O'Donovan, G. A. and Ingraham, J. L. (1965). Cold-sensitive mutants of *Escherichia coli* resulting from increased feedback inhibition. *Proc. Natl. Acad. Sci. USA.* **54,** 451–457.

Olsen, R. H. and Metcalf, E. S. (1968). Conversion of mesophilic to psychrophilic bacteria. *Science* **162,** 1288–1289.

Oppenheim, J., Scheinbuks, J., Biava, C. and Marcus, L. (1968). Polyribosomes in *Azotobacter vinelandii.* I. Isolation, characterization and distribution of ribosomes, polyribosomes and subunits in logarithmically growing *Azotobacter. Biochim. Biophys. Acta* **161,** 386–401.

Pace, B. and Campbell, L. L. (1967). Correlation of maximal growth temperature and ribosome heat stability. *Proc. Natl. Acad. Sci. USA* **57,** 1109–1116.

Pooley, H. M., Porres-Juan, J. M. and Shockman, G. D. (1970). Dissociation of an autolytic enzyme-cell wall complex by treatment with unusually high concentrations of salt. *Biochim. Biophys. Res. Commun.* **38,** 1134–1140.

Purohit, K. and Stokes, J. L. (1967). Heat labile enzymes in a psychrophilic bacterium. *J. Bacteriol.* **93,** 199–206.

Quist, R. G. and Stokes, J. L. (1969). Temperature range for formic hydrogenlyase induction and activity in psychrophilic and mesophilic bacteria. *Antonie van Leeuwenhoek* **35,** 1–8.

Quist, R. G. and Stokes, J. L. (1972). Comparative effect of temperature on the induced synthesis of hydrogenase and enzymes of the benzoate oxidation system in psychrophilic and mesophilic bacteria. *Can. J. Microbiol.* **18,** 1233–1239.

Reid, P. (1971). Isolation of cold sensitive-rifampicin resistant RNA polymerase mutants of *Escherichia coli. Biochem. Biophys. Res. Commun.* **44,** 737–744.

Schairer, H. U. and Overath, P. (1969). Lipids containing *trans*-unsaturated fatty acids change the temperature characteristic of thiomethylgalactoside accumulation in *Escherichia coli. J. Mol. Biol.* **44,** 209–214.

Scotti, P. D. (1968). A new class of conditional lethal mutants of bacteriophage T4D. *Mutat. Res.* **6,** 1–14.

Shaw, M. K. (1967). Effect of abrupt temperature shift on the growth of mesophilic and psychrophilic yeasts. *J. Bacteriol.* **93,** 1332–1336.

Shaw, M. and Ingraham, J. L. (1965). Fatty acid composition of *Escherichia coli* as a possible controlling factor of the minimal growth temperature. *J. Bacteriol.* **90,** 141–146.

Shaw, M., Marr, A. G. and Ingraham, J. L. (1971). Determination of the minimal temperature for growth of *Escherichia coli. J. Bacteriol.* **105,** 683–684.

Shing, Y. W., Akagi, J. M. and Himes, R. H. (1972). Thermolabile triose phosphate isomerase in a psychrophilic *Clostridium. J. Bacteriol.* **109,** 1325–1327.

Shing, Y. W., Akagi, J. M. and Himes, R. H. (1975). Psychrophilic, mesophilic, and thermophilic triosephosphate isomerases from three clostridial species. *J. Bacteriol.* **122,** 177–184.

Simensky, M. (1971). Temperature control of phospholipid biosynthesis in *Escherichia coli. J. Bacteriol.* **106,** 449–455.

Stevens, D. L. and Strobel, G. A. (1968). Origin of cyanide in cultures of a psychrophilic basidiomycete. *J. Bacteriol.* **95,** 1094–1102.

Stokes, J. L. and Larkin, J. M. (1968). Comparative effect of temperature on the oxidative metabolism of whole and disrupted cells of a psychrophilic and mesophilic species of *Bacillus. J. Bacteriol.* **94,** 95–98.

Szer, W. (1970). Cell-free protein synthesis at 0°. An activating factor from ribosomes of a psychrophilic microorganism. *Biochim. Biophys. Acta.* **213,** 159–170.

Taketa, K. and Pogell, M. (1965). Allosteric inhibition of rat liver fructose-1, 6-diphosphatase by adenosine-5'-monophosphate. *J. Biol. Chem.* **240,** 651–662.

Tai, P.-C. and Jackson, H. (1969). Growth and respiration of an obligate psychrophile, *Micrococcus cryophilus*, and its mesophilic mutants. *Can. J. Microbiol.* **15,** 1151–1155.

Tai, P.-C., Kessler, D. P. and Ingraham, J. L. (1969). Cold-sensitive mutations in *Salmonella typhimurium* which affect ribosome synthesis. *J. Bacteriol.* **97,** 1298–1304.

Tai, P.-C., Wallace, B. J., Herzog, E. L. and Davis, B. D. (1973). Properties of initiation-free polysomes of *Escherichia coli. Biochemistry,* **12,** 609–615.

Upadhyay, J. and Stokes, J. L. (1962). Anaerobic growth of psychrophilic bacteria. *J. Bacteriol.* **83,** 27–2075.

*Uydess, I. L. and Vishniac, W. V. (1976). Electron microscopy of Antarctic soil bacteria. *In* "Extreme Environments, Mechanisms of Microbial Adaptation" (Ed. M. R. Heinrich). Academic Press, New York and London.

Vinopal. R. T. (1972). "Cold-fitter mutants of *Salmonella typhimurium.*" Thesis, University California, Davis.

Ward, E. W. B. (1966). Preliminary studies of the physiology of *Sclerotinia borealis*, a highly psychrophilic fungus. *Can. J. Bot.* **44,** 237–246.

Ward, E. W. B. (1968). Temperature-induced changes in the hyphal morphology of the psychrophile *Sclerotinia borealis. Can. J. Bot.* **46,** 524–525.

Waskell, L. and Glaser, D. A. (1974). Mutants of *Escherichia coli* with cold-sensitive deoxyribonucleic acid synthesis. *J. Bacteriol.* **118,** 1027–1040.

Wehr, C. T., Waskell, L. and Glaser, D.A. (1975). Characteristics of cold-sensitive mutants of *Escherichia coli* K-12 defective in deoxyribonucleic acid replication. *J. Bacteriol.* **121,** 99–107.

Wilson, G. and Fox, C. F. (1971). Biogenesis of microbial transport systems: Evidence for coupled incorporation of newly synthesized lipids and proteins into membrane. *J. Mol. Biol.* **55,** 49–60.

Wilson, G., Rose, S. P. and Fox, C. F. (1970). The effect of membrane lipid unsaturation on glycoside transport. *Biochem. Biophys. Res. Commun.* **38,** 617–723.

Chapter 4

Microbial Life Under Pressure

R. E. MARQUIS and P. MATSUMURA

University of Rochester School of Medicine and Dentistry

I. Introduction 105
II. General mechanisms underlying biological pressure responses . . 110
A. Effects of pressure on biological equilibria 110
B. Effects of pressure on rates of biochemical reactions . . . 119
C. Conclusions 121
III. A few historical notes on microbial life under pressure . . . 122
IV. Microbial growth under pressure 126
V. Killing of microorganisms by pressure 135
VI. Pressure, pollution, petroleum microbiology and deep-sea mining . 138
VII. Biological effects of compressed gases 140
IX. Epilogue 143
Addendum 144
Acknowledgements 147
References 147

I. Introduction

Microbial barobiology had its origins in marine ecology, and much of the current effort in the field is focused on the role of hydrostatic pressure as an ecological factor affecting the distribution and activities of microorganisms in the deep sea. Pressure in the ocean increases by about one atm for every 10 m depth, due to the weight of the water column, to a maximum of some 1160 atm at the bottom of the Challenger Deep in the Pacific Ocean. The average pressure on the ocean floor is about 380 atm (Kinne, 1972), and high pressure coupled with ambient temperatures only slightly above 0°C produce very restrictive environments for life in the deep sea. Certainly, few of the bacteria described in Bergey's Manual can grow at 0 to 4°C and 380 atm pressure. Yet, samplings of all depths of the ocean, including the Challenger Deep, have yielded viable microorganisms. In fact, there is some evidence (ZoBell and Morita, 1957) that many of the bacteria retrieved from the deep ocean are obligately barophilic and grow only under pressure. Others are apparently only baroduric or pressure tolerant. It seems also that many of the bacteria recovered from the deep persist there in a state of suspended animation. Indeed, attempts to use hydrostatic pressure as a sterilizing agent (Timson and Short, 1965) have been hampered by so-called persistors that are not killed even by exposure to kilobar pressures. Some of the bacteria that inhabit the deep ocean do so as part of the normal flora of fish and other animals that have become adapted to life under considerable pressures.

Despite any adaptations that bacteria may have made to high pressure environments, microbial activity in the ocean depths appears to be very slow. One of the most striking demonstrations of slow growth of bacteria under deep-sea pressures and temperatures is that of Schwarz and Colwell (1975a) who grew cultures of the deep-sea isolate *Pseudomonas bathycetes* at 1000 atm and 3°C. The cultures showed lag phases of some four months, an exponential growth phase in which the generation time was 33 days and a stationary phase that occurred when the cultures had grown to only about one one-hundredth of the one atm density in terms of viable cells per ml. This very extended culture cycle is in contrast to the cycle at 37°C and one atm that lasts only a matter of a few hours with a generation time of less than one hour. Thus, growth under deep-sea conditions was nearly 1000 times slower than growth at near optimal conditions and appears to be highly inefficient in terms of yield. More recent experiments (Schwarz *et al.*, 1976) with bacteria isolated from the gut contents of deep-sea amphipods indicate that at least some of these bacteria can grow rapidly at 3°C and 750 atm, in contrast to bacteria isolated from deep water or sediment samples. Giant amphipods seem to be relatively abundant in the deep sea, and as the underwater photographs of Hessler *et al.* (1972) show, these animals are voracious in devouring dead fish that are used as bait to attract them. They probably serve a major function as scavengers in the abyssal-hadal depths. The bacteria within the guts of these animals presumably are also involved in the scavenging function. There seem to be many interesting associations of microorganisms with deep-sea animals, including the functional association of large masses of bacteria in the luminous organelles of deep-sea fish.

Although it does seem clear that there are microorganisms that can grow in the deep-sea environment and others that can persist there for long periods of time, there is still some question regarding whether or not there are truly barophilic bacteria that have become adapted to function only under deep-sea pressures. ZoBell and Morita (1957) found that deep-trench samples from the Pacific yielded more bacteria, as determined by the most-probable-numbers dilution test, when incubated at 1000 atm compared with one atm, but they were unable to subculture the apparently barophilic bacteria. Also, many bacteria show some barophilic character in that they grow better under pressures of 100 to 200 atm, or even as high at 300 to 350 atm (Kriss and Mitskevich, 1967), than at one atm.

In bodies of fresh water, maximum pressures are considerably less than those in the ocean. At the nadir of the deepest lake in the world, Lake Baykal in the Soviet Union, the pressure is only some 160 atm. The maximum pressure in the Great Lakes is only about 40 atm at the bottom of Lake Superior. However, these low pressures are not without biological effect. For example, microorganisms that contain gas vacuoles are extremely barosensitive. Walsby (1971, 1972) found that the gas vacuoles of *Halobacterium* and of photosynthetic bacteria can be collapsed by externally applied pressures of only about 6 atm. In fact, the vacuoles of *Halobacterium* were completely collapsed by only 2 atm. The positioning of certain blue-green bacteria in the water column appears to be related to metabolic activities (Walsby, 1971; Dinsdale and Walsby, 1972). At high intensities of light the turgor pressure of the cells increases, presumably due to increased photosynthetic metabolism, and a fraction of the gas vacuoles collapses to reduce the buoyancy of the organism. The cells then tend to settle to regions of lower light intensity. Increased hydrostatic pressure encountered during settling would also act to collapse the vacuoles and to speed up the descent. Apparently it is necessary for the organisms to synthesize new vacuoles in order to regain their former buoyancy. The extreme sensitivity to pressure here is due to the presence of a gas phase in the cells.

Extreme barosensitivity can also be associated with nonoptimal environmental conditions. For example, most organisms are most barotolerant at temperatures slightly above the optimum temperature for growth at one atm. At higher or lower temperatures, their barotolerance is decreased. Thus, for example, a bacterium such as *Escherichia coli* can grow at pressures as high as 500 atm at temperatures of 30 to 40°C. However, at a temperature of 9°C or at 49°C it is very much inhibited by only 50 to 100 atm. Similar extreme sensitivites are associated also with growth at pH extremes. For example, growth of *E. coli* can be stopped completely by only 100 atm pressure at 30°C when the pH of the growth medium is lowered to 5·2. In all, it seems that in many relatively shallow aquatic environments pressure may be an important ecological factor simply because growth conditions are not optimal for many of the organisms that are deposited there, or even for some of the resident flora.

Another natural environment in which high pressure prevails is in deep oil or sulphur wells. Pressure increases with depth in the earth at an average rate of about 0·1 atm m^{-1}. Temperature also increases with depth, by an average of about 0·014°C m^{-1}, but with major discontinuities in the temperature profile. ZoBell (1958) cultured thermophilic, sulphate-reducing bacteria from cores taken at depths greater than 3500 m in oil and sulphur wells where the pressure was some 400 atm and temperatures ranged from 60 to 105°C.

In addition, microorganisms living in other underground deposits, e.g. those associated with coal (Jaschhof and Schwartz, 1969), may be subjected to high ambient pressure due to the weight of the earth's crust. In these situations also, high pressure is coupled with high temperature.

Microorganisms are probably also subjected to high pressures in certain types of industrial equipment. Industrial fermentations are commonly run under a few atm positive pressure, and many chemical processes are carried out under much higher pressures. To our knowledge, there is no information on the effects of pressure on microbial activities in these situations.

In addition to the ecological focus of microbial barobiology, there is a great deal of interest in the use of pressure for basic and applied microbiology. Some of the most informative experiments in physiology or molecular biology involve perturbing the system under study, observing its response to perturbation and its response to withdrawal of the perturbing influence. Hydrostatic pressure is one of the most general of the perturbing forces in a category with temperature as an intrinsic parameter of any particular system. In recent years it has become apparent (Marquis, 1976) that the ideal gas law cannot be readily applied to condensed systems such as living cells, and that responses to pressure changes are not simply the inverse of responses to temperature changes. Therefore, pressure-sensitive and pressure-resistant mutant organisms should be unique in relation to heat-sensitive or cold-sensitive ones.

In general terms it can be said that pressure enhances those reactions or processes that are accompanied by a volume decrease, and inhibits those that are accompanied by a volume increase. Thus, if one knows how pressure affects some particular reaction or process, it is possible to calculate the volume change that accompanies it. This basic information regarding volume change is then useful in attempts to decipher reaction mechanisms. Moreover, a knowledge of the volume change allows one to calculate the pressure-volume work associated with the process or reaction, which is simply $P\Delta V$, where P is the pressure and ΔV is the volume change. In all, it seems that studies of the effects of pressure on biological reactions should yield as much information as have studies of the effects of temperature, and the results of both types of studies should allow for a detailed analysis of the mechanisms of biological reactions.

Interest in microbial barobiology extends to the highly practical. For example, pressure has great potential in modifying the physiology of industrially important microorganisms, as a sterilizing or disinfecting agent, especially for labile materials, and as a modifying factor for industrial enzymatic processes, including those that make use of immobilized enzymes. In recent years, there has also been some interest in the effects of pressure on the normal flora of man. It appears that man can work at higher pressures than was previously thought possible. Simulated dives conducted in the laboratory indicate that man may be able to work under conditions equivalent to those at a depth of some 5000 ft of sea water, where the pressure would be about 150 atm. It is reasonable to think that this pressure would not be without effect on microorganisms in the respiratory tract, especially since it is established mainly with inert gases that have narcotic effects at high pressures. In fact, man's ability to work at this high pressure may depend on a balance between the narcotic effects of inert gases and the stimulatory effects of pressure (Macdonald, 1975a, 1975b).

Of necessity, this article will differ in basic outlook from many of the others in this book, largely because we are as yet unsure that any organisms have undergone specific adaptations to life under pressure. Obviously then, it would be impossible to discuss the nature of these adaptations. That some bacteria are highly baroduric seems at the present time to be purely fortuitous and not related to specific adaptation. In fact, Kriss (1962) was able to isolate baroduric bacteria from garden soil. However, the effects of pressure on biopolymers are of the same general types as those of heat, cold and high ionic strength. Therefore, one might expect that certain amino acid substitutions in proteins would result in products that would maintain their native conformations better under pressure than at one atm. Thus, it seems likely that obligate barophilic bacteria do exist, or could be constructed by mutation, and current efforts at deep-sea sampling are likely to yield such organisms for study.

As primary interests of the authors are in microbial physiology rather than in marine biology, initial interests in barobiology were concerned with how pressure inhibits bacterial growth. This article will reflect our interests in that it will be focused mainly on the physiological, biochemical and molecular mechanisms underlying pressure responses of microorganisms. Articles with more emphasis on marine biology can be found in books on barobiology that have been published over the last few years, notably Zimmerman (1970), Brauer (1972), and Sleigh and Macdonald (1972). In addition, "Marine Ecology" (Kinne, 1972) contains contributions by Kinne, Morita, Vidaver and Flugel on pressure as an environmental factor in the ocean. Macdonald's book (1975b) contains excellent reviews of many aspects of barobiology, including microbiological ones.

In this review, pressures will usually be given in atmospheres since this unit is so easily appreciated by biologists in general. One standard atmosphere $= 1{\cdot}033\ \mathrm{kg\ cm^{-2}} = 1{\cdot}013\ \mathrm{bar} = 1{\cdot}013 \times 10^5\ \mathrm{Newtons\ m^{-2}}$.

II. General Mechanisms Underlying Biological Pressure Responses

It seems reasonable in terms of general orientation to review briefly the overall mechanisms by which pressure can affect living cells before going on to consider specific responses. Just how do cells sense changes in hydrostatic pressure? In our review, we shall distinguish between effects of pressure on biochemical equilibria and those on rates of reaction. Both types of effects are important in barobiology. The two fundamental relationships that are used to describe the effects are:

$$\left(\frac{\partial \ln K}{\partial P}\right)_T = -\Delta V/RT \tag{1}$$

$$\left(\frac{\partial \ln k}{\partial P}\right)_T = -\Delta V^{\ddagger}/RT \tag{2}$$

where K is the equilibrium constant, k is the reaction rate constant, P is the pressure, ΔV is the volume change of reaction, $\Delta V^{\ddagger}$ is the apparent volume change of activation, R is the gas constant, and T is the Kelvin temperature. Derivations of these equations are presented in standard treatments of high pressure chemistry, e.g. Weale (1967).

A. Effects of Pressure on Biological Equilibria

Most chemical reactions result in a volume change because the sum of the partial molar volumes of products is generally not equal to that of reactants. The volume change arises primarily from changes in the reacting species; for example, breakage of a covalent bond results in a volume increase. In addition, there may be changes in the solvent or suspending medium that result in volume changes; one of the most common is the electrostriction of water in the vicinity of ions that results in contraction. The volume change here is the ΔV of reaction that appears in equation (1). It can be determined in two ways —by assessing the pressure response of the system, or more directly, by use of a dilatometer. One of the most common types of dilatometer is the so-called Carlsberg vessel, which consists of a bifurcated glass reaction chamber connected to a calibrated capillary tube with an etched scale. Access to the chamber is through a three-way stopcock. Reactants are placed in the bifurcations, and the remainder of the vessel is filled with an inert fluid such as kerosene or heptane. After temperature equilibration, generally in a water bath with temperature control to a few hundredth of a Kelvin degree, the amount of kerosene or heptane in the vessel is adjusted so that there is a meniscus in the etched part of the capillary. The vessel is then closed off, inverted, and the movement of the meniscus is recorded as the reaction proceeds. Much more elaborate dilatometers can be constructed for special purposes, and there are many designs described in the literature, for example, in the papers by Katz and Ferris (1966), Christensen and Cassel (1967), Marquis and Fenn (1969) and by Jaenicke (1971). Obviously, it is also possible to calculate ΔV from measurements of changes in density of a system during the course of a chemical reaction. With a knowledge of the ΔV of reaction, it is possible to accurately predict the effect of pressure on the equilibrium constant by the use of equation (1). Conversely, if one knows the effect of pressure on the equilibrium constant, it is possible to accurately calculate the volume change.

Volume changes for selected reactions are presented in Table 1. The changes for simple, monomeric reactions are relatively small, less than about 30 ml mol^{-1}. Changes for cooperative, polymeric reactions may be an order of magnitude or more larger. For example, Infante and Baierlein (1971) found that the ΔV for association of ribosomal subunits is some 500 ml mol^{-1}, as indicated by the pressure sensitivity of dissociation. Volume changes of this magnitude clearly indicate cooperative phenomena. Thus, it appears that ribosomal subunit association involves a series of coupled "monomeric" reactions and that this coupling leads to exquisite barosensitivity.

1. Pressure Denaturation of Biopolymers

Probably the most thoroughly studied examples of cooperative reactions in biochemistry are biopolymer denaturations. Starting with the early work of Bridgman (1914), there have been many studies of protein denaturation by pressure. Recently, Zipp and Kauzmann (1973) took advantage of the differential light absorption of the heme moiety in native versus denatured metmyoglobin for such a study. They found that pressure up to about 3500 kg cm^{-2} (1 atm = 1·033 kg cm^{-2}) had essentially no effect. Then, over the relatively narrow range from about 3500 to 5000 kg cm^{-2} the protein underwent cooperative, reversible denaturation. Other examples of cooperative denaturation have been compiled by Suzuki and Taniguchi (1972) in their review article; their plots show pressures for 50% denaturation that range from about 3500 atm for β-lactoglobulin at pH 5·2 and 30°C to about 9000 atm for α-amylase at pH 5·8 and 30°C. Accurate determinations of the volume changes accompanying protein denaturation have yielded a vexing dilemma for protein chemists. The volume changes are not very large, and this finding is upsetting to current views of the forces that give conformational stability to proteins.

It is generally considered that hydrophobic interactions play major roles in determining native conformations of proteins. The disruption of hydrophobic interactions is commonly associated with a volume decrease of some 10 to

Table 1
Measured volume changes (ΔV) of some biologically important reactions

Reaction	ΔV (ml mol^{-1})	References
$H_2O \rightarrow H^+ + OH^-$	21·3	Bodanszky and Kauzmann (1962)
$HPO_4^{2-} + H^+ \rightarrow H_2PO_4^-$	24·0	Johnson *et al.* (1954)
Protein-COO^- + H^+ $\rightarrow$ protein-COOH	11·0 (average)	Kauzmann *et al.* (1962)
Diglycine-NH_2 + H^+ $\rightarrow$ diglycine-NH_3^+	−4·4	Kauzmann *et al.* (1962)
Imidazole + H^+ $\rightarrow$ imidazole H^+	−1·1	Kauzmann *et al.* (1962)
Protein-NH_3^+ + OH^- $\rightarrow$ protein-NH_2 + H_2O	16–18	Kauzmann *et al.* (1962)
Glucose $\rightarrow$ 2 lactic acid	11·8 (lactic acid)	Marquis and Fenn (1969)
Lactic acid $\rightarrow$ lactate anion + H^+	−11·7	Marquis and Fenn (1969)
Arginine + $4H_2O$ $\rightarrow$ ornithine + CO_2 + $2NH_4OH$	−22·8 (NH_4OH)	Marquis and Fenn (1969)
Native DNA $\rightarrow$ denatured DNA	0–2·7 (base pairs)	Chapman and Sturtevant (1969), Gunter and Gunter (1972)
Native metmyoglobin $\rightarrow$ denatured metmyoglobin	−98	Katz *et al.* (1973)
Gel-to-liquid-crystal transition of distearoyl-L-α-lecithin	20·5	Melchior and Morowitz (1972)
Allosteric binding of threonine by the aspartokinase-homoserine dehydrogenase I of *E. coli*	+40 ml 10^5 g protein	Wampler and Katz (1974)
Aggregation of poly-L-valyl-ribonuclease	203	Kettman *et al.* (1965)
Flagellin $\rightarrow$ flagellar filaments	157	Gerber and Noguchi (1967)
Ribosomal subunits $\rightarrow$ ribosome	500	Infante and Balerlein (1971)
Microtubulin $\rightarrow$ microtubules	90–400	Salmon (1975a, b)

20 ml mol^{-1} (Kauzmann, 1959). It is thought that, when nonpolar groups are transferred from a hydrophobic medium such as the interior of a globular protein to an aqueous medium, there is an ordering of water. The ordered water assembly is referred to as a clathrate or an "iceberg". Clathrate water differs from ordinary ice in being more rather than less dense than fluid water. Therefore, disruption of hydrophobic interactions results in contraction, and pressure acts to disrupt hydrophobic interactions. One would expect that denaturation of a protein with disruption of large numbers of hydrophobic interactions would be accompanied by a large decrease in volume of some hundreds of ml mol^{-1}.

Denaturation may also result in the breaking of electrostatic bonds in amphoteric proteins. This bond breakage would result in exposure of charged groups to the aqueous environment. The water around the groups would experience electrostriction, and there would be a decrease in volume, generally in the range of 1 to 11 ml mol^{-1} of charged groups.

Denaturation may also involve breakage of hydrogen bonds. However, this process would result in a volume increase of about 3 or 4 ml mol^{-1}. Any covalent bond breakage would also result in an increase in volume. For example, breakage of a carbon–carbon bond generally results in an increase in volume of about 12 ml mol^{-1}. Reversible denaturation does not usually involve breakage of covalent bonds.

A number of authors (Tanford, 1968; Brandts *et al.*, 1970; Zipp and Kauzmann, 1973) have pointed out that the volume change for protein denaturation should be large and negative because disruption of hydrophobic and electrostatic interactions should far overshadow hydrogen bond breakage. Attempts to identify other processes that might cause compensatory volume increases have proved unsuccessful. It seems that something must be askew in concepts of protein conformation, and Zipp and Kauzmann (1973) suggested the possibility that conformational stability may not in fact be achieved primarily through hydrophobic interactions.

Another, less traumatic explanation for the discrepancy between measured and expected volume changes for protein denaturation has come from the recent work of Bøje and Hvidt (1972) and Hvidt (1975). They pointed out that the chemical groups in a polymer are in one sense in very concentrated solution because of the close proximities of neighbours. With model compounds, they demonstrated that disruption of hydrophobic interactions in concentrated solutions results in an increase rather than a decrease in volume. Therefore, they proposed that the breaking of hydrophobic interactions during protein denaturation results in small, positive changes in volume. The small, negative volume changes that have been measured presumably reflect the net result of increases in volume for hydrophobic and hydrogen bond breakage, and decreases in volume for disruption of electrostatic interactions. Certainly, this is an exciting aspect of pressure studies that has resulted in a major re-evaluation of basic notions of biopolymer chemistry.

Denaturation of native DNA molecules also results in rather small changes in volume (Weida and Gill, 1966; Gunter and Gunter, 1972; Chapman and Sturtevant, 1969), but the changes are positive ones so that pressure stabilizes DNA in its native state. Pressure has been found (Hedén *et al.*, 1964; Weida and Gill, 1966; Suzuki *et al.*, 1971, 1972) to antagonize the denaturing effect of high temperature and to raise the "melting" temperature by a few degrees. The effects of heat on native DNA clearly indicate that denaturation is a cooperative reaction. Presumably, the small volume change is due to compensating increases and decreases in volume associated with various components of the overall reaction.

Most studies of protein denaturation are carried out with pressures of some thousands of atm rather than with deep-sea pressures of less than 1160 atm. In fact, pressures below 1000 atm are often thought of as stabilizing the native conformations of proteins as indicated, for example, by the extensive work of Johnson and Eyring (1970) on the effects of pressure on bioluminescence. There seems to be a general feeling that protein denaturation does not play a major role in growth inhibition by pressures that normally occur in the biosphere. However, it seems that this view might not be an entirely reasonable one. As part of his study of pressure denaturation of chymotrypsin, Hawley (1971) constructed constant-free-energy plots on a temperature-pressure plane. Denaturation can be considered as a two-state process in which the native state (N) is in equilibrium with the denatured state (D). The equilibrium constant for the reaction is then equal to $(D)/(N)$, and the free energy change is equal to $-RT\ln(D)/(N)$. Hawley's plots for zero free energy change show conditions under which $(D) = (N)$. They turned out to be closed elliptical figures with an enclosed area indicating conditions under which the native state is favoured and an outer area indicating conditions which favour the denatured form. Zipp and Kauzmann (1973) extended this sort of analysis in their study of metmyoglobin denaturation by including a pH axis so that a three-dimensional envelope figure was obtained. If we follow a contour of this envelope at constant temperature, say 20°C, to find points for zero free energy change at various pH values, we find that at pH 6, $(D) = (N)$ at about 4300 kg cm^{-2}. At pH 5, $(D) = (N)$ at about 3000 kg cm^{-2}; while at pH 4, the same equilibrium condition is attained at only 900 kg cm^{-2}. In other words, at this low pH, deep-sea pressures are denaturing for metmyoglobin. If we consider the situation at deep-sea temperatures of only a few degrees centigrade, the denaturing pressure at pH 4 is only a few hundred atm. There are probably few niches in the deep sea where the pH is as low as 4, especially since the water is buffered with bicarbonate and other buffer ions. However, the main point to be made here is that deep-sea pressures can be denaturing for proteins when other physical and chemical parameters are not optimal for maintaining the native conformational state.

Matsumura *et al.* (1974) described the interactions of pressure and pH in limiting bacterial growth. Their results are similar to those of protein denaturation studies. The net effect was that the pH range for growth was progressively narrowed as the growth pressure was increased. For the other point of view, it appeared that acid or alkaline growth conditions increased the barosensitivities of the bacteria studied. It has been found that pressure usually reduces growth yields as well as growth rates. Matsumura *et al.* (1974) found that at least part of this effect for many bacteria is due to increased sensitivity under pressure to metabolic acids that accumulate in enclosed culture vessels. Neutralization of these acids increased growth yields, but they still tended to be less than those of one atm cultures because of other effects of pressure on growth efficiency.

The similarity between the pressure-temperature-pH responses of microbial growth and those of protein denaturation seems more than just fortuitous. There presumably are many situations in which the failure of an organism to grow under pressure can be traced to reversible denaturation of enzymes or other cell proteins. Admittedly, these situations may be nonoptimal ones, but then optimal growth conditions are probably more the exception than the rule in natural environments.

2. Lipid Phase Changes

During the past few years, it has become apparent that the lipids in biological membranes undergo phase changes in response to changes in temperature. Functioning membranes seem to have fluid, liquid-crystalline interiors. Chilling causes a transition to the gel phases with resulting impairment of function. The temperature at which the phase transition takes place is related to the chemical nature of the membrane lipids. Membranes that are rich in unsaturated or branched-chain fatty acids have lower transition temperatures than do membranes that are rich in saturated, straight-chain fatty acids. In at least some organisms, the fatty acid composition of the membrane appears to be a major determinant of the range of temperatures over which growth can occur, as shown for example, by the work of McElhaney (1974) with *Acholeplasma laidlawii.*

Obviously, pressure also affects the phase state of membrane lipid aggregates and the temperatures at which phase transitions take place. However the effects are not major ones, and the volume changes accompanying lipid phase transitions are not large. For example, Melchior and Morowitz (1972) measured ΔV values of 11·8, 14·1 and 20·5 ml mol^{-1}, respectively, for gel to liquid crystal transitions of dimyristoyl-, dipalmitoyl- or distearoyl-L-α-lecithin in micellar or vesicular structures. Moreover, the relatively wide temperature ranges over which single membrane phase transitions take place suggest that they are relatively noncooperative, unlike the melting of DNA molecules. It appears also that the interaction of salts with charged lipid results in a volume change. The data presented by Plachy *et al.* (1974) indicate that the average density of mixed polar *Halobacterium* lipids is greater in 0·1 M $MgCl_2$ solution than in 4·0 M NaCl solution. It can be calculated that a shift from the first to the second would result in a volume increase of somewhat less than 10 ml mol^{-1}—again, a relatively small volume change. Pressure favours tightly packed gel phases of membrane lipids, and Trudell *et al.* (1974) found that 136 atm pressure, established with helium gas, produced 3 to 5°C increases in the temperatures at which mixed phosphatidyl choline bilayers underwent solid to liquid phase transitions. It seems almost axiomatic that pressure should have effects on minimal, optimal and maximal growth temperatures in situations in which these temperatures are determined by the physical states of membrane lipids. Pressure is known to shift optimal growth temperatures slightly upward (ZoBell, 1956; Johnson, 1957), and part of this shift may be due to the action of pressure in favouring condensed phases. The shift is a relatively small one of only a few degrees, but of course even a small shift could make a major difference in a natural environment where competition for food is keen.

3. *Polymer Aggregations*

In contrast to the small volume changes that accompany lipid phase transitions and the apparent noncooperativity of the process, large volume changes are associated with many types of polymeric aggregation (Table 1). The extreme pressure sensitivity of these aggregation processes has presented problems in interpretation of sedimentation patterns in the ultracentrifuge where pressures as high as 1950 atm can be generated (van Diggelen *et al.*, 1973).

Quite apart from any problems facing experimentalists who use the ultracentrifuge, living organisms may have dramatic adverse reactions to pressure due to disaggregation of structures such as microtubules. For example, pressures of 400 to 600 atm have major effects on protozoa and sea urchin eggs. These cells become immobilized (Regnard, 1884; Certes, 1884b; Ebbecke, 1935a, 1935b; Kitching, 1957a; Young *et al.*, 1972), phagocytosis and pinocytosis are blocked (Marsland and Brown, 1936; Zimmerman and Rustad, 1965), axopodia of heliozoans are collapsed (Kitching, 1957b; Tilney *et al.*, 1966), galvanotaxis of *Tetrahymena* is reversed (Murakami and Zimmerman, 1970), the cells round up (Marsland, 1970), and the mitotic apparatus is "frozen" (Pease, 1941, 1946). Many, if not all, of these effects appear to be due to pressure-induced disaggregation of microtubules. Tilney *et al.* (1966) showed by means of electron microscopy that the collapse of axopodia is associated with reversible loss of microtubules. Kennedy and Zimmerman (1970) found that immobilization of *Tetrahymena pyriformis* cells is associated with microtubule disintegration. Pressurization of *Arbacia* embryos at 408 atm leads to complete loss of cytoplasmic microtubules (Tilney and Gibbins, 1969), and Salmon *et al.* (1976) found that exposure of HeLa cells to 680 atm resulted in nearly complete depolymerization of astral and interpolar microtubules. Moreover, agents that stabilize microtubular structure, such as D_2O, have been found (Marsland, 1970) to antagonize inhibition of mitosis by pressure; while agents such as colchicine, that inhibit microtubule formation, potentiate the effect of pressure. A strong case can therefore be built for microtubular disaggregation as a primary cause of many of the adverse responses of animal cells to pressure. Recently, Salmon (1975a, 1975b) and Inoue *et al.* (1975) have demonstrated that microtubules are highly sensitive to pressure *in vivo* and that the apparent volume change for assembly is 90 to 400 ml mol^{-1} tubulin. Work by O'Connor *et al.* (1974) indicates, however, that some types of microtubules, specifically neuronal ones, are pressure resistant. Heath (1975) reported similar findings for fungal spindle microtubules, and Forer and Zimmerman (1976) indicate that the isolated mitotic apparatus of sea urchin zygotes contains pressure-sensitive nonmicrotubular elements that are involved in pressure induced losses in birefringence. Also, Salmon *et al.* (1976) found that kinetochore fibre microtubules in HeLa cells are relatively pressure resistant. Presumably animals that live at great depths in the ocean must have microtubules that are not disaggregated by pressure.

The largest volume change listed in Table 1 is that for dissociation of ribosomal subunits, some 500 ml mol^{-1}. Infante and Baierlein used the ribosomes of unfertilized, sea urchin eggs for their study, but similar results were obtained by van Diggelen *et al.* (1971, 1973) for prokaryotic ribosomes. Schulz *et al.* (1976) estimated the volume change for dissociation of *E. coli* ribosomes to be -240 ml mol^{-1}. They found that the pressure effect was reversible, even after dissociation at about 1500 atm, and that added Mg^{2+} had large effects on barosensitivity, enhancing or diminishing it depending on concentration. There is some difficulty in relating the *in vitro* responses of ribosomes in the ultracentrifuge to pressure responses of whole organisms. It has been shown by a number of research groups (Landau, 1966, 1970; Pollard and Weller, 1966; Albright, 1969; Zimmerman, 1971) that protein synthesis is more barosensitive than RNA or DNA synthesis. Recent detailed studies of the barosensitivities of the various steps in the process of protein synthesis indicate that the most sensitive step is the charging of tRNA (Hardon and Albright, 1974) or the binding of tRNA to polysomes (Pope *et al.*, 1975b). Polysomes have been found to be stable in whole cells and cell free extracts at pressures that completely inhibit protein synthesis (Schwarz and Landau, 1972a; Pope *et al.*, 1975a). Pope *et al.* (1975a) and Smith *et al.* (1975) have prepared hybrid ribosomes for *in vitro* protein synthetic systems from *E. coli* and pseudomonads. Protein synthesis in *E. coli* is relatively sensitive to pressure, while in the pseudomonads, it is relatively resistant. The resistance was found to be associated with the 30S ribosomal subunit and not with the 50S subunit or the soluble factors.

In all, it seems that the pressure sensitivity of protein synthesis cannot be readily related to the pressure sensitivity of the subunit dissociation reaction. It is possible that the ribosomes become stabilized when they are in the polysome form. It is also possible that inorganic ions may have stabilizing effects and that the blend of ions used for *in vitro* protein synthesis protects 70S ribosomes against the dissociative action of pressure.

The variance between predicted or expected responses based on knowledge of the behaviour of sub-cellular fractions and the actual responses of intact, living organisms is not confined to protein synthesis. Gerber and Noguchi (1967) found that formation of flagellar filaments from flagellin monomers has a ΔV of 157 ml mol^{-1} of flagellin. Thus, pressure acts to disaggregate the filaments. However, Meganathan and Marquis (1973) found that pressures as high as 612 atm did not cause disaggregation of the flagella of intact bacteria, although pressures as low as 100 to 200 atm did inhibit formation of new flagella. Apparently some factor(s) stabilizes the flagellum *in vivo* against the dissociating action of pressure.

A few years ago, Penniston (1971) developed the interesting thesis that pressure stimulates monomeric enzymes saturated with substrate but inhibits multimeric ones. The proposed basis for the effect on multimeric enzymes was pressure-induced dissociation of subunits. Penniston used data obtained with a variety of monomeric and multimeric enzymes to support his view, and subsequent work by Dicamelli *et al.* (1973) showed clearly that pressure does indeed cause tryptophane synthetases of *E. coli* and *Salmonella typhimurium* to dissociate into subunits that could be separated in the pressure field of the ultracentrifuge. It appears from work with still other enzymes that Penniston's hypothesis is not generally applicable but that it is valid for a number of enzymes. Actually, one would expect that dissociation of some multimeric enzymes would result in a volume increase instead of a decrease, and therefore, pressure could be a stabilizing influence. We have found (Matsumura and Marquis, 1975) that pressure actually stimulates the multimeric, membrane ATPase of *Streptococcus faecalis* at high substrate concentrations. Clearly dissociation and inactivation must not be determining the response of this enzyme to pressure.

There are other polymer aggregation reactions in which the effects of pressure on equilibria seem to determine the pressure response of intact organisms. For example, we have found (Marquis and Keller, 1975) in an extension of Landau's initial work (Landau, 1967) on the effects of pressure on the *lac* operon that it appears that the equilibrium between inducer plus operator-repressor and the inducer-operator-repressor complex determines the pressure response in *E. coli.* Pressure is inhibitory to derepression and it slows the rate at which β-galactosidase is synthesized after addition of inducer. Moreover, it also reduces the extent of derepression, and under pressures of 300 or 400 atm, cells never do become fully derepressed. The derepression process has been found (Barkley *et al.*, 1975) to be more complex than was originally believed, but still the pressure effects can be viewed in terms of shifts in equilibria that determine just how much repressor is bound in the active form to the operator.

B. Effects of Pressure on Rates of Biochemical Reactions

Many of the responses of living organisms to pressure are due to pressure-induced changes in rates of biochemical reactions rather than to any effects on equilibria. Living organisms are never at thermodynamic equilibrium, and there is constant exchange of energy and of various materials between the cell and its environment. Moreover, the whole array of biochemical reactions within a cell is an elaborately interlaced network so that no one reaction is completely independent of any other one. Therefore, the effects of pressure on reaction rates in cells are very complex, and in many ways it is a distorting abstraction to consider the effect of pressure on some isolated reaction.

Equation (2) is generally used to describe the effect of pressure on reaction rate. The pertinent volume change is the apparent volume of activation or $\Delta V^{\ddagger}$. (The development of the activated-complex concept is traced in Section III.) Unfortunately, $\Delta V^{\ddagger}$ can be determined in only one way—from a study of the effects of pressure on reaction rate. There is currently no means for measuring the partial molar volumes of transient, activated complexes. Equation (2) implies that $\Delta V^{\ddagger}$ is independent of pressure. More often than not, especially for biological processes, this is not so. Differences in compressibilities of reactants and activated complexes result in a nonconstant $\Delta V^{\ddagger}$. In complex biological systems the nature of the reaction may change with changes in pressure, and very often there are complications related to inactivation of enzymes by pressure. Also, it is clear that $\Delta V^{\ddagger}$ is not related to ΔV, even though the processes that lead to changes in volume are of the same type for both. In other words, there is a change in volume during the course of a reaction or during the course of formation of activated complexes because various bonds are broken or formed and because of changes in solvent interactions. However, the volume change for formation of the activated complex is not equal to the volume change of reaction, nor need it have the same sign. For example, the ΔV for protein hydrolysis is negative, in large part due to the increase in the numbers of charged groups during the process. However, pressure generally slows enzyme-catalysed proteolysis. Clearly, then, although the ΔV of reaction is negative, $\Delta V^{\ddagger}$ is positive.

Any consideration of the effect of pressure on biological reaction rates necessarily centers on pressure enzymology, the general aspects of which have been developed by Laidler (1951) and also in relation to control mechanisms (Hochachka *et al.*, 1972). Only a few of the main points will be considered here; the book by Laidler and Bunting (1973) should be consulted for more details.

The simplest sort of enzyme catalysed reaction can be considered as a two-step process,

$$E + S \underset{k_{-1}}{\overset{k_1}{\rightleftarrows}} ES \overset{k_2}{\rightarrow} E + P \tag{3}$$

where E is the enzyme, S is the substrate, ES is the enzyme-substrate complex and P is the product. The k symbols refer to reaction rate constants. The two steps here are formation of an enzyme-substrate complex or substrate binding and conversion of the complex to products. The later reaction is generally considered to occur in two stages—conversion of the enzyme-substrate complex to an activated state, and decay of the activated state to enzyme plus product. For most reactions, the later stage is very rapid, and so the rate-limiting stage is the formation of the activated complex. Therefore, the pertinent parameter in regard to pressure effects is the volume change associated with activation. The $\Delta V^\ddagger$ determined by use of equation (2) with data obtained at high substrate concentrations is usually the $\Delta V^\ddagger$ for this activation process. However, when the substrate is present in low concentrations, as is often the case in cells, the situation is more complex in that the initial binding of substrate may be rate limiting. Laidler (1951) considered two cases, the one for which $k_{-1} \gg k_2$ and the one for which $k_{-1} \ll k_2$. In the first case, $\Delta V^\ddagger$ determined by use of equation (2) is equal to the volume of reaction for formation of the enzyme-substrate complex plus the volume change associated with formation of the activated complex. For the second case, the more usual one in practice, the $\Delta V^\ddagger$ determined by use of equation (2) is the volume of activation for the binding of substrate to enzyme. Thus, at low substrate concentrations it is common for the pressure response to be related primarily to the binding reaction, while at high substrate concentration, the pressure response is related to the volume change for conversion of the enzyme-substrate complex to the activated complex. Obviously, more complicated reactions can be considered, but the important point here is that pressure effects on the rates of enzyme-catalysed reactions may be related to substrate binding or to activated complex formation or to both. A detailed analysis of each particular case is required in deciding which step in the process is mainly responsible for the observed effect. $\Delta V^\ddagger$ for binding may be opposite in sign from $\Delta V^\ddagger$ for activation. For example, we have found that pressure inhibits the membrane ATPase of *S. faecalis* at low substrate concentrations but stimulates it at high ATP levels. Apparently, for this reaction, the $\Delta V^\ddagger$ for substrate binding is positive, but the $\Delta V^\ddagger$ for formation of the activated enzyme-substrate complex is negative. There are many other examples in the literature of reactions with opposite signs for the two activation volumes.

Hochachka and his colleagues have studied enzymes isolated from midwater and abyssal fish in an attempt to define the nature of adaptive changes in enzyme structure that would result in catalysts capable of functioning at abyssal pressures. They concluded (Hochachka *et al.*, 1972) that "it is not $\Delta V^\ddagger$ or the activation energy, but it is enzyme-substrate, enzyme-cofactor, enzyme-modulator affinities which are most carefully tailored during pressure adaptation of enzymes." It should be noted that the $\Delta V^\ddagger$ referred to here is specifically the volume change for conversion of the enzyme-substrate complex to the activated complex and not necessarily the $\Delta V^{\ddagger}$ of equation (2). Hochachka *et al.* feel then that the substrate binding step is of primary importance in natural situations where substrate levels may be low and that binding of cofactors and modulating ligands may also be of major importance.

One of the enzymes they studied in some detail was fructose diphosphatase isolated either from rainbow trout or from the abyssal, rat-tail fish. They found that pressure stimulated both enzymes at test pH values above about 7·7 and at high substrate concentrations. However, at low substrate concentrations, the rat-tail enzyme was still stimulated, but the trout enzyme was inhibited. Thus, it appeared that pressure caused an increase in the Michaelis constant for the trout enzyme but a decrease for the rat-tail enzyme. Similar behaviour was noted with regard to the Michaelis constant for binding of the cofactor Mg^{2+}. The responses of the enzymes to the regulatory inhibitor, adenosine monophosphate, also indicated that the rat-tail enzyme was better adapted to function under pressure, especially at high concentrations of adenosine monophosphate. Clearly, many more examples should be studied in detail before any generalizations can be formulated. There are many enzymes from nonabyssal organisms that are stimulated by pressure at low substrate concentration, for example, lysozyme (Neville and Eyring, 1972), and so the behaviour of the rat-tail enzyme could possibly be fortuitous and not a true reflection of any pressure adaptation.

C. Conclusions

From the foregoing discussion it is apparent that for an interpretation of pressure responses of living organisms in biochemical terms it is necessary to consider the effects of pressure on both chemical equilibria and reaction rates. The best examples of responses to pressure in which equilibrium effects predominate are those caused by pressure-induced dissociation of specific polymeric aggregates, such as microtubules or multimeric enzymes. The denaturing action of pressure for proteins seems also to be important in many natural situations, especially if other environmental parameters such as temperature, pH or ionic strength are not optimal. Certainly, studies of isolated proteins demonstrate clearly that pressure denaturation can occur well below the kilobar level. The effects of pressure on most catabolic or biosynthetic processes in cells appear to be due primarily to pressure-induced changes in reaction rate. Most biological reactions are slowed by pressures of, say, 300 or more atm. However, many other reactions are speeded by pressure, and in fact, one might imagine that these differences in barosensitivities of the myriad of reactions in a cell would result in major upsets of the metabolic regulatory circuits of pressurized cells.

III. A Few Historical Notes on Microbial Life Under Pressure

We have no intention here of presenting any very thorough chronological review of the development of barobiology. Material of an historical nature can be found in the book by Johnson *et al.* (1954), in the review by Cattell (1936) and in the review by Fenn (1970) of Regnard's monograph. However, it does seem that there are a few topics that can profitably be considered from an historical perspective.

The first barobiological studies were those of Certes (1884) and Regnard (1884). The recovery of living specimens from depths as great as 8200 m during the Challenger Expedition in 1872 put to rest permanently any previously held notions that the ocean depths were sterile. Actually, some seven years prior to the voyage of the Challenger a species of living coral that was previously known only from fossils had been found encrusted on a broken telegraph cable that had been at a depth of 2000 m between Sardinia and Algeria. Subsequent exploratory voyages of the Travailler and the Talisman resulted in recovery of still more living specimens from the deep, including microorganisms that were recovered from water and sediment samples taken by Certes with the help of Milne-Edwards. Certes' (1884a) first note to the Academie des Sciences in Paris describing these findings was presented by Milne-Edwards. Its publication date is only five days prior to that of Regnard's initial presentation (Regnard, 1884) to the Société de Biologie. However, Regnard indicated that Certes' note had forced him to present his own results somewhat prematurely, before ". . . ils formeraient un corps de doctrine complet." Certes' paper focused mainly on the isolation of bacteria from the deep and the problems involved in excluding contaminating organisms. It is only at the end of the paper that he refers to experiments carried out in Pasteur's laboratory with a modified Cailletet press. Certes' view was clearly that the bacteria he recovered from the deep were in a state of suspended animation.

> En résumé, dès à present, il est légitime d'admettre que, dans les grandes profondeurs de l'Océan, l'eau et les sédiments renferment des germes qui, malgré l'énorme pression qu'ils ont à supporter, ne perdent pas la faculté de se multiplier, lorsqu'on les place dans les conditions de milieu et de température favorables.

Regnard also referred to life in the deep as "la vie latente." Later, Certes (1884b) seems to have changed his mind somewhat when he found that bacteria collected at a depth of some 5000 m were more barotolerant in their growth than were the terrestrial species he tested. More recent tests of deep-sea samples by ZoBell (1952), ZoBell and Morita (1957) and Kriss (1962) have generally confirmed Certes' finding of pressure-tolerant bacteria. However, even today, we still have the question of whether there are bacteria specifically adapted for growth under deep-sea pressures. The existence of a more barotolerant bacterial population in the deep may simply reflect a screening process that selects for the more barotolerant organisms that sink to the ocean floor. In fact, Kriss (1962) reported isolation of more barotolerant bacteria from garden-soil samples than from deep-sea mud. Certainly, the isolation of obligately barophilic bacteria would answer this question in an unequivocal way.

Certes and Regnard independently assessed the effects of pressures below 1000 atm on a variety of organisms and physiologic processes. They were able to show that pressures of hundreds of atm retarded many biological processes but accelerated others. Bacterial growth in natural or concocted media was an example of the former type of process in that pressures of 300 to 600 atm significantly slowed growth. Enzymes in saliva proved to be more resistant to the inhibitory action of pressure than were intact organisms, but processes such as fermentation and putrefaction by intact cells or cell extracts were slowed significantly by pressure. More extensive work by ZoBell and Johnson (1949), ZoBell and Oppenheimer (1950) and many other workers has shown that there are very few bacteria that can grow at pressures above 600 atm at their optimal growth temperature. Indeed, there are bacteria that are completely inhibited in their growth by pressures of 200 atm or less.

The barobiological interests of Certes and Regnard seem to have been related to their scientific associations. Certes carried out much of his work in Pasteur's laboratory and was primarily interested in prokaryotes and fungi. Regnard had Paul Bert as his mentor, and although he used microorganisms for his studies because of the ease with which they could be manipulated, he was very much oriented to studies of the effects of pressure on animals, on muscle and on nerve. He was an ingenious inventor and even devised an apparatus for observing through quartz windows the effects of pressure on muscle and on microscopic animals. As was the custom in science at the time, he had showings of some of his experiments for interested spectators. He summarized his barobiological work in a monograph entitled "Recherches Expérimentales sur les Conditions Physiques de la Vie dans les Eaux" published in 1891.

Subsequent developments in high-pressure technology led first to the attainment of pressures of about 12,000 atm in apparatus with self-tightening seals for closures and moving pistons, then to 50,000–60,000 atm with use of tungsten carbide pistons, and finally, to 100,000 atm with an apparatus that had a piston and cylinder within another piston and cylinder that created a supporting envelope of compressed fluid. Bridgman was prominent in this work and he describes much of its development in his book (1949) on high-pressure physics. Certainly, one of the major accomplishments of high-pressure technology was the formation of diamond from graphite at pressures of 45,000 to 90,000 atm and temperatures of 1500 to 3000°K.

Clearly, the types of apparatus needed to produce very high pressures are not routinely used in barobiology. The simple sorts of apparatus for pressurizing biological materials to about 2000 atm have been described in some detail by Morita (1967, 1970). The usual sort of apparatus used in biological laboratories consists of a hydraulic hand-pump connected by means of high pressure tubing to a cylinder. High-pressure valves can be inserted in the system to allow for closing off any part of it. Many types of pressure cylinders or barokams are used. The most common type consists simply of a hollowed out tube of 300 series stainless steel with an overlapping threaded cap. Oppenheimer and ZoBell (1952) modified the basic design of Johnson and Lewin (1946) by introducing O-ring seals and a threaded male fitting on the cap that will take a high-pressure needle valve with a female fitting. The valve can then be used to close off the barokam so that it can be disconnected from the pump assembly without loss of pressure. Many modifications of this basic design have been developed—quartz windows can be added, electrical leads can be passed through the cap into the main chamber, stirring shafts can be passed through the cap, etc.

When apparatus became available for generating kilobar pressures in the laboratory, barobiologists were lured into turning much of their attention to the effects of these nonbiospheric pressures. In fact, in the period between the turn of the century and the early 1930s, much of the focus of barobiology was on kilobar effects and the use of pressure as a disinfecting agent. Kilobar pressures are lethal for cellular microorganisms, including bacterial endospores, and are inactivating for viruses. A review of the early work on the lethal and denaturing actions of kilobar pressures appears in the book by Johnson *et al.* (1954). Since kilobar pressures denature proteins, it is not very surprising that they kill cells, although denaturation cannot always be identified as the cause of death. For example, in pressure killing of coliphage T_4, compression appears to trigger normal contraction of the tail sheath and abortive release of DNA (Solomon *et al.*, 1966). However, the results of denaturation studies at very high pressures were applied widely in interpretations of the effects of deep-sea pressures on organisms, so much so that Johnson *et al.* (1954) were led to make the following admonition.

> In the meantime the accumulated data with respect to the action of very high pressures has given rise to a widely prevalent view that the biological effect of high hydrostatic pressure in general is that of denaturing proteins or destroying life through protein denaturation or other poorly understood mechanisms. Nothing is farther from the truth.

It is interesting that some twenty years after Johnson *et al.* (1954) made this statement we are moving toward the more compromising middle-ground described in the previous section. Protein denaturation does seem to be involved in many of the biological effects of deep-sea pressures, especially for organisms that are growing in nonoptimal conditions. Certainly, some of the effects of pressure seem to be due to upsets in equilibria—such as that between microtubulin and microtubules in eukaryotic cells. However, many of the effects are not due to protein denaturation or to other shifts in equilibria, but as Johnson *et al.* (1954) described, are due to changes in reaction rates.

The development of a theoretical basis for considering the effects of pressure on reaction rates was closely intertwined with that for temperature effects. On the basis of experimental determinations of changes in reaction rate with changes in temperature, Arrhenius formulated his well known equation,

$$k = Ae^{-\mu/RT}$$

where k is the reaction velocity or rate, μ is the activation energy or temperature characteristic, R is the gas constant, T is the Kelvin temperature and A is a constant. The temperature characteristic has been found to be equal to $\Delta H - RT$, where H is the enthalpy of activation. In the 1901 edition of his book on physical and theoretical chemistry, van't Hoff presented a similar equation for pressure effects, based on experimentally determined changes with increasing pressure in the rates of hydrolysis of sucrose and methylacetate. The equation he gives, with a citation to earlier work by Planck, is,

$$\frac{\mathrm{dlog}}{\mathrm{d}p} K = \frac{\Delta v}{2T}$$

Here v is the pertinent volume change (approximately $- \Delta V^{\ddagger}$ in modern terminology) and K is equal to the ratio of the reaction rates at one pressure and at another, or as van't Hoff wrote it, k_{II}/k_{I}.

The potential importance of pressure in industrial processes was established in some detail by work such as that of Fawcett and Gibson (1934) who found that compression resulted in rate increases for a variety of organic chemical reactions. Of course, high pressures had been used previously in industry for the Haber ammonification process. Fawcett and Gibson considered the theoretical implications of their findings and stated

> The velocity would be influenced by pressure if the potential energy gained by the system as a result of the isothermal compression is available as part of the activation energy necessary for reaction.

Evans and Polanyi (1935) gave some substance to this notion in their consideration of the effects of pressure on reaction rates. They equated the probability of the transition state with the equilibrium constant for the formation of the transition state or activated complex from the initial state. They then indicated that the change in this equilibrium constant with change in pressure is related exponentially to V/RT, where V is the volume change accompanying formation of the transition state, or in the notation that is now commonly used, $\Delta V^{\ddagger}$. Evans and Polanyi pointed out that V may arise from changes in the reacting molecules themselves as they pass from the initial to the transition state and from changes in the solvent. The decay of the transition state to products is generally very rapid with an average lifetime for the activated complex of about 10^{-13} seconds. The reaction rate can then be directly related to the probability of formation of the transition state, and pressure can act to enhance or reduce this probability.

During the same year in which the paper by Evans and Polanyi appeared Eyring (1935a, b) presented his theory on the role of activated complexes in determining absolute rates of reaction. He and Polanyi had previously constructed potential surface diagrams for gas-phase reactions from which it was possible to estimate, at least roughly, activation energies. Eyring was then able to calculate the probability of the activated state or transition state by use of statistical mechanics. This probability multiplied by the decomposition rate of the activated complex yielded the specific reaction rate.

Subsequently, Johnson and Eyring successfully applied the theory of absolute reaction rates to the interpretation of the effects of pressure on biological reactions, including complex reactions such as bioluminescence and growth. Certainly, their efforts stand as major accomplishments in the field. They have recently reviewed their work in the book (1970) edited by Zimmerman.

IV. Microbial Growth Under Pressure

One of the primary observations of Certes and Regnard in the 1880s was that pressure slows microbial growth. Their work has been extensively elaborated, and it is now known that there are relatively barosensitive microorganisms that cannot grow at pressures much above 200 atm, relatively barotolerant ones that can grow at pressures up to 550 atm, and a few highly barotolerant ones that can grow at pressures as high as 1000 atm or so. The highest pressure at which growth has been reported (ZoBell's work cited by Vallentyne in 1963) to occur is some 1400 atm. Why is there such a wide range of barotolerance and just how does pressure inhibit growth? In this section, we shall briefly consider tentative answers to these questions that have been proposed by a number of investigators. We do not intend to review in detail the effects of pressure on various microbial biochemical reactions or physiological processes. For more detailed information, readers may wish to consult reviews by ZoBell (1964, 1970), Morita (1967, 1972) or Marquis (1976).

In considering the growth barotolerance of any particular microorganism, we can define cardinal pressures—the maximum, minimum and optimum. Maximum pressures for growth are very much dependent on the growth environment. For example, we have found (Marquis *et al.*, 1971; Marquis and ZoBell, 1971) that the maximal growth pressure for homofermentative, lactic-acid *S. faecalis* at 25 to 30°C depends very much on the source of ATP for growth. Cultures in a tryptone-yeast-extract medium showed no growth at pressures higher than about 200 atm when pyruvate was used as fuel source. However, growth occurred at pressures as high as about 450 atm when ribose was used as fuel, and at pressures as high as 550 atm when glucose, galactose, maltose or lactose was used as fuel. With these latter sugars, growth could occur at pressures as high as 750 atm if the media were supplemented with 50 mM Mg^{2+} or Ca^{2+} ions. Thus, it is apparent that a single species of microorganism can exhibit a remarkably wide range of barotolerance depending on environmental conditions.

Inorganic salts can enhance the barotolerance of many bacteria, and there appears to be some specificity in the enhancing reactions. For *S. faecalis*, only Mg^{2+} and Ca^{2+} are effective. Palmer and Albright (1970) and Albright and Henigman (1971) found that NaCl was the most effective salt for enhancing barotolerance of *Vibrio marinus* and other marine bacteria. Other salts were less effective, and those with divalent cations were at the lower end of the effectiveness scale. Mg^{2+} and Ca^{2+} are effective in enhancing the acid-resistance of many types of bacteria under pressure (Matsumura *et al.*, 1974), and they enhance barotolerance in part by extending the effective pH range for growth under pressure. Recently, Pope *et al.* (1976b) reported that growth and protein synthesis of the halophile, *Halobacterium salinarium*, are unusually resistant to pressure and are only reduced to 50% of the one atm rate at 1000 atm.

Changes in growth temperature also result in major changes in barotolerance. Maximum tolerance is generally at a temperature a few degrees higher than the optimum. Temperatures higher than this result in greatly decreased barotolerance. Below the optimum temperature, increases in temperature increase barotolerance.

Clearly, the ideal gas law cannot be applied to predict pressure-temperature interactions on growth of microorganisms in condensed aqueous phases. The cell itself is a mixture of an aqueous phase and a series of solid phases. Pressure-temperature interactions can be interpreted in terms of effects on the types of chemical bonds that are important in maintaining biological structure. Hydrogen bonds in model systems are stabilized by pressure increases but disrupted by temperature increases. Hydrophobic interactions are usually disrupted by pressure increases up to about 1000 to 3000 atm but are stabilized by further increases (Suzuki and Taniguchi, 1972). It was pointed out in a previous section that low pressures may stabilize hydrophobic interactions in concentrated regions such as the centre of a globular protein. Temperature increases tend to stabilize hydrophobic interactions, but above about 60°C, further increases are disruptive. Electrostatic interactions of the attractive type in polymers are disrupted by increases in pressure, whereas increases in temperature may stabilize or disrupt depending on the particular chemical groups involved. Repulsive electrostatic interactions are enhanced by increased pressure because the charged states of ionizable groups are favoured under pressure. Again, increased temperature may enhance or reduce repulsive interactions depending on the specific groups involved. In all, it is clear that increases in temperature and pressure may act in concert or antagonistically depending on the situation and the types of bonds that are important. Certainly, the effects of high pressures cannot generally be reversed simply by raising the temperature. Moreover, a review of the effects of the two parameters on various chemical bonds indicates that, for example, pressure denaturation of a protein would involve a different sequence of bond scissions and formations than would cold denaturation or heat denaturation.

Temperature adaptations seem to be essentially without effect on barotolerance. We have determined temperatures at which barotolerance is maximal for a psychrophile, *V. marinus*, a number of mesophiles, and a thermophile, *Bacillus stearothermophilus* (Marquis, 1976). For each organism, barotolerance was maximal at a temperature slightly above the optimum growth temperature. Studies with *V. marinus* and other cold-tolerant bacteria have indicated that the ability to grow at low temperature is not associated with an ability to grow at high pressure.

Other environmental factors such as E_h and ionic strength must also affect barotolerance, but little is known of their specific effects. Moreover, there are difficulties in designing experiments in which only one environmental factor is varied at a time. Thus, for example, pressure affects the pH of any medium because it affects dissociation reactions. Therefore, when the pressure is increased, the pH is changed. Also, pressure markedly potentiates the inhibitory effect of acids and bases for microorganisms (Matsumura *et al.*, 1974). Interactions of environmental factors are in general not of the additive type, but instead, antagonisms and synergisms are commonplace.

It seems that minimal growth pressures for most microorganisms must be in the negative pressure range, except of course, for barophiles, if they exist. These pressures can be produced in the laboratory by a variety of methods (Hayward, 1971) including the pulling apart of platens separated by a thin film of viscous liquid. Negative pressures occur naturally in plants. For example, Scholander *et al.* (1965) measured sap tensions in vascular plants and reported values of from -5 to -60 atm.

Our recent studies on optimum pressures for microbial growth suggest that many bacteria may grow best at pressures somewhat higher than one atm. For example, Fig. 1 shows growth curves of *S. faecalis* in a complex medium composed of tryptone, yeast extract and ribose. The growth temperature was 45°C. It is apparent that growth at 100 atm was more rapid and resulted in significantly higher yields than did growth at 1 atm. 200 atm was only slightly stimulatory, and in other experiments, we have found that 300 atm was markedly inhibitory. 50 and 150 atm pressure were somewhat less stimulatory than was 100 atm. The growth temperature here, 45°C, is slightly

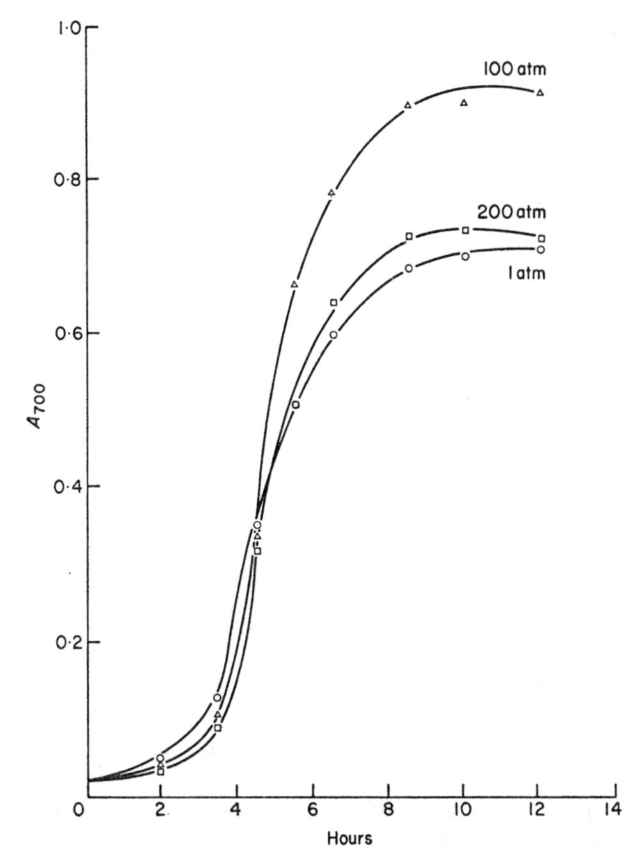

Fig. 1. Growth of *S. faecalis* 9790 in tryptone-yeast-extract-ribose medium at 45°C.

above the optimum, but the stimulatory effects of pressure are maximal at this temperature. The important point here is that growth rates and yields at 100 atm pressure and 45°C are higher than those attainable at any temperature at one atm, as we have previously reported for *E. coli* (Marquis, 1976). Low pressures appear to be stimulatory for other bacteria also—*V. marinus* (Morita, 1975) and a marine pseudomonad (Kriss and Mitskevich, 1967).

At the lower extremes of the temperature range for growth, bacteria become highly barosensitive. Figure 2 shows growth curves for *E. coli* in complex medium at 9°C and 1, 50 or 100 atm pressure. It is apparent that even relatively low pressures have significant inhibitory effects at low temperature. The extreme sensitivity exhibited here should be of ecological importance in bodies of fresh water such as the Great Lakes, in which the maximum pressure

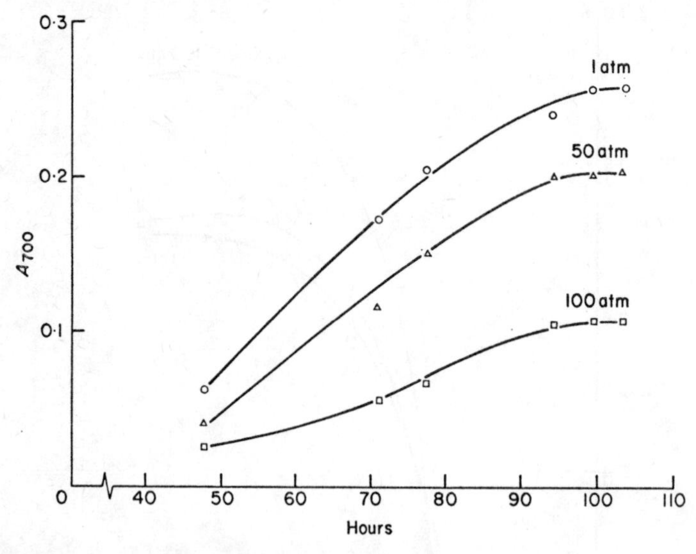

Fig. 2. Growth of *E. coli* B in trypticase-soy broth plus 0·1% (w/v) KNO_3 at 9°C

in Lake Superior is about 40 atm. Certainly, bacteria such as *E. coli* end up in large numbers in the Great Lakes, and the water temperature is usually low. Therefore, it seems reasonable that pressure would be a factor limiting growth of *E. coli*, at least in the deeper parts of the lakes. However, pressure effects are probably more important in regard to the resident flora since many of the organisms are not psychrophiles but are cold-tolerant mesophiles. It does seem that a reasonable case can be constructed for pressure as an ecologically important factor in fresh as well as salt water. Low pressures are also probably of major ecological importance in the continental shelf regions of the ocean.

What determines the maximum growth pressure for a particular micro-organism? Currently, there are many answers offered to this question, and in fact, it seems they may all be correct for certain organisms. Initially, ZoBell and Cobet (1964) found that some bacteria grew as long filaments under pressure with a sparsity of cross walls. Their results clearly indicated that cell division is more barosensitive in some bacteria than is cell growth. Many other bacteria do not show this differential sensitivity. *E. coli* B is especially liable to form filaments under pressure, and ZoBell and Cobet (1964) found that the bacterium also became relatively poor in DNA, relatively rich in RNA and showed little change in protein content per unit of cell mass. Initially, then, it seemed as if DNA synthesis might be unusually barosensitive.

Subsequent investigations with *E. coli* (Pollard and Weller, 1966; Landau, 1966, 1967; Yayanos and Pollard, 1969) showed that protein synthesis was more severely inhibited by pressure than was nucleic acid synthesis and that previously initiated rounds of chromosome replication could be completed under pressure but new rounds were not started, presumably because of inhibition of synthesis of initiation or termination proteins. Protein synthesis in *E. coli* was stimulated by low pressures, unaffected by a pressure of about 400 atm, and totally inhibited by pressures in excess of about 670 atm at 37°C. The maximum growth pressure for *E. coli* in complex media is about 550 atm, and somewhat lower in minimal media. When filamentous *E. coli* B cells growing under pressure are returned to one atm conditions, they undergo cell division and appear normal again. Therefore, at least for short periods of exposure, pressure causes no irreversible damage. In fact, Yayanos (1975) found that cell separation after division of *E. coli* was actually stimulated by a pressure of about 600 atm. Protozoa also have been found (Macdonald, 1967a, b) to be more severely inhibited in cell division than in cell mass increase by pressure. Zimmerman and Laurence (1975) found that *Tetrahymena pyriformis* cells could be synchronized in their cell division by subjecting them to a series of pressure pulses. At the present time, the relationship between the barosensitivity of cell division and that of protein synthesis is not clear. There is speculation of a connection between the two but no very firm evidence. Moreover, Kriss *et al.* (1969) found that growth of a marine pseudomonad under pressure resulted in cells with reduced RNA and polysaccharide contents per mass unit but no reduction in DNA or protein. For this bacterium, RNA synthesis seems to be more barosensitive than is protein synthesis.

In general, biosynthetic reactions are more severely inhibited by pressure than are catabolic reactions. This difference in sensitivity and the relatively high sensitivity of protein synthesis led Pope and Berger (1973, Berger, 1974) to propose that the inhibitory effects of pressure on microbial growth are due to inhibition of protein synthesis. This view has also been proposed by Albright (1975). Detailed studies of the effects of pressure on various steps in the process of protein synthesis (Albright, 1969; Hildebrand and Pollard, 1972; Pope *et al.*, 1975a, 1975b; Schwarz and Landau, 1972a, 1972b; Smith *et al.*, 1975) have led to the conclusion that the ribosome is highly important in barosensitivity and that the most sensitive step in the process is probably binding of charged RNA to polysomes. However, the work of Albright and coworkers (Arnold and Albright, 1971; Hardon and Albright, 1974) suggests that under at least some circumstances the most barosensitive step is charging of tRNA.

Swartz *et al.* (1974) found that incorporation of labelled amino acids into the trichloroacetic-acid-insoluble fraction of the highly barotolerant bacterium, *Pseudomonas bathycetes*, or of *P. fluorescens*, was markedly less sensitive to pressure than was incorporation by *E. coli*, even though *P. fluorescens* is not a highly barotolerant organism. Pope *et al.* (1975a) took advantage of this difference to identify the components of the systems that were most important in determining the pressure response. They prepared hybrid systems for *in vitro* protein synthesis with parts from *E. coli* and parts from the pseudomonads. The source of soluble factors was of no consequence for barotolerance, but the source of ribosomes did matter. Even finer analysis (Smith *et al.*, 1975) revealed that it was the 30S subunit of the ribosome that was of prime importance. It appears also that the ionic environment has an effect on the workings of ribosomes under pressure. *E. coli* ribosomes are relatively barosensitive at high (150 mM Na^+ plus 60 mM Mg^{2+}) or low (0 mM Na^+ plus 16 mM Mg^{2+}) ion levels. *P. fluorescens* ribosomes are relatively barotolerant at both ion levels, but *P. bathycetes* ribosomes are barotolerant only at the higher level (Smith *et al.*, 1976). Certainly, this work is bringing knowledge of the bases for barotolerance to the molecular level, and in the near future, it should allow for some interesting genetic experiments.

It was mentioned previously that the ability of *S. faecalis* to grow under pressure in complex media is related to its fuel source. The recent work of Matsumura (1975) has provided a basis for interpreting this phenomenon. He found that barotolerance could be related directly to the rates at which ATP could be supplied from various fuel sources. When ATP was supplied at a faster rate, the bacterium was better able to resist the inhibitory action of pressure. It was also found (Marquis *et al.*, 1971; Matsumura, 1975) that the demand for ATP for growth is increased under pressure, as reflected by lower cell yields per mole of ATP utilized for growth. Pressure has relatively small effects on catabolism and ATP production but has more significant effects on growth and ATP utilization. In effect, the cells find themselves under pressure in a situation of slightly diminished ATP supply but greatly increased ATP demand. The increase in demand appears to be due to stimulation by pressure of the membrane ATPase of *S. faecalis*, and cell yields per mol of ATP could be increased by adding 0·01 mM dicyclohexylcarbodiimide, an ATPase inhibitor, to growth media.

It appears that in some instances pressure inhibition of transport of solutes into cells can limit growth. Matsumura (1975) describes a reasonably clear-cut example for *S. faecalis*. Lactose is transported into this organism by means of a phosphotransferase system (Citti *et al.*, 1965). At one atm, uptake and hydrolysis of the sugar are rate-limiting, and growth in lactose medium is slower than is growth in media with lactose replaced by glucose or galactose. Hydrolysis is not greatly affected by pressure, but uptake is slowed. Therefore, the effect of pressure on transport is directly reflected by reduced growth rates.

Morita (1975) has developed the view, primarily in relation to psychrophilic bacteria, that the primary site of pressure damage is at the cell membrane. This view is based on the results of studies of amino acid or uracil incorporation and catabolism by psychrophiles at one atm and under pressure. In other bacteria, pressure can stimulate amino acid uptake (Schwarz and Landau, 1972a), and so the effect of pressure on transport reactions is not always inhibitory. Increased pressure also stimulates uptake of β-galactosides by *E. coli* (Schlamm and Daily, 1971; Marquis and Keller, 1975).

However, we have found, somewhat surprisingly, that pressure has a major effect on the potassium content of growing *S. faecalis* cells. The data in Fig. 3 show that the K^+ content of cells growing in a complex medium at 400 atm and 30°C is much less throughout the entire culture cycle than that of cells growing in the same medium at one atm. Reduced K^+ content does not seem to be due to the slower growth of the cells under pressure, because when cells are slowed in their growth by transfer to 15°C medium, they do not show reductions in K^+ content, even when the growth rate is slower than that at 400 atm and 30°C. The effect shown in Fig. 3 is a dramatic one. Loss of K^+ could possibly be one cause of reduced growth under pressure since protein synthesis requires K^+ as a cofactor and low intracellular K^+ levels result in a slowing of synthesis (Harold and Baarda, 1968).

The ability of an organism to grow under pressure in a situation that requires formation of adaptive enzymes may be limited by the barosensitivity of the adaptive process. For example, Marquis and Keller (1975) extended a previous study by Landau (1967) on the effects of pressure on derepression of the *lac* operon in *E. coli*. This adaptive process is relatively barosensitive, and when uninduced *E. coli* B cells are inoculated into minimal medium with lactose as sole source of carbon and fuel for growth, they are significantly more sensitive to pressure than are previously induced cells. This increased sensitivity is indicated by very long lag phases prior to growth, slower growth when it does occur and a significantly lower maximal growth pressure. Induction of penicillinase synthesis in *Bacillus licheniformis* proved also to be relatively barosensitive. However, a number of other adaptive systems were relatively insensitive to pressure and adaptation to new substrates occurred as readily under pressure as at one atm (Marquis and Keller, 1975).

The long-term survival of protozoa under pressure appears to be related to the effect of pressure on microtubules. Pressure interferes with the normal mitotic process because it shifts the microtubulin–microtubule equilibrium in the direction of the monomer (Salmon, 1975a, 1975b; Salmon *et al.*, 1976; Inoue *et al.*, 1975). A pressure of 400 atm is lethal for *Tetrahymena* or for amoebae (Marsland and Brown, 1936). It also inhibits mitosis and disrupts microtubules. Therefore, it seems that microtubular disruption is sufficient, if not necessary, to cause death.

On review, it seems that maximum growth pressures may be determined by the barosensitivities of many physiological processes—protein synthesis, ATP production and utilization, transmembrane solute transport, regulatory functions, assembly of microtubules or maintenance of the native states of proteins. The process that is paramount in terms of the maximum growth pressure seems to depend on the particular organisms and the particular growth conditions.

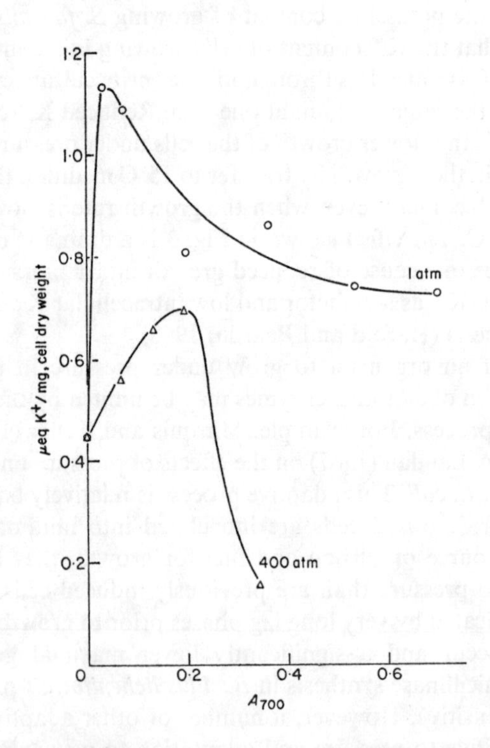

Fig. 3. Potassium content of *S. faecalis* 9790 cells growing at 30°C and one or 400 atm pressure. The abscissa indicates the optical densities of the cultures from which cells were harvested. The growth medium contained tryptone, yeast extract and glucose. Cells were harvested by centrifugation in the cold, washed once with cold deionized water and resuspended in a known volume of 6 N HCl solution. The suspension was then centrifuged and the K^+ content of the supernatant fluid assayed by use of an atomic absorption spectrophotometer.

V. Killing of Microorganisms by Pressure

Superficially, the lethal effects of high pressure for microorganisms are similar to those of heat. However, the thesis was developed elsewhere in this chapter that pressure effects are basically different from temperature effects. Even though proteins can be denatured by high pressure, by heat or by cold, the pathway to the denatured state is different for each agent. But of course, the physiologic effect may be the same in each case—the protein becomes non-functional. Killing by high pressure is similar to heat killing in regard to dose-effect relationships. There is generally a threshold pressure below which no killing occurs. Then, at lethal pressures, both intensity of pressure and time of exposure are important in determining the outcome of the treatment. Moreover, the hierarchy of heat resistance among various microorganisms is similar to the hierarchy of pressure resistance. Thus viruses are simpler in structure and organization than are cellular microorganisms and they are generally also more resistant to the lethal actions of heat and pressure. Unfortunately, there are relatively few studies in which pressure sensitivities of viruses and cellular organisms are compared under the same conditions, or even studies from which it is possible to calculate death rate values. Rutberg (1964a, 1964b) compared *E. coli* and two of its phages, T2 and T4. At 37°C and 2000 atm, the exponential death rate constant for the bacteria was about 0·125 min^{-1} compared with lower values of 0·045 and 0·033 min^{-1}, respectively, for the phages. Coliphage T5 was found to be intermediate in sensitivity between T2 and T4. Later Solomon *et al.* (1966) found that the rate of killing of T4 phage at 25°C was less than 0·1 min^{-1} at pressures below about 3400 atm but increased exponentially at higher pressures. Inactivation followed first-order kinetics at all pressures tested with an optimum temperature for inactivation of about 25°C in a test range from 0 to 50°C. Electron micrographs of pressurized phage suspensions revealed that most of the killed viruses had contracted tail sheaths, and many had empty heads. The change was similar to that caused by urea or exposure of the virus to host cell walls.

Pressure can also upset the lysogenic state between host and prophage. A pressure of 2000 atm for 5 min at 37°C was found to induce *E. coli* B to liberate virulent phage particles (Rutberg and Heden, 1960), which appeared in electron micrographs to be morphologically similar to coliphage T5 and possibly P1 (Edebo and Rutberg, 1960). Subsequent experiments by Rutberg (1964c) showed that 290 atm pressure could induce phage lambda production by lysogenic *E. coli* K.

As might be expected, phages show a relatively wide spectrum of barosensitivities. Rautenshtein and Muradov (1966) found that four phages isolated from a polylysogenic strain of *Actinomyces levoris* were extremely barosensitive and could be inactivated by only 500 atm for 30 min at 28°C. Phages from another actinomycete, *A. olivaceus* were more resistant and were essentially unaffected by 700 atm for 30 min. *A. levoris* spores could be induced to produce the four phages by pressures as low as 200 atm; whereas higher pressures of 500 or 700 atm were required to obtain even minimal induction of *A. olivaceus* spores. A 30 min exposure to 700 atm was not lethal for spores of either organism.

Animal and plant viruses can also be inactivated by pressure. Basset *et al.* (1938) used 45 min exposure times and found that staphylococcal phages were inactivated at 2000 to 3000 atm, herpes virus at 3000 atm; vaccinia, rabies, foot-and-mouth-disease and fowl-pox viruses at 4000 to 5000 atm; and equine encephalomyelitis virus at 7000 atm. Pressure inactivation of tobacco mosaic virus has been studied in some detail (Basset *et al.*, 1938; Lauffer and Dow, 1941). The process appeared to be a cooperative one in that pressures of 5000 to 6000 atm were without effect while pressures of 7500 to 9000 atm inactivated in only a few minutes, and the volume change appears to be very large. During inactivation, the virus protein was coagulated, and viral nucleic acid was released. However, inactivation was found (Lauffer and Dow, 1941) to precede coagulation, which seems to be secondary to virus disintegration.

The relative resistance of viruses to pressure makes one wonder if these organisms might not be able to survive for very long periods in the depths of the ocean where processes that ordinarily lead to their destruction are slowed. Perhaps there could even be some hazard in handling deep-sea sediment samples.

The killing of vegetative bacterial cells by pressure is remarkable in that it occurs at such low pressures, and also because there is so little information on its basic mechanisms. Commonly, pressures only somewhat above the maximum growth pressure are lethal for bacteria. These pressures kill the cells directly; death does not come about simply because the cells cannot grow. For example, we have found that *Arthrobacter crystallopoietes* cells in deionized water, in sea water or in dilute phosphate buffer are rapidly killed by pressures of only about 1000 atm even though the organisms survive in the nongrowing state for weeks at one atm. Typical exponential death rate constants ranged from about 0·185 h^{-1} in deionized water at 35°C to about 0·055 h^{-1} in 10 mM NaCl solution at 4°C. Killing of *A. crystallopoietes* cells by pressure was lower at 4°C than at room temperature. Thus, the temperature-pressure interaction here appears to be opposite to that for growth. We have found that other bacteria also are protected from pressure killing by being chilled. Kim and ZoBell (1971) found just the opposite effect of chilling for *E. coli* and *Serratia marinorubra*; killing by 1000 atm was more rapid at 4°C than at 8·5, 15 or 25°C. We found also that salts are protective, but that protection against pressure killing appears to be more related to ionic strength changes than to specific ion effects. Morita and Becker (1970) reported similar protective effects of salts. Pressure and salts both affect electrostatic interactions in biopolymers, and so it is not difficult to imagine ways in which they might interact antagonistically.

The extensive tables of ZoBell and Johnson (1949) and of Oppenheimer and ZoBell (1952) indicate that pressures as low as 200 atm can be lethal for many bacteria. Quantitative determinations of killing rates have not been very numerous, but as the values presented in the preceding paragraph indicate, the values obtained are highly dependent on the conditions of the experiment. In general, spores are more pressure resistant than are vegetative cells, with one rather interesting anomaly. Relatively low pressures can trigger germination of bacterial spores (Clouston and Wills, 1969, 1970; Gould and Sale, 1970, 1972; Sale *et al.*, 1970), and the resulting germinated forms are then much more barosensitive—about as sensitive as vegetative bacteria. For example, spores of *Bacillus pumilus* can be killed by pressures as low as 610 atm that trigger germination (Clouston and Wills, 1969), even though spores of many *Bacillus* organisms can withstand pressures in excess of 12,000 atm (Larson *et al.*, 1918; Basset and Macheboeuf, 1932) that inhibit germination. In effect, then, relatively low pressures may be lethal for spores while higher pressures are not. Moreover, Sale *et al.* (1970) found that germination and killing by pressure were greater at high temperatures than at lower ones and that NaCl or $CaCl_2$ was protective for *Bacillus coagulans* spores.

Eukaryotic cells also are killed by high pressures, and moreover, they appear to be significantly more barosensitive on the average than are prokaryotic cells. Hite *et al.* (1914) found that *Saccharomyces cerevisiae* was more readily killed by pressure than was *Serratia marcescens*. Pressures that caused death in one hour were, respectively, about 2040 and 2720 atm. Hill and Morita (1964) found that pressures of 600 to 1000 atm at 27°C killed mycelia and zoospores of the fresh-water mould *Allomyces macrogynus*, and they proposed that the demonstrated inhibition of mitochondrial respiration by pressure was one of the contributing factors in death. Marsland and Brown (1936) found that *Amoeba proteus* and *Amoeba dubia* were still more barosensitive in that they were killed by exposure to 450 atm for about one hour.

It is remarkable that so little is known about the details of pressure killing of microorganisms. The damage seems to be relatively subtle in the sense that pressure killed bacterial or fungal cells do not appear under the phase microscope to be grossly damaged. Moreover, there do seem to be potential uses for hydrostatic pressure as a disinfecting agent, especially since bacteria and fungi can be killed at pressures that do not denature many labile biopolymers. In 1914, Hite *et al.* found that pressures of about 6800 atm at room temperature for seven days did not inactivate enzymes in milk but did render the milk sterile. They also found that pears, peaches, grape juice and cider could be sterilized by somewhat lower pressures, but vegetables and berries spoiled after pressure treatment. More recently, Timson and Short (1965) attempted to sterilize milk with pressure but persistors, primarily *Bacillus* and *Micrococcus* organisms, were not completely eliminated even at about 10,000 atm. However, it seems that pressurization was certainly as effective as pasteurization in reducing bacterial counts of milk. Baross *et al.* (1975) found that persistors in sewage samples stored at 500 or 1000 atm at 4°C were *Arthrobacter*/*Corynebacterium* spp.

VI. Pressure, Pollution, Petroleum Microbiology and Deep-Sea Mining

There is currently a great deal of apprehension concerning marine pollution and a legitimate fear that indiscriminate dumping, especially of toxic substances, may lead to massive upsets of ocean communities and possibly to a global disaster. At least part of this fear comes from knowledge that the low temperatures and high pressures of deep-sea environments are highly inhibitory for many microbial processes. *In situ* measurements of the degradation of organic matter in the deep sea (Jannasch *et al.*, 1971; Jannasch and Wirsen, 1973) have done little to diminish this fear. Life is relatively abundant in the sediments of the ocean, as indicated by high counts of viable bacteria, but it seems that vital activities proceed very slowly there and that bacteria cannot become adapted to function rapidly at high pressures and low temperatures. Morita (1976) suggested that the slow metabolic activities of deep-sea organisms may be beneficial in terms of their long-term survival in nutrient-poor environments.

Schwarz *et al.* (1974a, b, 1975) have considered specifically the problem of ocean pollution by petroleum products. They found that mixed flora obtained from the sediment-water interfaces of cores from the Atlantic Ocean off the Florida coast was capable of using *n*-hexadecane as a sole carbon source on primary culture in a two-phase medium with salts and 0·37% (v/v) *n*-hexadecane. Subsequent subculturing resulted in selection of a population with increased capacity to use the alkane. The net conclusion of the studies of Schwarz *et al.* is that hydrocarbons, or at least *n*-hexadecane, can be degraded under the *in situ* conditions of the deep sea but that the process is a slow one that requires a mixed flora.

Bacteria may be active also in degradation of organic materials in coal, which is often under considerable pressure because of the weight of overlying strata. In fact, Beck and Poschenrieder (1957) proposed that pressure resistance could be used as a criterion for identifying the autochthonous flora of lignite. However, Jaschhop and Schwartz (1969) found both sensitive and tolerant bacteria in lignite and concluded that both types of bacteria were autochthonous. Of the 24 strains they isolated, 14 grew over a four-day period at 400 atm and 30°C, seven did not grow but were not killed, and three were killed. At 300 atm 18 grew, five survived without growing and one was killed. All the isolates grew at 200 atm. The most barosensitive bacterium was *Vibrio percolans*, and the ones that were killed at 400 atm, but not 300 atm, included *Alcaligenes* and *Bacillus*. However, none of the organisms was very active in degrading the organic components extracted from lignite with water or alkaline lignin or humic acids. Therefore, there remained some question concerning whether or not these bacteria actually are of much significance in transforming lignite *in situ*.

The slow pace of microbial activities in the deep sea appears in some cases at least to be an advantage in regard to human endeavours. Willingham and Quinby (1971) found that ingot iron coupons corroded faster in association with marine, sulphate-reducing bacteria at 200 atm and 20°C than at one atm. However, corrosion was markedly slowed to a level even below that of sterile control specimens at 600 atm. In other words, at a depth of 6000 m in the ocean, pressure would be highly inhibitory to corrosion of an iron rig, but at 2000 m corrosion would be enhanced by pressure. Tests carried out with coupons of 316 stainless steel indicated little or no corrosion at one, 200 or 600 atm. In contrast, tests with E.C. Grade aluminium coupons indicated that 200 and 600 atm pressure significantly inhibited corrosion, and so presumbly, would be protective for aluminium installations in the deep sea. The corrosion behaviour of high-strength materials in sea water has been reviewed by Kirk (1972) but he did not consider pressure effects.

Ferromanganese nodules and crusts are found in sufficient quantities on the ocean floor, especially in the Pacific, to have become the objects of deep-sea mining operations in the past few years. There has been a proposal that these deposits are types of bacterial stromatolites (Ehrlich, 1975) and that Mn (II)-oxidizing bacteria play a major role in their formation and growth. (See Chapter 10.) Representative oxidizing bacteria isolated from nodules have been found (Ehrlich, 1975) to be capable of growth at deep-sea temperatures and pressures, e.g. 4°C and 340 atm, and of Mn (II) oxidation at pressures as high as 476 atm. The oxidation process was very slow under pressure at low temperature, but as Ehrlich (1975) points out, accretion of new materials by nodules is also very slow—about 0·01 mm increase in diameter per thousand years. Manganese oxidizing bacteria include *Arthrobacter* and Gram negative rods. The nodules have been found to contain also Gram negative, MnO_2-reducing bacteria, which are active at deep-sea pressures (Ehrlich, 1974a). The reducing activity is inducible, and induction can occur at pressures as high as 476 atm (Ehrlich, 1974b).

VII. Biological Effects of Compressed Gases

In experiments on the effects of hydrostatic pressure on microorganisms, care is generally taken to exclude air or other gases from the vessels to be pressurized. Anaerobic organisms are outrightly killed by the oxygen in air and even aerobic or facultative organisms live only in a balanced truce with oxygen. The gas is an important and often necessary substrate for respiration, and in eukaryotic cells, it is a required nutrient for processes such as synthesis of monounsaturated fatty acids. However, oxygen can also be lethal and the development of mechanisms for destroying toxic free radicals and peroxides from oxygen metabolism was a prerequisite to the evolution of life on the primitive earth from a primarily anaerobic existence to a primarily aerobic one. Current work on detoxification reactions is focused on superoxide dismutase, the enzyme that catalyses the conversion of superoxide radicals to hydrogen peroxide, which can then be degraded through the actions of catalases or peroxidases (Fridovich, 1974, 1975). If air is trapped in a vessel that is subsequently pressurized, it dissolves slowly and the contents of the vessel are then exposed to hyperbaric oxygen. This exposure may not be a problem for an aerobic or facultative organism if the amount of trapped air is small. In fact, the more common problem with pressurized cultures is to supply sufficient oxygen to satisfy respiratory demands. Techniques for supplying oxygen under pressure without exposing cells to hyperbaric oxygen have been reviewed recently (Marquis, 1976).

There is currently a great deal of interest in determining the nature of the lesions induced in cells by hyperbaric oxygen. Gottlieb (1971) has reviewed the information for microorganisms, and Fridovich (1974, 1975) has extensively reviewed the general subject of oxygen toxicity and the role of superoxide dismutase as a protection against toxicity. ZoBell and Hittle (1967) found that hydrostatic pressure has a marked potentiating effect on oxygen toxicity for aerobic and facultative bacteria. For example, they found that approximately one atm of oxygen (36 $\mu g\ O_2\ ml^{-1}$) at one atm hydrostatic pressure had, if anything, an enhancing effect on growth of *E. coli*, *Bacillus subtilis* or *Bacillus megaterium* in nutrient broth. However, at 100 atm hydrostatic pressure, this same level of O_2 was lethal for all three species, none of which is inhibited by 100 atm hydrostatic pressure alone. We have recently attempted to determine the molecular basis for this remarkable potentiating effect of pressure. Our first thought was that pressure must be inhibitory for superoxide dismutase. However, Stephen Thom has found (unpublished observations) that pressures of up to 600 atm have no effect whatever on superoxide dismutase activities of *S. faecalis* or *E. coli*. It is possible that pressure may enhance production of free radicals, that it may stabilize them or that it may enhance the damage they do, but certainly it does not seem to impair superoxide dismutase.

We had previously found (Fenn and Marquis, 1968) that anesthetic gases also enhance oxygen toxicity for *S. faecalis* and that this enhancement is independent of any pressure effect. It seems then that there may be some common factor in the inhibitory effects of anesthetic gases, oxygen and pressure. We found also that anesthetic gases are effective not just for animals, but for bacterial cells as well. The effectiveness was assessed in terms of a slowing of growth. Testing of a number of gases indicated the following potency series: xenon and nitrous oxide > argon > nitrogen. Helium at gas pressures up to 41 atm proved to be impotent. Kenis (1971) later found that higher pressures of helium, about 100 atm, were inhibitory for growth of a wide range of bacteria. In contrast, Schlamm *et al.* (1974) found that 68 atm of helium in the presence of 0·2 atm of oxygen stimulated growth of *E. coli* in minimal medium. The stimulation appeared in growth curves as a reduction in lag time with no change in exponential growth rate or culture yield. The shortened lag appeared to be the result of enhanced iron uptake, and helium had no effect on cultures supplemented with the natural iron chelator, 2,3-dihydroxybenzoylserine, or its precursor, 2,3-dihydroxybenzoic acid. Hydrostatic pressure of 68 atm was ineffective as a stimulant, and so the effect seemed to be due to helium rather than to pressure. Previous studies had indicated that 68 atm of helium also increased the maximum velocity of β-galactoside uptake by *E. coli* but did not change the Michaelis constant for the reaction (Daily and Schlamm, 1972). This same pressure of helium also enhanced the susceptibility of *Staphylococcus aureus* to colistimethate but decreased susceptibility to penicillin, cephalothin, vancomycin and tetracycline (Schlamm and Daily, 1972).

Anesthetic gases have been found (Buchheit *et al.*, 1966) to slow growth of *Neurospora crassa*. Partial pressures for 50% reduction in growth rate were: 0·8 atm Xe, 1·6 atm Kr, 3·8 atm Ar, 35 atm Ne and about 300 atm He. Hydrostatic pressure was probably a major factor in the inhibitory effect of helium but not those caused by the other gases. In this particular study, nitrogen proved not to be very potent and was ranked with helium in the potency series. The relative lipid solubilities of the gases could be related directly to their potencies, as is the case for narcotic potencies in animals, and this relationship is one of the main bases for the proposal that the gases act by dissolving the lipid phase of the cell membrane, thereby distorting the membrane, and resulting in reversible disfunction. Potency is also directly related to molecular size and to polarizability. Anesthetic gases can cause reversible alterations in the structures of globular proteins (Balasubramanian and Wetlaufer, 1966), and their inhibitory action may well be related to their denaturing effects rather than, or in addition to, their membrane expanding effects. Pauling (1961) had proposed previously that the gases act by inducing clathrate formation and changes in cellular water structure. However, it has been difficult to mesh this theory with experimental observations. For the moment, it is necessary to say that there is little or no detailed information on the biochemical basis for growth inhibition by anesthetic gases.

Anesthetic gases also slow the growth of HeLa cells in monolayer culture (Bruemmer *et al.*, 1967). The potency series obtained was N_2O and xenon $>$ Kr $>$ Ar $\gg$ Ne or He. Exposure to 7·2 atm of Xe caused irreversible damage to the cells and this result differed from our findings of only reversible slowing of growth for *S. faecalis* (Fenn and Marquis, 1968). Xe also inhibited the attachment of HeLa cells to glass.

Recently, Macdonald (1975) has completed a study of the effects of high pressures of helium or hydrogen and high hydrostatic pressures on cell division of *Tetrahymena pyriformis* in mass culture or microdrop culture. He has previously found (Macdonald, 1967a, b) that 175 atm hydrostatic pressure inhibits preparation for cell division more severely than the actual cleavage phase of division or mass increase. In fact, hydrostatic pressure is sufficiently selective in its action so that it can be used to phase *Tetrahymena* cells in culture in relation to the cell cycle (Zimmerman, 1975). Hydrostatic pressures greater than 100 atm were found by direct microscopic observations (Macdonald, 1975) to inhibit cell division, but helium or hydrogen at the same pressures were less inhibitory. In other words, the gases seemed to antagonize the inhibitory effect of pressure. Helium was more inhibitory than was hydrogen. In microdrop cultures at 100 atm, growth curves were almost identical for cells under hydrogen or helium gas pressures and those under purely hydrostatic pressure. At 175 atm, cells under hydrogen grew best, but cells under helium grew better than those under hydrostatic pressure alone, and these differences were accentuated at 250 atm.

The antagonism between pressure and anesthetic gases may be important in the success of programmes aimed at increasing the depth to which human saturation dives are possible. Experiments at the Institute for Environmental Medicine at the University of Pennsylvania have shown that man can work effectively at 1600 ft of sea water (pressure of c. 50 atm), and the U.S. Navy has planned dives to 2000 ft (pressure of c. 62 atm). Macdonald (1975) suggests that the physiologic limit for human diving may be as high as 150 atm. It has been found in animal experiments, e.g. those of Brauer *et al.* (1971), that convulsions due to high pressures of 50 to 100 atm can be avoided by use of anesthetic gases. Simple models for the antagonism based on expansion of hydrophobic regions by anesthetic gases and antagonistic compression by pressure have been developed (Lever *et al.*, 1971; Miller, 1972). However, the so-called high pressure nervous syndrome is complex and clearly involves many elements, only some of which are reduced by anesthetic gases. Much work remains before we obtain any very clear picture of the interactions of hydrostatic pressure, oxygen and anesthetic gases in their effects on biological systems.

IX. Epilogue

A review of current developments in microbial barobiology suggests that there should be major advances during the next decade in our understanding of the molecular aspects of the subject and in our appreciation of pressure as an ecological force in the biosphere. Recent studies of protein denaturation by pressure have shaken the foundations of protein chemistry and have led to fundamental questioning of the roles of hydrophobic interactions in maintaining the native conformations of proteins. Moreover, it has become apparent that biopolymers can be denatured by biospheric pressures of less than 1000 atm when other conditions are not optimal for maintenance of native states. For eukaryotic organisms, the disaggregating effects of pressure for microtubules seem paramount in determining barotolerance. It now seems that the responses to pressure of prokaryotes also may be largely determined by changes in the balances between native and denatured, or aggregated and dispersed, states of biopolymers. Therefore, the most pertinent volume changes for interpreting barophysiologic responses in molecular terms may be ΔV, or volumes of reaction, rather than activation volumes, $\Delta V^{\ddagger}$.

The extreme barosensitivities of microorganisms in the nonoptimal conditions of natural environments suggests that pressure may be a potent ecological force even in relatively shallow marine or freshwater niches and that major ecologically important effects of pressure do not occur only in the deep ocean. In fact, it seems that the search for highly barotolerant or barophilic microorganisms, which has had only very limited success, has lured microbial barobiologists from studies of the biological effects of relatively low pressure.

It seems also that more should be done to develop the practical aspects of microbial barobiology. Pressure has much to offer as a means of preserving labile materials. Moreover, it could be used for processes such as milk preservation in which complete sterilization is generally not required. Development of these uses of pressure requires that we acquire more than our current meagre knowledge of the mechanisms by which pressure kills microbes. The finding that one atm is not optimal for growth of many microorganisms also opens up new possibilities for the use of pressure in industrial microbiology.

Another very important area that has been only marginally explored to date is the medical use of pressure. High pressure oxygen is used fairly commonly for treatment of diseases such as clostridial myonecrosis, although three atm is considered to be about the maximum safe limit. It is now clear that man can function under relatively high pressure, possibly as high as 150 atm, and that this capacity is due at least in part to an antagonism between high pressure and narcotic gases. The way seems open to consider pressure itself as a therapeutic or as an adjuvant to other therapeutic procedures. In fact, preliminary experiments (Dole *et al.*, 1975) suggest that high-pressure hydrogen may be an effective anticancer agent.

Addendum

Since the manuscript of this chapter was first submitted, there has been substantial progress in microbial barobiology. The quest for highly barotolerant or barophilic, deep-sea bacteria has taken a major step forward with the development of apparatus for sampling, retrieval and transfer of undecompressed, deep-sea samples (Taylor and Jannasch, 1976; Jannasch *et al.*, 1976; Jannasch and Wirsen, 1977a, 1977b). The results obtained to date indicate that the bacteria in the deep are not specifically adapted to perform better under pressure than are shallow water or terrestrial species, and the conclusion seems unavoidable that microbial processes in the depths of the ocean proceed very slowly. Schwarz and Colwell (1975b) used decompressed-recompressed sediment samples to come to the same conclusion. They also concluded that a large fraction of the metabolic activity of bacteria growing under pressure is associated with maintenance processes, and this conclusion can be related to the inefficient growth under pressure described by Matsumura and Marquis (1977). Wirsen and Jannasch (1975) have estimated a high ratio of respiration to biosynthesis for marine bacteria under pressure. In addition, they found that psychrophilic bacteria were better able to function under the deep-sea conditions of pressure and low temperature than were mesophiles. However, even the psychrophiles were very barosensitive at low temperature, and substrate incorporation was inhibited to a major extent by pressures as low as 20 atm at 8°C.

As a result of a study of the barosensitivities of protein synthetic systems from a variety of bacteria, Pope *et al.* (1976a) concluded that the ability of bacteria to synthesize protein at a deep-sea pressure of 680 atm cannot be related to environmental origin, physiological type or taxonomic group. Landau *et al.* (1977) extended previous studies on the effects of pressure on *in vitro* protein synthesis to establish clearly that the 30S ribosomal subunit is of paramount importance in determining barotolerance and to show that the subunit from ribosomes of *Pseudomonas bathycetes* is pressure resistant when isolated in the presence of 150 mM KCl but not 150 mM NaCl.

Arcuri and Ehrlich (1977) have shown that pressure can modify the effects of heavy metal ions on bacterial growth, but in a complex manner.

Morita (1976) has reviewed the subject of microbial survival with attention to cold, hyperbaric environments. He points out that mesophilic bacteria seem better able to survive under pressure at 4°C than are psychrophilic bacteria and that it should not be so surprising that deep-sea isolates are not psychrophiles. In fact, it seems that the most common bacteria isolated from deep-sea sediments are mesophilic members of the genus *Bacillus*. Of course, a distinction has to be made between mere survival and growth in a compressed environment. Morita also considers pressure-temperature interactions as they affect solute uptake, solute utilization and growth, but with the ideal gas law as a guide. Temperature does not always act antagonistically to pressure in affecting bacterial processes, in large measure because of the condensed state of the cell. Also presented is an interesting preliminary description of a vibrio, Ant-300, isolated from Antarctic water which becomes remarkably small and able to pass through 0·4 μm porosity filters when starved. A subsequent paper (Novitsky and Morita, 1977) presents a more detailed description of this bacterium which can survive starvation conditions for more than a year. Unfortunately, the organism is relatively barosensitive and pressure hastens its demise during starvation.

The production and survival of bacterial endospores under pressure have been considered in two publications. One, a thesis by J. D. Hauxhurst (1976), indicates that, for a variety of *Bacillus* organisms, the maximum pressure for sporulation is some 100 to 200 atm less than the maximum growth pressure. Murrell and Wills (1977) have extended previous work on pressure-induced germination of spores. On the basis of thermodynamic parameters for the process, including large, negative $\Delta V^{\ddagger}$ values, they concluded that germination involves a conformational change and also hydration or viscosity reduction in the spore core. It seems that pressure will prove to be a useful antimicrobial agent for preservation of food and medical supplies.

Since the first writing of this chapter, O'Brien (1976) has reported that growth of *Bacillus cereus* is enhanced in the negative pressure environment of a lens-plate arrangement in which culture fluid is stressed by pulling up on the lens part of the couple. The maximum negative pressure achieved was relatively low, only about $1{\cdot}28 \times 10^{-6}$ atm, but a 22% increase in growth was assessed.

Molecular studies of biological responses to pressure have led to changes in basic purview. For example, Low and Somero (1975) have discovered a curious linear correlation between the effects of salts on V_{max}, the maximum reaction rate in an enzyme catalysed reaction, and $\Delta V^{\ddagger}$. They interpret this correlation in terms of changes in water organization accompanying conformational changes in catalysis. Li *et al.* (1976) have proposed that the volume changes calculated for protein denaturation based on pressure sensitivity may be gross underestimates because of a plurality of denatured forms. In other words, denaturation appears on the basis of fluorescence studies to be a multistep process without a high degree of cooperativity, and so the volume change for any one step is not the volume change for the overall process. Certainly this view of a noncooperative, multistage process offers another solution to the dilemma posed by the small volume changes that have been found to accompany protein denaturation by pressure.

Macdonald and Miller (1976) have reviewed the subject of biological membranes under pressure with emphasis on eukaryotic membranes. They point out that there is very little basic information with which to interpret the role of pressure-induced membrane changes in complex processes such as the high-pressure nervous syndrome. Certainly, interpretations in terms of changes in membrane molecules are not currently possible, despite the advances in the study of synthetic membranes described by Trudell (1976).

The responses of microtubules to pressure appear also to be highly complex, more so than was previously thought. For example, Engelborghs *et al.* (1975) found the response of rat-brain microtubules to pressure was highly temperature dependent and that above 25°C only small effects were detectable. They also showed that pressure-temperature effects were primarily on nucleation rather than on propagation.

The successes achieved in deep-diving programmes have increased interest in the biological effects of compressed gases, including their actions on prokaryotic cells. Wild (1977) has described an increase in the resistance of staphylococci to penicillin due to 68 atm of helium, nitrogen or mixtures of He and N_2. The gases enhanced the induction of penicillinase. Argon was peculiar in that it has just the opposite effect. Wild had previously reported (1976) that hyperbaric, normoxic gases reduced reversion frequencies for tryptophan auxotrophs of *Salmonella typhimurium* treated with alkylating agents, but they increased the lethality of nitrosoguanidine. Enfors and Molin (1975) found that germination of *Bacillus cereus* endospores is extremely sensitive to compressed gases. The order of potency for anaesthetic gases was $N_2O > Ar > N_2 > He$. CO_2 also was highly effective in inhibiting germination, but O_2 was only a relatively weak inhibitor. We have repeated and confirmed these observations. However, we have also found that there is a hierarchy of sensitivity to anaesthetic gases among bacteria, and that growth of certain bacteria, such as *Staphylococcus aureus* strain H, is as sensitive to the gases as is endospore germination. However, spores may prove useful for future research on the mechanisms of action of anaesthetic gases.

Recently, Mathis and Brown (1976) have concluded from assays of intracellular ATP concentrations that the toxicity of 4·2 atm of oxygen for *E. coli* in a minimal medium is not due to any lack of ATP resulting from respiratory inhibition since growth stops well before there is any major reduction in ATP pools. Boehme *et al.* (1976) found that the protective effects of yeast extract for *E. coli* exposed to hyperbaric oxygen in minimal medium are due to the presence of ten amino acids. They feel that the toxic effects of oxygen can be interpreted in terms of oxygen sensitivities of enzymes in the biosynthetic pathways for amino acids.

Certainly, it seems that bacteria should be of considerable use in defining the basic mechanisms of toxicity of compressed gases. These gases affect all cells, and it is likely that there are universal targets, even though some cells may have highly sensitive, highly specialized targets. Microbial studies have formed the leading edge of molecular biology in the past three decades and they should hold the same position in the development of hyperbaric biology and medicine.

Acknowledgements

We wish to thank Diana Marquis for reading the manuscript critically. The work of the authors was supported by the U.S. Office of Naval Research.

References

Albright, L. J. (1969). Alternate pressurization-depressurization effects on growth and net protein, RNA and DNA synthesis by *Escherichia coli* and *Vibrio marinus*. *Can. J. Microbiol.* **15,** 1237–1240.

Albright, L. J. (1975). The influence of hydrostatic pressure upon biochemical activities of heterotrophic bacteria. *Can. J. Microbiol.* **21,** 1406–1412.

Albright, L. J. and Henigman, J. F. (1971). Seawater salts—hydrostatic pressure effects upon cell division of several bacteria. *Can. J. Microbiol.* **17,** 1246–1248.

Arcuri, E. J. and Ehrlich, H. L. (1977). Influence of hydrostatic pressure on the effects of the heavy metal cations of manganese, copper, cobalt, and nickel on the growth of three deep-sea bacterial isolates. *Appl. Environ. Microbiol.* **33,** 282–288.

Arnold, R. M. and Albright, L. J. (1971). Hydrostatic pressure effects on the translation stages of protein synthesis in a cell-free system from *Escherichia coli. Biochim. Biophys. Acta,* **238,** 347–354.

Balasubramanian, D. and Wetlaufer, D. B. (1966). Reversible alteration of the structure of globular proteins by anesthetic agents. *Proc. Natl. Acad. Sci. USA.* **55,** 762–765.

Barkley, M. D., Riggs, A. D., Jobe, A. and Bourgeois, S. (1975). Interaction of effecting ligands with *lac* repressor and repressor-operator complex. *Biochemistry* **14,** 1700–1712.

Baross, J. A., Hanus, F. J. and Morita, R. Y. (1975). Survival of human enteric and other sewage microorganisms under simulated deep-sea conditions. *Appl. Microbiol.* **30,** 309–318.

Basset, J. and Macheboeuf, M. A. (1932). Étude sur les effets biologiques des ultrapressions: résistance des bactéries, des diastases et des toxines aux pressions très élevées. *Compt. Rend. Acad. Sci., Paris,* **195,** 1431–1438.

Basset, J., Gratia, A., Macheboeuf, M. and Manil, P. (1938). Action of high pressures on plant viruses. *Proc. Soc. Exp. Biol. Med.* **38,** 248–251.

Beck, T. and Poschenrieder, H. (1957). Drucktoleranz, ein Kriterium für den autochthonen Charakter der Braunkohlenmitroflora. *Zentbl. Bakteriol.*, Abt. II, **110,** 534–539.

Berger, L. R. (1974). Enzyme kinetics, microbial respiration, and active transport at increased hydrostatic pressure. *Rev. Phys. Chem. Japan*, Special Issue, 639–642.

Bodanszky, A. and Kauzmann, W. (1962). The apparent molar volume of sodium hydroxide at infinite dilution and the volume change accompanying the ionization of water. *J. Phys. Chem.* **66,** 177–179.

Boehme, D. E., Vincent, K. and Brown, O. R. (1976), Oxygen and toxicity inhibition of amino acid biosynthesis. *Nature, Lond.* **262,** 418–420.

Bøje, L. and Hvidt, A. (1972). Volume effects in aqueous solutions of macromolecules containing non-polar groups. *Biopolymers* **11,** 2357–2364.

Brandts, J. F., Oliveira, R. J. and Westort, C. (1970). Thermodynamics of protein denaturation. Effect of pressure on the denaturation of ribonuclease A. *Biochemistry* **9,** 1038–1047.

Brauer, R. W. (1972). "Barobiology and the Experimental Biology of the Deep Sea." North Carolina Sea Grant Program. Chapel Hill, North Carolina.

Brauer, R. W., Way, R. O., Jordan, M. R. and Parrish, D. E. (1971). Experimental studies on the high pressure hyperexcitability syndrome in various mammalian species. "Proceedings of the Fourth Underwater Physiology Symposium." (Ed. C. J. Lambertsen), pp. 487–500. Academic Press, New York and London.

Bridgman, P. W. (1914). The coagulation of albumen by pressure. *J. Biol. Chem.* **19,** 511–512.

Bridgman, P. W. (1949). "The Physics of High Pressure." G. Bell, London.

Bruemmer, J. H., Brunetti, B. B. and Schreiner, H. R. (1967). Effects of helium group gases and nitrous oxide on HeLa cells. *J. Cell. Physiol.* **69,** 385–392.

Buchheit, R. G., Schreiner, H. R. and Doebbler, G. F. (1966). Growth responses of *Neurospora crassa* to increased partial pressures of the noble gases and nitrogen. *J. Bacteriol.* **91,** 622–627.

Cattell, M. (1936). The physiological effects of pressure. *Biol. Rev.* **11,** 441–476.

Certes, A. (1884a). Sur la culture, a l'abri des germes atmosphériques, des eaux et des sediments rapportes par les expeditions du Travailleur et du Talisman. *Compt. Rend. Acad. Sci., Paris,* 690–693.

Certes, A. (1884b). Note relative a l'action des hautes pressions sur la vitalité des micro-organismes d'eau douce et l'eau de mer. *Compt. Rend. Soc. Biol.* **36,** 220–222.

Chapman, R. E. and Sturtevant, J. M. (1969). Volume changes accompanying the thermal denaturation of deoxyribonucleic acid. I. Denaturation at neutral pH. *Biopolymers* **7,** 527–537.

Christensen, R. G. and Cassel, J. M. (1967). Volume changes accompanying collagen denaturation. *Biopolymers* **5,** 685–689.

Citti, J.E., Sandine, W.E. and Elliker, P.R. (1965). β-Galactosidase of *Streptococcus lactis*. *J. Bacteriol.* **89,** 937–942.

Clouston, J. G. and Wills, P. A. (1969). Initiation of germination and inactivation of *Bacillus pumilus* spores by hydrostatic pressure. *J. Bacteriol.* **97,** 684–690.

Clouston, J. G. and Wills, P. A. (1970). Kinetics of initiation of germination of *Bacillus pumilus* spores by hydrostatic pressure. *J. Bacteriol.* **103,** 140–143.

Daily, O. P. and Schlamm, N. A. (1972). Effect of hyperbaric atmospheres on β-galactoside transport in *Escherichia coli*. *Can. J. Microbiol.* **18,** 1162–1164.

Dicamelli, R. F., Balbinder, E. and Lebowitz, J. (1973). Pressure effects on the association of the α and β_2 subunits of tryptophan synthetase for *Escherichia coli* and *Salmonella typhimurium*. *Arch. Biochem. Biophys.* **155,** 315–324.

Dinsdale, M. T. and Walsby, A. E. (1972). The interrelations of cell turgor pressure, gas-vacuolation and buoyancy in a blue-green alga. *J. Exp. Bot.* **23,** 561–570.

Dole, M., Wilson, F. R. and Fife, W. P. (1975), Hyperbaric hydrogen therapy: A possible treatment for cancer. *Science* **190,** 152–154.

Ebbecke, U. (1935a). Das Verhalten von Paramecien unter den Einwirkung hohen Druckes. *Pflügers Arch. Ges. Physiol.* **236,** 658–661.

Ebbecke, U. (1935b). Uber die Wirkungen hoher Drucke auf marine Lebewesen. *Pflügers Arch. Ges. Physiol.* **236,** 648.

Edebo, L. and Rutberg, L. (1960). Electron-microscopical observations on a pressure-induced phage. *J. Ultrastr. Res.* **4,** 89–91.

Ehrlich, H. L. (1974a). Response of some activities of ferromanganese nodule bacteria to hydrostatic pressure. In "Effect of the Ocean Environment on Microbial Activities," (Eds R. R. Colwell and R. Y. Morita), pp. 208–221. University Park Press, Baltimore.

Ehrlich, H. L. (1974b). Induction of MnO_2-reductase activity under different hydrostatic pressures at 15°C. *Abst. Ann. Meeting Am. Soc. Microbiol.* p. 47.

Ehrlich, H. L. (1975). The formation of ores in the sedimentary environment of the deep sea with microbial participation: The case for ferromanganese concretions. *Soil Sci.* **119,** 36–41.

Enfors, S. O. and Molin, N. (1975). Inhibition of germination in *Bacillus cereus* spores by high gas pressure. *In* "Spores VI" (Eds P. Gerhardt, R. N. Costilow and H. L. Sadoff), pp. 506–512. American Society for Microbiology, Washington, D.C.

Engelborghs, Y., Heremans, K. A. H. and Hoebeke, J. (1975). The effect of pressure on neuronal microtubules. *In* "Microtubules and Microtubule Inhibitors" (Eds M. Borger and M. de Barbander), pp. 59–66. North-Holland Publishing Co.

Evans, M. G. and Polanyi, M. (1935). Application of the transition-state method to the calculation of reaction velocities. *Trans. Faraday Soc.* **31,** 875–894.

Eyring, H. (1935a). The activated complex in chemical reactions. *J. Chem. Phys.* **3,** 107–115.

Eyring, H. (1935b). The activated complex and the absolute rate of chemical reactions. *Chem. Rev.* **1,** 65–77.

Fawcett, E. W. and Gibson, R. O. (1934). The influence of pressure on a number of organic reactions in the liquid phase. *J. Chem. Soc.* (*Lond.*) 386–395.

Fenn, W. O. (1970). A study of aquatic life from the laboratory of Paul Bert. A review of "La Vie dans les Eaux," by Paul Regnard, Paris, 1891. *Resp. Physiol.* **9,** 95–107.

Fenn, W. O. and Marquis, R. E. (1968). Growth of *Streptococcus faecalis* under high hydrostatic pressure and high partial pressures of inert gases. *J. Gen. Physiol.* **52,** 810–824.

Forer, A. and Zimmerman, A. M. (1976). Spindle birefringence of isolated mitotic apparatus analyzed by pressure treatment. *J. Cell Sci.* **20,** 309–327.

Fridovich, I. (1974). Superoxide dismutase. *Adv. Enzymol.* **41,** 35–97.

Fridovich, I. (1975). Oxygen: Boon or bane. *Am. Sci.* **63,** 54–59.

Gerber, B. R. and Noguchi, H. (1967). Volume change associated with the G-F transformation of flagellin. *J. Mol. Biol.* **26,** 197–210.

Gottlieb, S. F. (1971). Effect of hyperbaric oxygen on microorganisms. *Ann. Rev. Microbiol.* **25,** 111–152.

Gould, G. W. and Sale, A. J. H. (1970). Initiation of germination of bacterial spores by hydrostatic pressure. *J. Gen. Microbiol.* **60,** 335–346.

Gould, G. W. and Sale, A. J. H. (1972). Role of pressure in the stabilization and destabilization of bacterial spores. *In* "The Effects of Pressure on Living Organisms" (Eds M. A. Sleigh and A. G. Macdonald), pp. 147–157. Academic Press, New York and London.

Gunter, T. E. and Gunter, K. K. (1972). Pressure dependence of the helixcoil transition temperature for polynucleic acid helices. *Biopolymers* **11,** 667–678.

Hardon, M. J. and Albright, L. J. (1974). Hydrostatic pressure effects on several stages of protein synthesis in *Escherichia coli. Can. J. Microbiol.* **20,** 359–365.

Harold, F. M. and Baarda, J. R. (1968). Effects of nigericin and monactin on cation permeability of *Streptococcus faecalis* and metabolic capabilities of potassium-depleted cells. *J. Bacteriol.* **95,** 816–823.

Hauxhurst, J. D. (1976). Lytic enzymes and the production and stability of bacterial endospores at increased hydrostatic pressure. Ph.D. Thesis, University of California.

Hawley, S. A. (1971). Reversible pressure-temperature denaturation of chymotrypsinogen. *Biochemistry* **10,** 2436–2442.

Hayward, T. J. (1971). Negative pressure in liquids: Can it be harnessed to serve man? *Am. Sci.* **59,** 434–443.

Heath, I. B. (1975). The effect of antitubule agents on the growth and ultrastructure of the fungus *Saprolegnia ferax* and their ineffectiveness in disrupting hyphal microtubules. *Protoplasma* **85,** 147–176.

Hedén, C.-G. (1964). Effects of hydrostatic pressure on microbial systems. *Bacteriol. Rev.* **28,** 14–29.

Hedén, C.-G., Lindahl, T. and Toplin, I. (1964). The stability of deoxyribonucleic acid solutions under high pressure. *Acta Chem. Scand.* **18,** 1150–1158.

Hessler, R. R., Isaacs, J. D. and Mills, E. L. (1972). Giant amphipod from the abyssal Pacific Ocean. *Science* **175,** 636–637.

Hildebrand, C. E. and Pollard, E. C. (1972). Hydrostatic pressure effects on protein synthesis. *Biophys. J.* **12,** 1235–1250.

Hill, E. P. and Morita, R. Y. (1964). Dehydrogenase activity under hydrostatic pressure by isolated mitochondria obtained from *Allomyces macrogynus. Limol. Oceanogr.* **9,** 243–298.

Hite, B. H., Giddings, N. J. and Weakly, C. E. (1914). The effect of pressure on certain microorganisms encountered in the preservation of fruits and vegetables. *W. Va. Agric. Exp. Stn. Bull.* **146,** 1–67.

Hochachka, P. W., Moon, T. W. and Mustafa, T. (1972). The adaptation of enzymes to pressure in abyssal and midwater fishes. *In* "The Effects of Pressure on Living Organisms" (Eds M. A. Sleigh and A. G. Macdonald), pp. 175–195. Academic Press, New York and London.

Hvidt, A. (1975). A discussion of pressure-volume effects in aqueous protein solutions. *J. Theor. Biol.* **50,** 245–252.

Infante, A. A. and Baierlein, R. (1971). Pressure-induced dissociation of sedimenting ribosomes: Effect on sedimentation patterns. *Proc. Natl. Acad. Sci. USA* **68,** 1780–1785.

Inoué, S., Fuseler, J., Salmon, E. D. and Ellis, G. W. (1975). Functional organization of mitotic microtubules. Physical chemistry of the *in vivo* equilibrium system. *Biophys. J.* **15,** 725–744.

Jaenicke, R. (1971). Volume changes in the isoelectric heat aggregation of serum albumin. *Eur. J. Biochem.* **21,** 110–115.

Jannasch, H. W. and Wirsen, C. O. (1973). Deep-sea microorganisms: *in situ* response to nutrient enrichment. *Science* **180,** 641–643.

Jannasch, H. W. and Wirsen, C. O. (1977a). Microbial life in the deep sea. *Sci. Am.* **236,** 42–52.

Jannasch, H. W. and Wirsen, C. O. (1977b). Retrieval of concentrated and undecompressed microbial populations from the deep sea. *Appl. Environ. Microbiol.* **33,** 642–646.

Jannasch, H. W., Eimhjellen, K., Wirsen, C. O. and Farmanfarmaian, A. (1971). Microbial degradation of organic matter in the deep sea. *Science* **171,** 672–675.

Jannasch, H. W., Wirsen, C. O. and Taylor, C. D. (1976). Undecompressed microbial populations from the deep sea. *Appl. Environ. Microbiol.* **32,** 360–367.

Jaschhof, H. and Schwartz, W. (1969). Untersuchungen zur Geomikrobiologie der Braunkohle. II. Verhalten von Bakterienstammen aus Braunkohle gegenüber höheren hydrostatischen Drucken und ihre biochemische. *Akt. ivität Z. Allg. Mikrobiol.* **9,** 347–361.

Johnson, F. H. (1957). The action of pressure and temperature. *In* "Microbial Ecology" (Eds R. E. O. Williams and C. C. Spicer), pp. 134–167. Cambridge University Press, Cambridge.

Johnson, F. H. and Eyring, H. (1970). The kinetic basis of pressure effects in biology and chemistry. *In* "High Pressure Effects on Cellular Processes" (Ed. A. M. Zimmerman), pp. 1–44. Academic Press, New York and London.

Johnson, F. H. and Lewin, I. (1946). The disinfection of *E. coli* in relation to temperature, hydrostatic pressure and quinine. *J. Cell. Comp. Physiol.* **28,** 23–45.

Johnson, F. H., Eyring, H. and Polissar, M. J. (1954). *In* "The Kinetic Basis of Molecular Biology." Wiley, New York. An updated version of this book has been published (1974) under the title "The Theory of Rate Processes in Biology and Medicine" with B. J. Stover as coauthor.

Katz, S. and Ferris, T. G. (1966). Dilatometric study of the interactions of bovine serum albumin with urea. *Biochemistry* **5,** 3246–3253.

Katz, S., Crissman, J. K. and Beall, J. A. (1973). Structure-volume relationships of proteins. *J. Biol. Chem.* **248,** 4840–4845.

Kauzmann, W. (1959). Some factors in the interpretation of protein denaturation. *Adv. Protein Chem.* **14,** 1–63.

Kauzmann, W., Bodanszky, A. and Rasper, J. (1962). Volume changes in protein reactions. II. Comparison of ionization reactions in proteins and small molecules. *J. Am. Chem. Soc.* **84,** 1777–1788.

Kenis, P. R. (1971). Effects of high pressure helium on bacterial growth. *Bact. Proc.* 57.

Kennedy, J. R. and Zimmerman, A. M. (1970). The effects of high hydrostatic pressure on the microtubules of *Tetrahymena pyriformis*. *J. Cell Biol.* **47,** 568–576.

Kettman, M. S., Hishikawa, A. H., Morita, R. Y. and Becker, R. R. (1965). Effect of hydrostatic pressure on the aggregation reaction of poly-L-valyl-ribonuclease. *Biochem. Biophys. Res. Commun.* **22,** 262–267.

Kim, J. and ZoBell, C. E. (1971). Death rates of barophobic bacteria at deep-sea pressure and temperature. *Bact. Proc.* 57.

Kinne, O. (1972). Pressure, general introduction. *In* "Marine Ecology" (Ed. O. Kinne), Vol. 1, part 3, pp. 1323–1360. Wiley-Interscience, London and New York.

Kirk, W. W. (1972). Corrosion behavior of high-strength materials in sea water. *In* "Barobiology and the Experimental Biology of the Deep Sea" (Ed. R. W. Brauer), pp. 302–320. North Carolina Sea Grant Program, Chapel Hill.

Kitching, J. A. (1957a). Effects of high hydrostatic pressures on the activity of flagellates and ciliates. *J. Exp. Biol.* **34,** 494–510.

Kitching, J. A. (1957b). The effects of high hydrostatic pressures on *Actinophrys sol* (Heliozoa). *J. Exp. Biol.* **34,** 511–525.

Kriss, A. E. (1962). *In* "Marine Microbiology (Deep Sea)" pp. 91–107. Wiley-Interscience, New York and London.

Kriss, A. E. and Mitskevich, I. N. (1967). Effect of the nutrient medium on the tolerance of barotolerant bacteria to high pressures. *Mikrobiologiya* **36,** 203–206.

Kriss, A. E., Mitskevich, I. N. and Cherni, N. E. (1969). Changes in ultrastructure and chemical composition of bacterial cells under the influence of high hydrostatic pressure. *Mikrobiologiya* **38,** 88–95.

Laidler, K. J. (1951). The influence of pressure on rates of biological reactions. *Arch. Biochem.* **30,** 226–236.

Laidler, K. J. and Bunting, P. S. (1973). "The Chemical Kinetics of Enzyme Action." Oxford University Press.

Landau, J. V. (1966). Protein and nucleic acid synthesis in *Escherichia coli*: Pressure and temperature effects. *Science* **153,** 1273–1274.

Landau, J. V. (1967). Induction, transcription and translation in *Escherichia coli*: A hydrostatic pressure study. *Biochim. Biophys. Acta* **149,** 506–512.

Landau, J. V. (1970). Hydrostatic pressure on the biosynthesis of macromolecules. *In* "High Pressure Effects on Cellular Processes" (Ed. A. M. Zimmerman), pp. 45–70. Academic Press, New York and London.

Landau, J. V., Smith, W. P. and Pope, D. H. (1977). Role of the 30S ribosomal subunit, initiation factors, and specific ion concentration in barotolerant protein synthesis in *Pseudomonas bathycetes*. *J. Bacteriol.* **130,** 154–159.

Larson, W. P., Hartzell, T. B. and Diehl, H. S. (1918). The effects of high pressures on bacteria. *J. Infect. Dis.* **22,** 271–284.

Lauffer, M. A. and Dow, R. B. (1941). The denaturation of tobacco mosaic virus at high pressures. *J. Biol. Chem.* **140,** 509–518.

Lever, M. J., Miller, K. W., Paton, W. D. M. and Smith, E. B. (1971). Pressure reversal of anaesthesia. *Nature, Lond.* **231,** 368–371.

Li, T. M., Hook, J. W., Drickamer, H. G. and Weber, G. (1976). Plurality of pressure-denatured forms in chymotrypsinogen and lysozyme. *Biochemistry* **15,** 5571–5580.

Low, P. S. and Somero, G. N. (1975). Protein hydration changes during catalysis. A new mechanism of enzymic rate-enhancement and ion activation/inhibition of catalysis. *Proc. Natl. Acad. Sci., USA.* **72,** 3305–3309.

Macdonald, A. G. (1967a). The effect of high hydrostatic pressure on the cell division and growth of *Tetrahymena pyriformis*. *Expl. Cell Res.* **47,** 569–580.

Macdonald, A. G. (1967b). Delay in the cleavage of *Tetrahymena pyriformis* exposed to high hydrostatic pressure. *J. Cell. Physiol.* **70,** 127–130.

Macdonald, A. G. (1975a). The effect of helium and of hydrogen at high pressure on the cell division of *Tetrahymena pyriformis* W. *J. Cell. Physiol.* **85,** 511–528.

Macdonald, A. G. (1975b). Physiological aspects of deep sea biology. Cambridge University Press.

Macdonald, A. G. and Miller, K. W. (1976). Biological membranes at high hydrostatic pressure. *In* "Biophysical and Biochemical Perspective in Marine Biology" (Eds J. Sargent and D. Mallins), pp. 27–57. Academic Press, New York and London.

Marquis, R. E. (1976). High-pressure microbial physiology. *Adv. Microbial Physiol.* **14**, 159–241.

Marquis, R. E. and Fenn, W. O. (1969). Dilatometric study of streptococcal growth and metabolism. *Can. J. Microbiol.* **15**, 933–940.

Marquis, R. E. and Keller, D. M. (1975). Enzymic adaptation by bacteria under pressure. *J. Bacteriol.* **122**, 575–584.

Marquis, R. E. and ZoBell, C. E. (1971). Magnesium and calcium ions enhance barotolerance of streptococci. *Arch. Mikrobiol.* **79**, 80–92.

Marquis, R. E., Brown, W. P. and Fenn, W. O. (1971). Pressure sensitivity of streptococcal growth in relation to catabolism. *J. Bacteriol.* **105**, 504–511.

Marsland, D. (1970). Pressure-temperature studies on the mechanism of cell division. *In* "High Pressure Effects on Cellular Processes" (Ed. A. M. Zimmerman), pp. 259–312. Academic Press, New York and London.

Marsland, D. and Brown, D. E. S. (1936). Amoeboid movement at high hydrostatic pressure. *J. Cell. Comp. Physiol.* **8**, 167–178.

Mathis, R. R. and Brown, O. R. (1976). ATP concentration in *Escherichia coli* during oxygen toxicity. *Biochim. Biophys. Acta* **440**, 723–732.

Matsumura, P. (1975). The physiologic bases for streptococcal barotolerance. Ph.D. thesis, University of Rochester.

Matsumura, P. and Marquis, R. E. (1975). Adenosine triphosphate supply and barotolerance of streptococci. *Abst. Ann. Meeting Am. Soc. Microbiol.* 191.

Matsumura, P. and Marquis, R. E. (1977). Energetics of streptococcal growth inhibition by hydrostatic pressure. *Appl. Environ. Microbiol.* **33**, 885–892.

Matsumura, P., Keller, D. M. and Marquis, R. E. (1974). Restricted pH ranges and reduced yields for bacterial growth under pressure. *Microbial Ecol.* **1**, 176–189.

McElhaney, R. N. (1974). The effect of alterations in the physical state of the membrane lipids on the ability of *Acholeplasma laidlawii* B to grow at various temperatures. *J. Mol. Biol.* **84**, 145–157.

Meganathan, R. and Marquis, R. E. (1973). Loss of bacterial motility under pressure. *Nature, Lond.* **246**, 525–527.

Melchior, D. L. and Morowitz, H. J. (1972). Dilatometry of dilute suspensions of synthetic lecithin aggregates. *Biochemistry* **11**, 4558–4562.

Miller, K. W. (1972). Inert gas narcosis and animals under high pressure. *Symp. Soc. Exp. Biol.* **26**, 363–378.

Morita, R. Y. (1967). Effects of hydrostatic pressure on marine microorganisms. *Oceanogr. Mar. Biol. Ann. Rev.* **5**, 187–203.

Morita, R. Y. (1970). Application of hydrostatic pressure to microbial cultures. *In* "Methods in Microbiology" (Eds J. R. Norris and D. W. Ribbons), Vol. 2, pp. 243–257. Academic Press, London and New York.

Morita, R. Y. (1972). Pressure. Bacteria, fungi and blue-green algae. *In* "Marine Ecology" (Ed. O. Kinne), Vol. 1, part 3, pp. 1361–1388. Wiley-Interscience, New York and London.

Morita, R. Y. (1975). Psychrophilic bacteria. *Bacteriol. Rev.* **39**, 144–167.

Morita, R. Y. (1976). Survival of bacteria in cold and moderate hydrostatic pressure environments with special reference to psychrophilic and barophilic bacteria.

In "The Survival of Vegetative Microbes" (Eds T. R. G. Gray and J. R. Postgate), pp. 279–298. Cambridge University Press.

Morita, R. Y. and Becker, R. R. (1970). Hydrostatic pressure effects on selected biological systems. *In* "High Pressure Effects on Cellular Processes" (Ed. A. M. Zimmerman), pp. 71–83. Academic Press, New York and London.

Murakami, T. H. and Zimmerman, A. M. (1970). A pressure study of galvanotoxis in *Tetrahymena. In* "High Pressure Effects on Cellular Processes" (Ed. A. M. Zimmerman), pp. 139–153. Academic Press, New York and London.

Murrell, W. G. and Wills, P. A. (1977). Initiation of *Bacillus* spore germination by hydrostatic pressure: Effect of temperature. *J. Bacteriol.* **129,** 1272–1280.

Neville, W. M. and Eyring, H. (1972). Hydrostatic pressure and ionic strength effects on the kinetics of lysozyme. *Proc. Natl. Acad. Sci. USA* **69,** 2417–2419.

Novitsky, J. A. and Morita, R. Y. (1977). Survival of a psychrophilic marine vibrio under long-term nutrient starvation. *Appl. Environ. Microbiol.* **33,** 635–641.

O'Brien, W. J. (1976). Effects of capillary penetration and negative pressure at sites of caries susceptibility. *In* "Microbial Aspects of Dental Caries" (Eds H. M. Stiles, W. J. Loesche and T. C. O'Brien), pp. 387–400. Information Retrieval, Washington, D.C.

O'Connor, T. M., Houston, L. L. and Samson, F. (1974). Stability of neuronal microtubules to high pressure *in vivo* and *in vitro*. *Proc. Natl. Acad. Sci. USA* **71,** 4198–4202.

Oppenheimer, C. H. and ZoBell, C. E. (1952). The growth and viability of sixty-three species of marine bacteria as influenced by hydrostatic pressure. *J. Marine Res.* **11,** 10–18.

Palmer, F. E. (1961). The effect of moderate hydrostatic pressures on the mutation rate of *Serratia marinorubra* to streptomycin resistance. M.Sc. thesis, University of California.

Palmer, D. S. and Albright, L. J. (1970). Salinity effects on the maximum hydrostatic pressure for growth of the marine psychrophilic bacterium, *Vibrio marinus*. *Limnol. Oceanogr.* **15,** 343–347.

Pauling, L. (1961). A molecular theory of general anesthesia. *Science* **134,** 15–21.

Pease, D. C. (1941). Hydrostatic pressure effects upon the spindle figure and chromosome movement. I. Experiments on the first mitotic division of *Urechis* eggs. *J. Morphol.* **69,** 405–441.

Pease, D. C. (1946). Hydrostatic pressure effects upon the spindle figure and chromosome movement. II. Experiments on the mitotic divisions of *Tradescantia* pollen mother cells. *Biol. Bull.* **91,** 145–169.

Penniston, J. T. (1971). High hydrostatic pressure and enzyme activity: Inhibition of multimeric enzymes by dissociation. *Arch. Biochem. Biophys.* **142,** 322–332.

Plachy, W. Z., Lanyi, J. K. and Kates, M. (1974). Lipid interactions in membranes of extremely halophilic bacteria. I. Electron spin resonance and dilatometric studies of bilayer structure. *Biochemistry* **13,** 4906–4913.

Pollard, E. C. and Weller, P. K. (1966). The effect of hydrostatic pressure on the synthetic processes in bacteria. *Biochim. Biophys. Acta* **112,** 573–580.

Pope, D. H. and Berger, L. R. (1973). Inhibition of metabolism by hydrostatic pressure: What limits microbial growth? *Arch. Mikrobiol.* **93,** 367–370.

Pope, D. H., Smith, W. P., Swartz, R. W. and Landau, J. V. (1975a). Role of bacterial ribosomes in barotolerance. *J. Bacteriol.* **121,** 664–669.

Pope, D. H., Connors, N. T. and Landau, J. V. (1975b). Stability of *Escherichia coli* polysomes at high hydrostatic pressure. *J. Bacteriol.* **121,** 753–758.

Pope, D. H., Smith, W. P., Orgrinc, M. A. and Landau, J. V. (1976a). Protein synthesis at 680 atm: Is it related to environmental origin, physiological type, or taxonomic group? *Appl. Environ. Microbiol.* **31,** 1001–1002.

Pope, D. H., Smith, W. and Landau, J. V. (1976b). Effect of hydrostatic pressure on growth and macromolecular synthesis in *Halobacterium salinarium. Abst. Ann. Meeting Am. Soc. Microbiol.* 125.

Rautenshtein, Y. I. and Muradov, M. (1966). Effect of high hydrostatic pressure on some lysogenic strains of actinomycetes and on their moderate phages. *Mikrobiologiya* **35,** 571–576.

Regnard, P. (1884). Note sur les conditions de la vie dans les profondeurs de la mer. *Compt. Rend. Soc. Biol.* **36,** 164–168.

Regnard, P. (1891). "Recherches Expérimentales sur les Conditions Physiques de la Vie dans les Eaux." Masson et Cie, Paris.

Rutberg, L. (1964a). On the effects of high hydrostatic pressure on bacteria and bacteriophage. 1. Action on the reproducibility of bacteria and their ability to support growth of bacteriophage T2. *Acta Pathol. Microbiol. Scand.* **61,** 81–90.

Rutberg, L. (1964b). On the effects of high hydrostatic pressure on bacteria and bacteriophage. 2. Inactivation of bacteriophages. *Acta Pathol. Microbiol. Scand.* **61,** 91–97.

Rutberg, L. (1964c). On the effects of high hydrostatic pressure on bacteria and bacteriophage. 3. Induction with high hydrostatic pressure of *Escherichia coli* K lysogenic for bacteriophage lambda. *Acta Pathol. Microbiol. Scand.* **61,** 98–105.

Rutberg, L. and Hedén, C.-G. (1960). The activation of prophage in *E. coli* B by high pressure. *Biochem. Biophys. Acta* **2,** 114–116.

Sale, A. J. H., Gould, G. W. and Hamilton, W. A. (1970). Inactivation of bacterial spores by hydrostatic pressure. *J. Gen. Microbiol.* **60,** 323–334.

Salmon, E. D. (1975a). Pressure-induced depolymerization of spindle microtubules. II. Thermodynamics of *in vivo* spindle assembly. *J. Cell Biol.* **66,** 114–127.

Salmon, E. D. (1975b). Spindle microtubules: Thermodynamics of *in vivo* assembly and role in chromosome movement. *Ann. N.Y. Acad. Sci.* **253,** 383–406.

Salmon, E. D., Goode, D., Maugel, T. K. and Bonar, D. B. (1976). Pressure-induced depolymerization of spindle microtubules. III. Differential stability in HeLa cells. *J. Cell Biol.* **69,** 443–454.

Schlamm, N. A. and Daily, O. P. (1971). Uptake of β-galactosides by *Escherichia coli* at elevated atmospheric pressure. *J. Bacteriol.* **105,** 1202–1204.

Schlamm, N. A. and Daily, O. P. (1972). Effect of elevated atmospheric pressure on penicillin binding by *Staphylococcus aureus* and *Streptococcus pyogenes. Antimicrob. Agents Chemother.* **3,** 147–151.

Schlamm, N. A., Perry, J. E. and Wild, J. R. (1974). Effect of helium gas at elevated pressure on iron transport and growth of *Escherichia coli. J. Bacteriol.* **117,** 170–174.

Scholander, P. F., Hammel, H. T., Bradstreet, E. and Hemmingsen, E. A. (1965). Sap pressure in vascular plants. *Science* **148,** 339–346.

Schulz, E., Lüdemann, H.-D. and Jaenicke, R. (1976). High pressure equilibrium studies on the dissociation-association of *Escherichia coli* ribosomes. *FEBS Lett.* **64,** 40–43.

Schwarz, J. R. and Colwell, R. R. (1975a). Macromolecular biosynthesis in *Pseudomonas bathycetes* at deep-sea pressure and temperature. *Abst. Ann. Meeting Am. Soc. Microbiol.* 162.

Schwarz, J. R. and Colwell, R. R. (1975b). Heterotrophic activity of deep-sea sediment bacteria. *Appl. Microbiol.* **30,** 639–649.
Schwarz, J. R. and Landau, J. V. (1972a). Hydrostatic pressure effects on *Escherichia coli*: Site of inhibition of protein synthesis. *J. Bacteriol.* **109,** 945–948.
Schwarz, J. R. and Landau, J. V. (1972b). Inhibition of cell-free protein synthesis by hydrostatic pressure. *J. Bacteriol.* **112,** 1222–1227.
Schwarz, J. R., Walker, J. D. and Colwell, R. R. (1974a). Growth of deep-sea bacteria on hydrocarbons at ambient and *in situ* pressure. *In* "Developments in Industrial Microbiology," Vol. 15, pp. 239–249. AIBS, Washington, D.C.
Schwarz, J. R., Walker, J. D. and Colwell, R. R. (1974b). Deep-sea bacteria: Growth and utilization of hydrocarbons at ambient and *in situ* pressure. *Appl. Microbiol.* **28,** 982–986.
Schwarz, J. R., Walker, J. D. and Colwell, R. R. (1975). Deep-sea bacteria: Growth and utilization of *n*-hexadecane at *in situ* temperature and pressure. *Can. J. Microbiol.* **21,** 682–687.
Schwarz, J. R., Yayanos, A. A. and Colwell, R. R. (1976). Metabolic activities of the intestinal miroflora of a deep-sea invertebrate. *Appl. Environ. Microbiol.* **31,** 46–48.
Sleigh, M. A. and Macdonald, A. G. (1972). "The Effects of Pressure on Organisms." Academic Press, New York and London.
Smith, W., Pope, D. and Landau, J. V. (1975). Role of bacterial ribosome subunits in barotolerance. *J. Bacteriol.* **124,** 582–584.
Smith, W., Landau, J. V. and Pope, D. (1976). Specific ion concentration as a factor in barotolerant protein synthesis in bacteria. *J. Bacteriol.* **126,** 654–660.
Solomon, L., Zeegen, P. and Eiserling, F. A. (1966). The effects of high hydrostatic pressure on coliphage T-4. *Biochim. Biophys. Acta* **112,** 102–109.
Suzuki, K. and Taniguchi, U. (1972). Effect of pressure on biopolymers and model systems. *In* "The Effects of Pressure on Living Organisms" (Eds M. A. Sleigh and A. G. Macdonald), pp. 103–124. Academic Press, New York and London.
Suzuki, K., Miyosawa, Y. and Taniguchi, Y. (1971). The effect of pressure on deoxyribonucleic acid. *J. Biochem., Tokyo* **69,** 595–598.
Suzuki, K., Taniguchi, Y. and Miyosawa, Y. (1972). The effect of pressure on the absorption spectra of DNA and DNA-dye complex. *J. Biochem., Tokyo* **72,** 1087–1091.
Swartz, R. W., Schwarz, J. R. and Landau, J. V. (1974). Comparative effects of pressure on protein and RNA synthesis in bacteria isolated from marine sediments. *In* "Effect of the Ocean Environment on Microbial Activities" (Eds R. R. Colwell and R. Y. Morita), pp. 145–159. University Park Press, Baltimore.
Tanford, C. (1968). Protein denaturation. *Adv. Protein Chem.* **23,** 121–282.
Taylor, C. D. and Jannasch, H. W. (1976). Subsampling technique for measuring growth of bacterial cultures under high hydrostatic pressure. *Appl. Environ. Microbiol.* **32,** 355–359.
Tilney, L. G. and Gibbins, J. R. (1969). Microtubules in the formation and development of the primary mesenchyme in *Arbacia punctulata.* II. An experimental analysis of their role in development and maintenance of cell shape. *J. Cell Biol.* **41,** 227–250.
Tilney, L. G., Hiramoto, Y. and Marsland, D. (1966). Studies on the microtubules of Heliozoa. III. A pressure analysis of the role of these structures in the formation and maintenance of the axopodia of *Actinosphaerium nucleofilum* (Barrett). *J. Cell Biol.* **29,** 77–95.

Timson, W. J. and Short, A. J. (1965). Resistance of microorganisms to hydrostatic pressure. *Biotech. Bioeng.* **7,** 139–159.
Trudell, J. R. (1976). The effect of high pressure on phospholipid bilayer membranes. *In* "Extreme Environments" (Ed. M. P. Heinrich), pp. 349–353. Academic Press, London and New York.
Trudell, J. R., Payan, D. G., Chin, J. H. and Cohen, E. N. (1974). The effect of pressure on the phase diagram of mixed dipalmitoyl-dimyristoyl-phosphatidylcholine bilayers. *Biochim. Biophys. Acta* **373,** 141–144.
Vallentyne, J. R. (1963). Environmental biophysics and microbial ubiquity. *Ann. N.Y. Acad. Sci.* **108,** 342–352.
van Diggelen, O. P., Oostrom, H. and Bosch, L. (1971). Association products of native and derived ribosomal subunits of *E. coli* and their stability during centrifugation. *FEBS Lett.* **19,** 115–120.
van Diggelen, O. P., Oostrom, H. and Bosch, L. (1973). The association of ribosomal subunits of *Escherichia coli* 2. Two types of association products differing in sensitivity to hydrostatic pressure during centrifugation. *Eur. J. Biochem.* **39,** 511–523.
van't Hoff, J. H. (1901). Vorlesungen über theoretische und physikalische Chemie. *In* "Chemische Dynamik," Vol. 1, 2nd ed., 236.
Walsby, A. E. (1971). The pressure relationships of gas vacuoles. *Proc. R. Soc.* B. **178,** 301–326.
Walsby, A. E. (1972). Gas-filled structures providing buoyancy in photosynthetic organisms. *In* "The Effects of Pressure on Living Organisms" (Eds M. A. Sleigh and A. G. Macdonald), pp. 233–250. Academic Press, New York and London.
Wampler, E. and Katz, S. (1974). Threonine inhibition of the aspartokinase-homoserine dehydrogenase-l complex of *Escherichia coli*: Dilatometric studies. *Biochem. Biophys. Acta* **365,** 414–417.
Weale, K. E. (1967). "Chemical Reactions at High Pressures." Spon, London.
Weida, B. and Gill, S. J. (1966). Pressure effect on deoxyribonucleic acid transition. *Biochim. Biophys. Acta* **112,** 179–181.
Wild, J. R. (1976). Enhancement of base pair substitution induced by alkylating mutagens in simulated hyperbaric diving environments. *Mutat. Res.* **38,** 259–270.
Wild, J. R. (1977). Induction of staphylococcal β-lactamase in response to low concentrations of methicillin under simulated diving environments. *Can. J. Microbiol.* **23,** 116–121.
Willingham, C. A. and Quinby, H. L. (1971). Effects of hydrostatic pressures on anaerobic corrosion of various metals and alloys by sulfate-reducing marine bacteria. *Dev. Ind. Microbiol.* **12,** 278–284.
Wills, P. A. (1975). Inactivation of *B. pumilus* spores by combination hydrostatic pressure-radiation treatment of parenteral solutions. *In* "Radiosterilization of Medical Products" pp. 45–61. International Atomic Energy Agency, Vienna.
Wirsen, C. O. and Jannasch, H. W. (1975). Activity of marine psychrophilic bacteria at elevated hydrostatic pressures and low temperatures. *Mar. Biol.* **31,** 201–208.
Yayanos, A. A. (1975). Stimulatory effect of hydrostatic pressure on cell division in cultures of *Escherichia coli. Biochim. Biophys. Acta* **392,** 271–275.
Yayanos, A. A. and Pollard, E. C. (1969). A study of the effects of hydrostatic pressure on macromolecular synthesis in *Escherichia coli. Biophys. J.* **9,** 1446–1482.
Young, P. G., Young, A. D. and Zimmerman, A. M. (1972). Action of hydrostatic pressure on sea urchin cilia. *Biol. Bull.* **143,** 256–264.

Zimmerman, A. M. (1970). "High Pressure Effects on Cellular Processes." Academic Press, New York and London.
Zimmerman, A. M. (1971). High-pressure studies in cell biology. *Int. Rev. Cytol.* **30,** 1–47.
Zimmerman, A. M. and Laurence, H. L. (1975). Induction of division synchrony in *Tetrahymena pyriformis.* A pressure study. *Exp. Cell Res.* **90,** 119–126.
Zimmerman, A. M. and Rustad, R. C. (1965). Effects of high pressure on pinocytosis in *Amoeba proteus. J. Cell Biol.* **25,** 397–400.
Zipp, A. and Kauzmann, W. (1973). Pressure denaturation of metmyoglobin. *Biochemistry* **12,** 4217–4228.
ZoBell, C. E. (1956). Effects of temperature and pressure on microbial activity. *Bacteriol. Rev.* **20,** 262.
ZoBell, C. E. (1958). Ecology of sulfate reducing bacteria. *Prod. Month.* **22,** 12–29.
ZoBell, C. E. (1964). Hydrostatic pressure as a factor affecting the activities of marine microbes. *In* "Recent Researches in the Fields of Hydrosphere, Atmosphere and Nuclear Geochemistry," pp. 83–116. Maruzen, Tokyo.
ZoBell, C. E. (1970). Pressure effects on morphology and life processes of bacteria. *In* "High Pressure Effects on Cellular Processes" (Ed. A. M. Zimmerman), pp. 85–130. Academic Press, New York and London.
ZoBell, C. E. and Cobet, A. B. (1964). Filament formation by *Escherichia coli* at increased hydrostatic pressures. *J. Bacteriol.* **87,** 710–719.
ZoBell, C. E. and Hittle, L. L. (1967). Some effects of hyperbaric oxygenation on bacteria at increased hydrostatic pressures. *Can. J. Microbiol.* **13,** 1311–1319.
ZoBell, C. E. and Hittle L. L. (1969). Deep-sea pressure effects on starch hydrolysis by marine bacteria. *J. Oceanogr. Soc., Japan* **25,** 36–47.
ZoBell, C. E. and Johnson, F. H. (1949). The influence of hydrostatic pressure on the growth and viability of terrestrial and marine bacteria. *J. Bacteriol.* **57,** 179–189.
ZoBell, C. E. and Morita, R. Y. (1957). Barophilic bacteria in some deep sea sediments. *J. Bacteriol.* **73,** 563–568.
ZoBell, C. E. and Oppenheimer, C. H. (1950). Some effects of hydrostatic pressure on the multiplication and morphology of marine bacteria. *J. Bacteriol.* **60,** 771–781.

Chapter 5

Microbial Life at High Temperatures: Ecological Aspects

M. R. TANSEY and T. D. BROCK

Indiana University and University of Wisconsin

I. Introduction 159
II. Prokaryotes 161
A. Photosynthetic prokaryotes 164
B. Chemolithotrophic bacteria 166
C. Methanogenic bacteria 167
D. Nonphotosynthetic bacteria 167
III. Eukaryotes 168
A. Fungi 168
B. Algae 186
C. Lichens 187
D. Protozoa 187
Addendum 189
Acknowledgements 193
References 194

I. Introduction

A wide variety of natural and man-made habitats of high temperature exist, including erupting volcanoes (1000°C), dry-steam fumaroles (up to 500°C), boiling or superheated springs (93 to 101°C, depending on the altitude), non-boiling hot springs (temperatures from near boiling down to ambient), sun-heated substrates such as soils, litter, rock (depending on colour and heat capacity, temperatures reach 60 to 70°C or higher), self-heating organic-rich materials such as compost piles, seaweed piles, and coal refuse piles (temperatures commonly up to 70°C, and if ignition occurs, above 100°C), hot-water heaters for domestic and industrial purposes (temperatures ranging from 55 to 80°C, occasionally higher), cooling waters from various industrial processes (from just above ambient to boiling), steam lines and steam-condensate lines of steam-heated buildings (condensate lines may reach temperatures above 60°C), and many other habitats. If other conditions are appropriate (pH, nutrients, etc.), certain nonphotosynthetic bacteria can grow in most of these habitats where the temperature is below 90°C, and a few bacteria can grow even up to the boiling point of water. This is not to say that all kinds of bacteria can grow at high temperatures, but the diversity is surprisingly large.

The photosynthetic prokaryotes (blue-green algae and photosynthetic bacteria) are unable to grow at temperatures as high as nonphotosynthetic prokaryotes, even if other environmental conditions (pH, light, nutrients) are appropriate, so that there is an upper temperature limit for photosynthetic life at about 70 to 73°C.

The upper temperature limit for eukaryotic microorganisms is even lower, approximately 60 to 62°C (Tansey and Brock, 1972), at which only a few species of fungi can grow. The upper temperature limits for eukaryotic algae and protozoa are slightly lower, and those for metazoans and higher plants are lower still. Although the exact upper limits for the various groups of living organisms cannot be precisely defined because of insufficient study, those which we consider to be based on satisfactorily thorough research are outlined in Table 1.

Table 1

Upper temperature limits for growth of various microbial groups

Group	Approximate upper temperature
Eukaryotic microorganisms	
Protozoa	56
Algae	55–60
Fungi	60–62
Prokaryotic microorganisms	
Blue-green algae (Cyanobacteria)	70–73
Photosynthetic bacteria	70–73
Chemolithotrophic bacteria	>90
Heterotrophic bacteria	>90

In the present review, we wish to consider only briefly the prokaryotic microorganisms living at high temperature, and concentrate primarily on the eukaryotes. The prokaryotes have been extensively reviewed in recent years

(Brock, 1967, 1969, 1970; Castenholz, 1969, 1973; Williams, 1975), and except for the discovery of several new genera (Brock *et al.*, 1972; Darland *et al.*, 1970), and a clearer ecological definition of the adaptation of bacteria to temperatures of 90°C and over (Brock *et al.*, 1971; Brock and Brock, 1971), the general conclusions given in the earlier reviews still stand. However, the eukaryotic microorganisms, especially the fungi, have not been reviewed extensively in recent years, and there are considerable new data about these groups. Thus, in the review we emphasize the eukaryotes, especially fungi.

II. Prokaryotes

The diversity of bacteria which have been cultivated at high or moderately high temperature is fairly large, although as the temperature is raised, the number of species able to grow drops markedly. We list the well-characterized genera and species of thermophilic bacteria in Tables 2 and 3. Several

Table 2

Prokaryotic microorganisms growing at high temperatures

Group, genus, species	Optimum temperature	Maximum temperature	Remarks	References
Blue-green Algae (Cyanobacteria)				
Chroococcales				
Synechococcus lividus	63–67	74	one strain (others, lower optimum	Meeks and Castenholz (1971)
S. elongatus		66–70		Castenholz (1969)
S. minervae		60		Castenholz (1969)
Synechocystis aquatilus		45–50		Castenholz (1969)
Aphanocapsa thermalis		>55		Castenholz (1969)
Chamaesiphonales				
Pleurocapsa sp.		52–54		Castenholz (1969)
Oscillatoriales				
Oscillatoria terebriformis		53		Castenholz (1969)
O. amphibia		57		Castenholz (1969)
O. germinata		55		Castenholz (1969)
O. okenii		>60		Castenholz (1969)
Spirulina sp.		55–60		Castenholz (1969)
Phormidium laminosum		57–60		Castenholz (1969)
P. purpurasiens		46–47		Castenholz (1969)
Symploca thermalis		45–47		Castenholz (1969)
Nostocales				
Calothrix sp.		52–54		Castenholz (1969)
Mastigocladus laminosus		63–64		Castenholz (1969)
Photosynthetic bacteria				
Chlorobiaceae				
Chloroflexus aurantiacus	55	70-73		Bauld and Brock (1973); Pierson and Castenholz (1975a)
Chromatiaceae				
Chromatium sp.		57–60	natural observations only	Castenholz (1969)

Table 2 (*contd.*)

Group, genus, species	Optimum temperature	Maximum temperature	Remarks	References
Nonphotosynthetic bacteria				
Spore formers				
Bacillus acidocaldarius	60–65	70	acidophile	Darland and Brock (1971); Uchino and Doi (1967)
B. coagulans	37–45	60	acidophile	Belly and Brock (1974); Fields (1970); Gibson and Gordon (1974)
Bacillus sp.	55–60	70	hydrocarbon oxidizer	Mateles *et al.* (1967)
Bacillus YT-P	72	82	proteolytic	Heinen and Heinen (1972)
Bacillus YT-G	80	85	needs confirmation	Heinen (1971)
B. thermocatenulatus		78		Golovacheva *et al.* (1975)
B. stearothermophilus	50–65	70–75	versatile, widespread	Fields (1970); Gibson and Gordon (1974); Gordon *et al.* (1973)
B. licheniformis		50–55		Gibson and Gordon (1974)
B. pumilus		45–50		Gibson and Gordon (1974)
B. macerans		40–50		Gibson and Gordon (1974)
B. circulans		35–50		Gibson and Gordon (1974)
B. laterosporus		35–50		Gibson and Gordon (1974)
B. brevis		40–60		Gibson and Gordon (1974)
B. subtilis		55–70	aromatic, heterocyclic, alcohol utilizers (many strains)	Allen (1953)
B. sphaericus		65–70	carboxylic acid utilizers (several strains)	Allen (1953)
Clostridium thermosaccharolyticum	55	67		Hollaus and Slytr (1972); Smith and Hobbs (1974)
C. thermohydrosulfuricum		74–76	reduces sulphite	Hollaus and Sleytr (1972)
C. tartarivorum		67		Hollaus and Sleytr (1972)
C. thermocellum	60	68	cellulose digester	Breed *et al.* (1957)
C. thermoaceticum	55–60	65		Breed *et al.* (1957)
C. thermocellulaseum	55–60	65	cellulose digester	Breed *et al.* (1957)
Clostridium sp.	60	75	cellulose digester	Loginova *et al.* (1966)
Clostridium sp.	50–75		butyric formation	Loginova *et al.* (1962)

Table 2 (*contd.*)

Group, genus, species	Optimum temperature	Maximum temperature	Remarks	References
Clostridium (many species)	35–45		some pathogens	Smith and Hobbss (1974)
Desulfotomaculum nigrificans	55	70	sulphate reducer	Campbell (1974); Hollaus and Klaus-hofer (1973)
Lactic acid bacteria				
Streptococcus thermophilus	40–45	50		Deibel and Seeley (1974)
Lactobacillus thermophilus	50–63	65		Gaughran (1947)
Lactobacillus (*Thermo-bacterium*) *bulgaricus*	40	52·5		Orla-Jensen (1942)
Lactobacillus	30–40	53		Rogosa (1974)
Bifidobacterium thermophilum	46·5			Rogosa (1974)
Actinomycetes				
Streptomyces fragmentosporus	50–60			Henssen (1969)
S. thermonitrificans	45–50			Pridham and Tresner (1974)
S. thermoviolaceus	50	60		Pridham and Tresner (1974)
S. thermovulgaris		60		Pridham and Tresser (1974)
Pseudonocardia thermophila	40–50	60		Henssen (1974)
Thermoactinomyces vulgaris	60	70		Küster (1974)
T. sacchari	55–60	65		Küster (1974)
T. candidus		60		Kurup *et al.* (1975)
Thermomonospora curvata	50	65		Manachini *et al.* (1966)
T. viridis	50	60		Küster (1974); Manachini *et al.* (1966)
T. citrina	55–60	70–75		Manachini *et al.* (1966)
Microbispora thermodiastatica		55		Cross (1974)
M. aerata		55		Cross (1974)
M. bispora		60		Cross (1974)
Actinobifida dichotomica	50–58			Cross (1974)
A. chromogena	55–58			Cross (1974)
Micropolyspora caesia	28–45	55		Cross (1974)
M. faeni	50	60		Cross (1974)
M. rectivirgula	45–55	65		Cross (1974)
M. rubrobrunea	45–55	65		Cross (1974)
M. thermovirida	40–50	57		Cross (1974)
M. viridinigra	45–55	65		Cross (1974)
Actinoplanes				
Streptosporangium album var. *thermophilum*	50–55	70		Manachini *et al.* (1965)
Methane-producing bacteria				
Methanobacterium thermoautotrophicum	65–70	75		Zeikus and Wolfe (1972)
Sulphur-oxidizing bacteria				
Thiobacillus thiooxidans		55	acidophile	Fliermans and Brock (1972)

Table 2 (*contd.*)

Group, genus, species	Optimum temperature	Maximum temperature	Remarks	References
Sulphur-oxidizing bacteria—(*contd.*)				
Thiobacillus sp.	50	55–60		Williams and Hoare (1972)
T. thermophilica	55–60	80	spore-former (valid species?)	Egorova and Deryugina (1963)
Sulfolobus acidocaldarius	70–75	85–90	acidophile	Brock *et al.* (1972); Mosser *et al.* (1973)
Sulphate-reducing bacteria				
Desulfovibrio thermophilus	65	85		Rozanova and Khudyakova (1974)
Mycoplasma				
Thermoplasma acidophilum	59	65	acidophile	Belly *et al.* (1973); Darland *et al.* (1970)
Spirochete				
Leptospira biflexa var. *thermophila*		54		Hindle (1932)
Methane-oxidizing				
Methylococcus capsulatus	30–50	55		Leadbetter (1974)
Pseudomonads				
Hydrogenomonas thermophilus	50	60		McGee *et al.* (1967)
Gram negative aerobes (uncertain affiliation)		85		
Thermomicrobium roseum	70–75	85		Jackson *et al.* (1973)
Thermus aquaticus	70	79		Brock and Freeze (1969)
T. flavus	70–75	80		Rozanova and Khudyakova (1974)
T. (*Flavobacterium*) *thermophilus*	70	85		Oshima and Imahori (1974)
T. ruber	60	80		Loginova and Egorova (1975)
Thermus X-1	69–71			Ramaley and Hixson (1970)

precautionary notes about these tables: (1) we do not vouch for the validity of all of the taxa; (2) the upper temperature limits for certain species may be found after more work to be either higher or lower than those listed; (3) it is likely that not all naturally occurring thermophilic bacteria are listed, because a concerted search for diverse types of thermophilic bacteria has not been made.

A. Photosynthetic Prokaryotes

It now seems well established that there is an upper temperature limit for photosynthetic life around 70 to 73°C. This limit has been reviewed by Brock (1967), and Castenholz (1969, 1973), and some recent work on two of the high temperature organisms, *Synechococcus lividus* (a blue-green alga) and *Chloroflexus aurantiacus* (a photosynthetic bacterium) have been reported

Table 3

Upper temperature limits for genera of nonphotosynthetic bacteria

Neutral pH	Maximum
Thermomicrobium	85
Thermus	85
Bacillus	82
Bacillus stearothermophilus	75
Methanobacterium	75
Thermomonospora	70–75
Clostridium	74–76
Desulfotomaculum	70
Thermoactinomyces	70
Streptosporangium	70
Micropolyspora	65
Streptomyces	60
Pseudonocardia	60
Microbispora	60
Hydrogenomonas	60
Leptospira	54
Lactobacillus	53
Acid pH	
Sulfolobus	85–90
Bacillus acidocaldarius	70
Thiobacillus	55
Thermoplasma	59

See Table 2 for details and references.

(Meeks and Castenholz, 1971; Bauld and Brock, 1973; Pierson and Castenholz, 1974a, b). The inability of photosynthetic organisms to live at temperatures above 70 to 73°C has not received any biochemical explanation, but it may be because of their inability to construct a functional photosynthetic membrane system. Since at least one autotroph, *Sulfolobus acidocaldarius*, is able to grow at considerably higher temperatures (see below), the temperature-sensitive lesion is presumably not in the CO_2-fixation system.

The conclusion concerning the upper temperature limit for photosynthesis is based on observations in the natural world, rather than those in culture. Conclusions about evolutionary adaptations to extreme environments are incomplete if they are based on observations of cultures alone, because if the cultures were isolated under one condition, they might not be able to grow at another (presumably more extreme) condition. To define environmental limits, one must establish by extensive field observations where organisms are able to grow; if we assume that there has been sufficient time for evolutionary processes to occur, failure of organisms to occupy a habitat is due to physico-chemical limitations. The same reasoning has been used to deduce the upper temperature for eukaryotic life (Tansey and Brock, 1972). (See Mitchell, 1974, for a contrasting view of upper temperature limits.) Blue-green algae that have a growth temperature optimum of over 45°C have a high fidelity to hot spring habitats (or artificially heated waters) in temperate or colder climates, but maintenance populations of some species may become established elsewhere in warmer areas, such as waters of the Everglades National Park (which reach 35–40°C) (Jackson and Castenholz, 1975).

Based on extensive field data, Fraleigh and Wiegert (1975) have recently proposed a model explaining successional change in a standing crop of thermal blue-green algae. Gross productivity and changes in the standing crop of biomass and chlorophyll *a* were measured during ecological succession in the community. Increasing concentration of chlorophyll *a* per unit biomass as succession proceeded in the thermal ecosystems contrasted with the decreasing chlorophyll concentration found in other ecosystems. The model for successional change was used to predict algal growth and was based on algal density, free CO_2 concentrations in the water, and day length; predictions agreed closely with actual field data for biomass increase during succession.

B. Chemolithotrophic Bacteria

Reduced sulphur compounds are common constituents of geothermal habitats; it should not be surprising to find thermophilic sulphur-oxidizing microorganisms. Two organisms, *Thiobacillus thiooxidans* and *Sulfolobus acidocaldarius*, have been found in acidic habitats (Fliermans and Brock, 1972; Brock *et al.*, 1972). *Sulfolobus* is the most thermophilic autotroph available in pure culture, being able to grow at temperatures up to 85 to 90°C. Very few sulphur-oxidizing bacteria able to grow at neutral pH under thermophilic conditions have been described. A spore-forming sulphur-oxidizing thermophile, *Thiobacillus thermophilica*, was isolated by Egorova and Deryugina (1963). The *Thiobacillus* isolated by Williams and Hoare (1972) grew only up to 60°C. Almost certainly, more thermophilic neutral pH thiobacilli (or other sulphur-oxidizers) exist, as sulphur crystals deposited in springs at temperatures close to boiling often are colonized by large numbers of bacteria (Brock, unpublished observations), but cultures have not been isolated. Attempts to isolate the sulphur bacteria living at temperatures over 90°C in Boulder Spring (Brock *et al.*, 1971) were unsuccessful, although the ecological studies showed clearly that these bacteria were able to function at high temperatures. Specifically, CO_2 fixation by the Boulder Spring bacteria required the presence of sulphide, and a temperature optimum around 90°C was found. Further work on the Boulder Spring bacteria was abandoned when *Sulfolobus* was isolated, since this latter organism was considerably more amenable to ecological and cultural studies (Mosser *et al.*, 1973; Brock *et al.*, 1972; Shivvers and Brock, 1973; Mosser *et al.*, 1974a; Mosser *et al.*, 1974b; Bohlool and Brock, 1974; Brock and Mosser, 1975). Natural populations of *Sulfolobus* also oxidize ferrous iron, with a temperature optimum similar to that for the oxidation of elemental sulphur (Brock *et al.*, 1976).

C. Methanogenic Bacteria

Methanogenic thermophiles can be readily isolated, even from sewage sludge (Zeikus and Wolfe, 1972), and also exist as a component of hot spring algal mats (Zeikus, personal communication). These bacteria can grow completely chemolithotrophically, and should provide interesting material for evolutionary studies.

D. Nonphotosynthetic Bacteria

It is among the nonphotosynthetic bacteria that most work has been done, and these bacteria are also of potential industrial importance. Although it is well established that certain bacteria thrive even in boiling water (Bott and Brock, 1969; Brock *et al.*, 1971; Brock and Brock, 1971; Brock, 1967), cultures have been obtained only of organisms able to grow at somewhat lower temperatures. The highest temperature at which it has been possible to grow a bacterial culture continuously and reproducibly is about 85°C (Brock *et al.*, 1972; Heinen, 1971). The reason that the bacteria from boiling water have not been cultured may be trivial. It is quite difficult to arrange in the laboratory for aerated continuous flow systems at the boiling point without evaporation, and it is under analogous conditions that these organisms grow in nature.

Although much new work on enzymes of thermophilic heterotrophic bacteria has been published, little new ecological work has been done. The genus *Thermus*, described by Brock and Freeze (1969) as the first non-sporulating extreme thermophile, has been isolated by a number of workers, and several new species have been described (Table 2). Most hot spring waters probably have small amounts of organic matter, so that populations of bacteria can be maintained even in the absence of reduced sulphur compounds. Attempts to culture heterotrophic bacteria at temperatures of 90°C have so far failed, although in at least one spring (Pool A, in Yellowstone Park), the temperature optimum of the resident population is about 90°C (Brock and Brock, 1971).

At temperatures of 70°C or below, the blue-green algae which are able to grow and form mats excrete sufficient organic matter to support the growth of some heterotrophic bacteria (Bauld and Brock, 1974), and a diverse heterotrophic flora exists, although it has been little studied. Microscopic examination of blue-green algal mats reveals, in addition to the blue-green algae, primarily Gram negative rods similar to *Thermus aquaticus*, and this organism can be readily enriched from algal mats by incubation at 70°C (Brock and Freeze, 1969). Interestingly, the ubiquitous thermophile *Bacillus stearothermophilus* almost never appears in 70°C enrichments, although if an enrichment temperature of 55°C is used, it can be routinely isolated from the same mats (Brock, unpublished observations).

A variety of other thermophilic heterotrophs, both aerobic and anaerobic, are listed in Table 2. The involvement of thermophilic actinomycetes in self-heating of hay and compost is well established and is discussed in the section on fungi which follows.

III. Eukaryotes

A. Fungi

1. Introduction

Approximately 67 different species or varieties of true fungi grow at 50°C or above (Table 4). Of the approximately 50,000 different species of fungi

Table 4

Thermotolerant and thermophilic fungi and their cardinal temperatures[a]

	Cardinal temperature (°C)		
	Minimum	Optimum	Maximum
Zygomycetes			
Absidia corymbifera (Cohn) Sacc. and Trott.	14	40	50
Mortierella turficola (Hayes, 1969)	25	47	55
Mucor miehei Cooney and Emerson	24–25	35–45	55–57
M. pusillus Lindt	20–27	35–55	55–60
Mucor sp. I	25	45	56
Mucor sp. II	25	42–45	55
Rhizomucor sp.	25–30	45–53	60–61
Rhizopus arrhizus (Nilsson, 1973b)	>15	40–45	55
Rhizopus cohnii Berl. and de Toni	10	42	55
R. microsporus v. Tiegh.	12	40	50
Rhizopus sp. I	16	40–45	55
Rhizopus sp. II	16	40	50
Rhizopus sp. III	28	48–50	60
Rhizopus sp. A	25	45–52	60
(See Yamamoto, 1930a, b, for description and comparison of thermophilic species of *Rhizopus*)			

Table 4 (*contd.*)

	Cardinal temperature (°C)		
	Minimum	Optimum	Maximum
Ascomycetes			
Allescheria terrestris Apinis stat. conid. *Cephalosporium* sp.	22	42–45	55
Byssochlamys verrucosa Samson and Tansey stat. conid. *Paecilomyces* (Samson and Tansey, 1975)	15	40	55
Chaetomium britannicum Ames	[b]	[b]	[b]
C. thermophile La Touche	25–27	50	58–61
C. thermophile var. *coprophile* Cooney and Emerson	25–28	45–55	58–60
C. thermophile var. *dissitum* Cooney and Emerson	25–28	45–50	58–60
C. virginicum Ames	[b]	[b]	[b]
Chaetomium sp. A	14	40	50
Chaetomium sp. (Eicker, 1972)	>20	40–43	⩾50
Emericella nidulans stat. conid. *Aspergillus* (Evans, 1971b; Saëz, 1966)	10	35–37	51
Hansenula polymorpha (Levine and Cooney, 1973; Cooney *et al.*, 1975a, b)	—	37–42	50
Myriococcum albomyces Cooney and Emerson (= *Melanocarpus albomyces* (Cooney and Emerson) v. Arx (v. Arx, 1975))	25–26	37–45	55–57
Sphaerospora saccata Evans	12	35–42	50
Talaromyces byssochlamydoides Stolk and Samson stat. conid. *Paecilomyces* (Stolk and Samson, 1972; Samson, 1974)	*c.* 25	40–45	>50
T. emersonii Stolk stat. conid. *Penicillium*	25–30	40–45	55–60
T. leycettanus Evans and Stolk stat. conid. *Penicillium*	18	42	55
T. thermophilus Stolk stat. conid. *Penicillium*	25–30	45–50	57–60
Thermoascus aurantiacus Miehe	20–35	40–46	55–62
T. crustaceus (Apinis and Chesters) Stolk stat. conid. *Paecilomyces*	20	37	55
T. thermophilus (Sopp) v. Arx stat. conid. *Polypaecilum* (Arx, 1970)	>20	—	>50
Thielavia australiensis Tansey and Jack stat. conid. *Chrysosporium* (Tansey and Jack, 1975)	20	—	⩾50
T. thermophila Fergus and Sinden stat. conid. *Sporotrichum* (= *Chrysosporium fergusii* von Klopotek, 1974)	20	45	56
Thielavia sp. A	16	40	52

Table 4 (*contd.*)

	Cardinal temperature (°C)		
	Minimum	Optimum	Maximum
Basidiomycetes			
Coprinus sp.	*c.* 20	*c.* 45	55
Phanerochaete chrysosporium Burds. (Burdsall and Eslyn, 1974)[c]	⩽12	36–40	50
Deuteromycetes			
Acremonium alabamensis Morgan-Jones (Morgan-Jones, 1974)	⩽25	—	>50
Acrophialophora fusispora (Saksena) Samson	14	40	50
Aspergillus candidus (Christensen, 1957; Christensen and Kaufman, 1974)	10–15	45–50	50–55
A. fumigatus Fres.	12–20	37–43	52–55
Calcarisporium thermophile Evans (= *Calcarisporiella thermophila* (Fergus) de Hoog (Hoog, 1974))	16	40	50
Cephalosporium (Mills and Eggins, 1970; Brown *et al.*, 1974)	—	—	⩾50
Chrysosporium sp. A [See also Basidiomycetes]	25	40–45	55
Humicola grisea var. *thermoidea* Cooney and Emerson	20–24	38–46	55–56
H. insolens Cooney and Emerson	20–23	35–45	55
H. lanuginosa (Griff. and Maubl.) Bunce	28–30	45–55	60
H. stellata Bunce	<24	40	50
Malbranchea pulchella var. *sulfurea* (Miehe) Cooney and Emerson	25–30	45–46	53–57
Paecilomyces puntonii (Vuillemin) Nannfeldt (Eicker, 1972)	>30	—	>50
P. variotii Bainier (Samson, 1974)	*c.* 5	35–40	*c.* 50
Paecilomyces spp. Group b, 7 str.	<30	45–50	55–60
Paecilomyces spp. Group c, 6 str.	<30	45–50	55–60
Paecilomyces spp. Group d, 4 str.	<30	45–50	55–60
Penicillium argillaceum Stolk, Evans and Nilsson (Stolk *et al.*, 1969; Minoura *et al.*, 1973a; Evans, 1971b)	*c.* 15	*c.* 35	*c.* 50
Penicillium sp. A	20	42	55
Penicillium sp. (Eicker, 1972)	>20	*c.* 45	>50
Scolecobasidium sp. A (= *Diplorhinotrichum gallopavum* W. B. Cooke) (= *Dactylaria gallopava* (W. B. Cooke) Bhatt and Kendrick)	14–16	40	50–52
Sporotrichum thermophile Apinis (= *Chrysosporium thermophilum* (Apinis) von Klopotek, 1974)	18–24	40–50	55
Stilbella thermophila Fergus	*c.* 25	35–50	*c.* 55

Table 4 (*contd.*)

	Cardinal temperature (°C)		
	Minimum	Optimum	Maximum
Thermomyces ibadanensis Apinis and Eggins	31–35	42–47	60–61
Torula thermophila Cooney and Emerson	23	35–45	58
Torulopsis candida (Andrusenko and Lakhonina, 1969)	—	⩽36–37	⩾50
Tritirachium sp. A (= *Nodulisporium cylindroconium* de Hoog (Hoog, 1973))	16	40	55
Mycelia sterilia			
Burgoa–Papulaspora sp. (Tansey, 1973)	⩽20	—	53
Papulaspora thermophila Fergus	29–30	*c.* 45	52

[a] Identity presented as published; entries without dates are from Crisan, 1973. Temperatures indicated as a range of temperatures are derived from values reported by several investigators or were originally reported in this manner

[b] Not determined but other data indicate probable ability to grow at 50°C or above

[c] It is unclear how many valid taxa are encompassed by organisms reported variously as *Sporotrichum pulverulentum*, *S. pruinosum*, *Sporotrichum* sp., *Chrysosporium lignorum*, *C. pruinosum*, *Chrysosporium* sp., *Ptychogaster*, and most recently *Phanerochaete chrysosporium*. For discussion of the status of these taxa, see Hofsten and Hofsten (1974); Eriksson and Rzedowski (1969); Smith and Ofosu-Asiedu (1972); Shields (1969b); Bergman and Nilsson (1971); Semeniuk and Charmichael (1966); Klopotek (1974); Eriksson and Petersson (1975); and Burdsall and Esyln, (1974)

(Ainsworth, 1968), these taxonomically diverse few have in common many ecological characteristics associated with the ability to grow at high temperatures. We shall adopt Cooney and Emerson's (1964) definition of thermophilic fungi as those which have a maximum temperature for growth at or above 50°C and a minimum temperature for growth at or above 20°C, and thermotolerant fungi as those with maxima near 50°C but minima well below 20°C. Alternative terms for fungi which grow at high temperatures are discussed by Crisan (1959, 1964), Cooney and Emerson (1964), Emerson (1968), and Apinis (1963a). The number of known species of these fungi is expanding rapidly and can be expected to continue to do so as upper temperature limits for growth of known species are determined, and especially as new isolations are made by the increasingly large number of workers who are deliberately seeking them. Variations in cardinal temperatures for growth are common and are often dependent on the chemical environment of the fungus and its cultural history (Fergus, 1964; Mahoney, 1972; Christensen, 1957; Prodromou and Chapman, 1974; Tendler *et al.*, 1967; Tansey, unpublished; Streets and Ingle, 1972). Nevertheless, major changes in cardinal growth temperatures through alterations of the chemical environment have not yet been reported for thermotolerant or thermophilic fungi.

A major problem in interpreting published data concerning the ecology of thermotolerant and thermophilic fungi is the frequently incomplete or inaccurate identification of taxa. Ecologically important traits such as cardinal temperatures for growth and ability to degrade particular substrates differ significantly among species which are culturally and morphologically similar enough to be confused. Compounding this unfortunate situation is the chaotic state of taxonomy and nomenclature of many species, the great and often unrecognized variation among strains of a single taxon, and the rapidly increasing number of new taxa. Progress in the study of the ecology of these fungi would be aided if binomials were accompanied by the name of the author whose concept of the taxon is accepted, and if representative isolates were carefully chosen, numbered, and deposited in a culture collection.

2. *Habitats*

Much is known about occurrence of thermotolerant and thermophilic fungi in composts, piles of hay, stored grains, and wood chip piles, all of which may self-heat to spontaneous ignition. These fungi are believed to be significant contributors to self-heating in each case. In the relatively brief period of time since publication of Cooney and Emerson's (1964) monograph on thermophilic fungi, numerous publications have appeared which describe the occurrence of thermotolerant and thermophilic fungi in a wide variety of habitats. A stimulus for these studies has been the realization that hot habitats can have a fungal population which might be responsible in part for heating and biodeterioration, and which might include toxigenic or pathogenic species. If enrichments and isolations in these studies were done at an elevated temperature much less than 50°C, several thermotolerant and a few thermophilic species were usually detected; the nearer the temperature was held to 50°C, the greater was the proportion of thermophiles isolated. These studies have rarely proved that the fungi found were growing in the hot habitat in question, but if quantitative data concerning the number of colony forming units of each species were presented, and if the species identifications were accurate, it is sometimes possible to infer that growth has occurred and has been accompanied by an increase in temperature and biodeterioration.

It is hazardous to conclude that thermotolerant and thermophilic fungi are optimally adapted to growing at high temperatures in habitats outside of laboratory pure cultues. These fungi have usually been sought in hot habitats; these habitats have therefore provided the preponderance of data concerning occurrence and possible growth. Comparative data from unheated habitats, especially for thermophilic species, are few. Nevertheless, the data which follow tend to support the contention that fungi which can grow at high temperatures in pure culture, and especially the thermophilic fungi which can grow only at high temperatures, do grow at high temperatures in natural and man-made hot habitats, and that the ability to do so is a dominant characteristic of these species.

In order to demonstrate that habitats suitable for growth of fungi at high temperatures are common, widespread, and of long existence, we shall review the habitats in which thermophilic fungi occur; in most of the reports cited, thermotolerant fungi were also found. In our review we shall emphasize reports which have appeared since Cooney and Emerson's book (1964). Cooney and Emerson reviewed earlier reports and presented new data concerning occurrence of thermophilic fungi. Habitats included soil, guayule rets, hay and grass, garden compost, manure, leaf mould, mushroom compost, herbivore dung, cacao seeds and husks, straw, grain, nesting material of birds, warm-blooded animals, paper mill slimes, municipal compost, and dust from mouldy hay. In each of these cases, the substrate is heated or is derived from a heated substrate. Other sources were cardboard and cordage, wood, and unheated plant debris.

a. Composts. The description by Fergus (1964) of thermophilic and thermotolerant fungi and actinomycetes isolated from mushroom compost, coupled with Cooney and Emerson's monograph of the same year, opened a period of increased interest in the study of thermophilic fungi. These publications attracted the attention of diverse workers to the presence and possible activity of thermophilic fungi in composts and other heated organic matter, and provided information needed for their isolation and identification. Cooney and Emerson (1964) review the earlier literature on the involvement of thermotolerant and thermophilic fungi in composting, especially the important work by Waksman and his colleagues in confirming the microbial nature of the composting process.

The cultivated mushroom *Agaricus bisporus* is grown on compost; this compost has traditionally been made from a mixture of straw and horse manure, with corn cobs often added. Municipal garbage has also been developed as an ingredient of mushroom compost (Gerrits, 1974; Franz, 1972). Preparation of mushroom compost is usually divided into two stages: a first phase occurring outdoors in piles, and a second phase occurring indoors in the wooden trays, beds, etc., which will eventually be inoculated with the spawn of *A. bisporus*. Self-heating occurs in both phases, although steam heat is often used during the second phase to assist in controlling temperature. Temperatures of about 70 to 80°C occur during the first phase and are required for ultimately productive compost. Aerobic conditions must be maintained, usually by restricting the cross-sectional dimensions of the pile and by repeated turnings. Temperatures of 50 to 60°C are maintained during the second phase. Variations in preparation procedures occur, but the principle and purpose are the same: microbial conversion of a non-selective substrate into a selective substrate suitable for growth and fruiting of *A. bisporus*. The heat pasteurizes the compost, killing most insect and fungus pests. Simple carbohydrates are preferentially utilized leaving a large proportion of the more complex substances (e.g. cellulose, hemicellulose, and lignin) which then may serve as substances for growth of *A. bisporus*. The presence of soluble carbohydrates at the end of composting is undesirable, because they seem to encourage the appearance of competing fungi (Hayes and Randle, 1968).

Fergus (1964) isolated thermophilic fungi from mushroom compost collected during the second phase of heating. This initial report and subsequent studies (Fergus, 1971; Fergus and Sinden, 1969) showed that mushroom compost is a rich source of thermophilic fungi and yielded several interesting new species. Other descriptive reports of thermophilic fungi originating from mushroom compost include those by Evans, 1968; Craveri *et al.*, 1966; Cooney and Emerson, 1964; Wood, 1955; Löhr and Olsen, 1969; and Stolk and Samson, 1972.

Population levels of thermophilic microorganisms, including fungi, during composting and later, during the period of growth and fruiting of *A. bisporus*, have been studied by Hayes, 1969; Hayes and Randle, 1968; Fordyce, 1970; Cailleux, 1973; and Staněk, 1972. Hayes (1969) found a carefully ordered, changing population of microorganisms during composting. With traditional mixes of wheat straw and horse manure, the early increase and later decrease in temperature during the first phase of composting was paralleled by similar changes in numbers of thermophilic bacteria, but not of actinomycetes and fungi. During the second phase, actinomycetes were more numerous than other microorganisms. Adding sucrose to the compost mixture (30 lb per ton on the seventh day) decreased the numbers of thermophilic actinomycetes, increased those of thermophilic bacteria, and increased the number of colony-forming units the of thermotolerant fungus *Aspergillus fumigatus*. The greater production of mushrooms obtained on compost prepared with supplementary sucrose may have been due to conservation of the nutrients, including cellulose, which are of value for fruiting of *Agaricus bisporus* (Hayes and Randle, 1968), or may have been due to the increased growth of thermophilic bacteria (Hayes, 1969; Chanter and Spencer, 1974). These bacteria would be less likely to be cellulolytic than would be the fungi and actinomycetes having similar heat tolerances; in addition, their dead cells might provide growth factors usable by the growing mycelium of *A. bisporus*. During composting soluble nitrogen is converted into insoluble nitrogen, in part of the form of nitrogen-containing components of thermophilic microorganisms; the growing mycelium of *A. bisporus* then metabolizes this nitrogen source (Gerrits *et al.*, 1967).

Mushroom composting offers an opportunity for experimental manipulation of thermophilic microbial populations. This can result in better understanding of the role of thermophilic fungi in composting, including their interactions with one another and their effects on the mesophilic *A. bisporus* (Renard and Cailleux, 1973). Commercial improvement of the composting process also motivates this research. Alteration of chemical and physical properties of the substrate is the experimental approach most commonly taken (San Antonio, 1966; Hayes, 1969; Hayes and Randle, 1968; Gerrits, 1970, 1972). Compost mixtures have also been inoculated with selected populations of one or more species of thermophilic fungi or bacteria (Pope *et al.*, 1962; Townsley, 1974); the results suggest that advantages (more mushrooms, shorter composting time and prevention of invasion by undesirable species) may be gained by doing this.

Poincelot (1972, 1974) presents useful reviews of the principles and practices of composting in municipal or garden composts. They emphasize the contribution of thermophilic microorganisms, including fungi, to the heating and decomposition of composting materials. In the 1974 article, Poincelot emphasizes that the limiting factor in composting is the long time required when large amounts of cellulose are present (as is often the case). It is suggested that production of cellulase by some species may be repressed by interactions with others, and that deliberate alteration of the microbial population may eliminate this repression. The goal of composting is rapid production of an odour-free, heat-pasteurized material which has become biologically stabilized and will decay only slowly because the readily utilized nutrients have been significantly decreased (mineralized).

Municipal refuse contains plastics which do not decay during composting. Mills *et al.* (1971), Mills and Eggins (1974) and Eggins and Mills (1971) found that thermophilic fungi can grow on a wide range of plasticizers as their sole source of carbon. With the aim of developing a process for converting waste plastics to protein, chemical degradation followed by fermentation has been investigated (Brown *et al.*, 1974; Mills and Eggins, 1970); chemical oxidation of polyethylene yields substrates suitable for use as carbon sources for the growth of several species of thermophilic fungi. The conversion of cellulose wastes to microbial protein using thermophilic fungi has also been investigated (Eriksson and Larsson, 1975; Barnes *et al.*, 1972). Many species of thermophilic fungi are highly cellulolytic (Fergus, 1969, Almin *et al.*, 1975; Romanelli *et al.*, 1975; Tansey, 1971b; Malik and Eggins, 1972; Chapman *et al.*, 1975; Knösel, 197?; Stutzenberger *et al.*, 1970; Eriksson and Petersson, 1975), and some attack lignin (Eslyn *et al.*, 1975; Tansey *et al.*, 1977). Thermophilic fungi are also suitable sources of xylanase (Matsuo *et al.*, 1975).

Instructive studies by Chang (1967) and Chang and Hudson (1967) on thermophilic microorganisms in experimental wheat straw composts showed that before composting there was a high population of mesophilic fungi and a much lower one of thermophiles. Both were killed off in the central region of the compost during the maximum heating phase, after which the thermophilic population increased rapidly during the plateau period above 50°C, and persisted after the compost cooled. Studies of populations of mesophilic and thermophilic fungi, bacteria, and actinomycetes showed an initial phase of activity of mesophiles, producing heat to raise the temperature to about 40°C, at which they are suppressed and the thermophiles grow vigorously. Above 60°C, thermophilic fungi become inactive and further microbial heat production is due solely to bacteria and actinomycetes.

Species of fungi usually appeared in a particular succession in samples taken from the centre of the compost. The ability to use complex carbon sources and the ability to thrive at high temperatures were the two important characteristics of successful colonizers of composts. The order of succession of individual species was governed by their individual growth temperatures and nutritional capabilities. For example, the thermophile *Mucor pusillus* occurred only early in the compost period. It could use only simple carbon compounds, but neither cellulose nor hemicellulose (cf. Somkuti, 1974); its failure to reinvade after the period of temperatures which were high enough to kill the fungi was probably due to lack of available carbon sources. The non-cellulolytic thermophile *Humicola lanuginosa* did persist, probably because it is capable of growing commensally with other thermophilic fungi (Hedger and Hudson, 1974).

The ability or inability of thermophilic fungi to survive the high temperatures commonly encountered in composts is of interest, especially in mushroom compost where resumption of growth might be harmful. Fergus and Amelung (1971) found that thermophilic fungi grown on compost for 42 h (simulating certain mushroom compost procedures) were not able to survive subsequent exposure to higher temperatures in the range which would commonly be encountered in composts. Many of the species could not produce their sexual spores (which might be more resistant than asexual spores) in the brief period allowed. Determination of heat resistance of mature sexual spores would aid in interpretation of data concerning occurrence and activity of these fungi in many habitats, including sun-heated soils. Data concerning longevity of spores in fluctuating temperature regimes are needed, as are tests at different relative humidities (Celerin and Fergus, 1971). Additional data on thermal resistance are given by Fergus (1971). Exoenzymes of thermophilic fungi, including cellulases, are stable and active at temperatures above those which the vegetative phase and some spores can tolerate, and enzymatic conversions in composts may continue after fungal development has been terminated by high temperatures (Knösel and Rész, 1973).

Other reports of thermophilic fungi in composts in addition to those reviewed by Cooney and Emerson (1964) include Löhr and Olsen, 1969; Awao and Otsuka, 1974; Arima *et al.*, 1972; Bai and Rao, 1966; Evans, 1968; Hedger, 1975; Hashimoto *et al.*, 1972; Stolk, 1965; Hedger and Hudson, 1970; Stolk and Samson, 1972; Okazaki and Iizuka, 1970; Maheshwari, 1968; Eriksen, 1974; Klopotek, 1974; Kane and Mullins, 1973; Taha *et al.*, 1968; Malik and Sandhu, 1973; and Gray, 1970. In general, thermophilic fungi may be isolated from heated composts. Home compost piles are a convenient source of thermophile-rich inocula for selective enrichment of thermophilic fungi having particular characteristics of interest (Tansey, unpublished). Succession of species usually, but not always, occurs. High temperatures, acidity, and low levels of O_2 may limit growth of thermophilic fungi in composts. The threshold concentrations of oxygen that allowed measurable growth to occur in pure culture were determined by Deploey (1970; Deploey and Fergus, 1975) for 12 common species of thermophilic fungi. Several species produced measurable growth at O_2 concentrations as low as 0·2 and 0·3%; traces of growth occurred at 0·05% O_2 for four species (see also Kane and Mullins, 1973).

A discussion of the beneficial activity of thermophilic fungi in composts would be incomplete without mentioning retting of guayule, discussed in Cooney and Emerson (1964). The shrub, guayule (*Parthenium argentatum*) was a potential source of rubber for the United States during World War II. The latex is improved by retting (microbial decomposition of plant materials including resins), resulting in rubber with greater tensile strength. The retting process is analogous to composting, and several species of thermophilic fungi able to decompose resins were isolated from the rets. Perhaps additional processes can be developed in which specific microbial processes occurring during self-heating fermentations can be used to carry out chemical transformations using little modern technology or investment.

b. Stored hay. Interest in the role which microorganisms play in the spontaneous heating and ignition of stored hay prompted the first detailed account of thermophilic fungi (Miehe, 1907; Cooney and Emerson, 1964). Continued interest in the subject has mostly concerned determination of factors affecting the heating process, identification and counting of microbial species present at different stages of heating, and correlation of these species with incidents and antigens of human and animal disease.

During the first few days of storage, damp hay heats spontaneously. Gregory *et al.* (1963) found that baled hay with an initial moisture content of 40 to 60% reached maximum temperatures of 60 to 70°C; the amount of moisture chiefly determined the degree of heating and the species of microorganisms which occurred. Typically, the initial heating period is accompanied by a general increase in the numbers of microorganisms and a succession of increasingly thermophilic fungi. Microbial growth and heating eventually abate as the hay dries out, and the hay then slowly cools to ambient temperature. Stacked hay responds much the same as baled hay but may become somewhat hotter and remain hot for a long time. Extensive studies of mouldy hay and straw associated with the incidence of farmer's lung disease show that there is association of thermophilic and thermotolerant fungi and actinomycetes, and only hay and straw in this category contain antigens of farmer's lung disease (Lacey, 1974b).

Experimental studies of hay in Dewar flasks confirm that the supply of air and moisture determine the maximum temperature reached and the nature of the microbial populations which will occur; adequate aeration and water contents above 40% ensure heating and a large population of thermophiles (Festenstein *et al.*, 1965; Festenstein, 1966). In a detailed study and review of the effects of aeration on microbial heating of hay, Rész (1968, 197?) found that microbial heating of hay having a water content of more than 50% could not be prevented by reasonable amounts of forced aeration, but that aeration of hay of lower water content could prevent microbial heating. Heating of hay beyond 70°C is due to non-biological exothermic chemical reactions (Festenstein, 1971; Currie and Festenstein, 1971; Hussain, 1973, 197?).

c. Stored grain. Stored grain is subject to loss in quantity and quality, due in large part to growth of fungi in and on the grain. Metabolism of mesophilic fungi and insects in grain produces heat and moisture which allow growth of a succession of thermotolerant and thermophilic fungi and bacteria. Continued accumulation of metabolic heat raises the temperature to approximately 75°C; non-biological processes may further raise the temperature to ignition under certain conditions. Causes and prevention of this process have been critically discussed by Christensen and Kaufmann (1969, 1974). Because the growth of mesophilic and thermotolerant fungi can raise the temperature of stored grain to the point where it is badly damaged (c. 55°C), the neglect of thermophilic fungi until recently is understandable. Cooney and Emerson (1964) cited a single report of a thermophilic fungus from stored grain; since then a growing number of publications have appeared which describe the abundance of thermophilic fungi in self-heated stored grains. The toxigenic (Davis *et al.*, 1975; Donovan, 1971; Berestets'kiï *et al.*, 1974; Kurbatskaya, 1974; Austwick, 1974; Yamazaki *et al.*, 1975) and pathogenic potential of these species, whatever their role in economically significant heating and spoilage may be, should ensure their further study.

Thermophilic fungi can be isolated from standing parts of plants in the field (Apinis, 1963b, 1972; Apinis and Chesters, 1964) and from freshly harvested grain (Clarke *et al.*, 1966), but are virtually absent from the kernels of barley and wheat collected before harvest (Flannigan, 1974).

Thermophilic fungi are therefore regarded as storage fungi: although they are present on grain while still in the field, they do not seem to damage grain prior to storage. Just as for other storage fungi, the major source of inoculum of thermophiles is undoubtedly the air, dust, and debris of the storage area itself (Lacey, 1971a). As heating of the stored grain progresses, a distinct succession of microbial species occurs, culminating in large populations of thermotolerant fungi, especially *Absidia corymbifera*, *Aspergillus fumigatus*, and *A. candidus*, and the thermophiles *Humicola lanuginosa* and *Mucor pusillus* (Clark *et al.*, 1967, 1969; Lacey, 1971a). The view that thermophilic fungi are initially present as superficial contaminants of grain, and grow and invade the grain kernel only under heated storage conditions, is supported by additional descriptions of the species which occur on and in stored grain (Mulinge and Apinis, 1969; Mulinge and Chesters, 1970; Flannigan, 1969, 1970, 1972, 1974; Flannigan and Dickie, 1972; Flannigan and Sellars, 1972). Taking the broadest possible view of the subject, Sinha and co-workers (Sinha *et al.*, 1973; Sinha and Wallace, 1965, 1973) are studying the complex relationships of biotic and abiotic factors involved in spoilage of grain. By integrating a large body of data, including such variables as the population dynamics of mites, insects, and microflora, they hope to provide predictive models for grain spoilage, including self-heating. Monitoring conditions of individual grain storage units, including mycological examination to determine population levels of thermotolerant and thermophilic fungi, provides an adequate and feasible basis for management of grain storage units at present (Christensen and Kaufmann, 1969, 1974).

Additional reports of thermophilic fungi in stored grain include those of Okafor (1966) and Awao and Mitsugi (1973) concerning self-heated maize; Lacey (1972) concerning heated grain stored in underground pits similar to those excavated on Iron Age sites in England; Festenstein *et al.* (1965), who measured farmer's lung hay (actinomycete) antigen in self-heated barley and oats grain; and Mehrotra and Basu (1975), who found low counts of fungi capable of growth at 55°C. Taber and Pettit (1975) have isolated thermophilic fungi from peanut kernels (seeds).

d. Wood chip piles. Thermophilic fungi participate in self-heating and deterioration in wood chip piles. More than 100 million tons (fresh weight) of pulpwood chips are stored and used annually as a raw material for manufacture of paper and related products; of this, more than three million tons are lost, and it is predicted that this loss of fibre will increase with the increasing tendency to store pulpwood in chip form (Ofosu-Asiedu and Smith, 1973a; Smith, 1973) and the increasing ratio of sapwood to heartwood in available pulpwood (Cowling *et al.*, 1974). Additional financial losses are caused by colour and chemical changes which increase processing costs or reduce product quality, and spontaneous ignition of chip piles (Schmidt, 1969; Hajny *et al.*, 1967). The temperature within a newly built wood chip pile usually increases rapidly to about 60°C within the first few weeks of storage, levels off for a variable period of time, then slowly declines over a period of a few months; in some piles temperatures increase to spontaneous ignition (Hajny, 1966). Respiration of wood parenchyma cells, and direct chemical oxidations, probably release sufficient heat to cause new chip piles to attain the high temperatures usually observed (Feist *et al.*, 1971; Springer *et al.*, 1971). Microbial metabolic heat may also contribute significantly to both initial and subsequent heating of wood chip piles (Feist *et al.*, 1973a, b). As the temperature in the centre of the pile rises, a succession of micro-organisms occurs, culminating in a large population of thermophilic fungi; these species can cause significant loss of wood substances and discolouration of chips (Nilsson, 1965, 1973a, b; Bergman, 1974; Bergman and Nilsson, 1966, 1967, 1968, 1971; Tansey, 1971a, b; Lundström, 1972, 1973, 1974; Ofosu-Asiedu and Smith, 1973a, b; Shields, 1966, 1967, 1969a, b, 1970; Fergus, 1969; Smith, 1973; Smith and Ofosu-Asiedu, 1973; Greaves, 1971, 1975). The heat alone can cause serious loss in wood pulp yields and pulp properties (Feist *et al.*, 1973a, b). Effective chemical treatment of wood chips to prevent losses due to heating and microbial activity are apparently possible, but the economic feasibility of the various available treatments has not yet been confirmed under the prolonged, diverse, and large-scale conditions of actual commercial storage (Cowling *et al.*, 1974; Hulme and Shields, 1973; Springer *et al.*, 1973a, b). In the absence of chemical treatment, deterioration can be reduced by appropriate management (storage on a first-on, first-off basis) (Smith and Hatton, 1971; Assarson and Bergman, 1972), and prolonged storage outdoors must be avoided if major losses in pulp yield and quality are to be avoided (Hatton and Hunt, 1972).

e. Soil. Widespread occurrence of thermophilic fungi in soils has been established (Taber and Pettit, 1975; Awao and Mitsugi, 1973; Awao and Otsuga, 1973, 1974; Crisan, 1959; Craveri *et al.*, 1967; Hashimoto *et al.*, 1972; Evans, 1968, 1971a; Ward and Cowley, 1972; Mil'ko and Belyakova, 1967; Minoura *et al.*, 1973a, b; Larcade, 1967; Kurata *et al.*, 1966; Khasanov and Mirhodzhayev, 1969; Apinis and Chesters, 1964; Eggins and Malik, 1969; Apinis, 1963a, b, 1964, 1965, 1972; Andrusenko and Lakhonina, 1969; Zakharchenko and Zhernova, 1971; Malik and Eggins, 1972; Khasanov, 1968; Mouchacca, 1973). Reports of occurrence in tropical and subtropical soils include Tendler and Korman, 1965a, c; Udagawa *et al.*, 1973; Mahoney, 1972; and Huang, 1971. Tansey (unpublished) has also isolated thermophilic fungi from numerous samples of sun-heated mud collected in the Florida Everglades. They have also been reported from forest litter (Morgan-Jones,1974).

The effects of duration, temporal distribution, and degree of solar heating of soil on the ability of thermophilic fungi to grow and compete with mesophilic soil microorganisms have not been determined. Eggins *et al.* (1972) used immersion tubes to provide evidence that thermophilic fungi grow in sun-heated soils; passive entry of spores into the soil tubes (e.g. on soil mites or by capillary action) might occur, however. If spores were more common in sun-heated areas than in shaded sites (perhaps because exposed areas, being more open, might be more likely to be visited by herbivores which could deposit spore-containing faeces), this would give the incorrect impression that growth had occurred. In the summer, sun-heated bodies of water, mud, margins of ponds, and spring-watered road cuts in temperate latitudes are at temperatures above 40°C for fairly long periods of time (Mitchell, 1960, 1974; Dingfelder, 1962; Young and Zimmerman, 1956; Tansey, unpublished; Deacon and Minckley, 1974); thermophilic fungi do occur in these places (Tansey, unpublished). Direct proof that insulation provides sufficient heat for growth and competition of thermophilic fungi in soil is, however, still lacking, as are studies describing the activities of these fungi in soil under non-permissive temperatures for growth. Seasonality of occurrence and abundance of particular species of thermophilic fungi in soil is known, as well as some correlation with soil type (Apinis, 1963a, b, 1972; Evans, 1971c; Tansey, unpublished). The concentration of colony-forming units in soil is usually low.

Desert soils and rocks provide fascinating habitats for a variety of microorganisms, including fungi (Friedmann and Galun, 1974), but published studies of thermophilic fungi in desert soils are inconclusive. Tansey (unpublished) found low concentrations of thermophilic fungi to be present in sun-baked, dry desert soils of Arizona and New Mexico.

Typically, thermophilic fungi occupy habitats which contain great diversity of potential competitors and antagonists. Since most habitats in which they can grow develop high temperatures for only transient periods of time, much of the existence of a thermophilic fungus is spent in co-existence with organisms which can outgrow it under mesophilic conditions. This is especially true in soil. Among the possible competitive aids might be the ability to produce antibiotics; many thermophilic fungi produce antibiotics (Craveri *et al.*, 1972; Aragozzini *et al.*, 1970; Kitano *et al.*, 1975a, b; Kleupfel *et al.*, 1972; Rode *et al.*, 1947; Walter, 1969; Bai and Rao, 1966; Cavazzoni and Vivani, 1973; Somkuti and Walter, 1970; Tendler and Korman 1965a, b, c; Tendler *et al.*, 1967; Okazaki and Iizuka, 1971; Thakre and Johri, 1973, 1973–1974). Considering the insolation-mediated growth of a thermophilic fungus in soil, mud, etc., the following scenario is proposed: periods of rapid vegetative growth at permissive temperatures are followed by many hours or days at lower temperatures which allow little if any growth (consider for example the most common thermophilic fungus, *Humicola* (= *Thermomyces*) *lanuginosa*, which has a minimum temperature for growth of 28 to 30°C). During these periods secondary metabolism might be stimulated, generating quantities of antibiotics (Weinberg, 1974). The possibility that this is an accurate picture of life of a thermophilic fungus in sun-heated soil seems great enough to warrant screening of the ability of thermophilic fungi to produce antibiotics at temperatures well below their optimum growth temperatures and under conditions of fluctuating temperatures.

Once spores of thermophilic fungi form and mature, which may require temperatures considerably above the minimum temperature for growth of the species (Tansey, 1972; Chapman, E. S., 1974), they are not especially sensitive to temperatures below those necessary for growth, and often remain viable for periods of several years at mesophilic temperatures.

f. Miscellaneous habitats. Thermophilic fungi have been isolated from many habitats in the past few years. Most of these are heated habitats or have a past history of being heated. These include stored sun-heated yams (Coursey and Nwankwo, 1968); fementing cacao beans (Broadbent and Oyeniran, 1968; Hansen and Welty, 1970; Riley, 1965; Maravalhas, 1966); heated stacks of oil palm kernels (Apinis and Eggins, 1966; Eggins and Coursey, 1964, 1968; Oso, 1974a, b, c; Coursey, 1965; Hartley, 1967); bagasse (the residue of sugar cane after extraction of the juice) (Seabury *et al.*, 1968; Ramabadran 1967–1968 and personal communication; Lacey, 1967, 1968, 1974a; Stolk and Samson, 1972); green and burnt sugar cane (Bevan and Bond, 1971); tobacco and tobacco products (Tansey, 1975; Pounds and Lucas, 1972; Fletcher *et al.*, 1967); piles of seaweed (Nonomura, personal communication; ZoBell, 1959); peat (Nikitina *et al.*, 1970; Isachenko and Mal'chevskaya, 1936; Mal'chevskaya, 1939; the latter two from Loginova *et al.*, 1966); plant tissue or soil from vineyards and processing plants (King *et al.*, 1969); and timber (Morton and Eggins, 1975). Hot springs and geothermal soils, especially those which are acidic, and associated baths and wells contain thermophilic fungi (Tansey and Brock, 1971, 1972, 1973; Loginova *et al.*, 1962; Geitler, 1963; Hedger, 1974). *In situ* studies have shown that fungi grow and reproduce in acidic hot springs and geothermal soils (Belly *et al.*, 1973; Tansey and Brock, unpublished).

Birds' nests are a rich source of thermophilic fungi (Apinis and Pugh, 1967; Cooney and Emerson, 1964; Hubálek *et al.*, 1973); such fungi have also been isolated from birds' feathers (Minoura *et al.*, 1973a, b). In most nests the nesting bird is assumed to be the source of heat responsible for occurrence of thermophilic fungi, although the temperatures attained would be considerably below the optima for growth of most species. In contrast, the nests of incubator birds (Frith, 1962) are, in effect, self-heating compost piles, and thermophilic fungi are abundant in these nests (Tansey and Jack, 1975; Samson and Tansey, 1975; Tansey, unpublished). Nests of the American alligator are also a rich source of thermophilic fungi (Tansey, 1973 and unpublished); the relative amount and temporal distribution of microbial and solar heating of alligator nests is controversial.

Thermophilic fungi are commonly isolated from herbivore dung and manure (Cooney and Emerson, 1964; Donovan, 1971; Ramabadran, 1967–1968; Sumner *et al.*, 1969; Evans, 1968; Minoura *et al.*, 1973a; Löhr and Olsen, 1969; Tansey, unpublished; Seal and Eggins, 1972). Often, it is not possible to determine whether the material examined came from a pile of manure from cleaning barns and stables, or from individual deposits of faeces. This distinction is important, however, because alternative interpretations of both the origin of the fungi and the occurrence of elevated temperatures are possible: self-heated hay or feed grain, fermented mixtures of bedding and faeces, or sun-heated deposits of faeces. Moist, sun-heated piles of herbivore dung can maintain temperatures suitable for growth of thermophilic fungi for surprisingly long periods of time (Anderson and Coe, 1974; Tansey, unpublished) and might be a common and widespread habitat for growth of these fungi.

Thermophilic fungi of self-heated coal waste piles have been intensively studied in England (Evans, 1971a, b, c; Evans and Stolk, 1971; Sumner *et al.*, 1969; Sumner and Evans, 1971), and we have isolated them from acidic self-heated coal waste piles in the United States (Tansey and Brock, 1972, 1973, and unpublished). These piles are widespread (Stahl, 1964; Myers *et al.*, 1966; Campbell, 1972); the activities of thermophilic fungi deserve further study.

Some thermophilic fungi are pathogens of humans and other warm-blooded animals (Ainsworth and Austwick, 1959; Pore and Larsh, 1967; Hughes and Crosier, 1973; Austwick, 1972; Lacey, 1975b; Scholer, 1974; Stretton, 1975; Commonwealth Mycological Institute, 1975). The thermophile *Mucor pusillus* in particular is a pathogen, causing a variety of mycoses (Meyer and Armstrong, 1973; Meyer *et al.*, 1973; Cooney and Emerson, 1964). Thermotolerant fungi such as *Absidia corymbifera* (Nottebrock *et al.*, 1974; Meyer and Armstrong, 1973) and *Aspergillus fumigatus* (Haller and Suter, 1974; Austwick, 1963; Rippon, 1974; Jungerman and Schwartzman, 1972) are more frequently reported as pathogens than are thermophiles. The thermotolerant fungus *Dactylaria* (= *Diplorhinotrichum* = *Scolecobasidium*) *gallopava* has recently attracted attention as a cause of epidemics in young turkeys and chickens (Tansey and Brock, 1973; Blalock *et al.*, 1973; Waldrip *et al.*, 1974; Ranck *et al.*, 1974; Georg *et al.*, 1964). Although several illnesses are associated with the handling of thermophile-rich materials (e.g., bagassosis, mushroom worker's lung, farmer's lung), thermophilic actinomycetes and not fungi are usually credited with causing these illnesses (Lacey, 1971b; Edwards, 1972; Gregory and Lacey, 1963a, b; Pirie *et al.*, 1971; Gray *et al.*, 1969; Wiseman *et al.*, 1973; Gregory *et al.*, 1964; Chan-Yeung *et al.*, 1972), although these actinomycetes are loosely referred to as fungi in some publications.

Few studies have been made of the occurrence of thermophilic fungi in air other than in association with heated habitats. Hudson (1973) found a low concentration of colony-forming units in air sampled at Cambridge, England, and Lacey (1975) found a low concentration in air in rural England; it may be inferred that Hughes and Crosier (1973) also found a low concentration in air. In contrast, the concentration of thermophilic fungi in air associated with potentially heated habitats is quite high: in farm buildings where hay is being shaken (Lacey and Lacey, 1964); in hay sheds, silos, and at a mushroom farm (Lacey *et al.*, 1972); in a room in which mushroom compost trays were dumped (Fergus, 1964); associated with moist stored grain (Lacey, 1973; Lacey, 1971a) and mouldy hay and straw (Lacey, 1974b). Evans (1972) recovered many species from air sampled at a distance of 4 km from the heated coal waste piles in which he found an abundance of many of the same species. Other isolations are occasionally reported from air (Stolk *et al.*, 1969; Hedger, 1975).

The occurrence of thermophilic fungi in aquatic sediments has been reported and is attributed to sedimentation from surrounding terrestrial habitats (Tubaki *et al.*, 1974).

g. Habitats deserving investigation. Many habitats in which thermophilic fungi are likely to occur and grow have not yet been investigated or have received only cursory study. In view of the capacity of these fungi for causing biodeterioration and illness, their possible presence and activity in a wider variety of habitats deserves study. The differing ability of different species to produce heat (Nikitina *et al.*, 1970) emphasizes the practical significance of accurate determination of the species composition and factors required for growth of each species in habitats where heating occurs. Some habitats, especially aquatic ones, may prove on examination to be more suitable for experimental manipulations (especially for *in situ* growth studies) than those already known. New species are needed for industrial exploitation. There are several inherent advantages (Bellamy, 1974, 1975; Sorensen and Crisan, 1974; Cooney and Wise, 1975) in using thermophilic microorganisms for industrial fermentations, and the thermotolerant Basidiomycete, *Phanerochaete chrysosporium*, appears to be a likely candidate for production of microbial protein from a variety of polysaccharides (Hofsten and Rydén, 1975), as does the thermotolerant yeast *Hansenula polymorpha* grown on hydrocarbons (Cooney and Levine, 1975; Cooney *et al.*, 1975a, b). Investigation of natural habitats which have environmental conditions similar to those in which a fungus or its products (especially exoenzymes) will be used (e.g. highly acidic or basic solutions, or with high concentrations of particular substrates or other substances) might yield new species of useful thermophilic fungi which are optimally adapted to function under these conditions.

Habitats especially deserving exploration include sun-heated trees (Derby and Gates, 1966); tree stumps, logs, and slash (Schmidt and Wood, 1969; Spaulding, 1929; Gooding *et al.*, 1966); leaf surfaces and other parts of plants (Björkman *et al.*, 1972; Gates, 1973); and slime fluxes (which might well yield the long-sought yeasts capable of rapid growth and development at 50°C or above; Loginova 1960; Loginova *et al.*, 1966, 1973). Other habitats of interest are bat guano in caves (Harris, 1970); electrical equipment, especially in humid climates (Rychtera, 1971); underground high voltage power cables (their organic components and surrounding soil); influorescences of certain Araceae (Nagy *et al.*, 1972; Knutson, 1974; Meeuse, 1975); heating bales of wool (Walker, 1963; Walker and Williamson, 1957; Rothbaum, 1961; Dye, 1964; Dye and Rothbaum, 1964; Rothbaum and Dye, 1964); bodies of certain insects (Heinrich, 1974); molasses pulp pellets in bulk storage (Jorgensen and Gaddie, 1969); sun-heated lakes (Por, 1969; Anderson, 1958; Hudec and Sonnenfeld, 1974); sawdust piles; saunas (Salminen *et al.*, 1974); Red Sea deep hot brines (Degens and Ross, 1969; Ross *et al.*, 1973; Ross, 1972; Brewer *et al.*, 1971; Hunt *et al.*, 1967); kinetically heated fuel systems of supersonic aircraft (Scott and Hill, 1971; Hill, 1971; Hill *et al.*, 1975; Sheridan and Soteros, 1974); timber in overheated mines (Ioachimescu, 1973); fermented rice (*arroz requemado*) of the Ecuadorian Andes (van Veen and Steinkraus, 1970); soil in the immediate vicinity of the Trans-Alaska Pipeline (McCown, 1973); garbage dumps (Böttcher *et al.*, 1973); piles of flood debris (Arnold, 1962); heating and humidification systems of homes and buildings (Seabury *et al.*, 1973); and hot water taps (Brock and Boylen, 1973; Pask-Hughes and Williams, 1975).

Some power plant cooling pipes and effluents are warm enough to be suitable habitats for growth of thermophilic fungi, and such fungi have been isolated from steam line discharge sites (Tansey and Brock, 1972). Some cooling towers should be excellent places for growth of thermophilic fungi. Thermophilic bacteria and mesophilic fungi and their roles in biodeterioration and fouling of cooling towers have been studied (Eaton, 1072; Eaton and Jones, 1971; Sládečková, 1969); the thermophilic fungal flora of cooling towers is almost (Sumner *et al.*, 1969; Burdsall and Eslyn, 1974) unknown. Effluents from geothermal power plants may also provide a suitable and chemically varied habitat (Axtmann, 1975). Dark, sun-heated oil-soaked soils (which occur in oil fields and in other areas of spillage or natural leakage), asphalt spillage and wastes (Turner and Ahearn, 1970), and anthracite waste piles (Schramm, 1966) are further examples of materials which might serve as natural enrichments for thermophilic fungi having nutritional and enzymatic characteristics desired for commercial exploitation. Sun-heated soils could also provide a natural setting for testing of ecological models of populations in seasonal or fluctuating environments (Fretwell, 1972; Mountford, 1971). Copper ore leaching piles (which heat to 80°C; Beck, 1967) might seem an unlikely place to search for thermophilic fungi (in view of the common toxicity of copper for fungi), but these piles do contain thermophilic procaryotes (Brierly and Murr, 1973); mesophilic fungi which tolerate extremely high copper concentrations (and low pH values and high concentrations of salt) are known (Gould *et al.*, 1974).

Hughes and Crosier (1973) demonstrated that thermophilic fungi are at least transients in the mycoflora of humans; further studies with improved sampling and isolation techniques may demonstrate that thermophilic fungi are more prevalent as human parasites than currently known.

B. Algae

For fungi the temperatures and species chosen in delimitation of different temperature groups are arbitrary because there is a continuum of cardinal temperatures, but among eukaryotic algae, one species stands out from all others in being able to grow at extreme temperatures. *Cyanidium caldarium* has an upper limit of 55 to 60°C, based upon field observations of the natural habitats, $^{14}CO_2$ incorporation in nature, and growth in pure culture (Doemel and Brock, 1970). There is a large gap in upper limits between *C. caldarium* and other eukaryotic algae. High temperature strains of *Chlorella* which grow at about 42°C are known (Sorokin, 1967). Diatoms occur in many hot springs and are often the dominant eukaryotic alga at 30 to 40°C. Although diatoms have been observed in hot springs at nearly all temperatures (Mann and Schlichting, 1967; Stockner, 1967, 1968a, b; Kullberg, 1968, 1971; Sprenger, 1930), growth and viability data from culture studies are required to establish accurate temperature limits for each species, due to difficulties in measuring and interpreting habitat temperatures in hot springs. The maximum temperature for growth for any diatom in culture (unialgal, not axenic) is between 43 and 44°C; for the hot spring diatom *Achanthes exigua*, the optimum is 40°C. Cultures can survive at 45°C for at least a week, although most cells die (Fairchild and Sheridan, 1974). Members of other groups of eukaryotic algae occur in hot springs; a species of the green alga *Mougeotia*, for example, has been found along hot spring margins at 47°C. This is a possible measure of the tolerance of the organism, however, not of its ability to sustain growth at that temperature (Stockner, 1967).

Thus, although thermophilic eukaryotic algae do exist, *Cyanidium caldarium* is unique. From a human point of view, *C. caldarium* attracts special attention as a thermophile: it is the only eukaryotic alga which grows in habitats which feel painfully hot. Low pH and high temperature favour its growth and exclude other photosynthetic organisms; it is the only photosynthetic organism in habitats with pH less than 5 and temperatures greater than 40°C (Doemel and Brock, 1971), and it can grow at pH values as low as 0 (Allen, 1959). It exists in two distinct kinds of habitats; aquatic and terrestrial. Aquatic habitats include hot spring pools and drainways; terrestrial habitats include steam-drenched soil and rock and warm dry soil in solfatara areas. Acid hot spring drainways are often covered with mats of *C. caldarium*, the bacterium *Bacillus coagulans*, and the fungus *Dactylaria gallopava*; the bacterial and fungal components apparently obtain nutrients from *C. caldarium* (Belly *et al.*, 1973b). In soils, *C. caldarium* often occurs as a thin layer at 3 to 5 mm below the surface, probably because of the susceptibility of the alga to water stress (Smith and Brock, 1973). Surface soil is too dry for growth; deeper soil is presumably too dark.

The taxonomic position of *Cyanidium caldarium* has long been a matter for debate; it has recently been assigned to the red algae (Chapman, D. J., 1974).

C. Lichens

Exposed lichens experience as high a temperature as any vegetative phase of an organism outside of those in hot springs, and to the extent that the algal and fungal components of a lichen may be considered microorganisms, it is appropriate to ask whether lichens or the separated mycobionts or phycobionts are thermophilic. From the evidence available for intact thalli, and the more limited data for the separated symbionts, it appears that many lichens are thermoduric; the desiccated thalli are able to survive periods of elevated temperature but do not grow at these elevated temperatures. A few lichens have been reported from geothermal habitats, however, and the possibility of growth at relatively high temperatures cannot be excluded (Kappen, 1973).

D. Protozoa

Reports of protozoa occurring at high temperatures in hot springs must be interpreted with caution unless accompanied by culture studies or detailed descriptions of *in situ* temperatures and the methods used for their determination. Thus, the reported occurrence of the ciliates *Cothuria* at 63°C (Pax,

1951), *Chilodon* at 68°C (Dombrowski, 1961), and *Oxytricha fallax* at 56°C (Uyemura, 1936, 1937); amoebae at 50 to 52°C (Issel, 1910), and 51°C (Uyemura, 1936, 1937); and protozoa at 53 to 64·7°C (Dogiel, 1965) cannot be taken as proof that these protozoa are capable of growth and completion of their life cycles at these temperatures.

Kahan (1969) found four species of protozoa in hot springs in quite abundant numbers at 57 to 58°C. In culture, protozoa were able to reproduce at 56°C (*Cercosulcifer hamathensis*) and 55°C (*Vahlkampfia reichi*), and to survive for half an hour at 60°C (*V. reichi*) and 58°C (*C. hamathensis*); over a period of several weeks, *C. hamathensis* was experimentally adapted to live at temperatures of 58 to 59°C. The thermophilic protozoa isolated from saline (3% salinity) hot springs could reproduce at a wide range of salinity; e.g. a flagellate reproduced over a salinity range of 0·4 to 11% in culture. Kahan (1972) collected *Cyclidium citrullus* from hot springs at 50 to 58°C. In monoxenic culture the highest temperature at which some of the cells would multiply was 47°C; cells survived for a few days at 49°C.

Hindle (1932) cultured a hot spring amoeba in abundance at 53 to 54°C for a year. Phelps (1961) found that a hot spring strain of *Tetrahymena pyriformis* could be acclimated to grow at 41·2°C in culture.

There is some correlation of upper temperature limits and the ability to parasitize warm-blooded animals. Griffin (1972) found that virulent strains of the freshwater amoeboflagellate *Naegleria fowleri*, could grow at 45 or 46°C, whereas non-virulent strains of *Naegleria gruberi*, had upper limits which were considerably lower and could not grow at normal or elevated body temperatures. His results suggest that combined coliform and thermal pollution would stimulate the growth of *N. fowleri* in freshwater habitats. Thermal pollution occasionally involves temperatures as high as those discussed here (Gibbons and Sharitz, 1974), and subjects such as the effects of elevated temperatures on oxygen concentration, community function, species diversity, etc., are common areas of concern. Cairns and Lanza (1972) review this subject with emphasis on algae and protozoa, and Coutant and Talmage (1975) provide a detailed review of thermal effects on microorganisms and other aquatic organisms.

Protozoa commonly occur in sun-heated habitats where they may be exposed to elevated temperatures. Dingfelder (1962) found that several species of ciliates which occur in sun-heated shallow puddles remain active at up to 52°C, although he did not claim that growth could occur at these temperatures.

The most remarkable report of the ability of eukaryotes to live and grow at high temperatures remains Dallinger's account (Dallinger, 1887; see also Hindle, 1932) of experimentally increasing the growth temperature of three species of flagellates to 70°C by slowly increasing their incubation temperature during the course of almost seven years. The thermophilic strains produced died when returned to a permissive temperature (15·5°C) for the wild types of the species. Dallinger was a highly respected and honoured researcher when he reported these data. He reported his results in detail, and had the equipment and skills necessary for careful microscopic observation and illustration of his cultures. There is no doubt that protozoa can be experimentally adapted to high and low temperatures (Poljansky and Sukhanova, 1967; Irlina, 1967; Sukhanova, 1967), but the extremely high temperatures claimed by Dallinger have been questioned. An attempt should be made to repeat Dallinger's experiments; success would benefit several areas of biology, and would surely gain the researcher admiration for outstanding patience and determination.

Addendum

A newly described thermophilic fungus is *Thielavia heterothallica* von Klopotek, the sexual stage of *Chrysosporium thermophilum* (Apinis) von Klopotek (von Klopotek, A., 1976, *Arch. Microbiol.* **107,** 223–244). The new taxon *Chaetomium cellulolyticum* Chahal et D. Hawksw., has been called thermotolerant by its authors (Chahal, D. S. and Hawksworth, D. L., 1976, *Mycologia* **68,** 600–610), but does not grow at 50°C. *Chaetomium gracile* Udagawa has been shown to be thermotolerant (Millner, P. D., 1975, *Biologia* **21,** 39–73); further studies of the temperature relationships of this fungus will be reported (Tansey, M. R., submitted for publication). Extensive new temperature and pH data for thermophilic and thermotolerant fungi have been published (Rosenberg, S. L., 1975, *Can. J. Microbiol.* **21,** 1535–1540); neither *Chrysosporium pruinosum* (Gilman and Abbott) Carmichael ATCC 24782 nor *Sporotrichum pulverulentum* Novobranova ATCC 24725 (syn. *Chrysosporium pruinosum* and *C. lignorum*) grew at 50°C. A report that *Humicola* (*Thermomyces*) *lanuginosa* grows at temperatures above 62°C has been published (Ohtomo, T., Sugiyama, J. and Iizuka, H., 1975, *Trans. Mycol. Soc. Japan* **16,** 289–300). It is apparent from detailed examination of conidial development that *Torula thermophila* must eventually be transferred to a different genus (Ellis, D. H. and Griffiths, D. A., 1976, *Can. J. Microbiol.* **22,** 1102–1112).

Particularly interesting reports of isolations of thermophilic and thermotolerant fungi include *Phanerochaete chrysosporium* (as *Peniophora mollis*) from redwood in cooling towers (Duncan, C. G. and Lombard, F. F., 1965, *For. Serv. Res. Pap., U.S.D.A.*, WO-4); *Aspergillus fumigatus* from aircraft fuel (Scott, J. A. and Forsyth, T. J., 1976, *Int. Biodeterior. Bull.* **12,** 1–4); *Humicola* (as *Thermomyces*) *lanuginosa* from a geyser (Ontomo, T., Sugiyama, J. and Iizuka, H., 1975, *Trans. Mycol. Soc. Japan* **16,** 289–300); *Humicola* (as *Thermomyces*) *stellata* isolated from a human (American Type Culture Collection, 1974, "Catalog of Strains" (11th ed.) A.T.C.C., Rockville); several species from Bahamian soils (Gochenaur, S. E., 1975, *Mycopathologia* **57,** 155–164); *Aspergillus fumigatus* and *Humicola insolens* from Taiwan hot springs (Volz, P. A., 1976, *Phytologia* **33,** 154–163); a thermotolerant member of the *Rhizopus pseudochinensis* Yamazaki-*R. chinensis* Saito complex from sun-heated soil where it formed macroscopic colonies (Tansey, M. R. and Jack, M. A., 1977, *Mycologia,* **69,** 563–578); several species from soil (Huang, L. H. and Schmitt, J. A., 1975, *Mycotaxon* **3,** 55–80); and several species from Nigerian palm kernels (Kuku, F. O. and Adeniji, M. O., 1976, *Int. Biodeterior. Bull.* **12,** 37–41). Fourteen isolates of fungi which grew at 55°C, and several species of thermophilic bacteria, have been isolated from the slime which forms on paper manufacturing equipment which operates at elevated temperatures (45 to 60°C) (Loginova, L. G., Sergeeva, V. V. and Seregina, L. M., 1973, *Priklad. Biokhim. Mikrobiol.* **9,** 701–709). The ecology of fungi, including thermotolerant species, during long-term storage of wheat has been described (Wallace, H. A. H., Sinha, R. N. and Mills, J. T., 1976, *Can. J. Bot.* **54,** 1332–1343). What is apparently the first part of a multifaceted study of thermophilic fungi (Jodice, R., Ferrara, R., Scurti, J. C., Fiussello, N., Obert, F. and Cortellezzi, G. C., 1974–1975, *Allionia* **20,** 53–74) includes reports of isolation of thermophilic and thermotolerant fungi from poultry manure, pig sludge, fermented bark piles, and mushroom compost. Optimum temperatures for growth of the various taxa were measured, as was antibacterial activity, mouse toxicity, and ability to degrade cellulose and lignin. Heated effluents of nuclear production reactors have proven to be one of the richest sources of thermophilic and thermotolerant fungi discovered to date, and include the encephalitis causing species *Dactylaria gallopava* in great abundance (Tansey, M. R., unpublished). A study of thermophilic and thermotolerant fungi in sun-heated temperate soils (Tansey, M. R. and Jack, M. A., 1976, *Mycologia* **68,** 1061–1075; Jack, M. A. and Tansey, M. R., 1977, *Mycologia* **69,** 109–117; Tansey, M. R. and Jack, M. A., *Mycologia* **69,** 563–578) has quantitatively described populations of these fungi at regular intervals over a 15 month period; compared fluctuations in population levels to changes in various environmental parameters; measured growth, sporulation, and germination of spores incubated in sun-heated soil; and measured growth of these fungi in soil during competition with mesophiles under fluctuating temperature conditions. Cooling tower fungi have been reviewed (Eaton, R. A., 1976, *In* "Recent Advances in Aquatic Mycology" (Ed. E. B. G. Jones), pp. 359–387. Wiley, New York), but detailed examination of this potential habitat for thermophilic fungi apparently remains to be done.

The fungus *Phanerochaete chrysosporium* is being intensively studied for use in food production (Hofsten, B. v. and Rydén, A.-L., 1975, *Biotechnol. Bioeng.* **17,** 1183–1197; Hofsten, B. v., 1976, *In* "Food from Waste" (Eds G. G. Birch, K. J. Parker and J. T. Worgan), pp. 156–166. Applied Science Publishers Ltd, London). It has been shown that decomposition of lignin by *P. chrysosporium* requires a growth substance such as cellulose or glucose (Kirk, T. K., Connors, W. J. and Zeikus, J. G., 1976, *Appl. Environ. Microbiol.* **32,** 192–194). Swedish workers have developed a cellulase-less mutant strain of *Sporotrichum pulverulentum* (syn. *P. chrysosporium*) for possible use in microbiological pulping of wood (Ander, P. and Eriksson, K.-E., 1975, *Sven. Papperstidn.* **78,** 643–652), and have continued their studies on lignin degradation by this species (Ander, P. and Eriksson, K.-E., 1976, *Arch. Microbiol.* **109,** 1–8). Some of the isolates which are assigned to *P. chrysosporium* (*S. pulverulentum*) are thermotolerant, but others are decidedly not; it is possible that there is more than one species masquerading under this name.

Interest in enzymes of thermophilic fungi is increasing and has resulted in publication of studies of thermomycolase, the extracellular serine protease of *Malbranchea pulchella* var. *sulfurea* (Ong, P. S. and Gaucher, G. M., 1976, *Can. J. Microbiol.* **22,** 165–176; Stevenson, K. J. and Gaucher, G. M., 1975, *Biochem. J.* **151,** 527–542; Voordouw, G. and Roch, R. S., 1975, *Biochemistry* **14,** 4659–4666); protease of *Mucor miehei* (McBride-Warren, P. A. and Rickert, W. S., 1976, *Can. J. Biochem.* **54,** 382–388; Rickert, W. S. and McBride-Warren, P. A., 1976, *Can. J. Biochem.* **54,** 120–129); cellulase, amylase, and pectinase of *Talaromyces thermophilus* (Tong, C. C. and Cole, A. L. J., 1975, *Mauri Ora* **3,** 37–43); lipase of *Mucor miehei* (Peppler, H. J., Dooley, J. G. and Huang, H. T., 1976, *J. Dairy Sci.* **59,** 859–862; Huang, H. T. and Dooley, J. G., 1976, *Biotechnol. Bioeng.* **18,** 909–919); and α-galactosidase from "*Penicillium duponti*" (Arnaud, N., Bush, D. A. and Horisberger, M., 1976, *Biotechnol. Bioeng.* **18,** 581–585).

Additional studies of the nutrition of thermophilic fungi have been published (Deploey, J. J., 1976, *Mycologia* **68,** 190–194; Sahm, D. F. and Chapman, E. S., 1976, *Mycologia* **68,** 168–174). Fatty acids of the thermotolerant fungus *Sporotrichum thermophile* have been described (Dart, R. K., 1976, *Trans. Br. Mycol. Soc.* **66,** 532–533; Dart, R. K. and Stretton, R. J., 1976, *Trans. Br. Mycol. Soc.* **66,** 529–532). The effects of environmental factors and nutrients on percent germination of conidia of *Talaromyces thermophilus* have been described (Jahnke, S. E. and Chapman, E. S., 1975, *Mycologia* **67,** 1223–1228).

Newly described thermophilic bacteria include a hydrocarbon-utilizing obligate thermophile, *Thermomicrobium fosteri* (Phillips, W. E., Jr. and Perry, J. J., 1976, *Int. J. Systemat. Bacteriol.* **26,** 220–225); an unnamed ("K2") obligate thermophile which might be assigned to *Thermus* (Ramaley, R. F., Bitzinger, K., Carroll, R. M. and Wilson, R. B., 1975, *Int. J. Systemat.*

Bacteriol. **25,** 357–364); an obligate methylotroph, *Methylococcus thermophilus* (Malashenko, Yu. R., Romanovskaya, V. A., Bogachenko, V. N. and Shved, A. D., 1975; *Mikrobiologiya* **44,** 855–862; see also *Mikrobiologiya* **44,** 707–713); and *Thermothrix thioparus,* a facultatively anaerobic extreme thermophile from hot springs (Caldwell, D. E. and Laycock, J. P., 1976, *Abst. Ann. Meeting Am. Soc. Microbiol.* 1976, p. N21). A new family name, Chloroflexaceae, has been introduced for the phototrophic, gliding, filamentous bacteria containing chlorobium vesicles and bacteriochlorophylls c and a; the type genus is *Chloroflexus* (Trüper, H. G., 1976, *Int. J. Systemat. Bacteriol.* **26,** 74–75).

A scheme for identification of thermophilic actinomycetes which are associated with hypersensitivity pneumonitis has been published (Kurup, V. P. and Fink, J. N. 1975, *J. Clin. Microbiol.* **2,** 55–61). Thermophilic actinomycetes have been isolated from air conditioners, humidifiers, wood dust, house dust, mouldy grains, furnace filters and other places in surveys of the prevalence of species which might cause human illness (Kurup, V. P., Fink, J. N. and Bauman, D. M., 1976, *Mycologia* **68,** 662–666; Seabury, J., Becker, B. and Salvaggio, J., 1976, *J. Allergy Clin. Immunol.* **57,** 174–176). Aerobic (Smith, J. E., Jr., Young, K. W. and Dean, R. B., 1975, *Water Res.* **9,** 17–24) and anaerobic (Garber, W. F., Ohara, G. T., Colbaugh, J. E. and Raksit, S. K., 1975, *J. Water Pollut. Control Fed.* **47,** 950–961; Cooney, C. L. and Ackerman, R. A., 1975, *Eur. J. Appl. Microbiol.* **2,** 65–72) treatment of wastes by thermophilic bacteria is receiving increased attention, including analyses of the costs of use of these methods for disposal of wastes and for production of single cell protein (Surucu, G. A., Engelbrecht, R. S. and Chian, E. S. K., 1975, *Biotechnol. Bioeng.* **17,** 1639–1662; Finstein, M. S. and Morris, M. L., 1975, *Adv. Appl. Microbiol.* **19,** 113–151). Interest in utilization of guayule has been renewed and has resulted in a brief review of the process of retting of that latex-producing plant (Allen, P. J. and Emerson, R., 197?, *In* "An International Conference on the Utilization of Guayule" (Eds W. G. McGinnies and E. F. Haase), pp. 146–149. University of Arizona, Tucson).

Considerable interest exists in use of thermophiles for conversion of waste materials into animal feeds, as well as for increased efficiency of disposal of wastes. K. J. Seal and H. O. W. Eggins (1976, *In* "Food from Waste" (Eds Birch, G. G., Parker, K. J. and Worgan, J. T.), pp. 58–78. Applied Science Publishers Ltd, London) have used thermophilic fungi to upgrade pig manure. The thermotolerant fungus *Aspergillus fumigatus* is being developed for production of protein-enriched feed from cassava (Reade, A. E. and Gregory, K. F., 1975, *Appl. Microbiol.* **30,** 897–904); an asporogenous mutant has been created for this use.

The chemistry and biology of thermal streams has been reviewed by R. W. Castenholz and C. E. Wickstrom (1975, *In* "River Ecology" (Ed.

B. A. Whitton), pp. 264–285. Blackwell Scientific Publications, Oxford). C. E. Wickstrom and R. W. Castenholz (1975, *J. Phycol.* **11** (Suppl.), (Abst.)) concluded that ostracod grazing pressure determined the distribution of thermophilic blue-green algae in a hot spring and accounted for observed seasonal distribution anomalies. The types and distribution of obligate thermophilic bacteria were found to be similar in a thermal gradient resulting from man-made thermal pollution and the thermal gradients of two natural hot springs (Ramaley, R. F. and Bitzinger, K., 1975, *Appl. Microbiol.* **30,** 152–155). Laboratory and field studies of thermophilic bacteria of hot springs in the U.S.S.R. support the conclusion that thermophilic species grow more rapidly than do mesophiles, but have a shorter period of high rate of growth and as a result accumulate less biomass (Pozmogova, I. N., 1975, *Mikrobiologiya* **44,** 313–316 and 492–497). L. A. Egorova and L. G. Loginova (1975, *Mikrobiologiya* **44,** 938–942) have described the widespread distribution of the extremely thermophilic bacterium *Thermus* in hot springs and geothermal soils of Tadzhikistan (U.S.S.R.). Strains of *Bacillus stearothermophilus* isolated from geothermal areas in the Southern Urals were capable of utilizing phenol as a source of carbon and energy; the thermal gases of this region contained phenol, which is perhaps an explanation for growth of these bacteria in steam condensates on the inside surfaces of well tubing (Golovacheva, R. S. and Oreshkin, A. E., 1975, *Mikrobiologiya* **44,** 470–475). The extremely thermophilic acidophilic bacterium *Sulfolobus acidocaldarius* has been isolated from New Zealand hot springs (Bahlool, B. B., 1975, *Arch. Microbiol.* **106,** 171–174). *Thermoplasma acidophila* requires an 8 to 10 amino acid oligopeptide for growth; it is suggested that the occurrence of this extreme thermophile only in heated coal wastes (and not in hot springs) may be due to this requirement for a polypeptide, which is perhaps met by concentration and localization of the required growth factor by the carbon-containing material in the coal wastes (Smith, P. F., Langworthy, T. A. and Smith, M. R., 1975, *J. Bacteriol.* **124,** 884–892). Lipids of *S. acidocaldarius* and of *T. acidophila* are based on the same type of cyclic diether combining glycerol and one of a series of unusual C_{40} isoprenoid diols (de Rosa, M., Gambacorta, A. and Bu'Lock, J. D., 1976, *Phytochemistry* **15,** 143–145).

Acknowledgements

We thank E. V. Crisan for reading and commenting upon portions of the manuscript of this article. This work and previously unpublished research of M. R. T. were supported in part by National Science Foundation Postdoctoral Fellowship 40003 and research grant GB-39880. Research of T. D. B. relevant to this paper was supported by National Science Foundation research grant GB-35046.

References*

Ainsworth, G. C. (1968). *In* "The Fungi" (Eds G. C. Ainsworth and A. S. Sussman), Vol. 3, pp. 505–514. Academic Press, New York and London.

Ainsworth, G. C. and Austwick, P. K. C. (1959). "Fungal Diseases of Animals." Commonwealth Agricultural Bureaux, Farnham Royal, Buckinghamshire.

Allen, M. B. (1953). The thermophilic aerobic sporeforming bacteria. *Bact. Rev.* **17,** 125–173.

Allen, M. B. (1959). Studies with *Cyanidium caldarium*, an anomalously pigmented chlorophyte. *Arch. Mikrobiol.* **32,** 270–277.

Almin, K. E., Eriksson, K.-E. and Pettersson, B. (1975). Extracellular enzyme system utilized by the fungus *Sporotrichum pulverulentum* (*Chrysosporium lignorum*) for the breakdown of cellulose. 2. Activities of the five endo-1,4-*β*-glucanases towards carboxymethylcellulose. *Eur. J. Biochem.* **51,** 207–211.

Anderson, G. C. (1958). Some limnological features of a shallow saline meromictic lake. *Limnol. Oceanogr.* **3,** 259–270.

Anderson, J. M. and Coe, M. J. (1974). Decomposition of elephant dung in an arid, tropical environment. *Oecologia* **14,** 111–125.

Andrusenko, M. Ya. and Lakhonina, G. M. (1969). K izucheniyu drozhzheĭ, sposobnykh razvivat'sya na uglevodosoderzhashchikh sredakh pri povyshennykh temperaturakh. *Uzbek. Biol. Zh.* **1969,** 21–23.

Apinis, A. E. (1963a). Occurrence of thermophilous microfungi in certain alluvial soils near Nottingham. *Nova Hedwigia* **5,** 57–78.

Apinis, A. E. (1963b). *In* "Soil Organisms" (Eds J. Doeksen and J. van der Drift), pp. 427–438. North-Holland Publishing Company, Amsterdam.

Apinis, A. E. (1964). On fungi isolated from soils and *Ammophila* depris. *Kew Bull.* **19,** 127–131.

Apinis, A. E. (1965). *In* "Biosoziologie" (Ed. R. Tüxen), pp. 290–303. W. Junk, The Hague.

Apinis, A. E. (1972). Thermophilous fungi in certain grasslands. *Mycopathol. Mycol. Appl.* **48,** 63–74.

Apinis, A. E. and Chesters, C. G. C. (1964). Ascomycetes of some salt marshes and sand dunes. *Trans. Br. Mycol. Soc.* **47,** 419–435.

Apinis, A. E. and Eggins, H. O. W. (1966). *Thermomyces ibadanensis* sp. nov. from oil palm kernel stacks in Nigeria. *Trans. Br. Mycol. Soc.* **49,** 629–632.

Apinis, A. E. and Pugh, G. J. F. (1967). Thermophilous fungi of birds' nests. *Mycopathol. Mycol. Appl.* **33,** 1–9.

Aragozzini, F., Toppino, P. and Rindone, B. (1970). Su di un eumicete termofilo produttore di penicillinia. *Annal. Microbiol.* **20,** 44–56.

Arima, K., Liu, W.-H. and Beppu, T. (1972). Studies on the lipase of thermophilic fungus *Humicola lanuginosa*. *Agric. Biol. Chem.* **36,** 893–895.

Arnold, D. C. (1962). Bacterial decomposition and the origin of domestic fire. *Compost Sci.* **3,** 12–14.

Arx, J. A. von (1970). "The Genera of Fungi Sporulating in Pure Culture." Verlag von J. Cramer, Lehre.

Arx, J. A. von. (1975). On *Thielavia* and some similar genera of Ascomycetes. Studies in Mycology No. 8. Centraalbureau voor Schimmllcultures, Baarn, Netherlands.

* References marked with an asterisk show additional recent work not dealt with in the text.

Assarsson, A. and Bergman, Ö. (1972). *In* "Biodeterioration of Materials" (Eds A. H. Walters and E. H. Hueck-van der Plas), Vol. 2, pp. 472–480. Wiley, New York.

Austwick, P. K. C. (1963). *In* "Recent Progress in Microbiology" (Ed. N. E. Gibbons), pp. 644–651. University of Toronto Press, Toronto.

Austwick, P. K. C. (1972). The pathogenecity of fungi. *Symp. Soc. Gen. Microbiol.* **22,** 251–268.

Austwick, P. K. C. (1974). *In* "Aspergillosis and Farmer's Lung in Man and Animal" (Eds R. de Haller and F. Suter), pp. 58–60. Hans Huber Publishers, Bern, Stuttgart.

Awao, T. and Mitsugi, K. (1973). Notes on thermophilic fungi in Japan (1). *Trans. Mycol. Soc. Japan* **14,** 145–160.

Awao, T. and Otsuka, S. (1973). Notes on thermophilic fungi in Japan (2). *Trans. Mycol. Soc. Japan* **14,** 221–236.

Awao, T. and Otsuka, S. (1974). Notes on thermophilic fungi in Japan (3). *Trans. Mycol. Soc. Japan* **15,** 7–22.

Axtmann, R. C. (1975). Environmental impact of a geothermal power plant. *Science* **187,** 795–803.

Bai, M. P. and Rao, P. L. N. (1966). Thermophilic microorganisms. Part IV. Elaboration of malbranchins A & B by *Malbranchea pulchella. Indian J. Biochem.* **3,** 187–190.

Barnes, T. G., Eggins, H. O. W. and Smith, E. L. (1972). Preliminary stages in the development of a process for the microbial upgrading of waste paper. *Int. Biodeterior. Bull.* **8,** 112–116.

Bauld, J. and Brock, T. D. (1973). Ecological studies of *Chloroflexis*, a gliding photosynthetic bacterium. *Arch. Mikrobiol.* **92,** 267–284.

Bauld, J. and Brock, T. D. (1974). Algal excretion and bacterial assimilation in hot spring algal mats. *J. Phycol.* **10,** 101–106.

Beck, J. V. (1967). The role of bacteria in copper mining operations. *Biotechnol. Bioeng.* **9,** 487–497.

Bellamy, W. D. (1974). Single cell proteins from cellulosic wates. *Biotechnol. Bioeng.* **16,** 869–880.

Bellamy, W. D. (1975). *In* "Single-cell Protein II" (Eds S. R. Tannenbaum and D. I. C. Wang), pp. 263–272. MIT Press, Cambridge, Massachusetts.

Belly, R. T. and Brock, T. D. (1974). Widespread occurrence of acidophilic strains of *Bacillus coagulans* in hot springs. *J. Appl. Bact.* **37,** 175–177.

Belly, R. T., Bohlool, B. B. and Brock, T. D. (1973a). The genus *Thermoplasma. Ann. N.Y. Acad. Sci.* **225,** 94–107.

Belly, R. T., Tansey, M. R. and Brock, T. D. (1973b). Algal excretion of ^{14}C-labeled compounds and microbial interactions in *Cyanidium caldarium* mats. *J. Phycol.* **9,** 123–127.

Berestets'kiĭ, O. O., Patika, V. P. and Nadkernichniĭ, S. P. (1974). Fitotoksichni vlastivosti *Aspergillus fumigatus* Fresenius. *Mikrobiol. Zh.* (*Kiev*) **36,** 581–586.

Bergman, Ö. (1974). Thermal degradation and spontaneous ignition in outdoor chip storage. Res. Note R91, 36 pp. Institutionen för Virkeslära, Skogshögskolan, Stockholm.

Bergman, Ö. and Nilsson, T. (1966). Studier över utomhuslagring av tallvedsflis vid Lövholmens Pappersbruk. Res. Note R53, 40 pp. Institutionen för Virkeslära, Skogshögskolan, Stockholm.

Bergman, Ö. and Nilsson, T. (1967). Studier över utomhuslagring av aspvedsflis vid Hörnefors Sulfitfabrik. Res. Note R55, 60 pp. Institutionen för Virkeslära, Skogshögskolan, Stockholm.

Bergman, Ö. and Nilsson, T. (1967). Studier över utomhuslagring av björkvedsflis vid Mörrums Bruk. Res. Note R60, 56 pp. Institutionen för Virkeslära, Skogshögskolan, Stockholm.

Bergman, Ö. and Nilsson, T. (1971). Studies on outside storage of sawmill chips. Res. Note R71, 43 pp. Institutionen för Virkeslära, Skogshögskolan, Stockholm.

Bevan, D. and Bond, J. (1971). Micro-organisms in field and mill—a preliminary survey. *Proc. Qd. Soc. Sug. Cane Techol.* **38,** 137–143.

Björkman, O., Pearcy, R. W., Harrison, A. T. and Mooney, H. (1972). Photosynthetic adaptation to high temperatures: a field study in Death Valley, California. *Science* **175,** 786–789.

Blalock, H. G., Georg, L. K. and Derieux, W. T. (1973). Encephalitis in turkey poults due to *Dactylaria* (*Diplorhinotrichum*) *gallopava*—a case report and its experimental reproduction. *Avian Dis.* **17,** 197–204.

Bohlool, B. B. and Brock, T. D. (1974). Population ecology of *Sulfolobus acidocaldarius*. II. Immunoecological studies. *Arch. Microbiol.* **97,** 181–194.

Bott, T. L. and Brock, T. D. (1969). Bacterial growth rates above 90°C in Yellowstone hot springs. *Science* **164,** 1411–1412.

Böttcher, B., Kaffanke, K. and Möller, H.-W. (1973). Zum Problem der Temperaturverteilung in Mülldeponien. Hinweise zu Ursachen det Selbstenzündung. *Zentbl. Bakt. Parastikde. Abt.* I Orig. B **157,** 165–177.

Breed, R. S., Murray, E. G. D. and Smith, N. R. (1957). *In* "Bergey's Manual of Determinative Bacteriology" (7th ed.) 1094 pp. Williams and Wilkins, Baltimore.

Brewer, P. G., Wilson, T. R. S., Murray, J. W., Munns, R. G. and Densmore, C. D. (1971). Hydrographic observations on the Red Sea brines indicate a marked increase in temperature. *Nature, Lond.* **231,** 37–38.

Brierley, C. L. and Murr, L. E. (1973). Leaching: use of a thermophilic and chemautotrophic microbe. *Science* **179,** 488–489.

Broadbent, J. A. and Oyeniran, J. O. (1968). *In* "Biodeterioration of Materials" (Eds A. H. Walters and J. J. Elphick), pp. 693–702. Elsevier, Amsterdam, London and New York.

Brock, T. D. (1967). Life at high temperatures. *Science* **158,** 1012–1019.

Brock, T. D. (1969). Microbial growth under extreme conditions. *Symp. Soc. Gen. Microbiol.* **19,** 15–41.

Brock, T. D. (1970). High temperature systems. *Ann. Rev. Ecol. Systemat.* **1,** 191–220.

Brock, T. D. and Boylen, K. L. (1973). Presence of thermophilic bacteria in laundry and domestic hot-water heaters. *Appl. Microbiol.* **25,** 72–76.

Brock, T. D. and Brock, M. L. (1971). Temperature optimum of non-sulphur bacteria from a spring at 90°C. *Nature, Lond.* **233,** 494–495.

Brock, T. D. and Freeze, H. (1969). *Thermus aquaticus* gen. n. and sp. n., a non-sporulating extreme thermophile. *J. Bact.* **98,** 289–297.

Brock, T. D. and Mosser, J. L. (1975). Rate of sulfuric-acid production in Yellowstone National Park. *Bull. Geol. Soc. Am.* **86,** 194–198.

Brock, T. D., Brock, M. L., Bott, T. L. and Edwards, M. R. (1971). Microbial life at 90°C: the sulfur bacteria of Boulder Spring. *J. Bact.* **107,** 303–314.

Brock, T. D., Brock, K. M., Belly, R. T. and Weiss, R. L. (1972). *Sulfolobus:* a new genus of sulfur-oxidizing bacteria living at low pH and high temperature. *Arch. Mikrobiol.* **84**, 54–68.

Brock, T. D., Cooks, S., Petersen, S. and Mosse, J. L. (1976). Biogeochemistry and bacteriology of ferrous iron oxidation in geothermal habitats. *Geochim. Cosmochim. Acta.* **40**, 493–500.

Brown, B. S., Mills, J. and Hulse, J. M. (1974). Chemical and biological degradation of waste plastics. *Nature, Lond.* **250**, 161–163.

Burdsall, H. H., Jr. and Eslyn, W. E. (1974). A new *Phanerochaete* with a *Chrysosporium* imperfect state. *Mycotaxon* **1**, 123–133.

Cailleux, R. (1973). Mycoflore du compost destiné à la culture du champignon de couche. *Rev. Mycol.* **37**, 14–35.

Cairns, J., Jr. and Lanza, G. R. (1972). *In* "Water Pollution Microbiology" (Ed. R. Mitchell), pp. 245–272. Wiley, New York.

*Caldwell, D. E., Caldwell, S. J. and Laycock, J. P. (1976). *Thermothrix thioparus* gen. et sp. nov. a facultative anaerobic facultative chemolithotroph living at neutral pH and high temperature. *Can. J. Microbiol.* **22**, 1509–1517.

Campbell, A. R. (1972). *In* "Solid Waste Treatment and Disposal" (Ed. N. Y. Kirov), pp. 87–92. Ann Arbor Science Publishers, Ann Arbor, Michigan.

Campbell, L. L. (1974). *In* "Bergey's Manual of Determinative Bacteriology" (8th ed.) (Eds R. E. Buchanan and N. E. Gibbons), pp. 572–573. Williams & Wilkins, Baltimore.

*Castenholz, R. W. (1969). Thermophilic blue-green algae and the thermal environment. *Bact. Rev.* **33**, 476–504.

Castenholz, R. W. (1973). *In* "The Biology of Blue-green Algae" (Eds N. G. Carr and B. A. Whitton), pp. 379–414. Blackwell Scientific Publications, Oxford.

Castenholz, R. W. (1976). The effect of sulfide on the blue-green algae of hot springs. I. New Zealand and Iceland. *J. Phycol.* **12**, 54–68.

Cavazzoni, V. and Vivani, M. A. (1973). Attività *in vitro* della termozimocidina. *Ann. Microbiol.* **23**, 151–156.

Celerin, E. M. and Fergus, C. L. (1971). Effects of nutrients, temperature and relative humidity on germination and longevity of the ascospores of *Chaetomium thermophile* var. *coprophile*. *Mycologia* **63**, 1030–1045.

Chang, Y. (1967). The fungi of wheat straw compost. II. Biochemical and physiological studies. *Trans. Br. Mycol. Soc.* **50**, 667–677.

Chang, Y. and Hudson, H. J. (1967). The fungi of wheat straw compost. I. Ecological studies. *Trans. Br. Mycol. Soc.* **50**, 649–666.

Chanter, D. P. and Spencer, D. M. (1974). The importance of thermophilic bacteria in mushroom compost fermentation. *Sci. Hortic.* **2**, 249–256.

Chan-Yeung, M., Grzybowski, S. and Schonell, M. E. (1972). Mushroom worker's lung. *Am. Rev. Resp. Dis.* **105**, 819–822.

Chapman, D. J. (1974). Taxonomic position of *Cyanidium caldarium*. The Porphyridiales and Goniotrichales. *Nova Hedwig.* **25**, 673–682.

Chapman, E. S. (1974). Effect of temperature on growth rate of seven thermophilic fungi. *Mycologia* **66**, 542–546.

Chapman, E. S., Evans, E., Jacobelli, M. C. and Logan, A. A. (1975). The cellulolytic and amylolytic activity of *Papulaspora thermophila*. *Mycologia* **67**, 608–615.

Christensen, C. M. (1957). Deterioration of stored grains by fungi. *Bot. Rev.* **23**, 108–134.

Christensen, C. M. and Kaufmann, H. H. (1969). "Grain Storage. The Role of Fungi in Quality Loss" University of Minnesota Press, Minneapolis.

Christensen, C. M. and Kaufmann, H. H. (1974). *In* "Storage of Cereal Grains and Their Products" (Ed. C. M. Christensen), pp. 158–192. American Association of Cereal Chemists, St Paul, Minnesota.

Clarke, J. H., Hill, S. T. and Niles, E. V. (1966). Microflora of high-moisture barley in sealed silos. *Pest Infest. Res.* **1965,** pp. 13–14.

Clarke, J. H., Hill, S. T., Niles, E. V. and Howard, M. A. R. (1969). Ecology of the microflora of moist barley in "sealed" silos on farms. *Pest Infest. Res.* **1968,** 17.

Clarke, J. H., Niles, E. V. and Hill, S. T. (1967). Ecology of the microflora of moist barley. Barley in "sealed" silos on farms. *Pest Infest. Res.* **1966,** 14–16.

Commonwealth Mycological Institute. (1975). "Catalogue of the Culture Collection of the Commonwealth Mycological Institute." CMI, Kew.

Cooney, C. L. and Levine, D. W. (1975). *In* "Single-Cell Protein II" (Eds S. R. Tannenbaum and D. I. C. Wang), pp. 402–423. MIT Press, Cambridge, Massachusetts.

Cooney, C. L. and Wise, D. L. (1975). Thermophilic anaerobic digestion of solid waste for fuel gas production. *Biotech. Bioeng.* **17,** 1119–1135.

Cooney, C. L., Levine, D. W. and Snedecor, B. (1975a). Production of single-cell protein from methanol. *Food Technol.* **29,** pp. 33–42.

Cooney, C. L., Makiguchi, N. and Montgomery, M. (1975b). Effect of temperature and growth rate on viability and cell yield on methanol grown *Hansenula polymorpha. Abst. Ann. Meeting Am. Soc. Microbiol.* 196.

Cooney, D. G. and Emerson, R. (1964). "Thermophilic Fungi. An Account of Their Biology, Activities, and Classification." W. H. Freeman, San Francisco and London.

Coursey, D. G. (1965). Biodeterioratire processes in palm oil stored in West Africa. S.C.I. Monogr. No. 23, 44–56.

Coursey, D. G. and Nwankwo, F. I. (1968). Effects of insolation and of shade on the storage behaviour of yams in West Africa. *Ghana J. Sci.* **8,** 74–81.

Coutant, C. C. and Talmage, S. S. (1975). Thermal effects. *J. Water Pollut. Control Fed.* **6,** 1656–1711.

Cowling, E. B., Hafley, W. L. and Weiner, J. (1974). Changes in value and utility of pulpwood during harvesting, transport, and storage. TAPPI **57,** 120–123.

Craveri, R., Guicciardi, A., and Pacini, N. (1966). Distribution of thermophilic actinomycetes in compost for mushroom production. *Annal. Microbiol.* **16,** 111–113.

Craveri, R., Craveri, A. and Guicciardi, A. (1967). Ricerche sulle proprietà ed attività di eumiceti termofili isolati dal terreno. *Annal. Microbiol.* **17,** 1–30.

Craveri, R., Manachini, P. L. and Aragozzini, F. (1972). Thermozymocidin new antifungal antibiotic from a thermophilic Eumycete. *Experientia* **28,** 867–868.

Crisan, E. V. (1959). The isolation and identification of thermophilic fungi. M.Sc. thesis, Purdue University.

Crisan, E. V. (1964). Isolation and culture of thermophilic fungi. *Contr. Boyce Thompson Inst. Pl. Res.* **22,** 291–301.

Crisan, E. V. (1973). Current concepts of thermophilism and the thermophilic fungi. *Mycologia* **65,** 1171–1198.

Cross, T. (1974). *In* "Bergey's Manual of Determinative Bacteriology" (8th ed.) (Eds R. E. Buchanan and N. E. Gibbons), pp. 861–863. Williams & Wilkins Baltimore.

Currie, J. A. and Festenstein, G. N. (1971). Factors defining spontaneous heating and ignition of hay. *J. Sci. Fd. Agric.* **22,** 223–230.

Dallinger, W. H. (1887). The president's address. *J. R. Microsc. Soc.*, Ser. 2, 185–199.

Darland, G. and Brock, T. D. (1971). *Bacillus acidocaldarius* sp. nov., an acidophilic thermophilic spore-forming bacterium. *J. Gen. Microbiol.* **67,** 9–15.

Darland, G., Brock, T. D., Samsonoff, W. and Conti, S. F. (1970). A thermophilic, acidophilic mycoplasma isolated from a coal refuse pile. *Science* **170,** 1416–1418.

Davis, N. D., Wagener, R. E., Morgan-Jones, G. and Diener, U. L. (1975). Toxigenic thermophilic and thermotolerant fungi. *Appl. Microbiol.* **29,** 455–457.

Deacon, J. E. and Minckley, W. L. (1974). In "Desert Biology" (Ed. G. W. Brown, Jr.), Vol. 2, pp. 385–488. Academic Press, New York and London.

Degens, E. T. and Ross, D. A. (eds) (1969). "Hot Brines and Heavy Metal Deposits in the Red Sea. A Geochemical and Geophysical Account" Springer. Verlag, New York.

*Degryse, E. (1976). Bacterial diversity at high temperature. *In* "Enzymes and Proteins from Thermophilic Microorganisms" (Ed. H. Zuber), pp. 401–410. Birkhäuser Verlag, Basel.

Deibel, R. H. and Seeley, H. W., Jr. (1974). *In* "Bergey's Manual of Determinative Bacteriology" (8th ed.) (Eds R. E. Buchanan and N. E. Gibbons), 490–509. Williams & Wilkins, Baltimore.

Deploey, J. J. (1970). The growth and sporulation of thermophilic fungi and actinomycetes in atmospheres of oxygen and nitrogen. Ph.D. thesis, Pennsylvania St. University.

Deploey, J. J. and Fergus, C. L. (1975). Growth and sporulation of thermophilic fungi and actinomycetes in O_2-N_2 atmospheres. *Mycologia* **67,** 780–797.

Derby, R. W. and Gates, D. M. (1966). The temperature of tree trunks—calculated and observed. *Am. J. Bot.* **53,** 580–587.

Dingfelder, J. H. (1962). Die Ciliaten vorübergehender Gewässer. *Arch. Protistenk.* **105,** 509–658.

Doemel, W. N. and Brock, T. D. (1970). The upper temperature limit of *Cyanidium caldarium. Arch. Mikrobiol.* **72,** 326–332.

Doemel, W. N. and Brock, T. D. (1971). The physiological ecology of *Cyanidium caldarium. J. Gen. Microbiol.* **67,** 17–32.

Dogiel, V. A. (Revised by J. I. Poljanskij and E. M. Chejsin) (1965). "General Protozoology" (2nd ed.). Oxford University Press, Oxford.

Dombrowski, H. (1961). Methoden und Ergebnisse der Balneobiologie. *Ther. Gegen.* **100,** 442–449.

Donovan, D. L. (1971). Determination of aflatoxin production by thermophilic fungi isolated from feedlot manure. M.A. thesis, Ball St. University.

Dye, M. H. (1964). Self-heating of damp wool. I. The estimation of microbial populations in wool. *N.Z. J. Sci.* **7,** 87–96.

Dye, M. H. and Rothbaum, H. P. (1964). Self-heating of damp wool. II. Self-heating of damp wool under adiabatic conditions. *N.Z. J. Sci.* **7,** 97–118.

Eaton, R. A. (1972). Fungi growing on wood in water cooling towers. *Int. Biodeterior. Bull.* **8,** 39–48.

Eaton, R. A. and Jones, E. B. G. (1971). The biodeterioration of timber in water cooling towers. II. Fungi growing on wood in different positions in a water cooling system. *Mater. Org.* **6,** 81–92.

Edwards, J. H. (1972). The isolation of antigens associated with farmer's lung. *Clin. Exp. Immunol.* **11,** 341–355.

Eggins, H. O. W. and Coursey, D. G. (1964). Thermophilic fungi associated with Nigerian oil palm produce. *Nature, Lond.* **203,** 1083–1084.
Eggins, H. O. W. and Coursey, D. G. (1968). The industrial significance of the biodeterioration of oilseeds. *Int. Biodeterior. Bull.* **4,** 29–38.
Eggins, H. O. W. and Malik, K. A. (1969). The occurrence of thermophilic cellulolytic fungi in a pasture land soil. *Antonie van Leeuwenhoek* **35,** 178–184.
Eggins, H. O. W., Mills, J., Holt, A. and Scott, G. (1971). *In* "Microbial Aspects of Pollution" (Eds G. Sykes and F. A. Skinner), pp. 267–279. Academic Press, London and New York.
Eggins, H. O. W., Szilvinyi, A. von and Allsopp, D. (1972). The isolation of actively growing thermophilic fungi from insolated soils. *Int. Biodeterior. Bull.* **8,** 53–58.
Egorova, A. A. and Deryugina, Z. (1963). The spore-forming thermophilic thiobacterium *Thiobacillus thermophilica* Imschenetskii nov. sp. *Mikrobiologiya* **32,** 437–446.
Eicker, A. (1972). Occurrence and isolation of South African thermophilic fungi. *S. Afr. J. Sci.* **68,** 150–155.
Emerson, R. (1968). *In* "The Fungi" (Eds G. C. Ainsworth and A. S. Sussman), Vol. 3, pp. 105–128. Academic Press, New York and London.
Eriksen, J. (1974). Cellulases from a thermophilic compost fungus, *Chaetomium thermophilae. J. Gen. Microbiol.* **81,** vi.
Eriksson, K.-E. and Larsson, K. (1975). Fermentation of waste mechanical fibers from a newsprint mill by the rot fungus *Sporotrichum pulverulentum. Biotechnol. Bioeng.* **17,** 327–348.
Eriksson, K.-E. and Petersson, B. (1975), Extracellular enzyme system utilized by the fungus *Sporotrichum pulverulentum* (*Chrysosporium lignorum*) for the breakdown of cellulose. 1. Separation, purification and physico-chemical characterization of five endo-1, 4-β-glucanases. *Eur. J. Biochem.* **51,** 193–206.
Eriksson, K.-E. and Rzedowski, W. (1969). Extracellular enzyme system utilized by the fungus *Chrysosporium lignorum* for the breakdown of cellulose. I. Studies on the enzyme production. *Arch. Biochem. Biophys.* **129,** 683–688.
*Esch, G. W. and R. W. McFarlane (Eds) (1976). "Thermal Ecology II." Energy Research and Development Administration, 406 pp. Oak Ridge.
Eslyn, W. E., Kirk, T. K. and Effland, M. J. (1975). Changes in the composition of wood caused by six soft-rot fungi. *Phytopathology* **65,** 473–476.
Evans, H. C. (1968). British records. *Trans. Br. Mycol. Soc.* **51,** 587–588.
Evans, H. C. (1971a). Thermophilous fungi of coal spoil tips. I. Taxonomy. *Trans. Br. Mycol. Soc.* **57,** 241–254.
Evans, H. C. (1971b). Thermophilous fungi of coal spoil tips. II. Occurrence, distribution and temperature relationships. *Trans. Br. Mycol. Soc.* **57,** 255–266.
Evans, H. C. (1971c). Thermophilous fungi of coal spoil tips. III. Seasonal and spatial occurrence. *Trans. Br. Mycol. Soc.* **57,** 267–272.
Evans, H. C. (1972). Thermophilous fungi isolated from the air. *Trans. Br. Mycol. Soc.* **59,** 516–519.
Evans, H. C. and Stolk, A. C. (1971). *Talaromyces leycettanus* sp. nov. *Trans. Br. Mycol. Soc.* **56,** 45–49.
Fairchild, E. and Sheridan, R. P. (1974). A physiological investigation of the hot spring diatom, *Achnanthes exigua* Grün. *J. Phycol.* **10,** 1–4.
Feist, W. C., Springer, E. L. and Hajny, G. J. (1971). Viability of parenchyma cells in stored green wood. TAPPI **54,** 1295–1297.

Feist, W. C., Hajny, G. J. and Springer, E. L. (1973a). Effect of storing green wood chips at elevated temperatures. TAPPI **56**, 91–95.

Feist, W. C., Springer, E. L. and Hajny, G. J. (1973b). Spontaneous heating in piled wood chips—contribution of bacteria. TAPPI **56**, 148–151.

Fergus, C. L. (1964). Thermophilic and thermotolerant molds and actinomycetes of mushroom compost during peak heating. *Mycologia* **56**, 267–284.

Fergus, C. L. (1969). The cellulolytic activity of thermophilic fungi and actinomycetes. *Mycologia* **61**, 120–129.

Fergus, C. L. (1971). The temperature relationships and thermal resistance of a new thermophilic *Papulaspora* from mushroom compost. *Mycologia* **63**, 426–431.

Fergus, C. L. and Amelung, R. M. (1971). The heat resistance of some thermophilic fungi on mushroom compost. *Mycologia* **63**, 675–679.

Fergus, C. L. and Sinden, J. W. (1969). A new thermophilic fungus from mushroom compost: *Thielavia thermophila* spec. nov. *Can. J. Bot.* **47**, 1635–1637.

Festenstein, G. N. (1966). Biochemical changes during moulding of self-heated hay in Dewar flasks. *J. Sci. Food Agric.* **17**, 130–133.

Festenstein, G. N. (1971). Carbohydrates in hay on self-heating to ignition. *J. Sci. Food Agric.* **22**, 231–234.

Festenstein, G. N., Lacey, J., Skinner, F. A., Jenkins, P. A. and Pepys, J. (1965). Self-heating of hay and grain in Dewar flasks and the development of farmer's lung antigens. *J. Gen. Microbiol.* **41**, 389–407.

Fields, M. L. (1970). The flat sour bacteria. *Adv. Food Res.* **18**, 163–217.

Flannigan, B. (1969). Microflora of dried barley grain. *Trans. Br. Mycol. Soc.* **53**, 371–379.

Flannigan, B. (1970). Comparison of seed-borne mycofloras of barley, oats and wheat. *Trans. Br. Mycol. Soc.* **55**, 267–276.

Flannigan, B. (1972). *In* "Biodeterioration of Materials" (Eds A. H. Walters and E. H. Hueck-van der Plas), Vol. 2, 35–41. Wiley, New York.

Flannigan, B. (1974). Distribution of seed-borne micro-organisms in naked barley and wheat before harvest. *Trans. Br. Mycol. Soc.* **62**, 51–58.

*Flannigan, B. and Dickie, N. A. (1972). Distribution of micro-organisms in fractions produced during pearling of barley. *Trans. Br. Mycol. Soc.* **59**, 377–391.

*Flannigan, B. and Hui, S. C. (1976). The occurrence of aflatoxin-producing strains of *Aspergillus flavus* in the mould floras of ground spices. *J. Appl. Bacteriol.* **41**, 411–418.

Flannigan, B. and Sagoo, G. S. (1977). Degradation of wood by *Aspergillus fumigatus* isolated from self-heated wood chips. *Mycologia* **69**, 514–523.

Flannigan, B. and Sellars, P. N. (1972). Activities of thermophilous fungi from barley kernels against arabinoxylan and carboxymethyl cellulose. *Trans. Br. Mycol. Soc.* **58**, 338–341.

Fletcher, J. T., Lucas, G. B. and Welty, R. E. (1967). Thermophilic fungi and bacteria isolated from tobacco. *Phytopathology* **57**, 458–459.

Fliermans, C. B. and Brock, T. D. (1972). Ecology of sulfur-oxidizing bacteria in hot acid soils. *J. Bact.* **111**, 343–350.

Fordyce, C., Jr. (1970). Relative numbers of certain microbial groups present in compost used for mushroom (*Agaricus bisporus*) propagation. *Appl. Microbiol.* **20**, 196–199.

Fraleigh, P. C. and Wiegert, R. G. (1975). A model explaining successional change in standing crop of thermal blue-green algae. *Ecology* **56**, 656–664.

Franz, M. (1972). Municipal garbage into mushroom soil. *Compost Sci.* **13,** (6), 6–9.

Fretwell, S. D. (1972). "Populations in a Seasonal Environment" Princeton University Press, Princeton.

Friedmann, E. I. and Galun, M. (1974). *In* "Desert Biology" (Ed. G. W. Brown), Vol. 2, pp. 165–212. Academic Press, New York and London.

Frith, H. L. (1962). "The Mallee-fowl. The Bird That Builds an Incubator." Angus & Robertson, Sydney.

Gates, D. M. (1973). *In* "Temperature and Life" (Eds H. Precht, J. Christophersen, H. Hensel and W. Larcher), pp. 87–101. Springer Verlag, New York, Heidelberg, Berlin.

Gaughran, E. R. L. (1947). The thermophilic microorganisms. *Bact. Rev.* **11,** 189–225.

Geitler, L. (1963). Die angebliche Cyanophycee *Isocystis pallida* ist ein hefeartige Pilz (*Torulopsidosira*). *Arch. Mikrobiol.* **46,** 238–242.

Georg, L. K., Bierer, B. W. and Cooke, W. B. (1964). Encephalitis in turkey poults due to a new fungus species. *Sabouraudia* **3,** 239–244.

Gerrits, J. P. G. (1970). Inorganic and organic supplementation of mushroom compost. *MGA Bull.* **251,** 3–15.

Gerrits, J. P. G. (1972). The influence of water in mushroom compost. *Mushr. Sci.* **8,** 43–57.

Gerrits, J. P. G. (1974). Development of a synthetic compost for mushroom growing based on wheat straw and chicken manure. *Neth. J. Agric. Sci.* **22,** 175–194.

Gerrits, J. P. G., Bels-Koning, H. C. and Muller, F. M. (1967). Changes in compost constituents during composting, pasteurization and cropping. *Mushr. Sci.* **6,** 225–243.

Gibbons, J. W. and Sharitz, R. R. (1974). Thermal alteration of aquatic ecosystems. *Am. Sci.* **62,** 660–670.

Gibson, T. and Gordon, R. E. (1974). *In* "Bergey's Manual of Determinative Bacteriology" (8th ed.) (Eds R. E. Buchanan and N. E. Gibbons), pp. 529–550. Williams & Wilkins, Baltimore.

*Golovacheva, R. S. (1976). Thermophilic nitrifying bacteria from hot springs. *Mikrobiologiya* **45,** 377–379.

Golovacheva, R. S., Loginova, L. G., Salikhov, T. A., Kolesnikov, A. A. and Zaïtseva, G. N. (1975). Novyï vid termofil'nykh batsill—*Bacillus thermocatenulatus* nov. sp. *Mikrobiologiya* **44,** 265–268.

Gooding, G. V., Jr., Hodges, C. S., Jr. and Ross, E. W. (1966). Effect of temperature on growth and survival of *Fomes annosus*. *For. Sci.* **12,** 325–333.

Gordon, R., Haynes, W. and Pang, C. (1973). "The Genus *Bacillus*." U.S. Dep. Agr. Handbook No. 427, 283 pp.

Gould, W. D., Fujikawa, J. I. and Cook, F. D. (1974). A soil fungus tolerant to extreme acidity and high salt concentrations. *Can. J. Microbiol.* **20,** 1023–1027.

Gray, K. (1970). Research on composting in British universities. *J. Soil Ass.* **16,** 27–34.

Gray, R. L., Wenzel, F. J. and Emanuel, D. A. (1969). Immunofluorescence identification of *Thermopolyspora polyspora*, the causative agent of farmer's lung. *Appl. Microbiol.* **17,** 454–456.

Greaves, H. (1971). Biodeterioration of tropical hardwood chips in outdoor storage. TAPPI **54,** 1128–1133.

Greaves, H. (1975). Microbiological aspects of wood chip storage in tropical environments. *Austral. J. Biol. Sci.* **28,** 315–322.

Gregory, P. H. and Lacey, M. E. (1963a). Liberation of spores from mouldy hay. *Trans. Br. Mycol. Soc.* **46,** 73–80.

Gregory, P. H. and Lacey, M. E. (1963b). Mycological examination of dust from mouldy hay associated with farmer's lung disease. *J. Gen. Microbiol.* **30,** 75–88.

Gregory, P. H., Lacey, M. E., Festenstein, G. N. and Skinner, F. A. (1963). Microbial and biochemical changes during the moulding of hay. *J. Gen. Microbiol.* **33,** 147–174.

Gregory, P. H., Festenstein, G. N., Lacey, M. E., Skinner, F. A., Pepys, J. and Jenkins, P. A. (1964). Farmer's lung disease: the development of antigens in moulding hay. *J. Gen. Microbiol.* **36,** 429–439.

Griffin, J. L. (1972). Temperature tolerance of pathogenic and non-pathogenic free-living ameobas. *Science* **178,** 869–870.

Hajny, G. J. (1966). Outside storage of pulpwood chips—a review and bibliography. TAPPI **49,** (10), 97A–109A.

Hajny, G. J., Jorgensen, R. N. and Ferrigan, J. J. (1967). Outside storage of hardwood chips in the northeast. I. Physical and chemical effects. TAPPI **50,** 92–96.

Haller, R. de and Suter, F. (eds) (1974). "Aspergillosis and Farmer's Lung in Man and Animals." Hans Huber Publishers, Bern, Stuttgart, Vienna.

Hansen, A. P. and Welty, R. E. (1970). Microflora of raw cacao beans. *Mycopathol. Mycol. Appl.* **44,** 309–316.

Harris, J. A. (1970). Bat-guano cave environment. *Science* **169,** 1342–1343.

Hartley, C. W. S. (1967). "The Oil Palm." Longman Group Limited, London.

Hashimoto, H., Iwaasa, T. and Yokotsuka, T. (1972). Thermostable acid protease produced by *Penicillium duponti* K1014, a true thermophilic fungus newly isolated from compost. *Appl. Microbiol.* **24,** 986–992.

Hatton, J. V. and Hunt, K. (1972). Effect of prolonged outside chip storage on yield and quality of kraft pulps from *Picea glauca* and *Pinus contorta* chips. TAPPI **55,** 122–126.

Hayes, W. A. (1969). Microbiological changes in composting wheat straw/horse manure mixtures. *Mushr. Sci.* **7,** 173–186.

Hayes, W. A. and Randle, P. E. (1968). The use of water soluble carbohydrates and methyl bromide in the preparation of mushroom composts. *MGA Bull.* **218,** 81–102.

Hedger, J. N. (1974), *In* "Biodegredation et Humification I" (Eds G. Kilbertus, O. Reisinger, A. Mourey and J. A. Cancela da Fonseca), pp. 59–65. Pierron Sarreguemines.

Hedger, J. N. and Hudson, H. J. (1970). *Thielavia thermophilia* and *Sporotrichum thermophile*. *Trans. Br. Mycol. Soc.* **54,** 497–500.

Hedger, J. N. and Hudson, H. J. (1974). Nutritional studies of *Thermomyces lanuginosus* from wheat straw compost. *Trans. Br. Mycol. Soc.* **62,** 129–143.

Heinen, U. J. and Heinen, W. (1972). Characteristics and properties of a caldoactive bacterium producing extracellular enzymes and two related strains. *Arch. Mikrobiol.* **82,** 1–23.

Heinen, W. (1971). Growth conditions and temperature-dependent substrate specificity of two extremely thermophilic bacteria. *Arch. Mikrobiol.* **76** 2–17.

Heinrich, B. (1974). Thermoregulation in endothermic insects. *Science* **185,** 747–756.

Henssen, A. (1969). *Streptomyces fragmentosporus*, ein neuer thermophiler Actinomycet. *Arch. Mikrobiol.* **67,** 21–27.

Henssen, A. (1974). *In* "Bergey's Manual of Determinative Bacteriology" (8th ed.) (Eds R. E. Buchanan and N. E. Gibbons), pp. 746–747. Williams & Wilkins, Baltimore.

Hill, E. C. (1971). *In* "International Conference on Global Inpacts of Applied Microbiology, 3rd, Bombay, 1969" (Eds Y. M. Freitas and F. Fernandes), pp. 201–202. University of Bombay.

Hill, E. C., Thomas, A. R. and John, D. (1975). Thermophilic growth in aviation kerosene. *Int. Biodeterior. Bull.* **11,** iv.

Hindle, E. (1932). Some new thermophilic organisms. *J. R. Microsc. Soc.* **52,** 123–133.

Hofsten, B. v. and Hofsten, A. v. (1974). Ultrastructure of a thermotolerant Basidiomycete possibly suitable for production of food protein. *Appl. Microbiol.* **27,** 1142–1148.

Hofsten, B. v. and Rydén, A.-L. (1975). Submerged cultivation of a thermotolerant basidiomycete on cereal flours and other substrates. *Biotechnol. Bioeng.* **17,** 1183–1197.

Hollaus, F. and Klaushofer, H. (1973). Identification of hyperthermophilic obligate anaerobic bacteria from extraction juices of beet sugar factories. *Int. Sug. J.* **75,** 237–241, 271–275.

Hollaus, F. and Sleytr, U. (1972). On the taxonomy and fine structure of some hyperthermophilic saccharolytic clostridia. *Arch. Mikrobiol.* **86,** 129–146.

Hoog, G. S. de (1973). Additional notes on *Tritirachium. Persoonia* **7,** 437–441.

Hoog, G. S. de (1974). The genera *Blastobotrys*, *Sporothrix*, *Calcarisporium* and *Calcarisporiella* gen. nov. Studies in Mycology No. 7, Centraalbureau voor Schimmelcultures, Baarn, Netherlands.

Huang, L. H. (1971). Studies on soil microfungi of Nigeria and Dominica. Ph.D. thesis, Univ. Wisconsin.

Hubálek, Z., Balát, F., Toušková, I. and Vlk, J. (1973). Mycoflora of birds' nests in nest-boxes. *Mycopathol. Mycol. Appl.* **49,** 1–12.

Hudec, P. P. and Sonnenfeld, P. (1974). Hot brines on Los Roques, Venezuela. *Science* **185,** 440–442.

Hudson, H. J. (1973). Thermophilous and thermotolerant fungi in the air-spora at Cambridge. *Trans. Br. Mycol. Soc.* **60,** 596–598.

Hughes, W. T. and Crosier, J. W. (1973). Thermophilic fungi in the mycoflora of man and environmental air. *Mycopath. Mycol. Appl.* **49,** 147–152.

Hulme, M. A. and Shields, J. K. (1973). Treatments to reduce chip deterioration during storage. TAPPI **56,** 88–90.

Hunt, J. M., Hays, E. E., Degens, E. T. and Ross, D. A. (1967). Red Sea: detailed survey of hot-brine areas. *Science* **156,** 514–516.

Hussain, H. M. (1973). Ökologische Untersuchungen über die Bedeutung thermophiler Mikroorganismen für die Selbsterhitzung von Heu. *Z. Allg. Mikrobiol.* **13,** 323–334.

Hussain, H. M. (197?). *In* "Self heating of Organic Materials: Proceedings of the Symposium." (International Symposium, Utrecht, 1971, Delft), pp. 111–126.

Ioachimescu, M. (1973). Influenta temperaturii asupra creşterii şi dezvoltării ciupercilor izolate de pe lemnul din mină. *Stud. Cercet. Biol. Ser. Bot.* **25,** 167–170.

Irlina, I. S. (1967). *In* "The cell and Environmental Temperature" (Ed. A. S. Troshin), pp. 249–251. Pergamon Press, Oxford.

Issel, R. (1910). La Faune des Sources thermales de Viterbo, Int. Revue ges. *Hydrobiol. Hydrogr.* **3,** 178–180.

Jackson, J. E., Jr. and Castenholz, R. W. (1975). Fidelity of thermophilic blue-green algae to hot spring habitats. *Limnol. Oceanogr.* **20**, 305–322.

Jackson, T. J., Ramaley, R. F. and Meinschein, W. G. (1973). *Thermomicrobium*, a new genus of extremely thermophilic bacteria. *Int. J. Syst. Bacteriol.* **23**, 28–36.

*Jonckheere, J. de. 1977. Use of an axenic medium for differentiation between pathogenic and nonpathogenic *Naegleria fowleri* isolates. *Appl. Environ. Microbiol.* **33**, 751–757.

Jorgensen, J. D. and Gaddie, R. (1969). Decomposition of molasses pulp pellets in bulk storage. *J. Am. Soc. Sugar Beet Technol.* **15**, 277–281.

Jungerman, P. F. and Schwartzman, R. M. (1972). "Veterinary Medical Mycology." Lea & Febiger, Philadelphia.

*Kahan, D. (1969). The fauna of hot springs. *Verh. Int. Verein. Theor. Angew. Limnol.* **17**, 811–816.

Kahan, D. (1972). *Cyclidium citrullus* Cohn, a ciliate from the hot springs of Tiberias (Israel). *J. Protozool.* **19**, 593–597.

Kahan, D. and Sharon, R. (1976). Effect of temperature on growth, cell size, and free amino acid pool of the thermophilic ciliate *Cyclidium citrullus*. *J. Protozool.* **23**, 478–481.

Kane, B. E. and Mullins, J. T. (1973). Thermophilic fungi in a municipal waste compost system. *Mycologia* **65**, 1087–1100.

Kappen, L. (1973). *In* "The Lichens" (Eds V. Ahmadjian and M. E. Hale), pp. 311–380. Academic Press, New York and London.

Khasanov, O. (1968). O termofil'nykh gribakh v pochvakh Chirchik-Kelesskogo raïona Tashkentskoï oblasti. *UzSSR Fanalr Akad. Dokl., Dokl. A.N. UzSSR* **1968**, 53–54.

Khasanov, O. and Mirkhodzhaev, M. (1969). Vliyanie razlichnykh istochnikov azota na rasshcheplenie tsellyolozy termofil'nymi mikromitsetami. *Uzbek. Biol. Zh.* **1969**, 13–15.

King, A. D., Jr., Michener, H. D. and Keith, A. I. (1969). Control of *Byssochlamys* and related heat-resistant fungi in grape products. *Appl. Microbiol.* **18**, 166–173.

Kitano, K., Kintaka, K., Suzuki, S., Katamoto, K., Nara, K. and Nakao, Y. (1975a). Screening of microorganisms capable of producing β-lactam antibiotics. *J. Ferment. Technol.* **53**, 327–338.

Kitano, K., Kintaka, K., Katamoto, K., Nara, K. and Nakao, Y. (1975b). Occurrence of 6-aminopenicillanic acid in culture broths of strains belonging to the genera *Thermoascus, Gymnoascus, Polypaecilum* and *Malbranchea*. *J. Ferment. Technol.* **53**, 339–346.

Klopotek, A. von (1974). Revision der thermophilen *Sporotrichum*-Arten: *Chrysosporium thermophilum* (Apinis) comb. nov. und *Chrysosporium fergusii* spec. nov. = status conidialis von *Corynascus thermophilis* (Fergus und Sinden) comb. nov. *Arch. Microbiol.* **98**, 365–369.

Kluepfel, D., Bagli, J., Baker, H., Charest, M.-P., Kudelski, A., Sehgal, S. N. and Vézina, C. (1972). Myriocin, a new antifungal antibiotic from *Myriococcum albomyces*. *J. Antibiot.* **25**, 109–115.

Knösel, D. (197?). *In* "Self heating of Organic Materials: Proceedings of the Symposium." (International Symposium, Utrecht, 1971, Delft), pp. 139–146.

Knösel, D. and Rész, A. (1973). Pilze als Müllkompost. Enzymatischer Abbau von Pektin und Zelloluse durch wärmeliebende Spezies. *Städtehygiene* **24**, (6), 143–148.

Knutson, R. M. (1974). Heat production and temperature regulation in eastern skunk cabbage. *Science* **186,** 746–747.
Kullberg, R. G. (1968). Algal diversity in several thermal spring effluents. *Ecology* **49,** 751–755.
Kullberg, R. G. (1971). Algal distribution in six thermal spring effluents. *Trans. Am. Microsc. Soc.* **90,** 412–434.
Kurata, H., Ichinoe, M. and Naito, A. (1966). A thermophilic fungus, *Humicola lanuginosa* found in soil from Japan. *Trans. Mycol. Soc. Japan.* **7,** 99–100.
Kurbatskaya, Z. A. (1974). Vivchennya toksichnosti riznikh shtamiv *Aspergillus fumigatus*. *Mikrobiol. Zh.* (*Kiev*) **36,** 722–725.
Kurup, V. P., Barboriak, J. J., Fink, J. N. and Lechevalier, M. P. (1975). *Thermoactinomyces candidus*, a new species of thermophilic actinomycetes. *Int. J. Syst. Bacteriol.* **25,** 150–154.
Küster, E. (1974). *In* "Bergey's Manual of Determinative Bacteriology." (8th ed.) (Eds R. E. Buchanan and N. E. Gibbons), pp. 858–859. Williams & Wilkins, Baltimore.
Lacey, J. (1967). Airborne moulds and actinomycetes from grain in storage. *Rep. Rothamsted Exp. Stn.* **1966,** 133–134.
Lacey, J. (1968). Moulds and actinomycetes from stored crops. *Rep. Rothamsted Exp. Stn.* **1967,** 129–130.
Lacey, J. (1971a). The microbiology of moist barley storage in unsealed silos. *Ann. Appl. Biol.* **69,** 187–212.
Lacey, J. (1971b). *Thermoactinomyces sacchari* sp. nov., a thermophilic actinomycete causing bagassosis. *J. Gen. Microbiol.* **66,** 327–338.
Lacey, J. (1972). The microbiology of grain stored underground in Iron Age type pits. *J. Stored Prod. Res.* **8,** 151–154.
Lacey, J. (1973). Actinomycete and fungus spores in farm air. *J. Agric. Labour Sci.* **1,** 61–78.
Lacey, J. (1974a). Moulding of sugar-cane bagasse and its prevention. *Ann. Appl. Biol.* **76,** 63–76.
Lacey, J. (1974b). *In* "Aspergillosis and Farmer's Lung in Man and Animals" (Eds R. de Haller and F. Suter), pp. 16–23. Hans Huber Publishers, Bern, Stuttgart, Vienna.
Lacey, J. (1975a). Airborne spores in pastures. *Trans. Br. Mycol. Soc.* **64,** 265–281.
Lacey, J. (1975b). Potential hazards to animals and man from microorganisms in fodders and grain. *Trans. Br. Mycol. Soc.* **65,** 171–184.
Lacey, J. and Lacey, M. E. (1964). Spore concentrations in the air of farm buildings. *Trans. Br. Mycol. Soc.* **47,** 547–552.
Lacey, J., Pepys, J. and Cross, T. (1972). *In* "Safety in Microbiology" (Eds D. A. Shapton and R. G. Board), pp. 151–184. Academic Press, London and New York.
Larcade, R. J. (1967). Studies on the effects of temperature and nutrition on growth and morphogenesis of the thermophilic fungus *Thermoascus aurantiacus*. M.Sc. thesis, Adelphi University.
Leadbetter, E. R. (1974). *In* "Bergey's Manual of Determinative Bacteriology" (8th ed.) (Eds R. E. Buchanan and N. E. Gibbons), pp. 267–269. Williams & Wilkins, Baltimore.
Levine, D. W. and Cooney, C. L. (1973). Isolation and characterization of a thermotolerant methanol-utilizing yeast. *Appl. Microbiol.* **26,** 982–990.
Loginova, L. G. (1960). "Fiziologiya Eksperimental'no Poluchennykh Termofilnykh Drozhzheĭ." Izdatel'stovo Akademii Nauk SSSR, Moscow.

Loginova, L. G. and Egorova, L. A. (1975). Obligatno-termofil'nye bacterii *Thermus ruber* v gidrotermakh kamchatki. *Mikrobiologiya* **44,** 661–665.

Loginova, L. G., Kosmachev, A. E., Golovacheva, R. S. and Seregina, L. M. (1962). Issledovanie termofil'noï mikroflory gory Yangan-Tau na yuzhnom Urale. *Mikrobiologiya* **31,** 1082–1086.

Loginova, L. G., Golovacheva, R. S. and Shcherbakov, M. A. (1966a). Thermophilic bacteria forming active cellulolytic enzymes. *Mikrobiologiya* **35,** 796–804.

Loginova, L. G., Golovacheva, R. S. and Egorova, L. A. (1966b). "Zhizn' Mikroorganizmov pri Vysokikh Temperaturakh." Publishing House "Nauka," Moscow.

Loginova, L. G., Golovacheva, R. S., Golovina, I. G., Egorova, L. A., Pozmogova, I. N., Khokhlova, Yu. M. and Tsaplina, I. A. (1973). "Sovremennye Predstavleniya o Termofilii Mikroorganizmov." Publishing House "Nauka." Moscow.

Löhr, E. and Olsen, J. (1969). The thermophilic fungus *Humicola lanuginosa*. *Friesia* **9,** 140–141.

Lundström, H. (1972). Microscopic studies of cavity formation by soft rot fungi *Allescheria terrestris* Apinis, *Margarinomyces luteo-viridis* v. Beyma and *Phialophora richardsiae* (Nannf.) Conant. *Stud. For. Suec.* **98,** 1–18.

Lundström, H. (1973). Studies of the wood-decaying capacity of the soft rot fungi *Allescheria terrestris, Phialophora* (*Margarinomyces*) *luteo-viridis* and *Philaphora richardsiae*. Res. Note R87, 23 pp. Institutionen för Virkeslära, Skogshögskolan Stockholm.

Lundström, H. (1974). Studies on the physiology of the three soft rot fungi *Allescheria errestris, Phialophora* (*Margarinomyces*) *luteo-viridis* and *Phialophora richardsiae*. *Stud. For. Suec.* **115,** 1–42.

Maheshwari, R. (1968). Occurrence and isolation of thermophilic fungi. *Curr. Sci.* **37,** 277–279.

Mahoney, D. P. (1972). Soil and litter microfungi of the Galapagos Islands, Ph.D. thesis, University of Wisconsin.

Malik, K. A. and Eggins, H. O. W. (1972). Some studies on the effect of pH on the ecology of cellulolytic thermophilic fungi using a perfusion technique. *Biologia* **18,** 143–151.

Malik, K. A. and Sandhu, G. R. (1973). Some studies on the fungi of kallar grass (*Diplachne fusca* (L.) P. Beauv.) compost. *Pak. J. Bot.* **5,** 57–63.

Manachini, P., Ferrari, A. and Craveri, R. (1965). Forme termofile di Actinoplanaceae, Isolamento e ceratteristiche di *Streptosporangium album* var. *thermophilum*. *Annal. Microbiol.* **15,** 129–144.

Manachini, P., Craveri, A. and Craveri, R. (1966). *Thermomonospora citrina*, una nuova specie di Attinomicete Termofilo isolato dal suolo. *Annal. Microbiol.* **16,** 83–90.

Mann, J. E. and Schlichting, H. E., Jr. (1967). Benthic algae of selected thermal springs in Yellowstone National Park. *Trans. Am. Microsc. Soc.* **86,** 2–9.

Maravalhas, N. (1966). Mycological deterioration of cocoa beans during fermentation and storage in Bahia. *Int. Choc. Rev.* **21,** 375–378.

Mateles, R. I., Baruah, J. N. and Tannenbaum, S. R. (1967). Growth of a thermophilic bacterium on hydrocarbons: A new source of single-cell protein. *Science* **157,** 1322–1323.

Matsuo, M., Yasui, T. and Kobayashi, T. (1975). Production and saccharifying action for xylan of xylanase from *Malbranchea pulchella* var. *sulfurea* No. 48. *J. Agric. Chem. Soc. Japan*, **49**, 263–270.

McCown, B. H. (1973). *In* "Proceedings of the Symposium on the Impact of Oil Resource Development on Northern Plant Communities," pp. 12–33. Institute of Arctic Biology, Fairbanks, Alaska.

McGee, J. M., Brown, L. R. and Tischer, R. G. (1967) A high-temperature, hydrogen-oxidizing bacterium—*Hydrogenomonas thermophilus*, n. sp. *Nature, Lond.* **214,** 715–716.

Meeks, J. C. and Castenholz, R. W. (1971). Growth and photosynthesis in an extreme thermophile, *Synechococcus lividus* (Cyanophyta). *Arch. Mikrobiol.* **78,** 25–41.

Meeuse, B. J. D. (1975). Thermogenic respiration in aroids. *A. Rev. Plant Physiol.* **26,** 117–126.

Mehrotra, B. S. and Basu, M. (1975). Survey of the microorganisms associated with cereal grains and their milling fractions in India. Part I. Imported wheat. *Int. Biodeterior. Bull.* **11,** 56–63.

Meyer, R. D. and Armstrong, D. (1973). Mucormycosis—changing status. *CRC Crit. Rev. Clin. Lab. Sci.* **4,** 421–451.

Meyer, R. D., Kaplan, M. H., Ong, M. and Armstrong, D. (1973). Cutaneous lesions in disseminated mucormycosis. *J. Am. Med. Assoc.* **225,** 737–738.

Miehe, H. (1907). "Die Selbsterhitzung des Heus. Eine biologische Studie." Gustav Fischer, Jena.

Mil'ko, A. A. and Belyakova, L. A. (1967). Vidy roda *Mucor* s sharovidnymi sporangiosporami. *Mikrobiologiya* **36,** 111–120.

Mills, J. and Eggins, H. O. W. (1970). Growth of thermophilic fungi on oxidation products of polyethylene. *Int. Biodeterior. Bull.* **6,** 13–17.

Mills, J. and Eggins, H. O. W. (1974). The biodeterioration of certain plasticisers by thermophilic fungi. *Int. Biodeterior. Bull.* **10,** 39–44.

Mills, J., Barnes, T. G. and Eggins, H. O. W. (1971). *Talaromyces emersonii*—a possible biodeteriogen. *Int. Biodeterior. Bull.* **7,** 105–108.

Mills, J., Allsopp, D. and Eggins, H. O. W. (1972). *In* "Biodeterioration of Materials" (Eds A. H. Walters and E. H. Hueck-van der Plas), Vol. 2, pp. 227–232. Wiley, New York.

Minoura, K., Yokoe, M., Kizima, T. and Nehira, T. (1973a). Thermophilic filamentous fungi in Japan (1). *Trans. Mycol. Soc. Japan* **14,** 352–361.

Minoura, K., Ochi, K. and Nehira, T. (1973b). Thermophilic filamentous fungi in Japan (2). *Trans. Mycol. Soc. Japan* **14,** 362–366.

Mitchell,R. (1960). The evolution of thermiphilous [*sic*] water mites. *Evolution* **14,** 361–377.

Mitchell, R. (1974). The evolution of thermophily in hot springs. *Q. Rev. Biol.* **49,** 229–242.

Morgan-Jones, G. (1974). Notes on Hyphomycetes. V. A new thermophilic species of *Acremonium. Can. J. Bot.* **52,** 429–431.

Morton, L. H. G. and Eggins, H. O. W. (1975) Thermophilous fungi from timber. *Int. Biodeterior. Bull.* **11,** iv (Abst.).

Mosser, J. L., Mosser, A. G. and Brock, T. D. (1973). Bacterial origin of sulfuric acid in geothermal habitats. *Science* **179,** 1323–1324.

Mosser, J. L., Bohlool, B. B. and Brock, T. D. (1974a). Growth rates of *Sulfolobus acidocaldarius* in nature. *J. Bact.* **118,** 1075–1081.

Mosser, J. L., Mosser, A. G. and Brock, T. D. (1974b). Population ecology of *Sulfolobus acidocaldarius*. I. Temperature strains. *Arch. Microbiol.* **97,** 169–179.
Mouchacca, J. (1973). Les *Thielavia* des sols arides: espèces nouvelles et analyse générique. *Bull. Soc. Mycol. France* **89,** 295–311.
Mountford, M. D. (1971). Population survival in a variable environment. *J. Theor. Biol.* **32,** 75–79.
Mulinge, S. K. and Apinis, A. E. (1969). Occurrence of thermophilous fungi in stored moist barley grain. *Trans. Br. Mycol. Soc.* **53,** 361–370.
Mulinge, S. K. and Chesters, C. G. C. (1970). Ecology of fungi associated with moist stored barley grain. *Ann. Appl. Biol.* **65,** 277–284.
Myers, J. W., Pfeiffer, J. J., Murphy, E. M. and Griffith, F. E. (1966). Ignition and control of burning of coal mine refuse. U.S. Bur. Mines Rep. Invest. 6758. 24 pp.
Nagy, K. A., Odell, D. K. and Seymour, R. S. (1972). Temperature regulation by the inflorescence of philodendron. *Science* **178,** 1195–1197.
Nikitina, Z. I., Pchelintseva, T. P. and Kholler, V. A. (1970). Termogenez gribov, vydelennykh iz samorazogrevayushchegosya torfa. *Biol. Nauk. USSR* **13,** (8), 88–91.
Nilsson, T. (1965). Mikroorganismer i flisstackar. *Svensk Papp-Tidn.* **68,** 495–499.
Nilsson, T. (1973a). Micro-organisms in chip piles. Res. Note No. R83, 23 pp. Dept. For. Prod., R. Coll. For., Stockholm.
Nilsson, T. (1973b). Studies on wood degradation and cellulolytic activity of microfungi. *Stud. For. Suec.* **104,** 1–40.
Nottebrock, H., Scholer, H. J. and Wall, M. (1974). Taxonomy and identification of mucormycosis-causing fungi. I. Synonymity of *Absidia ramosa* with A. *corymbifera. Sabouraudia* **12,** 64–74.
Ofosu-Asiedu, A. and Smith, R. S. (1973a). Some factors affecting wood degradation by thermophilic and thermotolerant fungi. *Mycologia* **65,** 87–98.
Ofosu-Asiedu, A. and Smith, R. S. (1973b). Degradation of three softwoods by thermophilic and thermotolerant fungi. *Mycologia* **65,** 240–244.
Okafor, N. (1966). Thermophilic micro-organisms from rotting maize. *Nature, Lond.* **210,** 220–221.
Okazaki, H. and Iizuka, H. (1970). Lysis of living mycelia of a thermophilic fungus, *Humicola lanuginosa*, by the cell wall-lytic enzymes produced from a thermophilic actinomycete. *J. Gen. Appl. Microbiol., Tokyo* **16,** 537–541.
Okazaki, H. and Iizuka, H. (1971). On yeast cell lytic activity and β-1, 3-glucanase produced from thermophilic fungi. *J. Agr. Chem. Soc. Japan* **45,** 460–470.
Orla-Jensen, S. (1942). "The Lactic Acid Bacteria" (2nd ed.) 197 pp. Kobenhavn.
Oshima, T. and Imahori, K. (1974). Description of *Thermus thermophilus* (Yoshida and Oshima) comb. nov., a nonsporulating thermophilic bacterium from a Japanese thermal spa. *Int. J. Syst. Bactiol.* **24,** 102–112.
Oso, B. A. (1974a). Thermophilic fungi from stacks of oil palm kernels in Nigeria. *Z. Allg. Mikrobiol.* **14,** 593–601.
Oso, B. A. (1974b). Carbon source requirements for the thermophilic ascomycete *Chaetomium thermophile* var. *coprophile. Z. Allg. Mikrobiol.* **14,** 603–610.
Oso, B. A. (1974c). Utilization of lipids as sole carbon sources by thermophilic fungi. *Z. Allg. Mikrobiol.* **14,** 713–717.
Pask-Hughes, R. and Williams, R. A. D. (1975). Extremely thermophilic Gram-negative bacteria from hot tap water. *J. Gen. Microbiol.* **88,** 321–328.
Pax, F. (1951). Die Grenzen tierischen Lebens in mitteleuropäischen Thermen. *Zool. Anz.* **147,** 275–284.

Phelps, A. (1961). Studies on factors influencing heat survival of a ciliate, a mite, and an ostracod, obtained from a thermal stream. *Am. Zool.* **1**, 467.

Pierson, B. K. and Castenholz, R. W. (1974a). A phototrophic gliding filamentous bacterium of hot springs, *Chloroflexus aurantiacus*, gen. and sp. nov. *Arch. Microbiol.* **100**, 5–24.

Pierson, B. K. and Castenholz, R. W. (1974b). Studies of pigments and growth in *Chloroflexus aurantiacus*, a phototrophic filamentous gliding bacterium. *Arch. Microbiol.* **100**, 283–305.

Pirie, H. M., Dawson, C. O., Breeze, R. G., Selman, I. E. and Wiseman, A. (1971). Fog fever and precipitins to micro-organisms of mouldy hay. *Res. Vet. Sci.* **12**, 586–588.

Poincelot, R. (1972). The biochemistry and methodology of composting. Conn. Agr. Exp. Stn. Bull. 727. 38 pp.

Poincelot, R. (1974). A scientific examination of the principles and practice of composting. *Compost Sci.* **15**, (3), 24–31.

Poljansky, G. I. and Sukhanova, K. M. (1967). *In* "The Cell and Environmental Temperature" (Ed. A. S. Troshin), pp. 200–209 +258–261. Pergamon Press, Oxford.

Pope, S., Knaust, H. and Knaust, K. (1962). Production of compost by thermophilic fungi. *Mushr. Sci.* **5**, 123–126.

Por, F. D. (1969). Limnology of the heliothermal Solar Lake on the coast of Sinai (Gulf of Elat). *Verh. Int. Verein. Theor. Angew. Limnol.* **17**, 1031–1034.

Pore, R. S. and Larsh, H. W. (1967). First occurrence of *Thermoascus aurantiacus* from animal and human sources. *Mycologia* **59**, 927–928.

Pounds, J. R. and Lucas, G. B. (1972). Thermophilic fungi of tobacco. N. C. Agr. Exp. Stn., Raleigh Tech. Bull. No. 211. 24 pp.

Pridham, T. G. and Tresner, H. D. (1974). *In* "Bergey's Manual of Determinative Bacteriology." (8th ed). (Eds R. E. Buchanan and N. E. Gibbons), pp. 748–829. Williams & Wilkins, Baltimore.

Prodromou, M. C. and Chapman, E. S. (1974). Effects of nitrogen sources at various temperatures on *Papulospora* [sic] *thermophila. Mycologia* **66**, 876–880.

Qureshi, A. R. and Johri, B. N. (1972). Temperature relationship of some thermophilic fungi. *Bull. Bot. Soc. Univ. Sagar* **19**, 28–30.

Ramabadran, R. (1967–1968). The role of thermophilic micro-organisms in nature. *Annamalai Univ. Agr. Mag.* **8**, 74–76.

Ramaley, R. F. and Hixson, J. (1970). Isolation of a nonpigmented, thermophilic bacterium similar to *Thermus aquaticus. J. Bact.* **103**, 527–528.

Ranck, R. M., Jr., Georg, L. K. and Wallace, D. H. (1974). Dactylariosis—a newly recognized fungus disease of chickens. *Avian Dis.* **18**, 4–20.

Renard, Y. and Cailleux, R. (1973). Contribution à l'étude des microorganismes du compost destiné à la culture du champignon de couche. *Rev. Mycol.* **37**, 36–47.

Rész, A. (1968). Untersuchungen über den Mikroorganismenbesatz von belüftetem Heu. *Zentbl. Bakt. Parasitkde.* Atb. II. **122**, 597–634.

Rész, A. (197?). *In* "Self heating of Organic Materials; Proceedings of the Symposium." (International Symposium Organized by the Laboratory of Physical Technology, Technical University of Delft, and the Federation of Mutual Fire Insurance Companies, Utrecht, February 18–19, 1971, Delft), pp. 127–138.

Riley, J. (1965). Further studies on temperature rise in large stacks of cocoa. Nigerian Stored Prod. Res. Inst. A. Rep., Tech. Rep. No. 1, 21–23.

Rippon, J. W. (1974). "Medical Mycology." W. B. Saunders, Philadelphia, London and Toronto.

Rode, L. J., Foster, J. W. and Schuhardt, V. T. (1947). Penicillin production by a thermophilic fungus. *J. Bact.* **53,** 565–566.

Rogosa, M. (1974). *In* "Bergey's Manual of Determinative Bacteriology." (8th ed.) (Eds R. E. Buchanan and N. E. Gibbons), pp. 669–676. Williams & Wilkins, Baltimore.

Romanelli, R. A., Houston, C. W. and Barnett, S. M. (1975). Studies on thermophilic cellulolytic fungi. *Appl. Microbiol.* **30,** 276–281.

Ross, D. A. (1972). Red Sea hot brine area: revisited. *Science* **175,** 1455–1457.

Ross, D. A., Whitmarsh, R. B., Ali, S. A., Boudreaux, J. E., Coleman, R., Fleisher, R. L., Girdler, R., Manheim, F., Matter, A., Nigrini, C., Stoffers, P. and Supko, P. R. (1973). Red Sea drillings. *Science* **179,** 377–380.

Rothbaum, H. P. (1961). Heat output of thermophiles occurring on wool. *J. Bacteriol.* **81,** 165–171.

Rothbaum, H. P. and Dye, M. H. (1964). Self-heating of damp wool. Part III. Self-heating of damp wool under isothermal conditions. *N.Z. J. Sci.* **7,** 119–146.

Rozanova, E. P. and Khudyakova, A. I. (1974). Novji bessporovji termofil'nyi organizm, vostablivayushchii sut'faty, *Desulfovibrio thermophilus* nov. sp. *Mikrobiologiya* **43,** 1069–1075.

Rychtera, M. (1971). *In* "Deterioration of Electrical Equipment in Adverse Environments" (Ed. E. W. Firth) Daniel Davey and Company, Hartford.

Saëz, H. (1966). *Aspergillus nidulans* (Eidam) Winter, une espece thermophile commune, chez l'animal. *Bull. Mens. Soc. Linn. Lyon* **35,** 467–472.

Saiki, T., Kimura, R. and Arima, K. (1972). Isolation and characterization of extremely thermophilic bacteria from hot springs. *Agr. Biol. Chem.* **36,** 2357–2366.

Salminen, K., Talja, R., Vasenius, H. and Weckström, P. (1974). The fungous flora of the suana and the influence of certain disinfectants. *Zentbl. Bakt. ParasitKde* Abt. I. Orig. B **158,** 552–560.

Samson, R. A. (1974). *Paecilomyces* and some allied hyphomycetes. Studies in Mycology No. 6. Centraalbureau voor Schimmelcultures, Baarn, Netherlands.

Samson, R. A. and Tansey, M. R. (1975). *Byssochlamys verrucosa* sp nov. *Trans. Br. Mycol. Soc.* **65,** 512–514.

*Samson, R. A., Crisman, M. Jack and Tansey, M. R. (1977). Observations on the thermophilic ascomycete *Thielavia terrestris. Trans. Br. Mycol. Soc.* **69,** 417–**423.**

San Antonio, J. P. (1966). Effects of injection of nutrient solutions into compost on the yield of mushrooms (*Agaricus bisporus*). *Proc. Am. Soc. Hortic. Sci.* **89,** 415–422.

*Satyanarayana, T., Johri, B. N. and Saksena, S. B. (1977). Seasonal variation in mycoflora of nesting materials of birds with special reference to thermophilic fungi. *Trans. Br. Mycol. Soc.* **68,** 307–309.

Schmidt, F. L. (1969). Observations on spontaneous heating toward combustion of commercial chip piles. TAPPI **52,** 1700–1701.

Schmidt, R. A. and Wood, F. A. (1969). Temperature and relative humidity regimes in the pine stump habitat of *Fomes annosus. Can. J. Bot.* **47,** 141–154.

Scholer, H. J. (1974). *In* "Aspergillosis and Farmer's Lung in Man and Animal" (Eds R. de Haller and F. Suter), pp. 35–40. Hans Huber Publishers, Bern, Stuttgart, Vienna.

Schramm, J. R. (1966). Plant colonization studies on black wastes from anthracite mining in Pennsylvania. *Trans. Am. Philos. Soc. N. S.* **56,** 1–194.

Scott, J. A. and Hill, E. C. (1971). *In* "Microbiology 1971" (Ed. P. Hepple), pp. 25–41. Institute of Petroleum, London.

Seabury, J., Salvaggio, J., Buechner, H. and Kundur, V. G. (1968). Bagossois III. Isolation of thermophilic and mesophilic actinomycetes and fungi from moldy bagasse. *Proc. Soc. Exp. Biol. Med.* **129,** 351–360.

Seabury, J., Salvaggio, J., Domer, J., Fink, J. and Kawai, T. (1973). Characterization of thermophilic actinomycetes isolated from residential heating and humidification systems. *J. Allergy Clin. Med.* **51,** 161–173.

Seal, K. J. and Eggins, H. O. W. (1972). The role of micro-organisms in the biodegradation of farm animal wastes with particular reference to intensively produced wastes. A review. *Int. Biodeterior. Bull.* **8,** 95–100.

Semeniuk, G. and Carmichael, J. W. (1966). *Sporotrichum thermophile* in North America. *Can. J. Bot.* **44,** 105–108.

*Sheridan, R. P. (1976). Sun-shade ecotypes of bluegreen algae in a hot spring. *J. Phycol.* **12,** 279–285.

Sheridan, J. E. and Soteros, J. J. (1974). A survey of fungi in jet aircraft fuel systems in New Zealand. *Int. Biodeterior. Bull.* **10,** 105–107.

*Sheridan, R. P. and Ulik, T. (1976). Adaptive photosynthetic responses to temperature extremes by the thermophilic cyanophyte *Synechococcus lividus. J. Phycol.* **12,** 255–261.

Shields, J. K. (1966). Microbiological deterioration of piled wood chips stored outside. Can. Dep. For., Bi-mon. Res. Notes **22,** 5–6.

Shields, J. K. (1967). Microbiological deterioration in the wood chip pile. Dep. Publ. No. 1191, 29 pp. Forestry Branch (Canada).

Shields, J. K. (1969a). Inhibition of fungi in a softwood chip pile. Can. Dept. For., Bi-mon. Res. Notes **25,** 3–5.

Shields, J. K. (1969b). Microflora of eastern Canadian wood chip piles. *Mycologia* **61,** 1165–1168.

Shields, J. K. (1970). Brown-stain development in stored chips of spruce and balsam fir. TAPPI **53,** 455–457.

Shivvers, D. W. and Brock, T. D. (1973). Oxidation of elemental sulfur by *Sulfolobus acidocaldarius. J. Bact.* **114,** 706–710.

Sinha, R. N. and Wallace, H. A. H. (1965). Ecology of a fungus-induced hot spot in stored grain. *Can. J. Plant Sci.* **45,** 48–59.

Sinha, R. N. and Wallace, H. A. H. (1973). Population dynamics of stored-product mites. *Oecologia* **12,** 315–327.

Sinha, R. N., Yaciuk, G. and Muir, W. E. (1973). Climate in relation to deterioration of stored grain. A multivariate study. *Oecologia* **12,** 69–88.

Sládečková, A. (1969). Control of slimes and algae in cooling systems. *Verh. Int. Verein. Theor. Angew. Limnol.* **17,** 532–538.

Smith, D. W. and Brock, T. D. (1973). The water relations of the alga *Cyanidium caldarium* in soil. *J. Gen. Microbiol.* **79,** 219–231.

Smith, L. D. S. and Hobbs, G. (1974). *In* "Bergey's Manual of Determinative Bacteriology." (8th ed.) (Eds R. E. Buchanan and N. E. Gibbons), pp. 551–572. Williams & Wilkins, Baltimore.

Smith, R. S. (1973). Colonization and degradation of outside stored softwood chips by fungi. Res. Note No. R83, 16 pp. Dept. For. Prod., R. Coll. For., Stockholm.

Smith, R. S. and Hatton, J. V. (1971). Economic feasibility of chemical protection for outside chip storage. TAPPI **54,** 1638–1640.

Smith, R. S. and Ofosu-Asiedu, A. (1972). Distribution of thermophilic and thermotolerant fungi in a spruce-pine chip pile. *Can. J. For. Res.* **2,** 16–26.

Smith, R. S. and Ofosu-Asiedu, A. (1973). Degradation of arabinose in wood attacked by thermophilic fungi. Can. Dept. For., Bi-mon. Res. Notes **29,** 3–4.

Somkuti, G. A. (1974). Synthesis of cellulase by *Mucor pusillus* and *Mucor miehei. J. Gen. Microbiol.* **81,** 1–6.

Somkuti, G. A. and Walter, M. M. (1970). Antimicrobial polypeptide synthesized by *Mucor pusillus* NRRL 2543. *Proc. Soc. Exp. Biol. Med.* **133,** 780–785.

Sorensen, S. G. and Crisan, E. V. (1974). Thermostable lactase from thermophilic fungi. *J. Food Sci.* **39,** 1184–1187.

Sorokin, C. (1967). New high-temperature *Chlorella. Science* **158,** 1204–1205.

Spaulding, P. S. (1929). Decay of slash of northern white pine in southern New England. U.S.D.A. Tech. Bull. No. 132, 1–20.

Sprenger, E. (1930). Bacillariales aus den Thermen und der Umgebung von Karlsbad. *Arch. Protistenk.* **71,** 502–542.

Springer, E. L., Feist, W. C., Zoch, L. L., Jr. and Hajny, G. J. (1973a). Evaluation of chemicals for preserving wood chips using pile simulators. TAPPI **56,** 125–128.

Springer, E. L., Feist, W. C., Zoch, L. L., Jr. and Hajny, G. L. (1973b). New and effective chemical treatment to preserve stored wood chips. TAPPI **56,** 157.

Springer, E. L., Hajny, G. J. and Feist, W. C. (1971). Spontaneous heating in piled wood chips. II. Effect of temperature. TAPPI **54,** 589–591.

Stahl, R. W. (1964). Survey of burning coal-mine refuse banks. U.S. Bur. Mines Inform. Circ. 8209, 39 pp.

Staněk, M. (1972). Microorganisms inhabiting mushroom compost during fermentation. *Mushr. Sci.* **8,** 797–811.

Stockner, J. G. (1967). Observations of thermophilic algal communities in Mount Rainier and Yellowstone National Parks. *Limnol. Oceanogr.* **12,** 13–17.

Stockner, J. G. (1968a). The ecology of a diatom community in a thermal stream. *Br. Phycol. Bull.* **3,** 501–514.

Stockner, J. G. (1968b). Algal growth and primary productivity in a thermal stream. *J. Fish. Res. Bd. Canada* **25,** 2037–2058.

Stolk, A. C. (1965). Thermophilic species of *Talaromyces* Benjamin and *Thermoascus* Miehe. *Antonie van Leeuwenhoek* **31,** 262–276.

Stolk, A. C. and Samson, R. A. (1972). The genus *Talaromyces.* Studies on *Talaromyces* and related genera II. Studies in Mycology No. 2. Centraalbureau voor Schimmelcultures, Baarn, Netherlands.

Stolk, A. C., Evans, H. C. and Nilsson, T. (1969). *Penicillium argillaceum* sp. nov., a thermotolerant *Penicillium. Trans. Br. Mycol. Soc.* **53,** 307–311.

Streets, B. W. and Ingle, M. B. (1972). The effect of temperature on spore germination and growth of *Mucor miehei* in submerged culture. *Can. J. Microbiol.* **18,** 975–979.

Stretton, R. J. (1975). Experimentally induced mycetoma: species of *Sporotrichum and Sporothrix. Mycopathologia* **55,** 83–90.

Stutzenberger, F. J., Kaufman, A. J. and Lossin, R. D. (1970). Celluloytic activity in municipal solid waste composting. *Can. J. Microbiol.* **16,** 553–560.

Sukhanova, K. J. (1967). *In* "The Cell and Environmental Temperature" (Ed. A. S. Troshin), pp. 255–257. Pergamon Press, Oxford.

Sumner, J. L. and Evans, H. C. (1971). The fatty acid composition of *Dactylaria* and *Scolecobasidium*. *Can. J. Microbiol.* **17,** 7–11.

Sumner, J. L., Morgan, E. D. and Evans, H. C. (1969). The effect of growth temperature on fatty acid composition of fungi in the order Mucorales. *Can. J. Microbiol.* **15,** 515–520.

Taber, R. A. and Pettit, R. E. (1975). Occurrence of thermophilic microorganisms in peanuts and peanut soil. *Mycologia* **67,** 157–161.

Taha, S. M., Zayed, M. N. and Zohdy, L. (1968). Bacteriological and chemical studies in rice straw compost. III. Effect of ammoniacal nitrogen. *Zentbl. Bakt. ParasitKde* Atb. **II. 122,** 500–509.

Tansey, M. R. (1971a). Isolation of thermophilic fungi from self-heated, industrial wood chip piles. *Mycologia* **63,** 537–547.

Tansey, M. R. (1971b). Agar-diffusion assay of celluloytic ability of thermophilic fungi. *Arch. Mikrobiol.* **77,** 1–11.

Tansey, M. R. (1972). Effect of temperature on growth rate and development of the thermophilic fungus *Chaetomium thermophile*. *Mycologia* **64,** 1290–1299.

Tansey, M. R. (1973). Isolation of thermophilic fungi from alligator nesting material. *Mycologia* **65,** 594–601.

Tansey, M. R. (1975). Isolation of thermophilic fungi from snuff. *Appl. Microbiol.* **29,** 128–129.

Tansey, M. R. and Brock, T. D. (1971). Isolation of thermophilic and thermotolerant fungi from hot spring effluents and thermal soils of Yellowstone National Park. *Bact. Proc.* **1971,** 36.

Tansey, M. R. and Brock, T. D. (1972). The upper temperature limit for eukaryotic organisms. *Proc. Natl. Acad. Sci. USA* **69,** 2426–2428.

Tansey, M. R. and Brock, T. D. (1973). *Dactylaria gallopava*, a cause of avian encephalitis, in hot spring effluents, thermal soils and self-heated coal waste piles. *Nature, Lond.* **242,** 202–203.

Tansey, M. R. and Jack, M. A. (1975). *Thielavia australiensis* sp. nov., a new thermophilic fungus from incubator-bird (mallee fowl) nesting material. *Can. J. Bot.* **53,** 81–83.

*Tansey, M. R., Murrmann, D. N., Behnke, B. K. and Behnke, E. R. (1977). Enrichment, isolation, and assay of growth of thermophilic and thermotolerant fungi in lignin-containing media. *Mycologia* **69,** 463–476.

Tendler, M. D. and Korman, S. (1965a). Antibiosis among the thermophilic *Eumyces:* temperature and nutritional effect. *Fed. Proc.* (1) **24,** 403.

Tendler, M. D. and Korman, S. (1965b). Temperature and nutritional effects on the enzyme activity of antibiotic-producing thermophilic *Eumyces*. *Bact. Proc.* **1965,** 20.

Tendler, M. D. and Korman, S. (1965c). Physiological studies of thermophily among the *Eumyces*. *23rd Int. Congr. Physiol. Sci., Tokyo.*

Tendler, M. D., Korman, S. and Nishimoto, M. (1967). Effects of temperature and nutrition on macromolecule production by thermophilic Eumycophyta. *Bull. Torrey Bot. Club* **94,** 175–181.

Thakre, R. P. and Johri, B. N. (1973). In vitro activity of aureofungin, amphotericin B and hamycin on the mycellial growth of some thermophilic fungi. *Hindustan Antibiot. Bull.* **15,** 75–78.

Thakre, R. P. and Johri, B. N. (1973–1974). Influence of antibiotics on the swelling phase and spore germination of thermophilic fungi. *Hindustan Antibiot. Bull.* **16,** 109–114.

Townsley, P. M. (1974). Pure cultural industrial fermentation in the production of mushrooms. *Can. Inst. Food Sci. Technol. J.* **7,** 254–255.

Tubaki, K., Ito, T. and Matsuda, Y. (1974). Aquatic sediments as a habitat of thermophilic fungi. *Annal. Microbiol.* **24,** 199–207.

Turner, W. E. and Ahearn, D. B. (1970). *In* "Recent Trends in Yeast Research" (Ed. D. G. Ahearn), pp. 113–123. Georgia State University, Atlanta.

Uchino, F. and Doi, S. (1967). Acido-thermophilic bacteria from thermal waters. *Agri. Biol. Chem.* **31,** 817–822.

Udagawa, S., Furuya, K. and Horie, Y. (1973). Notes on some ascomycetous microfungi from soil. *Bull. Natl. Sci. Mus.* (*Tokyo*) **16,** 503–520.

Uyemura, M. (1936). Biological studies of thermal waters in Japan. IV. *Ecol. St.* **2,** 171.

Uyemura, M. (1937). Biological studies of thermal waters in Japan. V. *Rep. Jap. Sci.* A. **12,** 264.

van Veen, A. G. and Steinkraus, K. H. (1970). Nutritive value and wholesomeness of fermented foods. *Agric. Food Chem.* **18,** 576–578.

Waldrip, D. W., Padhye, A. A. and Ajello, L. (1974). Isolation of *Dactylaria gallopava* from broiler house litter. *Abst. A. Meeting Am. Soc. Microbiol.* **1974,** 143.

Walker, I. K. (1963). Spontaneous combustion of wool. *Wool Rec.* **104,** 45.

Walker, I. K. and Williamson, H. M. (1957). The spontaneous ignition of wool. I. The causes of spontaneous fires in New Zealand wool. *J. Appl. Chem.* **7,** 468–480.

Walter, M. C. (1969). Isolation and characterization of a peptide antibiotic synthesized by *Mucor pusillus.* M. Sc. thesis, Duquesne University.

Ward, J. E., Jr. and Cowley, G. T. (1972). Thermophilic fungi of some central South Carolina forest soils. *Mycologia* **64,** 200–205.

Weinberg, E. D. (1974). Secondary metabolism: control by temperature and inorganic phosphate. *Dev. Ind. Microbiol.* **15,** 70–81.

Williams, R. A. D. (1975). Caldoactive and thermophilic bacteria and their thermostable proteins. *Sci. Prog. Oxford* **62,** 373–393.

Williams, R. A. D. and Hoare, D. S. (1972). Physiology of a new facultatively autotrophic thermophilic *Thiobacillus. J. Gen. Microbiol.* **70,** 555–566.

Wiseman, A., Selman, I. E., Dawson, C. O., Breeze, R. G. and Pirie, H. M. (1973). Bovine farmers' lung: a clinical syndrome in a herd of cattle. *Vet. Rec.* **93,** 410–417.

Wood, J. L. (1955). *Mucor pusillus* Lindt from Maryland and some notes on its nutrition. *Proc. Pa. Acad. Sci.* **29,** 115–120.

Yamamoto, Y. (1930a). Ein Beitrag zur Kenntnis der Gattung *Rhizopus.* I. *J. Fac. Agr. Hokkaido Imp. Univ.* **28,** 1–102.

Yamamoto, Y. (1930b). Ein Beitrag zur Kenntnis der Gattung *Rhizopus.* II. *J. Fac. Agr. Hokkaido Imp. Univ.* **28,** 103–327.

Yamazaki, M., Fujimoto, H. and Kawasaki, T. (1975). Structure of a tremorgenic metabolite from *Aspergillus fumigatus,* fumitremorgin A. *Tetrahedron Lett.* **14,** 1241–1246.

Young, F. N. and Zimmerman, J. R. (1956). Variations in temperature in small aquatic situations. *Ecology* **37,** 609–611.

Zakharchenko, V. O. and Zhernova, I. I. (1971). Vidilennya termofil'nikh ta termotolerantnikh gribiz z gruntiv URSR. *Mikrobiol. Zh.* (*Kiev*) **33,** 705–706.
Zeikus, J. G. and Wolfe, R. S. (1972). *Methanobacterium thermoautotrophicum* sp. n., an anaerobic, autotrophic, extreme thermophile. *J. Bact.* **109,** 707–713.
ZoBell, C. E. (1959). Factors affecting drift seaweeds on some San Diego beaches. IMR Ref. No. 59–3. Institute of Marine Sciences, University of California, La Jolla.

Chapter 6

Microbial Life at High Temperatures: Mechanisms and Molecular Aspects

R. E. AMELUNXEN and A. L. MURDOCK

University of Kansas Medical Centre

I. Introduction 217
II. Classification and general characteristics 218
III. Proposed mechanisms of thermophily and related aspects: general considerations 220
A. Lipids and membrane involvements 220
B. Enzymes from cell-free extracts, transferable stabilizing factors, and rapid resynthesis 223
C. Macromolecular thermostability 226
D. Highly charged macromolecular environment 235
E. Allosterism, thermoadaptation and transformation 238
IV. Detailed consideration of selected proteins from obligate and caldoactive thermophiles 243
A. Glyceraldehyde-3-phosphate dehydrogenase 243
B. Ferredoxin 252
C. Thermolysin 258
D. Formyltetrahydrofolate synthetase 262
V. Conclusions 264
Acknowledgements 267
References 267

I. Introduction

It is evident from Chapter 5 that a great variety of thermophilic prokaryotic and eukaryotic organisms exist in nature. The authors have presented a very thorough classification of thermophilic microorganisms which is so necessary in evaluating all future work on thermophily. In this chapter it is our purpose to discuss and evaluate the various mechanisms and molecular aspects that have been presented in explaining thermophilic existence.

Of the wide variety of environmental stresses, certainly one of the most extreme is that of elevated temperature. It is natural for scientists to be curious and intrigued about microorganisms which not only tolerate but reproduce, often obligately, at temperatures which normally preclude life by destroying the necessary macromolecules. In recent years, a concerted effort has been made to elucidate molecular reasons for thermophily. It must be emphasized that thermophily undoubtedly encompasses an array of molecular mechanisms and cannot be explained by any one single attribute. Comparative physicochemical studies of thermophilic proteins with their counterparts from non-thermophilic systems are numerous and afford in many instances valid basis for a consideration of molecular mechanisms. It is our intent to first present an overview of work on thermophily followed by a detailed consideration of a few selected systems that have been well characterized.

II. Classification and General Characteristics

Most known bacterial species reproduce optimally at temperatures between 30°C and 45°C and are termed mesophiles. Bacteria which have optima for growth at about 20°C, and a range of growth from 0°C to 30°C are referred to as psychrophiles or psychrotrophes (Chapter 2; Stanier *et al.*, 1970). In considering thermophiles, three main groups have been proposed (Farrell and Campbell, 1969).

(1) Strict, or obligate thermophiles demonstrate optimal growth at 65°C to 70°C but do not grow below 40°C to 42°C.
(2) Facultative thermophiles have a maximal growth temperature between 50°C and 65°C but also are capable of reproducing at room temperature.
(3) Thermotolerant bacteria have growth maxima at 45°C to 50°C and also grow at room temperature.

It is not unusual for classification schemes to be fraught with some difficulties. As stressed in a review by Williams (1975), there are now many bacterial strains which are capable of growing at higher temperatures than a classical obligate thermophile such as *Bacillus stearothermophilus*. It becomes apparent that group (1) above has to be supplemented or re-defined to include organisms growing above 70°C.

Brock and Freeze (1969) were the first to describe the properties of a new genus and species, i.e. the Gram negative extreme thermophile, *Thermus aquaticus*. Strains of these yellow-pigmented, non-sporulating rods were isolated from a variety of sources, e.g. hot springs in Yellowstone National Park, hot tap water and thermally polluted streams. Ramaley and Hixson

(1970) isolated *Thermus* X-1 which is not pigmented, and Oshima and Imahori (1974) isolated an extreme thermophile, *Thermus thermophilus* HB8 (initially called *Flavobacterium thermophilum* HB8). Since many extreme thermophiles have growth optima above 70°C, Heinen and Heinen (1972) have introduced the new term "caldoactive" (caldus from the Latin meaning hot), to describe extremely thermophilic bacteria; the following subdivision has been proposed (Williams, 1975).

(1) Caldoactive bacteria—maximum growth temperature above 70°C, optimum above 65°C, and minimum above 40°C.
(2) Thermophilic bacteria—maximum growth temperature above 60°C, optimum above 50°C, minimum above 30°C.

Although it is not our purpose to delve into classification, it seems to us that group (1) above has validity, but only as an addition to the original classification of Farrell and Campbell (1969). Group (2) appears to ignore certain facultative thermophiles such as *Bacillus coagulans* KU which have a maximum growth temperature of about 58°C, and does not provide for the thermotolerant bacteria. Additional difficulties arise in the application of broad and rigid classifications in light of the recent reports of thermoadaptation wherein the cardinal growth temperatures can be altered with concomitant alterations in the enzyme levels and thermostability of certain enzymes. This important aspect will be discussed in Section III.E.

Babel *et al.* (1972) suggested that changes in the conformation of one or more essential enzymes control the cardinal growth temperatures. Based on this assumption, these authors define cardinal temperatures as follows: T_{min} is the minimal temperature for growth where a transition from a rigid inactive conformation to one of limited flexibility occurs; T_{opt} is the optimal growth temperature and defines the most favourable conformational state; at T_{max} there is the onset of conformational disruption and decreased enzymic activity, and above this temperature growth ceases through thermal denaturation. In comparing the properties of nucleic acids, lipids and enzymes in defining T_{max}, it was concluded that a stronger case can be built for implicating the enzymes. Babel *et al.* (1972) attempted to explain T_{min} for growth using the thermophile, *B. stearothermophilus* ATCC 12980; the three cardinal temperatures for this organism are 37°C (T_{min}), 63°C (T_{opt}) and 72°C (T_{max}). In interpreting their data, the authors excluded many of the proposed causes for T_{min}, e.g. accumulation of toxic metabolic products due to an imbalance of enzymes with different temperature characteristics, feed-back inhibition, repression of enzyme synthesis, inhibition of membrane transport, membrane alterations and translational errors in protein synthesis. It was proposed that the DNA-dependent RNA polymerase is the major controlling factor for T_{min}. This enzyme from *B. stearothermophilus* ATCC 12980 demonstrated a sharp breaking point at 33°C with an activation energy of 65 kcal mol^{-1} below this temperature, and 13 kcal mol^{-1} above this temperature; this represents a significant increase in catalytic activity.

It should be emphasized that regardless of the ability of thermophilic microorganisms to reproduce at high temperatures, they do in general resemble their mesophilic counterparts in terms of carbon and nitrogen sources for growth, metabolic pathways, their existence as aerobes, anaerobes, facultative anaerobes, autotrophs and heterotrophs. However, the extreme thermophiles do not appear closely related to non-thermophilic isolates. Thermophiles occur in a wide (and often unusual) variety of environmental conditions including arctic glaciers, freshly fallen snow, thermal pools, desert sands, soil of temperate and tropical zones, milk and various foodstuffs.

Previous reviews concerned with various aspects of thermophily include: Gaughran (1947), Allen (1953), Koffler (1957), Brock (1967), Chapman (1967), Brandts (1967), Farrell and Rose (1967a, b), Campbell and Pace (1968), Friedman (1968), Castenholz (1969), Farrell and Campbell (1969), Brock (1970), Brock and Darland (1970), Singleton and Amelunxen (1973), Williams (1975), Ljungdahl and Sherod (1976).

III. Proposed Mechanisms of Thermophily and Related Aspects: General Considerations

A. Lipids and Membrane Involvements

It was first reported by Heilbrunn (1924) and Bělehrádek (1931) that thermophilic organisms had lipids with higher melting points than non-thermophiles. It was suggested that the melting temperature of cellular lipids might determine the upper limit for growth. Kaneda (1963) demonstrated that branched-chain fatty acids of 15 carbons are predominant in several mesophilic species of the genus *Bacillus*. Brock (1967) suggested that an increase in the percentage of saturated and branched-chain fatty acids with an increase in temperature could provide an organism with a more stable membrane. Fatty acids with 17, 18, and 19 carbon atoms were found to be the major constituents of some extreme thermophiles (Bauman and Simmonds, 1969). Daron (1970) studied the fatty acid composition of lipid extracts of a thermophilic *Bacillus* similar to *B. stearothermophilus*. Fatty acids containing 16 or 17 carbons comprised over 80% of the total produced; branched-chain fatty acids were at a higher level than normal fatty acids. Increasing the growth temperature from 40 to 60°C effected a three to four-fold increase in the ratio of normal to branched chain hexadecanoic acid and a decrease in the proportion of unsaturated fatty acids.

Studies on the effect of growth temperatures on the fatty acid composition of *T. aquaticus* have been reported by Ray *et al.* (1971a). In effecting a change in growth temperature (50 to 75°C in 5°C intervals], the proportions of monoenoic and branched-C_{17} fatty acids decreased, and the proportions of the higher-melting iso-C_{16} and normal-C_{16} fatty acids increased. In cells grown at 75°C, the total fatty acid content was 70% higher than in cells grown at 50°C; the largest increases were in the content of iso-C_{16} and normal-C_{16} fatty acids. In a subsequent paper, Ray *et al.* (1971b) demonstrated that on shift of growth from 50 to 75°C (5°C intervals), there was a two-fold increase in the level of phospholipids and carotenoids and a four-fold increase in glucolipids. The proportion of lipids within each class remained constant as the growth temperature increased. From their data on changes in cellular lipids, the authors suggested that lipids play a role as a molecular mechanism of thermophily and that perhaps the glucolipids increase the thermal stability of the membranes.

Chan *et al.* (1973) conducted spin-labelling studies on a facultative thermophile, *Bacillus* strain Tl (BTl). Using nitroxide stearate probes, it was observed that this fatty acid was incorporated into the membrane fraction. Cells grown at 55°C contained a higher percentage of iso-fatty acids than those grown at 37°C, whereas the latter contained a higher proportion of anti-iso-fatty acids. The membrane of cells grown at 37°C showed a rapid rate of fluidization with increasing temperature which is in accord with the higher melting points of iso-fatty acids. Electron spin resonance studies indicated that the membrane of cells grown at 55°C is an anisotropically rigid environment in which molecular motion is restricted. The potentially major role of proteins in controlling the membrane fluidity was not determined.

Wisdom and Welker (1973) reported that the protoplasts of *B. stearothermophilus* 1503-4R are resistant to osmotic rupture and that stability is maintained by divalent cations. Cells grown at elevated temperatures showed increased thermostability of the protoplasts. For cells grown at elevated temperature, a correlation was demonstrated between protoplast thermostability and a lower lipid content of the membrane of cells, and an increase in the membrane content of the cell and the protein content of the membrane. Based on the stability studies with NADH oxidase and alkaline phosphatase, the authors proposed that the membrane acts as an insulator to the transfer of heat from the environment which protects soluble enzymes from thermal denaturation. No details were given regarding such a mechanism.

The fact has been established that the lipid composition of thermophiles responds to growth temperature. However, there is little definitive information about the relationship of such changes to a generalized mechanism of thermophily, and since the membrane structure of bacteria is relatively unknown it is difficult to make any specific correlations. As an approach to this problem,

Esser and Souza (1974) attempted to correlate thermal death and membrane fluidity in *B. stearothermophilus.* Spin label experiments demonstrated that the dynamics of the lipid phase are unchanged at different growth temperatures. Such data indicate that the preservation of the physical state of the lipids in the membrane of *B. stearothermophilus* is a reason for changes in lipid composition induced by temperature. The authors commented that their data were not in complete accord with the interpretation of Chan *et al.* (1973) that the membrane of cells grown at higher temperatures is more rigid than at lower temperatures. In contrast, membrane rigidity was indicated only if comparisons were made at temperatures below the growth temperature; there were no discernible differences at the growth temperature. The authors proposed the following: (1) for the cytoplasmic membrane of *B. stearothermophilus* to function, lateral lipid phase separations appear necessary; (2) the temperature limit for thermophilic growth may be determined by the phase diagram boundary conditions for the total lipid mixture in the membrane; such conditions are dependent on the melting points of the individual lipids; (3) the T_{min}, T_{max}, and temperature boundaries for lipid phase separations can be predicted, but not the T_{opt} which probably depends on the intrinsic thermostability of other cellular macromolecules.

From the mechanistic viewpoint, the thermophilic acidophile, *Thermoplasma acidophilum*, is of much interest. This cell wall-less mycoplasma-like organism is obviously not a suitable model in analysing thermophily *per se* but there are some unusual properties associated with this most stable and rigid membrane. Smith *et al.* (1973) concluded from their studies that thermophily (growth at 59°C) may be related to the long isoprenol chains of the lipids and that acidophily (growth at pH 2) could be explained by the sole presence of ether lipids and by the drastic reduction in chargeable groups on the membrane proteins. Ruwart and Haug (1974) isolated the membrane from this organism. Although most neutral lipids were found to be esterified with fatty acids, the bulk of the glycolipid and phospholipid contained long chains in ether linkages. The amino acid composition showed relatively low levels of charged amino acids, and relatively high levels of half-cystines, suggesting that the protein is hydrophobic. It was concluded that the solubilization of *T. acidophilum* at elevated pH is probably related to the charge distribution of the amino acids. If the pH is higher than 4, carboxyl groups are ionized resulting in charge repulsion and destabilization of the membrane followed by solubilization. The effects are less than with mesophilic *Mycoplasma* since *T. acidophilum* has only half as many carboxyl and amide groups, and the membrane has fewer carboxyl and amide groups than the total cellular constituents. The authors speculated as to the ability of this organism to function both as an acidophile and a thermophile, with an upper limit of growth of 70°C. A high acidic amino acid content may be a requirement for thermophilic growth but a minimum number of ionizable groups is probably necessary for existence as an acidophile. Since it is known that hydrophobic interactions have a maximum stability at about 60°C, the authors felt that the highest temperature at which optimal growth at low pH could occur is 60°C.

A great deal of information has accumulated on membrane properties and regulation using *Escherichia coli* as a model system (Cronan and Gelmann, 1975). There are many possible analogies to the thermophilic systems, some of which are: (1) the ratio of saturated to unsaturated fatty acids that accumulate during inhibition of phospholipid biosynthesis is highly sensitive to temperature; e.g. the free fatty acid fraction from cells incubated at 15°C contain 10 times more unsaturated fatty acids than that of cells grown at 43°C; (2) the lipids of the membrane of *E. coli* show order–disorder phase transition as the temperature is increased; synthesis of membranes containing less than one-third the normal amount of such lipids results in cell death.

There seems to be little doubt that thermophiles are able to control the physical properties of the cytoplasmic membrane by manipulating its composition in response to temperature changes. However, this general mechanism also seems to operate in the mesophile, *E. coli*. It appears that the nature of the chemical modifications involved define the range of optimal function for the membrane, which undoubtedly defines the cardinal temperatures for growth. Like the intrinsically thermostable proteins found in obligate and caldoactive bacteria, the precise mechanism(s) involved in membrane stability remain(s) evasive. As pointed out by Cronan and Gelmann (1975), the molecular basis for regulation of lipid phase transition is one of the more important unstudied areas in membrane function.

B. Enzymes From Cell Free Extracts, Transferable Stabilizing Factors and Rapid Resynthesis

Early reports on the thermostability of enzymes from thermophilic organisms were concerned with crude cell free extracts. Militzer *et al.* (1949) demonstrated that malate dehydrogenase from an obligate thermophile was stable at 65°C for 2 h; by contrast this enzyme from the mesophile *B. subtilis* was rapidly inactivated at 65°C. The thermostability of 11 enzymes from cell free extracts of *B. stearothermophilus* and *B. cereus* were compared by Amelunxen and Lins (1968). In general, the thermophilic enzymes were significantly more resistant to heat inactivation than the homologous mesophilic enzymes. The glyceraldehyde-3-phosphate dehydrogenase from the thermophile exhibited thermostability as high as 90°C whereas the pyruvic kinase and glutamic-oxaloacetic transaminase showed inactivation at 70°C and were much closer to the range of inactivation exhibited by the mesophile. These relatively less stable enzymes could be involved in defining the maximum growth temperature for the organism.

Howell *et al.* (1969) carried out similar experiments with glycolytic enzymes and 6-phosphogluconate dehydrogenase from the thermophiles *Clostridium tartarivorum* and *C. thermosaccharolyticum* and verified the greater thermostability of thermophilic enzymes in cell free crude extracts than of those from the mesophile *C. pasteurianum.* Of interest is the fact that pyruvate kinase was also less stable in the thermophilic anaerobes than the other thermophilic enzymes.

Acetate synthesis in the mesophile *C. formicoaceticum*, and the thermophile *C. thermoaceticum*, involves a series of enzymes (O'Brien and Ljungdahl, 1972; Andreesen *et al.*, 1973). Cell free extracts of the thermophile catalyse acetate synthesis at 57°C (Ljungdahl *et al.*, 1965; Poston *et al.*, 1966) indicating that all of the required enzymes are quite thermostable; however, in the mesophile, an enzyme such as the methylenetetrahydrofolate dehydrogenase is much less thermostable (Moore *et al.*, 1974).

Early studies by Koffler (1957) and Koffler and Gale (1957) considered transferable protective factors within the cell that might impart thermostability to thermophilic enzymes. This possible mechanism of thermophily was tested by mixing cell free extracts from thermophiles with those from mesophiles and determining the extent of coagulation after heat treatment. The proteins of thermophilic extracts were significantly more stable (i.e., less precipitable) following heat treatment than those from mesophilic extracts, and experiments with mixtures of the two extracts argued against the presence of stabilizing factors in the thermophilic extract or labilizing factors in the mesophilic extract. Because of the low concentration of protein utilized in these earlier experiments, protein concentrations of about 20 mg ml^{-1} were used in re-evaluating the effect of heat treatment in mixture experiments (Amelunxen and Lins, 1968). On heating mixtures at either 60°C or 70°C for 10 min, the total protein loss correlated very closely with half the summation of the individual systems in agreement with the earlier conclusions regarding transferable stabilizing or labilizing factors. However, the above experiments do not exclude the possibility that stabilizing factors could be firmly bound moieties which are non-transferable.

Inherent difficulties are apparent in the above experiments since the causes for enzymic inactivation are subject to different interpretations. It is possible that the proteins could maintain their structural integrity upon heating but precipitate due to aggregation. As an example, in the purification of glyceraldehyde-3-phosphate dehydrogenase from *B. stearothermophilus* (Amelunxen, 1966) this enzyme remains completely in solution after heating crude extracts for 5 min at 80°C, but 70% of the total protein is precipitated. This could be interpreted to mean that 70% of the proteins are inactivated, but resuspension of the precipitate revealed that many of the enzymes are still active. However, no activity remained when mesophilic extracts were heated at lower temperatures and the precipitate resuspended (Amelunxen, unpublished).

From early data and the present state of knowledge of thermophilic enzymes, the concept of transferable protective factors seems untenable for the following reasons:

(1) lack of effect in mixture experiments of cell free crude extracts;
(2) retention of unchanged thermostability during purification, after reaching homogeneity, and for crystallizable enzymes on repeated recrystallization;
(3) ability to account quantitatively for the molecular weight based on the amino acid content, and the identity of molecular weights for homologous enzymes from mesophilic or thermophilic sources. As previously mentioned, stabilizing factors could conceivably be firmly bound but necessarily must be of low molecular weight.

Another proposal to explain thermophily states that rapid resynthesis of thermophilic macromolecules adequately counteracts thermal inactivation. Early work supporting this hypothesis has been reviewed by Allen (1953). Bubela and Holdsworth (1966a, b) have shown that the rates of protein and nucleic acid synthesis and turnover are higher in *B. stearothermophilus* than in *E. coli.* It should be pointed out that in sporeforming bacteria, a high rate of protein turnover is associated with sporulation which can pose complications in interpretation. As shown by Lodish (1969), the ribosomes of *B. stearothermophilus* differ from those of *E. coli* in their requirements for initiating polypeptide synthesis. By contrast the Gram negative, nonsporulating extremely thermophilic bacterium *Thermus thermophilus* HB8 (Ohno-Iwashita *et al.*, 1975) which is morphologically closer to *E. coli* than *B. stearothermophilus*, translates in a manner similar to *E. coli.*

A review by Friedman (1968) concluded that the protein synthesizing system of *B. stearothermophilus* is significantly more thermostable than that of mesophilic organisms. Irwin *et al.* (1973) demonstrated that the ribosomes from the thermophilic *C. tartarivorum* and *C. thermosaccharolyticum* are significantly more thermostable than those from the mesophile *C. pasteurianum* and the psychrophile *Clostridium* strain 69. It was suggested that the increased thermostability was due to ribosomal protein and not to rRNA.

Koffler (1957) was the first to point out that if thermophilic growth is due to a simple kinetic function of resynthesis then thermophiles should exhibit 16 times more activity during growth at 70°C than at 30°C, assuming a doubling of the rate for a 10°C increase in temperature. Brock (1967) plotted the optimal growth temperatures and growth rates of several thermophilic and mesophilic organisms in an Arrhenius fashion. Even though non-linearity was observed at higher temperatures and the data points were sparse, he concluded that thermophiles do not grow nearly as rapidly at their temperature optima as might be predicted on the basis of rapid resynthesis.

In evaluating the concept of rapid resynthesis, it is interesting to consider the generation times of the facultative thermophile, *B. coagulans* KU. Novitsky *et al.* (1974) using the growth rate constant found that this organism growing in a tryptone-glucose medium had a generation time of 28 min at 37°C and 23 min at 55°C. While much faster growth for this organism was obtained at both temperatures in antibiotic medium 3 (Crabb *et al.*, 1975), the generation time at 55°C was increased only by a factor of about 1·2 to 1·3 (consistent with above). Considering that the growth temperature spread involved is 18°C, it seems difficult to perceive that rapid resynthesis presents an adequate explanation for survival of this organism.

In assessing the role of rapid resynthesis as a mechanism of thermophily, no unequivocal conclusions can be reached based on the available data. The fact that the protein synthesizing apparatus and other macromolecules of obligate and caldoactive thermophiles show inherent thermostability is well established. This also seems to be inconsistent with the necessity for rapid resynthesis as a major controlling factor in thermophilic growth. It is possible that a better system for studying rates of resynthesis would be the caldoactive bacteria, which have significantly longer generation times than the classical obligate thermophile, *B. stearothermophilus*.

C. Macromolecular Thermostability

1. Introduction

From the data presented in reviews and numerous recent reports, the isolated macromolecules from thermophilic organisms, in general, exhibit a greater intrinsic thermostability than their mesophilic counterparts. This enhanced thermostability seems to be associated with macromolecules throughout the entire cell, and includes enzymes from the major metabolic pathways (Table 1), as well as ferredoxins (see Section IVB), ribosomes (Ljungdahl and Sherod, (1976), and flagellin (Singleton and Amelunxen, 1973) Although we are unaware of any reports on the characterization of membrane proteins or DNA-associated proteins, they would also be expected to show enhanced thermostability.

A point that needs to be made is that thermostability of proteins is not unique to organisms that grow at thermophilic temperatures. The classical example is myokinase which was shown to retain activity after heating at 100°C (Colowick and Kalckar, 1943). Myoglobin exhibits a transition temperature around 80°C, and avidin (Donovan and Ross, 1973), a tetramer, in the absence and presence of biotin has transition temperatures of 85 and 132°C, respectively. Phosphofructokinases from *C. pasteurianum* and *E. coli* have transition temperatures of approximately 70 and 60°C, respectively (Cass and Stellwagen, 1975). Hence, the major difference between mesophilic and thermophilic proteins is that all of the thermophilic proteins must function at thermophilic growth temperatures. This shows that caution should be exercised in looking for differences in the properties of mesophilic and thermophilic proteins, particularly those that do not show a large difference in thermostability between the thermophilic and mesophilic source.

2. *Measurement of Thermostability*

Since most of the proteins isolated from thermophiles have been enzymes, the thermostability generally has been studied by measuring the retention of catalytic function. In general a solution of the enzyme is heated at different temperatures for a fixed time of 5 to 10 min or aliquots are removed over a longer time period and then assayed for activity at a temperature where the enzyme is known to be stable. This type of data have clearly shown that the thermophilic proteins are more stable than their mesophilic counterparts but it is based on an irreversible change and provides little information on the cause of inactivation.*

The determination of the temperature of maximum activity also indicates enhanced thermostability, and the loss of all or most of the activity 10 to 15°C above this temperature is consistent with the denaturation of the protein. However, this may not represent intrinsic thermostability of the enzyme, since substrates and other components in the assay may have a stabilizing effect on the protein.

Besides greater resistance to thermal denaturation, several investigators use the marked resistance to denaturation by compounds such as urea, guanidine-HCl and sodium dodecyl sulphate as a measure of thermal stability. Again, such studies do not provide much information about the cause of the enhanced stability or prove that the enzyme is stable at the maximum growth temperature. In fact, careful examination of the data in this area does not show unequivocally that all proteins from thermophiles are intrinsically thermostable at the thermophilic growth temperature and particularly at the maximum growth temperature (Table 1). As an approach towards standardization of procedures in the evaluation of thermostability, we suggest that for an

* Inactivation, rather than denaturation, should be used to describe such losses in function unless measurements have been made to show that denaturation of the protein has occurred. Inactivation may be due to an alteration of a functional group, dissociation, a small conformational change or denaturation. Tanford (1968) defined denaturation as "a major change from the original native structure without alteration of the amino acid sequence".

Table 1
Thermostability and related properties of thermophilic enzymes[a]

Enzyme	Source[b]	Cardinal temperature (°C)[c]		Thermostability (°C)[d]		Arrhenius[e] plot	pI[b]	References
		T_{opt}	T_{max}	Intrinsic	Stabilizers			
Glucokinase	B.s.	50–65	70–75	65	0·2 M NaCl	na[b]	na	Hengartner and Zuber (1973)
Glucose-6-P isomerase	B.s.	50–65	70–75	55	subst. (<65)	—	na	Muramatsu and Nosoh (1971)
Fructose-1,6-diP aldolase	B.s.	50–65	70–75	50	na	+	na	Sugimoto and Nosoh (1971)
Phosphofructokinase	T.X-1	69–71	na	<70	na	na	na	Cass and Stellwagen (1975)
	T.a.	70	79	80	na	na	na	Hengartner *et al.* (1976)
Glyceraldehyde-3-P dehydrogenase	B.s.	50–65	70–75	>80	na	na	4·6	Ameluxen (1966), Singleton *et al.* (1969)
	T.a.	70	79	<95	na	na	4·8	Hocking and Harris (1973, 1976)
	T.t.	70	85	85	na	na	na	Fujita *et al.* (1976)
Triose-P isomerase	C.ts.	55	67	55	na	na	5·6	Shing *et al.* (1975)
Phosphoglycerate kinase	B.s.	50–65	70–75	60	na	na	4·9	Suzuki and Imahori (1974)
Enolase	B.s.	50–65	70–75	40	Mg^{2+}(60)	na	na	Boccu *et al.* (1976)
	T.X-1	69–71	na	74	Mg^{2+}(88)	na	na	Stellwagen and Barnes (1976)
	T.a.	70	79	88	Mg^{2+}(100)	—	na	Stellwagen and Barnes (1976)
6-P gluconate dehydrogenase	B.s.	50–65	70–75	60	na	+	na	Veronese *et al.* (1974)
NADP-isocitrate dehydrogenase	B.s.	50–65	70–75	55	subst. (<70)	na	4·5	Pearse and Harris (1973), Hibino *et al.* (1974)
	T.f.	70–75	80	>70	na	+	na	Saiki *et al.* (1976)

Malate dehydrogenase	B.s.	50–65	70–75	>60	na	na	na	Murphey *et al.* (1967)
Formyltetrahydrofolate synthetase	C.ta.	55–60	65	<60	NH_4^+,K^+(60)	na	na	Ljungdahl *et al.* (1970)
Glutamine synthetase	B.s.	50–65	70–75	<60	cofactors plus subst. (70)	+	4·6	Hachimori *et al.* (1974), Wedler and Hoffman (1974a, b)
Deoxythymidine kinase	B.s.	50–65	70–75	75	na	na	na	Kobayashi *et al.* (1974)
DNA-dependent RNA polymerase	B.s.	50–65	70–75	<60	na	na	na	Air and Harris (1974)
	T.a.	70	79	60	na	na	na	Air and Harris (1974)
	T.t.	70	85	>70	na	+	na	Date *et al.* (1975)
α-Amylase	B.s.	50–65	70–75	50	Ca^{2+}(<70)	na	9·3	Ogasahara *et al.* (1970), Pfueller and Elliott (1969)
	B.cd.	72	82	45	Ca^{2+}(70)	na	na	Heinen and Lauwers (1976)
Thermolysin	B.t.	50–60	na	<50	Ca^{2+}(80)	—	na	Dahlquist *et al.* (1976), Ohta *et al.* (1966)
Inorganic pyrophosphatase	B.s.	50–65	70–75	65	na	na	na	Hachimori *et al.* (1975)
Aminopeptidase I	B.s.	50–65	70–75	>80	na	—	na	Roncari and Zuber (1969)
Adenosine triphosphatase	B.s.	50–65	70–75	65	na	na	na	Hachimori *et al.* (1970)
Esterase	B.s.	50–65	70–75	50	na	+	na	Matsunaga *et al.* (1974)

[a] Only enzymes that appeared to be essentially homogeneous are included

[b] Abbreviations are: B.s. = *Bacillus stearothermohilus*; T.X-1 = *Thermus* X-1; T.a. = *Thermus aquaticus*; T.t. = *Thermus thermophilus*; C.ts. = *Clostridium thermosaccharolyticum*; T.f. = *Thermus flavus*; C.ta. = *Clostridium thermoaceticum*; B.cd. = *Bacillus caldolyticus*; B.t. = *Bacillus thermoproteolyticus*; na = not available; pl = isoelectric point

[c] T_{opt} and T_{max} values for all of the thermophiles listed were obtained from the chapter of Tansey and Brock in this text, except for *Bacillus caldolyticus* (Williams, 1975) and *Bacillus thermoproteolyticus* (Endo, 1962)

[d] The conditions that were used to measure thermostability were variable and other components in the system may have had an effect on some of the values listed for intrinsic thermostability

[e] A + means a break was observed in the Arrhenius plot and a — means no break occurred

enzyme to be considered intrinsically thermostable, a dilute solution of the apoprotein (i.e. free of cofactors, substrates, ions, etc.) should exhibit stability when heated in deionized water or a very dilute buffer for 5 to 10 min at the maximum growth temperature, and shown to retain at least 90% of its enzymic activity at this temperature. It would be most ideal if the appropriate homologous mesophilic protein also be evaluated under identical experimental conditions.

3. Physicochemical Properties of Thermophilic Proteins

Enhanced thermostability of most thermophilic proteins from caldoactive and obligate thermophiles is well established. However, the physicochemical characterization of thermophilic proteins shows that they have a great deal of homology with their mesophilic counterparts. In contrast to mesophilic proteins which usually denature below 60°C, thermophilic proteins appear to undergo small conformational changes without denaturation when heated from room temperature to 55 or 60°C. Surprisingly, there is very little information on the relation between conformation and catalytic activity of these proteins at the thermophilic temperatures. Amelunxen *et al.* (1970), were the first to show that the secondary structure of glyceraldehyde-3-phosphate dehydrogenase (GPDH) could change before the catalytic site was destroyed (see Section IVA). Matsunaga and Nosoh (1974) have clearly shown that a conformational change occurs in glutamine synthetase of *B. stearothermophilus*, well below the temperature required for inactivation. In the presence of cofactors and substrates, which have been shown to protect the enzyme from inactivation at 70°C for 5 h, they showed that the enzyme became susceptible to thermolysin digestion above 55°C. As pointed out by Ljungdahl and Sherod (1976) there is a break in the Arrhenius plot of many of the thermophilic enzymes (Table 1). Although these plots are based on kinetic measurements, and temperature may have an effect on substrate and kinetic factors, the type of break usually observed is consistent with an alteration of protein conformation (Han, 1972). One of the best examples that the broken Arrhenius plot is due to a conformational change is glutamine synthetase which shows a break at 58°C and susceptibility to proteolysis above 55°C.

(*a*) *Nonprotein stabilizing factors.* Table 1 shows that several of the isolated thermophilic enzymes are not stable at the optimum growth temperature, which suggests that additional stabilizing factors are needed. Although attempts to detect unique stabilizing factors have failed (see Section IIIB), it is possible that normal cellular components such as cofactors, substrates, membranes and the highly charged macromolecular intracellular environment could provide the additional stabilization needed. Unfortunately, few investigators have looked at the effects of these factors on thermostability. One of the best examples of substrate and modifier stabilization is with the glutamine synthetase from *B. stearothermophilus*. The enzyme rapidly loses activity at 65°C unless NH_4^+ and glutamate, or ATP and glutamate are present (Wedler and Hoffmann, 1974b) and has been shown to be stable for 5 h at 70°C in the presence of Mg^{2+}, glutamate and NH_4^+ (Hachimori *et al.*, 1974). This enzyme is also stabilized by feedback modifiers, particularly alanine, histidine, and CTP (Wedler and Hoffmann, 1974b). Although we are unaware of any reports on coenzyme stabilization of purified enzymes from thermophiles, pyridoxal phosphate does enhance the thermostability of a partially purified threonine deaminase preparation (Thomas and Kuramitsu, 1971). Calcium seems to be required for the stabilization of the extracellular hydrolases such as the α-amylases and thermolysin (see Section IVC). The GPDH from *B. coagulans* is the only highly purified enzyme that has been shown to require a highly charged environment for thermostability (see Section 111D).

(*b*) *Effect of temperature and water on secondary interactions*. In considering the three main types of secondary interactions (i.e. salt linkages, hydrogen bonding, and apolar interactions) that are associated with protein structure, the importance of water must be emphasized. The high dielectric constant and H-bonding capacity of water markedly decrease the strength of salt-linkages or H-bonds and provide enhancement of the apolar interactions which are usually referred to as hydrophobic bonds. Temperature will have a direct effect on these interactions and an indirect effect via the solvent. In going from mesophilic to thermophilic temperatures, the direct effect would be to weaken the salt-linkages and H-bonds, whereas increased temperature will decrease the dielectric constant of water which would enhance these interactions. By contrast, increasing the temperature would increase the apolar interactions which are considered to reach a maximum in water around 60°C. The net effect of an increased temperature should be a decrease in both salt-linkages and hydrogen bonding and an increase in hydrophobic interactions up to a temperature of 60°C. However, Brandts (1967) on the basis of a two state model has estimated that, in the case of chymotrypsinogen, stabilization due to hydrogen bonding would remain constant from 0 to 70°C and hydrophobic interactions for the large side-chains would increase rapidly up to 75°C and then slowly decrease.

(*c*) *Hydrophobicity and thermostability*. Since hydrophobic interactions would be enhanced at thermophilic temperatures, it was natural to conclude that thermophilic proteins might contain a higher content of apolar side-chains. Singleton and Amelunxen (1973) calculated the relative hydrophobicity, *NPS*, $H\phi_{ave}$, and *p* for 11 thermophilic and 39 nonthermophilic enzymes and found no correlation between thermostability and hydrophobicity. Bull and Breese (1973) combined the average residue volume with a parameter of hydrophobicity and found good correlation to the transition temperature of 14 nonthermophilic proteins. However, when they applied their parameter to two mesophilic and two thermophilic enolases, the most stable enzyme from *T. aquaticus* was predicted to be the most unstable. Singleton (1976) has made an exhaustive evaluation of all of the various parameters associated with hydrophobicity measurements from the amino acid composition of thermophilic and nonthermophilic proteins without any significant correlation of hydrophobicity to thermostability. As noted by Bigelow (1967), his parameter ($H\o_{ave}$) correlates well with *NPS* but not with *p*. The author related this to a difference in choosing which side-chains are polar, and which apolar. Also he recognized that $H\o_{ave}$ would not accurately measure the hydrophobic contributions to stability, but felt that it would be closely related to the actual value. Whether or not these three parameters are closely related to the actual hydrophobic contribution to thermostability, they are related to hydrophobicity and show that there is no difference between the mesophilic and thermophilic proteins. However, these parameters provide only relative values based on the amino acid composition, and do not exclude the possibility that enhanced thermostability of certain regions of the thermophilic proteins is the result of stabilization by additional hydrophobic interactions.

(*d*) *Hydrogen bonding and thermostability*. Barnes and Stellwagen (1973) compared the amino acid composition of enolase from two thermophilic and two mesophilic sources and found a positive correlation of the thermostability of these proteins with an increase in the number of residues they considered capable of hydrogen bonding. However, some of the residues that were included in this list, e.g. tryptophan, methionine, and cysteine would not be expected to participate in hydrogen bonding, and lysine was omitted whereas arginine was included. Suzuki and Imahori (1974) made a similar calculation for phosphoglycerate kinases and found a decreased hydrogen bonding potential for the thermophile. Cass and Stellwagen (1975) have looked at this parameter in phosphofructokinase from one thermophilic and two mesophilic bacteria and found no difference. In addition, a comparison of the percentage of various types of amino acids in α-amylases (Hasegawa *et al.*, 1976) and phosphoglycerate kinase (Suzuki and Imahori, 1974) does not show any significant difference between the mesophilic and thermophilic sources. As with the hydrophobicity parameters, these measurements only show that the overall potential hydrogen bonding capacity is the same in mesophilic and thermophilic proteins, but do not exclude the possibility that certain regions of thermophilic proteins are stabilized by additional hydrogen bonding.

(*e*) *Salt-linkages and thermostability*. Perutz and Raidt (1975) have attributed the enhanced thermostability of ferredoxins to an increased number of external salt-linkages. *Sundaram et al.* (1976) have suggested that the enhanced thermolability of malate synthetase in 0·2 M KCl is consistent with their proposal. The conclusions of Perutz and Raidt were derived from building models of several ferredoxins based on the known tertiary structure of the one from *Micrococcus aerogenes* (see Section IVB). Such a mechanism seems very unlikely since external salt-linkages should be very weak in an aqueous environment and be decreased even further in the presence of ions. The thermostability studies that they referred to (Devanathan *et al.*, 1969) were actually performed in 0·1 M phosphate buffer. It has also been shown that the number of charged groups within an *E. coli* cell is equivalent to 1 mol kg^{-1} (Damadian, 1973), which suggests that external salt-linkages probably would not be stable inside such a cell. However a salt-linkage or ion-dipole interaction would have considerable strength if shielded from the solvent.

(*f*) *Secondary structure and thermostability*. Stellwagen and Barnes (1976) have found a positive correlation between the thermostability of enolases and secondary structure and have proposed as a first approximation that the secondary structure of globular proteins, particularly the β-structure, determines their thermal transition temperatures. Singleton (1976) also calculated structural parameters from amino acid compositions and found no difference in the helical and turn content but estimated that a decrease might occur in the β-structure of the thermophilic proteins. Other investigators have estimated the helical and β-structure from spectropolarimetric measurements and found no difference between the mesophilic and thermophilic proteins (Hasegawa *et al.*, 1976; Hibino *et al.*, 1974; Suzuki and Imahori, 1974; Ogasahara *et al.*, 1970; Yoshida *et al.*, 1975; Fontana *et al.*, 1976). Thus, in most cases, the content of ordered secondary structure of homologous mesophilic and thermophilic proteins appears to be very similar.

(*g*) *Aggregation and thermostability*. Glutamine synthetase of *B. stearothermophilus* loses activity at 70°C more slowly at a protein concentration of 4 mg ml^{-1} than at one of 0·04 mg ml^{-1}. Because of this and evidence that the enzyme could aggregate, Wedler and Hoffmann (1974b) suggested that aggregation may provide a partial explanation for the thermostability of this enzyme. Again this does not seem to be a general way that thermophilic proteins would be stabilized. Except for the natural association of monomers to form native protomers, aggregation frequently leads to inactivation. Donovan and Ross (1973), in studies with the remarkably stable avidin found that some turbidity of the protein occurred around 70°C, suggesting that aggregation might be involved in the thermal transition. However, from their experimental data, they concluded that aggregation did not contribute to heat stability of the protein. In fact, thermophilic proteins appear to be more resistant to heat-induced aggregation as evidenced by the greater resistance to heat coagulation of the cytoplasmic proteins from thermophilic than of mesophilic bacteria (Koffler and Gale, 1957). Possibly the tendency towards low isoelectric points and the greater resistance to unfolding which would decrease the exposure of apolar regions of the molecule is related to this decreased tendency to aggregate.

(*h*) *Cysteine and thermostability*. Hocking and Harris (1976) have suggested that a decrease in cysteine content is associated with thermostability, particularly a decrease of ancillary sulphydryl groups that would be exposed on the surface of the protein and subject to oxidation. Their proposal is primarily based on the differences in the cysteine content of GPDHs (see Section IVA) and phosphofructokinases (Hengartner *et al.*, 1976). Other examples of thermophilic enzymes with fewer sulphydryl groups are phosphoglycerate kinase (Suzuki and Imahori, 1974), enolases (Barnes and Stellwagen, 1973), and adenosine triphosphatase (Yoshida *et al.*, 1975). Since there is a tendency for both mesophilic and thermophilic bacterial enzymes to contain less sulphur amino acids, caution should be exercised in comparing the content of these amino acids from bacteria with other sources. It is interesting that most of the enzymes listed are from caldoactive bacteria, particularly *T. aquaticus*. However, phosphofructokinase from *Thermus* X-1 (Cass and Stellwagen, 1975) contains 12 equivalents of cysteine compared to none in *T. aquaticus* (Hengartner *et al.*, 1976) eight in *B. stearothermophilus* (Hengartner *et al.*, 1976) and 17 in *C. pasteurianum* (Cass and Stellwagen, 1975). Although sulphydryl groups exposed on the surface of enzymes would be more subject to oxidation, the intracellular environment is reductive and should minimize an oxidative process. Possibly the cysteine content is more closely related to inactivation rather than structural resistance to denaturation. Considerations of sulphydryl content may be important in the extracellular enzymes such as α-amylases and thermolysin which do not contain cysteine or cystine and in *in vitro* studies of purified enzymes. Formyltetrahydrofolate synthetase from *C. thermoaceticum* contains 24 cysteine residues and has a relatively low thermostability (Table 1).

(*i*) *Cystine, calcium and thermostability*. Hsiu *et al.* (1964) in studies on the amylase from *B. subtilis* noted that cystine, which stabilizes the mammalian extracellular enzymes, was absent from bacterial enzymes and proposed that a metal ion formed intramolecular cross-links similar in function to disulphide bridges. The lack of cystine in bacterial extracellular hydrolases has been confirmed, and may be due to the absence of a well-defined organelle system such as the Golgi apparatus and secretory granules found in eukaryotes. Although these enzymes may be stabilized by calcium ions in other organisms than thermophiles, calcium ions seem to be essential to the thermostability of the α-amylase from *B. stearothermophilus* (Yutani, 1976) and *B. caldolyticus* (Heinen and Lauwers, 1976). However, in the case of thermolysin, the formation of additional calcium binding sites appears to be related to the enhanced thermostability (see Section IVC).

(*j*) *Acidic residues and thermostability.* With similar basic amino acid compositions to mesophiles (Singleton, 1976) and lower isoelectric points (Table 1), thermophilic enzymes appear to have more acidic residues. Unfortunately, due to incomplete data, Singleton (1976) had to group glutamine–glutamate and asparagine–aspartate in comparing the amino acid compositions of thermophilic and nonthermophilic proteins. In contrast to a basic group such as the amino group, the carboxyl group is well suited to enhance the thermostability of proteins, since temperature has little effect on its pK_a, whereas the polarity of the environment has a marked effect. Data on the helix–coil transition of polyglutamate in dioxane-water suggested that the pK_a's of the carboxyl groups are shifted to 6·5, 7·2, and 8·1 in 10, 30, and 50% dioxane, respectively (Iizuka and Yang, 1965). This is in agreement with the observation that acidic groups of proteins often titrate abnormally and may be found within the interior of the protein molecule. Hence an increase in carboxyl groups could enhance the thermostability by: (1) being protonated with the potential of forming two hydrogen bonds with another carboxyl group or other groups; (2) forming an ion–dipole interaction with the phenolic hydroxyl of tyrosine; (3) forming a salt-linkage; or (4) forming a bond with an ion such as calcium in thermolysin. Other than the involvement of acidic residues in forming the calcium binding sites of thermolysin, the involvement of these residues in thermostability must await further studies.

In conclusion, the physicochemical properties of the thermophilic proteins show that there are no unique or gross differences in the structure of the thermophilic proteins that would explain their enhanced stability. Thus the inherent thermostability must reside in a small number of changes which can only be elucidated by detailed studies of homologous mesophilic and thermophilic proteins (see Section IV).

D. Highly Charged Macromolecular Environment

The consensus of the recent reviews on thermophily is that thermophilic bacteria synthesize proteins that are intrinsically thermostable. In the previous section, many of the isolated enzymes from obligate and caldoactive bacteria were shown not to be stable at the optimum and/or maximum growth temperatures, although it is evident that they are more thermostable than their mesophilic counterparts. However, it was pointed out by Novitsky *et al.* (1974), in studies with the cell walls of the facultative thermophile, *B. coagulans* KU, that selected glycolytic enzymes from crude extracts of this organism grown at 37 and 55°C are as labile as those from mesophilic organisms. *B. coagulans* has a T_{opt} of 55°C and a T_{max} of 57 to 58°C, and growth at the mesophilic temperature of 37°C is abnormally rapid in comparison to a true mesophile. The instability of enzymes was confirmed in the case of glyceraldehyde-3-phosphate dehydrogenase (GPDH) for cells grown at both 37 and 55°C (Crabb *et al.*, 1975). The enzyme in crude extracts of the facultative thermophile was rapidly inactivated after 5 min at 50°C (Table 2). By contrast, the GPDH from *B. cereus* (Suzuki and Imahori, 1973a) is stable for

Table 2

Induced thermostability of glyceraldehyde-3-phosphate dehydrogenase from extracts of Bacillus coagulans *grown at 55 and 37°C*

Samples	% Activity remaining after 5 min[a]					
	37°C	50°C	55°C[b]	60°C[b]	65°C	70°C
55°C Extract						
No addition	97	19	0	0	0	0
$(NH_4)_2SO_4$	90	92	98	97	86	31
NaCl	102	99	95	99	68	17
37°C Extract						
No addition	70	29	0	0	0	0
$(NH_4)_2SO_4$	93	87	95	97	52	0
NaCl	93	98	95	89	31	0

[a] Controls for each of the samples above were kept at 4°C throughout the experiment; there was no change in the enzymic activity

[b] In other experiments, heat treatment was extended to 15 min; there was no significant additional loss in enzymic activity for either the 55 or 37°C extract

more than 40 min at 50°C, and the enzyme from rabbit muscle shows thermostability for 10 min at 60°C (Amelunxen, 1966). As shown in Table 2, thermostability above T_{max} could be conferred to the enzyme in crude extracts (at a protein concentration of approximately 25 mg ml^{-1}) by increasing the ionic strength of the extracts by a factor of 1·8; this corresponded to the addition of 8% $(NH_4)_2SO_4$ or 10% NaCl. Sodium chloride was included in the experiments because it has been shown to be a neutral salt (inert as either a stabilizer or destabilizer) with respect to protein stability (Von Hippel and

Schleich, 1969). In order to eliminate the effect of any factor(s) in the crude extract, these experiments have been repeated with the homogeneous enzyme and essentially the same results were obtained (Amelunxen and Singleton, 1976). Although we have not determined the minimum ionic strength required for thermostability of the enzyme, we know that the enzyme is unstable in 0·15 M ammonium sulphate at 4°C. From the elegant studies of Damadian (1973), the intracellular inorganic ion concentration of *E. coli* is approximately 260 mM and primarily consists of K^+ and NH_4^+. If *B. coagulans* has a similar composition, the inorganic ions would not provide sufficient ionic strength to stabilize the GPDH. However, the polyelectrolytes, such as proteins, phospholipids, and nucleic acids, make the anionic and cationic charge equivalent to 1 mol kg^{-1} (Damadian, 1973) which would provide an intracellular matrix capable of stabilizing enzymes such as the GPDH.

Based on this information, we propose that the concept of stabilization by macromolecular charge *in vivo* as a mechanism of thermophily in *B. coagulans* is feasible for the following reasons: (1) the obvious thermolability of the GPDH (Crabb *et al.*, 1975) and other glycolytic enzymes (Novitsky *et al.*, 1974) argues strongly against intrinsic thermostability, and therefore the ability of this facultative thermophile to reproduce at 55°C must involve another mechanism; (2) rapid resynthesis is inconsistent based on the relatively small difference exhibited in the generation time between cells grown at 37 and 55°C and would be an expensive mechanism from the energetic viewpoint; (3) in crude extracts of cells grown at either temperature, the charged macromolecules are present, but in the *in vitro* situation are unable to exert a complete protective effect on enzymes due to dilution. However, simulating the *in vivo* charged environment by the addition of $(NH_4)_2SO_4$ or NaCl results in complete protection of the enzyme; (4) with the isolated crystalline GPDH where no other macromolecules are present to interact, charge alone stabilized the enzyme; (5) the growing realization that the microbial cell is highly organized with little intracellular free water lends more support to the concept that a highly charged macromolecular environment is an important mechanism in thermostabilizing proteins within the intact cell. It is also probable that *in vivo* macromolecular charge suffices to stabilize relatively thermolabile enzymes in obligate and caldoactive bacteria. An example might be the lactate dehydrogenase from *B. caldolyticus* (Weerkamp and MacElroy, 1972).

As with any new proposed mechanism, we realize that further experimental documentation is necessary both qualitatively and quantitatively. In future experiments, the heat resistance of other enzymes from *B. coagulans* and their response to increased charge will be evaluated. Although we have shown that the addition of salts to crude extracts of *B. coagulans* can confer thermostability to GPDH by simulating intracellular charge, it is obvious that these salts are not present in the cell at the concentrations used *in vitro*. Therefore, in further studies, we will investigate the substitution of polyelectrolytes for salts as a means of estimating the minimal and optimal ionic environment required for stability at the thermophilic growth temperature.

Another interesting aspect in our studies of *B. coagulans* is the temperature dependent alteration in the protein population. In extracts of cells grown at 55°C, the proteins are completely resistant to heat coagulation up to 70°C which is well above the thermophilic growth temperature. However, extensive coagulation occurs at 50°C in extracts of cells grown at 37°C. Polyacrylamide gel electrophoresis has revealed differences in the protein profiles, and gels stained for carbohydrate showed that the majority of proteins in extracts of cells grown at 37°C are glycoproteins. Chemical analyses showed that the extracts of cells grown at 37°C contained three times more carbohydrate than those from cells grown at the thermophilic temperature. We are presently investigating whether or not this intriguing aspect has any relationship to the proposed mechanism of thermophily for *B. coagulans*.

E. Allosterism, Thermoadaptation and Transformation

That allosteric interactions are of major importance in metabolic regulation is now well established. Because of the unusual thermostability of enzymes from thermophilic bacteria, Brock (1967) suggested that this characteristic may be the result of a rigid and inflexible conformation. The concept of an inflexible conformation would appear to be incompatible with allosteric interactions where structural flexibility is a functional requirement. However, there are now many examples of thermophilic enzymes subject to allosteric control: Ljungdahl and Sherod (1976) have divided allosteric enzymes into two groups.

The first group consists of enzymes showing allosterism at both high and low temperatures; these are aspartokinase (Kuramitsu, 1968; Kuramitsu, 1970; Cavari *et al.*, 1972), threonine deaminase (Thomas and Kuramitsu, 1971), phosphofructokinase (Yoshida *et al.*, 1971; Yoshida, 1972) and fructose 1,6-diphosphatase (Yoshida and Oshima, 1971). Some properties of this group have been discussed previously (Singleton and Amelunxen, 1973).

The second group consists of those enzymes showing allosterism only at high temperatures: included in this group are the lactate dehydrogenase (Weerkamp and MacElroy, 1972), homoserine dehydrogenase (Cavari and Grossowicz, 1973), uridine kinase (Orengo and Saunders, 1972), and a ribonucleotide reductase (Sando and Hogenkamp, 1973). As pointed out by Ljungdahl and Sherod (1976), enzymes of the latter group may behave as Brock (1967) speculated, but inflexibility of conformation exists only at low temperatures. Hence, *in vivo* the conformation of thermophilic enzymes, in terms of response to allosteric effectors, appears to be as flexible as those from mesophiles.

Prior to a discussion of thermoadaptation *per se*, it seems appropriate to summarize the studies on aminopeptidases because of their probable relationship to the adaptive process. It has been reported that *B. stearothermophilus* contains a thermostable aminopeptidase I (API), but also two relatively thermolabile aminopeptidases, APII and APIII (Roncari and Zuber, 1969; Roncari and Zuber, 1970). Comparative studies using different strains of *B. stearothermophilus* demonstrated that the distribution of APs I, II and III showed interesting variations. Obligate thermophilic strains produce high levels of APs I and II, but little APIII; facultative strains contain comparable amounts of APs I, II, and III; and mesophilic mutants synthesize large amounts of APIII, but little API or II. It appears that the quantitative relationship among the three aminopeptidases depends on temperature and growth conditions. This is a unique example of what could be referred to as temperature-dependent isozymes. The thermostable API has been isolated in homogeneous form; in the metal-enzyme complex of API, the Co^{2+} is strongly bound at pH 8, and at a pH of less than 6, the apoenzyme is formed.

In a series of papers, detailed physicochemical studies including subunit analyses were described (Moser *et al.*, 1970; Stoll *et al.*, 1972; Roncari *et al.*, 1972; Stoll *et al.*, 1973; Balerna and Zuber, 1974; Stoll and Zuber, 1974). The molecular weight of API was found to be 400,000 which is very close to values reported for this enzyme from *E. coli* (Dick *et al.*, 1970), and from amino acid analyses it was suggested that the enzyme was markedly hydrophobic. The enzyme is composed of 12 subunits of two different types (α, β) which can combine in different ratios. Only the α-subunit is needed for degradation of neutral peptides, but dipeptides having an amino-terminal aspartic or glutamic acid require the β-subunit. Even though the subunits have different hydrolytic specificities, the molecular weight of the subunits is identical. The two subunits were observed to occur in varying proportions in API, leading to hybrids $\alpha_6\beta_6$, $\alpha_8\beta_4$ and $\alpha_{10}\beta_2$. Dissociation of API was accomplished but it was not known whether the components were monomers or dimers. Conditions are being sought to assess whether or not it is possible from the energetic viewpoint to prepare β-rich species ($\alpha_4\beta_8$, $\alpha_2\beta_{10}$, β_{12}) hybrids which are not found in extracts of *B. stearothermophilus*.

Aminopeptidase III synthesized in cultures of *B. stearothermophilus* grown at 37°C has also been purified to homogeneity and found to be present at a ten-fold higher level than in cells grown at 55°C. The molecular weight of APIII is 108,000, and the enzyme is much less thermostable than API. Interesting features of APIII include its specificity as an aminotripeptidase and its similarity to the aminotripeptidase from pig kidney (Chenoweth *et al.*, 1973). Although the specific functions of the high molecular weight API and the low molecular weight APII and APIII are unclear, the ability of an organism such as *B. stearothermophilus* to synthesize these enzymes in different ratios dependent on growth temperature is of importance in an evaluation of thermophily and thermoadaptation.

In considering thermoadaptation in the overall scheme of thermophily, many basic questions still remain. Studies in this area have been essentially quiescent for many years, but renewed interest through recent reports, prompts a critical re-evaluation. If thermoadaptation should prove to be a generalized phenomenon among thermophilic bacteria, then as indicated in a previous section, much caution will be needed in the interpretation of rigid classification schemes due to profound alterations in the cardinal growth temperatures by adaptive modification. In reviewing some of the early findings, Campbell (1954) demonstrated that an obligate thermophile could be grown at 36°C using the appropriate medium; this observation was confirmed subsequently by Long and Williams (1959). Bausum and Matney (1965) found that the facultative thermophile, *B. licheniformis* (strain Allen) growing at 37°C required a period of adaptation at an intermediate temperature to obtain growth at 55°C. Campbell (1955) demonstrated that the α-amylase from a facultative thermophile (*Bacillus*) grown at 55°C was thermostable, but the enzyme from the same organism grown at 37°C was thermolabile.

It was not until recently that thermoadaptation studies were reintroduced into our conceptual thinking about thermophily. Jung *et al.* (1974) have presented studies of thermoadaptation and metabolic differences in *B. stearothermophilus* grown at 55 and 37°C. By cultivation at the intermediate temperature of 46°C, these investigators converted not only a facultative strain of *B. stearothermophilus* (ATCC 7954), but also three obligate thermophilic strains (NCIB 8924, ATCC 7953 and 12977) into mesophilic variants and back again into their original thermophilic forms. To accomplish this conversion, brain-heart-infusion medium was used. Depending on the growth temperature, some striking metabolic differences were shown in cellular extracts. High activities of glyceraldehyde-3-phosphate dehydrogenase (GPDH) and alcohol dehydrogenase (ADH) were observed in extracts of cells grown at 55°C. However, no ADH was found in extracts of cells grown at 37°C and the level of GPDH was extremely low; in addition succinate dehydrogenase and isocitrate dehydrogenase (IDH) appeared to be higher. The authors suggested that during exponential growth at 55°C both respiratory and fermentative pathways could be used (depending on oxygen content) but at the top of the exponential phase, fermentation only is possible. In the mesophilic variant (growth at 37°C), energy is obtained mainly by aerobic respiration. The authors paid particular attention to the logical criticism that in their experiments they selected for a mesophilic contaminant or a temperature-sensitive mutant. The following criteria were submitted to show the species identity of the variants: (1) control of strain purity by single-colony picks: (2) direct transfers from 37 to 55°C and vice versa remained sterile; (3) morphological identity of endospores; (4) occurrence of thermostable API in both variants.

In a subsequent paper, Haberstich and Zuber (1974) studied thermoadaptation of enzymes in thermophilic and mesophilic cultures of *B. stearothermophilus* (NCIB 8924) and *B. caldotenax* YT-G. The experiments indicated that certain enzymes (glucokinase, glucose-6-phosphate dehydrogenase, GPDH and IDH) of both organisms are thermostable when cells are grown at thermophilic temperatures and thermolabile when cells are grown at mesophilic temperatures. In both organisms cultivated at intermediate temperatures, a mixture of thermostable and thermolabile enzyme is seemingly present as indicated by the intermediate stability curves. Differences were found between the two bacterial systems in the process of thermoadaptation. With *B. stearothermophilus*, adaptation occurs only when the intermediate temperature of 46°C is used; no growth could be obtained when cells were transferred directly. However, in *B. caldotenax* direct transfer and conversion can be made; adaptation occurs during the lag period and the length of the lag period depends on the difference between the temperature in the preculture and the culture. In *B. caldotenax* cultivated at 37°C, then at 70°C, it was proposed that thermolabile enzymes disappear by denaturation and thermostable enzymes are newly synthesized. In the converse experiment where *B. caldotenax* was cultivated at 70°C and then at 37°C, it was concluded that the thermostable enzymes are not denatured at 37°C, but disappear during exponential growth of the mesophilic cells. The authors presented arguments against spontaneously arising mesophilic or thermophilic mutants during thermoadaptation. Emphasis was also placed on the fact that thermoadaptation is dependent upon the culture medium. Only brain-heart-infusion medium was successful in demonstrating thermoadaptation, and in other media employed, the strains of *B. stearothermophilus* used were obligately thermophilic.

We agree with the importance of the following questions posed by Haberstich and Zuber (1974) regarding thermoadaptation: (1) does the genetic information for two enzyme forms derive from different genes as a function of temperature? (2) is only one gene involved with thermoadaptation occurring at a subsequent phase of protein synthesis? (3) are the completed enzymes undergoing modification by other enzymic reactions?

Recently, a comprehensive presentation of thermoadaptation studies has been presented (Frank *et al.*, 1976). We have made a few comments on these. In all cases the organisms used are sporeformers which can pose metabolic complications in any study of this type. More credence probably would be afforded if a thermophilic non-sporeformer (e.g. *Thermus*) were shown to exhibit enzymic and metabolic alterations dependent on the growth temperature. It seems to us that the shadow of contamination or spontaneously arising mutants in the thermoadaptation studies could be eliminated by genetically marking the strains prior to adaptation. In future studies using *B. stearothermophilus*, the nature of the component(s) in brain-heart-infusion medium that facilitate thermoadaptation should be clarified. Both defined and minimal media (liquid and solid) have been reported for *B. stearothermophilus* 1503 (Rowe *et al.*, 1975). Use of such media in combination with fractionated components of brain-heart-infusion broth could lead to identification of the active component(s). Of passing interest in this regard are the studies of Weerkamp and MacElroy (1972) involving lactate dehydrogenase from *B. caldolyticus*, a caldoactive bacterium showing growth up to 84°C. The enzyme obtained from cells grown in a minimal medium supplemented with brain-heart-infusion broth was more thermostable than from cells grown in the minimal medium only.

As stated by Williams (1975):

> it seems likely that comparisons of enzyme stability in cells of facultative thermophiles grown at high and low temperatures will be an area of research where advances will certainly be made, because the most direct comparisons will be possible, unclouded by strain variation in enzyme structure.

Studies in our laboratories are now involved with such a system, i.e. the facultative thermophile *B. coagulans* KU. In comparing the characteristics of this organism with the thermoadaptation studies of Jung *et al.* (1974) and Haberstich and Zuber (1974), we would like to reiterate a few points (for more details see Section IIID): (1) the GPDH is thermolabile (from crude extracts or in crystalline form) whether obtained from cells grown at 37 or 55°C; (2) there is no difference in the level or specific activity of this enzyme from cells grown at 37 or 55°C, and ADH is absent from cells grown at either temperature; (3) *B. coagulans* shows no lag for growth at either 37 or 55°C, shows no lag in temperature shift experiments (37 to 55°C and vice versa), requires no adaptation for growth at mesophilic or thermophilic temperature and exhibits generation times at 37 and 55°C that are unusually close; (4) the implications of elevated glycoprotein content in *B. coagulans* grown at 37°C remain to be elucidated; the possibility of contamination or mutation in this system has also been carefully considered and is also contraindicated.

Turning now to the topic of transformation, of much interest in the area of thermophily is the recent paper of Lindsay and Creaser (1975). Using *B. caldolyticus* (T_{opt} of 72°C) as the DNA donor, the authors reported transforming a mesophilic strain of *B. subtilis* so that it could grow at temperatures above 70°C. In these studies, an essential biosynthetic enzyme, the L-histidinol dehydrogenase (HDH) served as a well defined biochemical marker. Transformed cells of *B. subtilis* also contain a thermostable HDH similar to that present in the donor *B. caldolyticus*, whereas the original HDH from *B. subtilis* was inactive above 70°C. One transformant (HT-1) exhibited thermostability in activity curves for HDH at 100°C, and identical results were obtained whether the enzyme was evaluated in crude extracts or as a pure protein. However, the HDH from HT-1 is not identical to the enzyme from *B. caldolyticus* since it requires 2-mercaptoethanol for full activity. This requirement was interpreted as indicating a change in tertiary structure.

Though thermostable enzymes can be transformed, it is not easy to visualize conferring thermophilic growth characteristics to a mesophile by transformation. The authors address themselves to the latter problem by proposing that sufficient individual genes, each coding for a single enzyme, would have to be transferred to enable the recipient cells to grow at high temperatures, or a small number of genes which are capable of controlling a wide range of pleiomorphic effects would have to be transformed. Of the two possibilities, the latter was strongly favoured. As a further extension, it was proposed that changes in the protein synthesizing mechanism of thermophiles at the tRNA or ribosomal level could result in translationally altered enzymes which can function at high temperatures. It was concluded that such an alteration in the translational apparatus could lead to modification of a large number of enzymes in transformants.

While it is obvious that the mechanism involved in the transformation of mesophilic to thermophilic cells is yet to be clarified, it is equally obvious that an understanding of this important system should have a profound effect in future studies of thermophily.

IV. Detailed Consideration of Selected Proteins from Obligate and Caldoactive Thermophiles

A. Glyceraldehyde-3-Phosphate Dehydrogenase

1. Introduction

Glyceraldehyde-3-phosphate dehydrogenase (GPDH) is found in most living systems where it functions as a substrate-level oxidative phosphorylative enzyme in glycolysis. At the time this enzyme was crystallized in homogenous form from the obligate thermophile *B. stearothermophilus* 1503 and shown to be remarkably thermostable (Amelunxen, 1966), no intracellular enzyme from a thermophilic bacterium had been isolated in purified form. An updated purification scheme for crystallization of the enzyme has been reported (Amelunxen, 1975). Suzuki and Harris (1971) have described another procedure for the crystallization of the enzyme from *B. stearothermophilus* 1503. Recently the enzyme has been purified from two caldoactive bacteria, *T.*

aquaticus (Hocking and Harris, 1973) and *T. thermophilus* (Fujita *et al.*, 1976) and a facultative thermophile, *B. coagulans* KU, in our laboratory (unpublished). The fact that the enzyme has been well characterized from many other sources provides a basis for comparative physicochemical analyses.

2. *Thermostability*

In evaluating the thermostability of the crystalline enzyme from *B. stearothermophilus* (Amelunxen, 1966), crystals were dissolved in deionized water to a protein concentration of 10 to 11 mg ml^{-1}. Even when assayed without cysteine, little or no inactivation could be demonstrated after heat treatment for 10 min at 80°C, and only about 10% inactivation occurred at 90°C, clearly showing inherent molecular thermostability of the thermophilic enzyme. By contrast, this enzyme from rabbit muscle under identical experimental conditions was extensively inactivated after heat treatment at 70°C. Unfortunately the optimum temperature for activity could not be determined because of instability of assay components; marked absorbance increases in the absence of enzyme above 60°C were observed (Amelunxen, 1967). An Arrhenius plot of the data increased linearly up to 55°C, and then levelled off at 60°C. Since the activity was the same at 70°C when corrected for the non-enzymic absorbance (unpublished), the break in the plot may be due to instability of the assay components. Under the same experimental conditions, the reaction rate of the rabbit muscle enzyme began to decrease rapidly at 40°C and showed marked reduction in activity at 50 and 60°C.

A procedure has been described for crystallization of the thermophilic GPDH after removal of bound NAD^+ (Amelunxen and Clark, 1970); the thermostability of the apoenzyme is essentially the same as that of the holoenzyme. Hocking and Harris (1973) reported that the holoenzyme from *T. aquaticus* demonstrated a half-life at 98°C greater than 30 min with no appreciable difference observed for the apoenzyme. Though difficult to evaluate because of different experimental procedures employed, the thermostability recently reported for the GPDH from *B. stearothermophilus* (Hocking and Harris, 1976) is considerably lower than that observed consistently for many years in our laboratories. Recently, Fujita *et al.* (1976) crystallized the GPDH from *T. thermophilus* HB8 and demonstrated 20% inactivation after 10 min at 90°C (protein concentration of 2·1 mg ml^{-1} in Tris-HCl buffer pH 8·45). The percentage yield after two recrystallizations was 2·4% which is extremely low compared to other GPDHs.

3. *Physicochemical Properties*

The homogeneity of the GPDH from *B. stearothermophilus* has been demonstrated using several physical methods (Amelunxen, 1966; Singleton *et al.*, 1969; Amelunxen *et al.*, 1970). In subsequent studies, polyacrylamide gel electrophoresis (with and without SDS) at protein concentrations as high as 200 μg/gel failed to reveal any impurity in re-crystallized preparations (unpublished). The homogeneity of the enzyme from *T. aquaticus* (Hocking and Harris, 1973) and *T. thermophilus* (Fujita *et al.*, 1976) has also been verified.

Thus far, all of the NAD-dependent GPDHs appear to be tetramers with a molecular weight around 144,000 composed of four identical subunits. High speed sedimentation equilibrium of the enzyme from *B. stearothermophilus* in 0·1 M KCl and 5·0 M guanidine-HCl gave molecular weights of 148,700 and 36,000 ± 1200, respectively (Amelunxen *et al.*, 1970). The $S^{o}_{20,w}$ value for the thermophilic enzyme is 7·17 (Singleton *et al.*, 1969); this value is consistent with that of other GPDHs. The molecular weight of the enzyme from *T. aquaticus* (Hocking and Harris, 1976) and *T. thermophilus* (Fujita *et al.*, 1976) was reported to be 150,000 and 130,000 to 135,000 for the native enzyme, and 37,000 and 33,700 for the subunit, respectively.

Like the rabbit muscle enzyme (Murdock and Koeppe, 1964), there are 4 mol of firmly bound NAD^+ per mol of the enzyme from *B. stearothermophilus* (Singleton *et al.*, 1969; Suzuki and Harris, 1971). To our knowledge, all GPDHs isolated except the yeast enzyme have been found to contain the full complement of bound NAD^+. The irreversible substrate-induced inactivation of the thermophilic GPDH by NADH (Amelunxen, 1967) is similar to that of the rabbit muscle enzyme (Amelunxen and Grisolia, 1962) except that thermophilic temperatures are required. The optimal pH for activity is 10 (Amelunxen, 1967) in contrast to other GPDHs, which have optimal pH values ranging from 8·4 to 8·6. Fujita *et al.* (1976) reported an optimal pH of 8·3 for the enzyme from *T. thermophilus.*

In considering structural properties of the GPDH from *B. stearothermophilus,* Amelunxen *et al.* (1970) reported on the dissociation of the enzyme in 8 M urea and 5 M guanidine-HCl. As for nonthermophilic GPDHs, dissociation in the latter denaturant is immediate and irreversible. Hocking and Harris (1976) have reported that the enzyme from *T. aquaticus* could be reversibly denatured in 6 M guanidine-HCl. Possibly this is related to the low content of SH groups in this enzyme. However, in contrast to rapid and irreversible dissociation of the rabbit muscle enzyme in 8 M urea, the dissociation of the enzyme from *B. stearothermophilus* has the characteristics of a first order denaturation reaction. Remarkable resistance to the denaturant was observed at 30, 40, and 50°C under the experimental conditions employed. However, at thermophilic temperatures (55 and 60°C), the inactivation proceeded very rapidly. If assayed in the presence of cysteine at the termination of the incubation period, reactivation occurred at all temperatures but ranged from about 50% at 30°C, to about 20% at 60°C; activity could also be demonstrated in an assay system (containing cysteine) which was

8 M in urea. It should be pointed out that dissociation of the enzyme occurs after exhaustive dialysis in 8 M urea, as indicated by sedimentation velocity studies (unpublished). Marked resistance to urea and SDS has also been reported for the enzyme from *T. aquaticus* (Hocking and Harris, 1973) and *T. thermophilus* (Fujita *et al.*, 1976).

As shown in Table 3 the enzyme from *B. stearothermophilus* is extensively unfolded in both 8 M urea or 5 M guanidine-HCl. Based on calculated values

Table 3

Optical rotation parameters of glyceraldehyde-3-phosphate dehydrogenase from Bacillus stearothermophilus

Solvent[a]	$-a_0$	$-b_0$
H_2O, pH 6, 25 or 30°C	129 ± 14	134 ± 11
H_2O, pH 6, 55°C	119 ± 6	135 ± 8
H_2O, pH 6, 60°C	176 ± 10	206 ± 14
H_2O, pH 6, 30°C[b]	128 ± 10	211 ± 15
Buffer, 30°C[c]	148 ± 6	119 ± 9
7·8–8·3 M urea, 25 or 30°C[d]	526 ± 19	66 ± 11
8·3 M urea in H_2O, 60°C	462 ± 17	68 ± 8
5·1 M guanidine-HCl in buffer, 30°C	589 ± 18	64 ± 8
5·1 M guanidine-HCl in buffer, 60°C	540 ± 20	66 ± 9

[a] The solutions, except where noted, contained approximately 0·04 M $(NH_4)_2SO_4$
[b] After heating at 60°C for 16 h
[c] Buffer in all cases is 0·01 M imidazole—0·001 M EDTA, pH 7·4
[d] In H_2O, pH 6·0, or in 0·01 M imidazole—0·001 M EDTA, pH 7·4

of $(m')_\lambda$ at 300 and 400 nm, the extent of unfolding was close to that of a random coil. Although the enzyme is still catalytically active at 70°C, a conformational change has occurred by 60°C (Table 3). When the enzyme in water is heated from 30 to 55°C, the parameters a_0 and b_0 remained essentially the same. However, on heating to 60°C, the value of a_0 changed from −129 to −176 and that of b_0 from −134 to −206. Suzuki and Imahori (1973b) have confirmed this from CD spectra and have shown that considerably more change occurred at 65°C. Thus the molecule is not inflexible and some regions of the enzyme can unfold without destroying the catalytic site. This, in conjunction with the unpublished observation that the protein can be significantly digested by trypsin with minimal loss in activity, suggests that the thermostability may reside in a core of residues around the active site that resists denaturation.

The secondary structure of GPDH from thermophiles seems to be similar to that found in mesophiles. From spectropolarimetric measurements, Suzuki and Imahori (1973b) concluded that the enzyme from *B. stearothermophilus* was structurally very similar to the rabbit muscle enzyme. Fontana *et al.* (1976) in studies of the thermal properties of GPDH from *E. coli* concluded that a high content of β-structure is common to the GPDH from a number of sources.

Table 4 represents a comparative view of the amino acid composition of GPDHs over a wide phylogenetic spectrum. In general there is marked

Table 4

Amino acid composition of the subunit of nine glyceraldehyde-3-phosphate dehydrogenases[a]

Amino acid	T.a.[b]	T.t.	B.s.[c]	B.c.	E.c.	Y.[b]	E.g.[d]	L.m.[b]	P.m.[b]
Lys	22	19	24	25	26	26	26	28	26
His	10	8	9	7	6	7	6	5	11
Arg	16	17	16	13	12	11	12	9	10
Asx	34	31	39	40	45	37	30	32	38
Thr	22	23	19	24	28	23	19	20	22
Ser	13	12	18	14	14	24	35	25	19
Glx	23	26	28	27	22	22	23	24	18
Pro	12	6	11	10	9	13	15	12	12
Gly	25	37	25	32	31	29	44	30	32
Ala	42	39	40	36	35	33	33	32	32
Cys	1	1	2	2	4	2	2	5	4
Val	30	30	40	32	32	36	25	38	34
Met	7	5	5	7	8	6	4	10	9
Ile	22	18	19	21	17	19	15	18	21
Leu	31	33	27	21	20	22	19	18	18
Tyr	10	9	7	7	8	10	8	9	9
Phe	7	7	5	12	11	10	12	15	14
Trp	3	na	3	na	4	3	2	3	3
Total	332[e]	321	337	330	332	334	331	333	332
pI[a]	4·8	na	4·3–4·6	*c.* 5	na	*c.* 5	8·3	*c.* 7	*c.* 7

[a] Abbreviations and references are: T.a. = *Thermus aquaticus* (Hocking and Harris, 1976); T.t. = *Thermus thermophilus* (Fujita *et al.*, 1976); B.s. = *Bacillus stearothermophilus* (Singleton *et al.*, 1969); B.c. = *Bacillus cereus* (Suzuki and Imahori, 1973a); E.c. = *Escherichia coli* (D'Alessio and Josse, 1971); Y = yeast (Jones and Harris, 1972); E.g. = *Euglena gracilis* (Grissom and Kahn, 1975); L.m. = lobster muscle (Davidson *et al.*, 1967); P.m. = pig muscle (Harris and Perham, 1968); na = not available; pI = isoelectric point

[b] From amino acid sequence data

[c] Original amino acid composition data corrected to a subunit molecular weight of 36,000; from the amino acid sequence data, there are actually 333 residues (Harris, unpublished)

[d] NAD^+-dependent enzyme

[e] Two residues in the sequence were not identified (positions 303 and 304)

similarity in the composition and total residues of the enzyme from all sources. If one searches for trends, there appears to be an increased level of Leu and to a lesser extent Arg in all three thermophilic GPDHs. The increased Leu is accompanied by other changes so that the overall hydrophobicity of the thermophilic enzyme is similar to counterparts from mesophilic sources

(Singleton and Amelunxen, 1973). Even though the overall hydrophobicity of the enzyme appears not to be increased, Suzuki and Imahori (1973b) have presented evidence that certain hydrophobic regions are more stable. From fluorescence emission spectra and titration data, the thermophilic enzyme was found to have many buried tyrosine residues with a higher pK value (11·8), and the tryptophan residues appeared to be buried in a more hydrophobic region than those of the rabbit muscle enzyme.

Since the GPDH from *T. aquaticus* contains only one cysteine/subunit, in contrast to the lobster and pig muscle which contain five and four respectively, Hocking and Harris (1976) suggested that the reduced cysteine content may be related to thermostability (see Section IIIC). It is possible that the cysteine content is related to the relatively small enhanced thermostability of the enzyme from *T. aquaticus* and *T. thermophilus* over that of *B. stearothermophilus* (two cysteines/subunit), but this does not provide an adequate explanation for intrinsic thermostability since the GPDH from *B. cereus*, yeast and *E. gracilis* all contain two cysteines/subunit (see Table 4). However, as pointed out by Hocking and Harris (1976), only cysteines exposed to the surface of the molecule would be subject to oxidation. This may be related to the decreased thermostability of mesophilic GPDHs such as that from rabbit muscle which contains 16 sulphydryl groups/tetramer of which eight react rapidly with *p*-hydroxymercuribenzoate (PHMB) (Murdock and Koeppe, 1964). In the case of the GPDH from *B. stearothermophilus*, there is only partial reactivity of the eight sulphydryl groups of the native enzyme with PHMB (two to three groups/tetramer) or Ellman's reagent (four groups/tetramer). However, unlike the muscle enzyme the reaction with PHMB does not cause irreversible inactivation since full activity could be restored by the addition of cysteine (Amelunxen, 1967). Of interest would be a comparison of the reactivity of the sulphydryl groups in the enzyme from *T. aquaticus* and *T. thermophilus*.

Hocking and Harris (1976) also noted that there appeared to be a positive correlation between thermostability and acidic isoelectric points and cited as examples the thermophilic and muscle GPDHs (Table 4). Since the GPDH from *B. cereus* and yeast also have acidic isoelectric points, this trend is also questionable. However, these mesophilic enzymes are more stable than the muscle enzymes.

Since the molecular basis for thermostability of the thermophilic GPDHs could not be obtained from overall structural properties such as size, shape, and amino acid composition, more detailed structural studies have been pursued. The active-site peptide of this enzyme has been sequenced from 14 mesophilic species (Perham, 1969) and all but that from the lobster muscle enzyme have an identical sequence of 17 residues (1 in Table 5). The lobster muscle peptide has two very conservative substitutions of Val for

Table 5

Amino acid sequence of the active-site tryptic peptides of glyceraldehyde-3-phosphate dehydrogenase from several sources

	1	2	3	4	5	6	7	8	9	10	11	12	13	14	15	16	17	18	19	20
1	-Ser	-Leu	-Lys	Ile	-Val	-Ser	-Asn	-Ala	-Ser	-Cys-[a]	Thr	-Thr	-Asn	-Cys	-Leu	-Ala	-Pro	-Leu	-Ala	-Lys-
				↑																
2	-Asp	-Met	-Thr	-Val	-Val	-Ser	-Asn	-Ala	-Ser	-Cys-[a]	Thr	-Thr	-Asn	-Cys	-Leu	-Ala	-Pro	-Val	-Ala	-Lys-
3	-Ala	-*His*	-*His*	-Ile	-Val	-Ser	-Asn	-Ala	-Ser	-Cys-[a]	Thr	-Thr	-Asn	-Cys	-Leu	-Ala	-Pro	-*Phe*	-Ala	-Lys-
4	-Arg	*His*	-*His*	-Ile	-Ile	-Ser	-Asn	-Ala	-Ser	-Cys-[a]	Thr	-Thr	-Asn	-*Ser*	-Leu	-Ala	-Pro	-Val	-*Met*	-Lys-
		↑																		

1 17 residues; pig muscle (Harris and Perham, 1968)
2 20 residues; lobster muscle (Davidson *et al.*, 1967)
3 20 residues; *Bacillus stearothermophilus* (Bridgen *et al.*, 1972)
4 19 residues; *Thermus aquaticus* (Hocking and Harris, 1976)
[a] Active-site cysteine
↑ Cleavage site of trypsin

Ile or Leu (positions 4 and 18 in Table 5) and has the trypsin cleavage site three residues towards the N-terminus which gives a 20 residue peptide. The active-site peptide from *B. stearothermophilus* (Bridgen *et al.*, 1972) also has 20 residues and differs from the prototype sequence only in position 18 where Phe replaces a Leu residue (Val in the lobster muscle enzyme). However, the active-site peptide from *T. aquaticus* (Hocking and Harris, 1976) shows two substitutions, i.e. Met for Ala in position 19, and Ser for Cys in position 14. The two adjacent His residues (positions 2 and 3) occur in a variable portion of the protein chain (Olsen *et al.*, 1975), and represent a novel feature of the enzyme from *B. stearothermophilus* and from *T. aquaticus*. It will be interesting to see if these substitutions are conserved in other thermophilic GPDHs, particularly since His 3 represents a two base change in the genetic codon. In relating these substitutions to the three-dimensional structure of the lobster muscle enzyme (Olsen *et al.*, 1975), the two His residues are in contact with hydrophobic residues which suggests that His 3, which replaces hydrophilic residues Lys or Thr, may provide additional stabilization to this region of the molecule. It has been reported by Hocking and Harris (1976) that the amino acid sequence of the monomeric subunit of the GPDH from *B. stearothermophilus* and *T. aquaticus* show considerable homology (50 to 60%) with mesophilic GPDHs. Two findings of interest are the deletion of all but the essential SH groups (Cys-149) in the enzyme from *T. aquaticus*, and the replacement of Lys-183 which is known to be the site of acetyl transfer in the apoenzyme (Harris and Perham, 1968) by Arg in both thermophilic enzymes. In addition, a comparison of the amino acid sequence and three-dimensional structure of lobster muscle GPDH (Olsen *et al.*, 1975) with the sequence of the enzyme from *T. aquaticus* shows a number of substitutions that could alter the thermostability. Two segments that look interesting involve residues 121–125 and 280–282. The first of these is in a surface region of the molecule with a sequence Pro-Ser-Ala-Asp-Ala in the lobster muscle enzyme, and Lys-Gly-Glu-Asp-Ile in the enzyme from *T. aquaticus*. The second segment contains Ser-Ser-Asp in the lobster muscle enzyme, and Glx-Asx-Ile in the enzyme from *T. aquaticus*, and is in the region of the molecule that is associated with subunit side chain contacts. Substitutions such as these may be related to the differences in subunit interactions and overall thermostability of the enzyme.

It was also reported by Hocking and Harris (1976) that X-ray crystallographic studies in progress for the apo and holo enzyme from *B. stearothermophilus* closely resemble that of the lobster enzyme. In personal communication with Dr. Harris, we were informed that the sequence of the subunit of *B. stearothermophilus* has been completed and shows a high degree of homology with that from *T. aquaticus*. In addition, a molecular model based on X-ray structure at 2·9Å has been constructed for the GPDH from *B. stearothermophilus*. Papers concerned with detailed comparisons of the lobster muscle GPDH and the enzyme from *B. stearothermophilus* are in preparation by Harris and co-workers. Hopefully the three-dimensional analyses will clarify whether or not the substitutions in the active-site region are of significance in stabilizing the thermophilic enzymes, and will reveal other stabilizing regions in the total conformation.

4. *Immunological, Kinetic and Genetic Studies*

Sauvan *et al.* (1972) prepared antibodies against GPDH from *B. stearothermophilus*, rabbit muscle and yeast. The maximum inhibition of enzymic activity by homologous antiserum was 94% for the yeast enzyme, 96% for the thermophilic enzyme and 50% for the rabbit muscle enzyme. No cross inhibition was demonstrable with any of the enzymes and their heterologous antisera. Preincubation of the thermophilic enzyme with NAD^+ had no effect on inhibition by homologous antiserum but decreased antiserum inhibition of the yeast and rabbit muscle enzyme. No cross reactivity could be shown with any of the enzymes and their heterologous antisera using the quantitative precipitin test, immunodiffusion or immunoelectrophoresis.

K_m and V_{max} values for substrate and coenzyme have been determined for the thermophilic GPDH (Amelunxen *et al.*, 1974) at pH 8·6 and 10 (optimal) over a temperature range of 30 to 60°C. The marked thermostability of the enzyme afforded an opportunity to study the effect of temperature on kinetic parameters without the complications of enzymic inactivation. For glyceraldehyde-3-phosphate (G3P), the K_m at pH 8·6 was found to be $7{\cdot}1 \times 10^{-4}$ M at both 30 and 60°C; at pH 10, the K_m was $3{\cdot}5 \times 10^{-4}$ M at either 30 or

60°C. V_{max} values at pH 8·6 increased more than 3-fold in going from 30 to 60°C; however at pH 10, the change was negligible. For the GPDH from *T. aquaticus*, Hocking and Harris (1976) reported a K_m for G3P of $7{\cdot}5 \times 10^{-4}$ M (at either 25 or 35°C) with the V_{max} increasing more that 2-fold from 25 to 35°C. Fujita *et al.* (1976) calculated the K_m (G3P) for the enzyme from *T. thermophilus* and reported a value of 3×10^{-4} M at 30°C. Based on the fact that the K_m values reported for thermophilic enzymes are significantly higher than those from nonthermophilic sources, e.g. rabbit muscle enzyme (Velick, 1955), one could speculate that somehow thermophilic enzymes sacrifice catalytic efficiency for increased stability. However, in a report by Suzuki and Imahori (1973a) concerning the GPDH from the mesophile, *B. cereus*, a K_m value of $5{\cdot}1 \times 10^{-4}$ M (20°C) was obtained for G3P, a value which is quite close to those given above for the thermophilic enzymes.

A genetic approach for pursuing the molecular bases of thermophily and the metabolism of thermophiles was afforded by the isolation of mutants from *B. stearothermophilus* (Rowe *et al.*, 1973). Mutants blocked in fumarase, aconitase and alcohol dehydrogenase were obtained using the mutagen *N*-methyl-*N′*-nitro-*N*-nitrosoguanidine. Some properties of a double mutant (aconitase negative and alcohol dehydrogenase negative) and two fumarase negative mutants of *B. stearothermophilus* have been reported (Rowe *et al.*, 1976). In the double mutant, alcohol dehydrogenase was not synthesized whereas aconitase was synthesized but in inactive form. Rowe *et el.* (1975) have described both defined and minimal media (both liquid and solid) that support good growth of *B. stearothermophilus.* The solid media have been used to isolate several amino acid-requiring mutants of this organism. These media should greatly facilitate more detailed studies of the genetics and metabolism of *B. stearothermophilus* 1503.

From the accumulated data, it is apparent that the GPDH from thermophilic sources is the most completely characterized intracellular enzyme that exhibits high intrinsic thermostability. Physicochemical characterization has clearly demonstrated that the enhanced thermostability of the enzyme is not due to unique structural properties, but rather must reside in the specific alterations of the primary and tertiary structure of the polypeptide. Since the level of secondary structure seems to be the same for thermophilic or mesophilic GPDHs, the difference in stability must be associated with key substitutions of amino acids that enhance the stability of the secondary structure, and possibly of more importance, other critical regions of the molecule. Although there are minor differences in the amino acid composition, no conclusive relationship to thermostability is evident. However, with the forthcoming publication of the three-dimensional structure, a more concerted effort can be made to identify the substitutions that are involved in conferring thermostability.

B. Ferredoxin

1. Introduction

Ferredoxins are iron-sulphur proteins that are capable of electron transfer at very low redox potentials. They are of low molecular weight, comparable in size to polypeptides (such as insulin) rather than to enzymes. As such, ferredoxins are readily amenable to amino acid sequence analyses and hence to an evaluation of phylogenetic relationships among ferredoxins from a wide variety of sources. In recent years, ferredoxins have been detected in all living systems, from obligate anaerobic fermenting bacteria to higher plants and animals, and to date about 18 ferredoxins from bacteria, algae and plants have been sequenced.

Thermostable ferredoxin has been purified from the thermophiles *C. tartarivorum* and *C. thermosaccharolyticum* (Devanathan *et al.*, 1969) by modification of a scheme for purification of mesophilic ferredoxin (Buchanan *et al.*, 1963). A stable ferredoxin has also been purified from *B. stearothermophilus* (Mullinger *et al.*, 1975).

2. Thermostability

Devanathan *et al.* (1969) incubated ferredoxins from *C. tartarivorum* and *C. thermosaccharolyticum* at a protein concentration of 2·5 mg ml^{-1} at 70°C in 0·1 M potassium phosphate pH 7·4. After 1 h, the ferredoxin from *C. tartarivorum* retained about 85% of initial activity whereas the ferredoxin from *C. thermosaccharolyticum* retained more than 90% of the initial activity. Under the same experimental conditions, the ferredoxins from the mesophiles *C. acidi-urici* and *C. pasteurianum* retained 50% and 30% of their activity, respectively. The decrease in absorbance (A_{390}) of three ferredoxins at 70°C as a function of time demonstrated that heat inactivation may be due simply to loss of iron and sulphide from the molecule and not to irreversible denaturation; the rate of absorbance decrease for the ferredoxin from *C. tartarivorum* was slower than for the mesophilic counterparts. Hence all of the ferredoxins appeared to be fairly thermostable with the difference residing in the affinity for the iron and sulphide ions.

3. Physicochemical Properties

From almost all of the comparative analyses reported by Devanathan *et al.* (1969), such as molecular weight (6000), content of iron (8 mol), inorganic sulphide (8 mol) and cysteine (8 mol), and absorption spectra, the thermophilic ferredoxins show near identity to the mesophilic ferredoxins. The clostridial ferredoxins are acidic polypeptides and are composed of 55 amino acid residues (Table 6). A greater number of qualitative variations have been reported for mesophilic ferredoxins (Lovenberg *et al.*, 1963); in general, the

Table 6

Amino acid sequences of two mesophilic and two thermophilic ferredoxins

	1	2	3	4	5	6	7	8	9	10	11	12	13	14	15	16	17	18	19	20	21	22	23	24	25	26	27	28
C.p.	-Ala	Tyr	Lys	Ile	Ala	Asp	Ser	Cys	Val	Ser	Cys	Gly	Ala	Cys	Ala	Ser	Glu	Cys	Pro	Val	Asn	Ala	Ile	Ser	Gln	Gly	Asp	Ser-
C.a.	-Ala	Tyr	Val	Ile	Asn	Glu	Ala	Cys	Ile	Ser	Cys	Gly	Ala	Cys	Asp	Pro	Glu	Cys	Pro	Val	Asp	Ala	Ile	Ser	Gln	Gly	Asp	Ser-
C.t.	-Ala	His	Ile	Ile	Thr	Asp	Glu	Cys	Ile	Ser	Cys	Gly	Ala	Cys	Ala	Ala	Glu	Cys	Pro	Val	Glu	Ala	Ile	His	Glu	Gly	Thr	Gly-
C.ts.	-Ala	His	Ile	Ile	Thr	Asp	Glu	Cys	Ile	Ser	Cys	Gly	Ala	Cys	Ala	Ala	Glu	Cys	Pro	Val	Glu	Ala	Ile	His	Glu	Gly	Thr	Gly-

	29	30	31	32	33	34	35	36	37	38	39	40	41	42	43	44	45	46	47	48	49	50	51	52	53	54	55
C.p.	-Ile	Phe	Val	Ile	Asp	Ala	Asp	Thr	Cys	Ile	Asp	Cys	Gly	Asn	Cys	Ala	Asn	Val	Cys	Pro	Val	Gly	Ala	Pro	Val	Gln	Glu-
C.a.	-Arg	Tyr	Val	Ile	Asp	Ala	Asp	Thr	Cys	Ile	Asp	Cys	Gly	Ala	Cys	Ala	Gly	Val	Cys	Pro	Val	Asp	Ala	Pro	Val	Gln	Ala-
C.t.	-Lys	Tyr	Gln	Val	Asp	Ala	Asp	Thr	Cys	Ile	Asp	Cys	Gly	Ala	Cys	Gln	Ala	Val	Cys	Pro	Thr	Gly	Ala	Val	Lys	Ala	Glu-
C.ts.	-Lys	Tyr	Glu	Val	Asp	Ala	Asp	Thr	Cys	Ile	Asp	Cys	Gly	Ala	Cys	Glu	Ala	Val	Cys	Pro	Thr	Gly	Ala	Val	Lys	Ala	Glu-

Abbreviations are: C.p. = *Clostridium pastuerianum*; C.a. = *Clostridium acidi-urici*; C.t. = *Clostridium tartarivorum*; C.ts. = *Clostridium thermosaccharolyticum*

Boxed in areas represent amino acid substitutions between mesophilic and thermophilic ferredoxins; the inner boxes indicate substitutions between the thermophilic ferredoxins thermophilic ferredoxins are similar to the mesophilic ferredoxins, but do vary in that the thermophilic ferredoxins contain two histidine residues and an increased number of basic and acidic residues. In contrast to the ferredoxins from *C. tartarivorum* and *C. thermosaccharolyticum*, the ferredoxin from *B. stearothermophilus* (Mullinger *et al.*, 1975) shows some significant differences. Instead of two sets of 4Fe-4S clusters found in clostridial ferredoxins, that from *B. stearothermophilus* has only one 4Fe-4S cluster. The molecular weight of the ferredoxin from the aerobic thermophile is 7900; it contains 67 amino acid residues but does not have any His or Arg. Unfortunately, experiments concerned with thermostability were not reported.

Optical rotatory studies (Devanathan *et al.*, 1969) showed that clostridial ferredoxins have little secondary structure; i.e. there is no well defined α-helical or β-structure, which is consistent with the three-dimensional structure of the ferredoxin from *M. aerogenes* (Adman *et al.*, 1973).

The primary amino acid sequences of the ferredoxins from *C. tartarivorum* and *C. thermosaccharolyticum* have been determined (Tanaka *et al.*, 1971; Tanaka *et al.*, 1973). A comparison of the primary sequences of the thermophilic ferredoxins with those of the ferredoxins from *C. pasteurianum* and *C. acidi-urici* (Tanaka *et al.*, 1966; Rall *et al.*, 1969) is presented in Table 6. The ferredoxins are arranged in the table according to the range of thermostability observed, with the least stable on top. The only difference between the two thermophilic ferredoxins is the substitution of glutamate for glutamine in positions 31 and 44 (inner boxes) of the ferredoxin from *C. thermosaccharolyticum*. Tanaka *et al.* (1973) concluded that these two substitutions might be responsible for the increased thermostability of *C. thermosaccharolyticum* by allowing hydrogen bonding to other amino acid side chains; however, the authors did not consider the histidine at position 2 to be of significance in terms of thermostability since this mutation is also found in the thermolabile ferredoxin from *Peptostreptococcus elsdenii* (Yasunobu and Tanaka, 1973).

The boxed-in areas in Table 6 represent amino acid substitutions occurring between the mesophilic and thermophilic ferredoxins. These substitutions have been analysed in terms of the mutation values for interconversion of amino acid pairs. A mutation value of 1 represents one nucleotide change in a codon, 2 represents two nucleotide changes and 3 indicates no codon homology. Hence, a mutation value of 3 would represent the most drastic substitution. The mutation values are categorized in Table 7; for the purpose of this analysis, a mutation value of 2 will be considered as a major substitution since in general two nucleotide changes in the codon can affect the charge properties of amino acids, whereas one nucleotide change rarely does. In comparing the mutation values between the two mesophilic ferredoxins, there are four major substitutions at positions 5, 29, 42 and 45. The major substitutions occurring between the least stable ferredoxin from *C. pasteuria-*

Table 7

Mutation values for interconversion of amino acid pairs for mesophilic and thermophilic ferredoxins

Sequence position	[a]Between C.p. and C.a.	[a]Between C.p. and C.t., or C.ts.	[a]Between C.a. and C.t., or C.ts.
2	0	1	1
5	2	1	2
6	1	0	1
7	1	2	1
9	1	1	0
15	1	0	1
16	1	1	1
21	1	2	1
24	0	2	2
25	0	1	1
27	0	2	2
28	0	1	1
29	2	2	1
30	1	1	0
31[b]	0	2, 1	2, 1
32	0	1	1
42	2	2	0
44[b]	0	2, 1	2, 1
45	2	2	1
49	0	2	2
50	1	0	1
52	0	2	2
53	0	2	2
54	0	2	2
55	1	0	1

[a] Abbreviations are given in Table 6
[b] Only points of substitutions occurring between the thermophilic ferredoxins (see Table 6); the mutation value equals 1 between Glu and Gln

num and the thermophilic ferredoxins number 11 or 13 (due to variance at positions 31 and 44 of thermophilic ferredoxins). However, between the relatively more stable ferredoxin from *C. acidi-urici* and the thermophilic ferredoxins, there are seven or nine major substitutions. While presumptive, the data suggest that as the thermostability of the ferredoxin increases, the number of major substitutions also increase. Because the substitutions between the two mesophilic ferredoxins tend to complicate a mutational analysis with the thermophilic ferredoxins, a comparison of mutation values perhaps should involve only positions where homology exists between the mesophilic ferredoxins. As seen in Table 7, major substitutions occur at positions 24, 27, 31 (variable), 44 (variable), 49, and 52–54; of the two histidines in the thermophilic ferredoxins only His-24 represents a significant alteration. Thus there are six to eight major substitutions between mesophilic and thermophilic ferredoxins; in most cases these alterations involve charge differences and change from a hydrophilic to a hydrophobic amino acid or vice versa. The apparent overall effects of these changes are that the thermostable ferredoxins have approximately twice the number of charged amino acids as do the mesophilic ferredoxins. In addition, the ratio of basic to acidic residues is also increased in the thermophilic ferredoxins although they are more acidic than the mesophilic ferredoxins. From the standpoint of primary structure only, the observed genetic changes are major. The question now remains as to their significance in a three-dimensional analysis. Adman *et al.* (1973) using X-ray diffraction have resolved the three-dimensional structure of the ferredoxin from *Micrococcus aerogenes*. The validity of comparing the three-dimensional structure of *M. aerogenes* with the clostridial ferredoxins is somewhat questionable from the phylogenetic viewpoint, and there are also differences in the total number of amino acid residues. Nevertheless, as a starting point such analyses have been made by Tanaka *et al.* (1973). Using the model of Adman *et al.* (1973), it can be demonstrated that the glutamine or glutamic acid in positions 31 and 44 of the thermophilic ferredoxins can be hydrogen bonded to other amino acid side chains, which additional interactions could account in part for the thermostability.

Perutz and Raidt (1975) have also analysed the three-dimensional structure of the ferredoxin from *M. aerogenes* and attempted to visualize possible stereochemical reasons for the thermostability of the thermophilic ferredoxins. An atomic model of the ferredoxin from *M. aerogenes* was constructed with altered patterns of the side chains to correspond to the four different sequences of the clostridial ferredoxins. Associated with the formation of the complex between apoferredoxin and 8Fe-8S, there are two separate complexes formed (4Cys-4Fe-4S). In looking first at the primary sequence of the clostridial ferredoxins in Table 6, there are two clusters of Cys residues at positions 8, 11, 14 and 18 and at positions 37, 40, 43, and 47. In the primary sequence of *M. aerogenes*, the first cluster of Cys residues is identical to that of the clostridial ferredoxins, but in the second cluster the positions are 35, 38, 41 and 45 since the clostridial ferredoxins have an extra amino acid between residues 21 and 22 of *M. aerogenes* (Ala in all *Clostridium* sp.), and between residues 25 and 26 (Asp in mesophilic clostridia and Thr in thermophilic clostridia). The three-dimensional structure of *M. aerogenes* (Adman *et al.*, 1973) shows that both halves of the molecule are needed to form either cluster, i.e. 8, 11, 14, and 45 in the first, and 18, 35, 38, and 41 in the second. This is consistent with the general observation that most of the polypeptide backbone is essential to form the active clusters. However, this does not preclude the deletion of segments of the backbone in regions not essential for stabilization of the clusters subsequent to their formation.

From their simulated three-dimensional model of the clostridial ferredoxins Perutz and Raidt (1975) concluded that the increased stability of the thermophilic ferredoxins is due mainly to external salt bridges linking residues near the N-terminus to others near the C-terminus. Examples cited are His 2 to Gln 44 or Glu 44, and Glu 7 to Lys 53. Other salt bridges that might tend to shield the active clusters are His 24 to Glu 25 and Lys 29 to Glu 31. (We also suggest that His 24 may lend additional hydrophobic shielding to the active site cluster and add stability to the external loop formed in this region due to its hydrophobic character.) Hydrogen bonds gained in the thermophilic ferredoxins were listed as being His 2 to Cys 43 and Ser 10 to Tyr 30. As pointed out by Perutz and Raidt (1975), if the above alterations do contribute to increased thermostability, then differences between the heat stability of the different ferredoxins should decrease on weakening the salt bridges, e.g. with increasing pH or ionic strength. It was also concluded that the thermophilic ferredoxins do not consistently show more non-polar contacts than their mesophilic counterparts.

In commenting on some of the possible mechanisms of thermophily for the thermophilic ferredoxins, it seems clear that definitive reasons can most likely be reached only when the three-dimensional structure of mesophilic and thermophilic ferredoxins from the different *Clostridium* sp. is determined. However, the mechanism of thermostability may be difficult to assess if the major difference between the ferredoxins resides in the affinity for iron and sulphide ions. In assessing the ultimate structural reasons for the thermostability of the ferredoxins from *C. tartarivorum* and *C. thermosaccharolyticum*, consideration must also be given to the range of thermostability exhibited, which appears to be species-dependent. At this point, the comparisons while seemingly valid could well be misleading. It should also be stressed that, because of the very low molecular weight and absence of subunit structure, ferredoxins are not ideal models for arriving at structural conclusions concerning thermostability of high molecular weight proteins having quaternary structure.

If increased salt bridges do indeed furnish enhanced thermostability, experiments using increased pH and ionic strength to weaken these bridges could perhaps clarify the gradations of thermostability observed in the clostridial ferredoxins. However, Devanathan *et al.* (1969) evaluated the thermostability of the clostridial ferredoxins in 0·1 M potassium phosphate, pH 7·4; under these conditions there did not appear to be destabilization since significant inactivation was not observed.

C. Thermolysin

1. Introduction

Thermolysin is an extracellular metalloendopeptidase originally isolated and crystallized from *Bacillus thermoproteolyticus* by Endo (1962). The enzyme is mainly bacterial in origin, functions at neutral pH and is an endopeptidase directed toward peptide bonds in which the amino nitrogen is contributed by hydrophobic residues (Matsubara and Feder, 1971). With respect to specificity, thermolysin resembles pancreatic carboxypeptidase A (Pétra, 1970). The presence of bound Ca^{2+} in thermolysin is an important feature distinguishing it from carboxypeptidase A.

2. Thermostability

Ohta *et al.* (1966) thoroughly evaluated the thermostability of thermolysin. The enzymic reaction was initiated by the addition of the protease to substrate (carbobenzoxy-Gly-Pro-Leu-Ala-Pro) at various temperatures between 25 and 88°C. A linear Arrhenius plot was obtained between 25 and 80°C. Above 80°C, the affinity for substrate starts to decrease. The authors concluded that the mechanism of enzyme action is similar at both low and high temperatures and that the enzyme maintains its structure over a rather wide temperature range. However, in the above experiments, the possible protective effect of substrate could not be excluded.

Heat inactivation experiments were then performed by treating the enzyme at 80°C for 1 h in 50 mM Tris buffer, pH 7, containing 10 mM $CaCl_2$. Activity was measured at 45°C. Although the maximum velocity of untreated and treated thermolysin was the same, the K_m was considerably higher after heat treatment, suggesting that a slight conformational change had occurred. Since the intrinsic viscosity of the native and heat treated enzyme were essentially the same, this confirmed that a gross conformational change had not occurred (Ohta, 1967). Thermolysin heated at 80°C for 1 h was referred to as "modified enzyme".

3. Physicochemical Properties

Ohta *et al.* (1966) reported 19 abnormally ionizing tyrosine residues in the native enzyme and a Cotton effect as measured by circular dichroism; however, in the "modified enzyme", the abnormal ionization of tyrosines was not found (Ohta, 1967), and the Cotton effect completely disappeared. The pH stability curve showed marked differences at 20 versus 80°C (where the abnormally ionizing residues disappeared), and the enzyme was stable at both temperatures in the pH range where ionization did not occur. It was concluded that the 19 tyrosine residues stabilize the enzyme via hydrogen bonding of the phenolic groups and the hydrophobic bonds involving the aromatic ring. Ohta (1967) also reported that removal of Ca^{2+} by dialysis against EDTA causes loss of enzymic activity and a decrease in helical conformation; he emphasized the importance of Ca^{2+} in stabilization of thermolysin. In terms of other properties, thermolysin is quite similar to pancreatic carboxypeptidase (Pétra, 1970), e.g. pH profile, response to inhibition by chelating agents, molecular weight (35,000), amino acid composition (except for the absence of cysteine or cystine in thermolysin) and zinc content (1 g atom mol^{-1}). Denaturation of thermolysin at room temperature in 8 M urea could not be detected by difference spectra, measurement of flourescence spectra emitted by tryptophan residues, optical dispersion by peptide bonds or circular dichroism due to tyrosine residues (Ohta, 1967). Marked denaturation (as measured by optical density change) at 70°C was observed following combined treatment with heat and 8 M urea. The rate of denaturation was essentially independent of enzyme concentration but was dependent on pH, the presence of Ca^{2+}, and the ionic strength.

The amino acid sequence of thermolysin was determined by Titani *et al.* (1972). In agreement with the data of Ohta *et al.* (1966), amino acid analyses showed two methionyl residues and the absence of cysteine or cystine but instead of five tryptophan residues, only three were found by several different procedures. In the sequence of thermolysin (316 residues), unusual distributions of amino acid residues are obvious, e.g., no arginine or lysine occurs between residues 101 and 182 but 10 carboxyl residues can be found; the C-terminal residues (227–316) contain only four carboxyl groups but 10 basic residues. No significant structural homology could be demonstrated between thermolysin and bovine carboxypeptidase A (Bradshaw *et al.*, 1969) even though there are many other physicochemical similarities between the enzymes. From the amino acid sequence only, it was concluded that thermolysin represents a distinct class of enzymes unrelated to mammalian carboxypeptidases or any other protease of known structure. No unique feature of the sequence could be related to the thermostability of the enzyme.

Matthews *et al.* (1972a) constructed an electron density map of thermolysin which enabled them to identify many of the 316 residues, and provided an overall view of the conformation of the enzyme. The asymmetric shape of the molecule was interpreted to indicate that the surface to volume ratio is greater than that of a compact sphere. However, the molecule has the appearance of two spheres in contact. The lower half of the molecule was described as being a large hydrophobic core enclosed by the six helices in the *C*-terminal portion of the molecule. An improved electron density map was subsequently reported by Matthews *et al.* (1974). These authors concluded that there are no unusual characteristics in the thermolysin conformation that could explain its thermostability, with the possible exception of the four calcium binding sites.

In obtaining information about the active site and calcium binding sites of thermolysin, Matthews *et al.* (1972b) correlated the amino acid sequence data (Titani *et al.*, 1972) with X-ray analyses. Amino acid residues of the central region of the polypeptide chain are involved in formation of the active site which shows a deep cleft across the centre of the molecule with the zinc atom located at the bottom. Three zinc ligands were demonstrated to be His-142, His-146, and Glu-166. This combination is identical to that found in the zinc ligands in bovine carboxypeptidase A (Lipscomb *et al.*, 1969), showing similarity in the active site region of the two enzymes. However, the imidazole ring of a third His residue (His-231) is in front of the zinc atom of thermolysin, which represents a major structural difference compared to carboxypeptidase A.

The inhibition of thermolysin by ethoxyformic anhydride has been studied by Burstein *et al.* (1974), and has provided evidence that a histidine residue is essential for enzymic activity. Although this reagent reacts in proteins with several amino acid side chains, control experiments indicated that only one of these groups was responsible for loss of enzymic activity. Spectral changes during reactivation with hydroxylamine gave further proof that modification of a single residue resulted in activation. The competitive inhibitor Cbz-L-phenylalanine which binds to the active site of thermolysin, protected the enzyme from inhibition by ^{14}C-ethoxyformic anhydride, and prevented incorporation of a single ethoxyformyl group. Spectral changes during ethoxyformylation were found to be characteristic of modification of both histidine and tyrosine residues, but spectral changes related to tyrosine modification were not prevented by the competitive inhibitor. Additional evidence was also presented implicating the essential role of histidine in the active site of thermolysin.

In terms of calcium binding sites, two adjacent Ca^{2+} ions (double site) appear to be bound inside the molecule by chelation to five acidic groups and the other two single Ca^{2+} ions (single sites) are bound at exposed surface regions to aspartic acid residues (Matthews *et al.*, 1972b). Feder *et al.* (1971) demonstrated that removal of three or four of the Ca^{2+} ions known to bind to thermolysin resulted in loss of thermostability. In a detailed presentation of the structure of thermolysin, Colman *et al.* (1972) proposed that two Ca^{2+} ions in the proximity of the central cleft may stabilize the molecule by linking the two halves, each of which has a hydrophobic core which extends across the cleft.

Further studies of much importance on the role of calcium in conferring thermostability to thermolysin have recently been reported by Dahlquist *et al.* (1976). It was demonstrated that excess Ca^{2+} ions stabilize native holothermolysin against what appears to be a cooperative structural transition resulting in autolysis at temperatures above 50°C. Neither zinc (which binds at a site distinct from the four Ca^{2+} sites), nor terbium (bound at the double Ca^{2+} site), shows this stabilizing effect. Binding of terbium to sites other than the double Ca^{2+} site also results in stabilization, leading the authors to suggest that the transition resulting in autolysis involves a region of the molecule distinct from the active site and the double Ca^{2+} binding site. With a single terbium ion tightly bound at the double Ca^{2+} binding site, removal of the remaining two Ca^{2+} ions from their sites resulted in an irreversibly altered enzyme which retains about 40% of the native catalytic activity. However, thermolysin altered in this way denatures at low temperature and is not subject to thermostabilization by Ca^{2+} ions.

The authors interpreted their data as suggesting that the major role of Ca^{2+} in stabilizing native thermolysin from thermal denaturation is to afford protection of the region near one or both single Ca^{2+} binding sites from a cooperative conformational change which can result in irreversible modification and loss of thermostability.

The elegant and detailed studies of thermolysin make it the most highly characterized of thermophilic enzymes. It is significant that there are no unusual structural features of the protein *per se* which can account for its marked thermostability, except for its ability to bind Ca^{2+} ions in a very specific manner in terms of total conformation. It is clear that the apoenzyme is not thermostable (in contrast to apoenzymes derived from thermophilic intracellular enzymes, e.g. GPDH). It seems conclusive that the thermostability of thermolysin resides in the formation of calcium binding sites and that the two single sites represent critical areas in terms of thermostability. Such a system of stabilization has not been found to be the case for intracellular enzymes and may or may not be a general mechanism for extracellular enzymes. In considering mechanisms of thermophily, thermolysin is unique based on the present state of knowledge, but does not appear to be a particularly ideal model for studying the intrinsic molecular reasons for thermostability.

Stahl and Ljunger (1976) have proposed that calcium stabilizes cellular functions against the action of heat and that calcium uptake by *B. stearothermophilus* is a requirement for thermophilic growth. Except for extracellular enzymes such as thermolysin and α-amylase, we are not aware of other enzymes that show an absolute calcium requirement for thermostability. Although the presence of tightly bound Ca^{2+} could go undetected, Barnes and Stellwagen (1973) used elemental and colorimetric analyses to show that no metal ions are necessary for the thermostability of the enolase from *Thermus* X-1. Also, in the preparation and assay of crystalline GPDHs, 10^{-3} M EDTA is routinely used without any demonstrable enzymic inhibition.

D. Formyltetrahydrofolate Synthetase

1. Introduction

Formyltetrahydrofolate synthetase catalyses the synthesis of 10-formyltetrahydrofolate from formic acid, H_4-folate and ATP. The enzyme has been purified from pigeon liver (Greenberg *et al.*, 1955), *M. aerogenes* (Whiteley *et al.*, 1959), *C. acidi-urici* and *C. cylindrosporum* (Rabinowitz and Pricer, 1962), *C. thermoacetacum* (Sun *et al.*, 1969; Ljungdahl *et al.*, 1970) and *C. formicoaceticum* (O'Brien *et al.*, 1976). The enzyme from both the obligate thermophile *C. thermoaceticum* (Himes and Wilder, 1968; Ljungdahl and Wood, 1969) and the mesophile *C. formicoaceticum* (O'Brien and Ljungdahl, 1972) can incorporate CO_2 into the 5-methyl group of 5-methyltetrahydrofolate which is subsequently carboxylated to acetate (Rabinowitz and Pricer, 1957; Ljungdahl and Wood, 1969). The comparative properties of the synthetase from the various clostridial sources have been presented (Himes and Harmony, 1973; O'Brien *et al.*, 1976).

2. Thermostability

Formyltetrahydrofolate synthetase from *C. thermoaceticum* has been evaluated for thermostability under a variety of experimental conditions (Ljungdahl *et al.*, 1970). The effect of monovalent cations on the enzyme at different pH values was tested over a 30 min period at 60°C. Using 0·05 M Tris-HCl (pH 7·5) as the solvent in all cases, 0·01 M NH_4Cl and 0·01 M KCl stabilized the enzyme to a greater extent than did 0·01 M NaCl or the buffer alone. However, except in the presence of NH_4Cl, the enzyme exhibited significant inactivation after 30 min at 60°C. Using the above conditions but at pH 7·0, the enzyme was reasonably stable over the 30 min period. At 50°C in 0·01 M Tris-maleate buffer only, the enzyme showed stability between pH 6·0 and 8·5, but at pH 5·5 and 8·9, slow inactivation was observed; at 60°C inactivation was observed at any pH. However, at 60°C, the enzyme in Tris-maleate buffer in the presence of KCl showed very slow inactivation between pH 6·8 and 9·0 which was similar to the inactivation observed at 50°C in buffer alone. Potassium ions appeared to increase the thermostability of the synthetase between pH 6·8 and 9·0, but below pH 6·8 rapid inactivation occurred. The ion effect, while related to thermostabilization, did not affect increased enzyme activity.

Although the thermophilic synthetase demonstrated more thermostability than mesophilic counterparts, it is not as marked as shown for many other thermophilic enzymes.

3. Physicochemical Properties

The formyltetrahydrofolate synthetase from *C. thermoaceticum* has been characterized physicochemically by Brewer *et al.* (1970). Recently, O'Brien *et al.* (1976) have made a detailed comparison of this enzyme from *C. thermoaceticum* and three mesophilic clostridia. There are close similarities in all cases for such parameters as amino acid composition, molecular weight, $S^{o}_{20,w}$, $\bar{V}$, f/fo and fluorescence emission maximum. Although the molecular weight of the enzyme from all sources is 240,000, the enzyme from mesophilic sources requires potassium or ammonium ions for activity and removal of these ions results in dissociation into subunits (MacKenzie and Rabinowitz, 1971; Welch *et al.*, 1971). However, the synthetase from *C. thermoaceticum* is not dissociated on removal of the monovalent cations. The K_m values for the enzyme from all sources are essentially the same (O'Brien *et al.*, 1976). However, the synthetase from *C. cylindrosporum* has a broken K_m when plotted against temperature with a break at about 30°C (Himes and Harmony, 1973), compared to a break at about 45°C for the enzyme from the thermophile (Ljungdahl and Sherod, 1976). The Arrhenius plot for the thermophilic enzyme also shows a break at about 45°C. Even though the kinetic data suggest a conformational change, it was not detectable in optical rotatory studies. A break at 45°C in the absorbance at 295 nm versus temperature (Shoaf *et al.*, 1974) supported the possibility that a small conformational change does occur.

With regard to the subunit structure, Buttlaire *et al.* (1972) demonstrated that dissociation of the synthetase from *C. cylindrosporum* occurs at pH 5·3; the enzyme from *C. thermoaceticum* also dissociates at this pH (Ljungdahl and Sherod, 1976). However, in contrast to the mesophilic enzyme, the subunits from the thermophile cannot reassociate, indicating greater instability of the subunit; this rather unexpected observation suggested to the authors that peptide chain interactions stabilize the native tetrameric enzyme. In delving further into the possible reasons for the thermostability of the synthetase from *C. thermoaceticum*, O'Brien *et al.* (1976) examined the effect of deuterium oxide on the relative thermostability of the enzyme from *C. formicoaceticum* and *C. thermoaceticum*. Presumably, since D_2O forms stronger hydrogen bonds, it would tend to strengthen hydrophobic interactions in proteins and to weaken intraprotein hydrogen bonding. The results suggested to the authors that the thermophilic enzyme is stabilized by hydrophobic interactions although they commented that more experiments are needed to confirm their results.

Compared to the synthetase from mesophilic sources, it is of potential significance that the enzyme from *C. thermoaceticum* does not dissociate on removal of monovalent cations, and that the thermophilic subunits exhibit a relative instability preventing reassociation to the native enzymic form.

V. Conclusions

In evaluating potential mechanisms to explain thermophilic existence, the earlier proposals of non-protein stabilizing factors and rapid resynthesis cannot be given serious consideration since supportive evidence is either very weak or absent. That thermophilic proteins cannot be under allosteric control also is contraindicated by studies of many isolated thermostable enzymes. Thus the possibility that enzymes from thermophiles are inflexible and sacrifice effector control for thermostability is untenable. There seems to be little question that membrane stability is an important aspect of thermophilic function, and the necessity for further work in this understudied area must be emphasized.

At present, it is premature to evaluate the role of thermoadaptation and transformation in the overall scheme of thermophily. Eventually, it should be possible via thermoadaptation and/or transformation to fully characterize, e.g. an enzyme with varying levels of thermostability from the same microbial species. Such an experimental system should be capable of magnifying the subtle structural differences that are involved in conferring increased thermostability to thermophilic proteins.

The proposal that intracellular macromolecular charge is a functional mechanism of thermophily in the facultative thermophile *B. coagulans* KU is substantiated for the thermolabile glyceraldehyde-3-phosphate dehydrogenase from this organism. Whether or not this is a general mechanism in other facultative thermophiles remains to be seen. The possibility exists that the charged intracellular environment is also of importance in the survival of obligate and caldoactive thermophiles by stabilizing certain enzymes that show decreased thermostability (as measured *in vitro*) at the optimum or maximum growth temperature.

Among the various proposals that have been made in attempting to explain the ability of thermophilic microorganisms to reproduce at high temperatures, there is no doubt that caldoactive and obligate thermophiles synthesize macromolecules that have sufficient intrinsic molecular stability to withstand increased thermal stress. Although the intrinsic thermostability of thermophilic proteins is greater than their mesophilic counterparts, frequently the isolated protein does not exhibit thermostability at the optimum growth temperature. Since naturally occurring cellular components such as metal ions, substrates, modifiers, and ionic strength can enhance thermostability, no unusual factors are probably necessary for intracellular stability. However, caution should be exercised in evaluating the intrinsic thermostability of proteins. Usually stability is determined by showing that a particular thermophilic protein is more resistant to heat inactivation than its mesophilic counterpart, with little attention given to the optimum and maximum growth temperatures of the organism or to other components in the system. In addition, the cause(s) for the inactivation of the mesophilic protein is (are) frequently ignored. If the mesophilic protein were inactivated due to a minor rather than a major structural alteration, a comparison of its structure with that of the thermophilic protein could lead to erroneous conclusions. Hence, in studies on the molecular mechanisms of thermophily, a careful characterization of the holo- and/or apoprotein from both the thermophilic and mesophilic sources is essential.

In determining the molecular basis for thermophily, the selection of the protein is important in relating thermostability to polypeptide structure. As pointed out by Hocking and Harris (1976), the difference between the free energy of stabilization of thermophilic and mesophilic proteins appears to be small, which suggests that it is preferable to select a specific thermophilic protein where the difference in intrinsic thermostability is great compared to the homologous protein from the proper mesophilic source. For example, it may not be surprising that the three-dimensional structure of thermolysin does not show any unusual interactions of stabilization since the apoenzyme has about the same thermostability as other proteases; the difference is in the additional Ca^{2+} ion binding sites. On the other hand, all of the apoferredoxins seem to have considerable thermostability with the major difference being in their affinity for the iron and sulphide ions.

Even if an ideal protein is selected, the enhanced thermostability may be related to only a few key substitutions in the primary structure. In this regard mention should be made of the work of Langridge (1968). In studies with the β-galactosidase gene of *E. coli*, 56 amber mutations were suppressed by crossing into a stock containing the *sup D* suppressor gene. Stability at 57°C was tested for the resultant enzymes, which differed only in the position of the inserted serine into the protein chain at the point of the nonsense mutation. Some of the suppressed enzymes were very thermolabile when compared to the normal enzyme. Other experiments indicated that the degree of stability was characteristic of a particular position in the polypeptide chain of the amino acid substitution, and independent of the amino acid inserted. Such data emphasize the dramatic effect that a single amino acid substitution can have on molecular thermostability.

Physicochemical examination of thermophilic proteins show that there are no unusual structural properties associated with the enhanced thermostability. In other words, thermophilic proteins have similar helical, β-structure, and hydrophobic content and are essentially homologous physicochemically to their mesophilic counterparts. Based on our knowledge of protein structure, the helices and β-structure would be expected to be more stable than many other regions of the molecule due to hydrogen bonding and the additional stabilization from side-chain interactions. These regions may be potentially thermostable in mesophilic proteins. Chen and Lord (1976) using laser Raman spectroscopy to follow the thermal unfolding of ribonuclease A found that a substantial amount of the helical and β-structure remained at 70°C. In addition, the hydrophobic regions of both mesophilic and thermophilic proteins should be stable at most thermophilic growth temperatures; although hydrophobic interactions would decrease in stability above 60°C, they would still be greater throughout the thermophilic growth range than at mesophilic temperatures. Since the unfolding of globular proteins has several cooperative features, the difference in thermostability between thermophilic and mesophilic proteins may reside only in a few regions which would lead to denaturation of the rest of the molecule. The N-terminal and C-terminal regions may be especially important in this regard, in addition to other critical regions near the surface of the molecule. The detailed studies on the thermal denaturation of ribonuclease A (Burgess and Scheraga, 1975) serve as an example of the variability in thermostability of different regions of a protein molecule. From 35 to 45°C, the region between residues 17 to 25 is unfolded without inactivating the enzyme. This segment is on the surface of the molecule and contains five neutral hydrophilic groups and two alanines. Such a segment would readily interact with water but have little interaction with the hydrophobic interior. Possibly the substitution of a residue with the ability for interaction with the hydrophobic interior would stabilize this segment. The C-terminal region (residues 104 to 124) does not unfold until approximately 60°C, and possibly would remain intact if other regions of the molecule had not unfolded first. This region does not have any ordered secondary structure but is rich in apolar groups, contains α, β, and γ carboxyl groups, and forms a compact loop. Removal of the four residues from the C-terminus, which contains a valine, and the α and β carboxyl groups, essentially inactivates the enzyme. The N-terminus is one of the helical regions, and is even more stable, but the stability is dependent upon the interactions of its side-chains with the hydrophobic interior. Again the side-chains associated with these interactions are hydrophobic and acidic. Hence, it is conceivable to visualize the homologous mesophilic and thermophilic protein folding in a cooperative manner to produce compact molecules with similar secondary and tertiary structures containing many regions that are thermostable in both molecules but differing only in the stability of a few key regions. For example, thermolysin increases its thermostability from approximately 50 to 80°C by stabilizing one or two regions with Ca^{2+} ions.

Since a molecular model based on X-ray structure has been constructed for the glyceraldehyde-3-phosphate dehydrogenase from *B. stearothermophilus*, more information on potential structural alterations of this enzyme should be available in the near future. It is anticipated that X-ray crystallography will serve as the crucial tool for observing structural differences between mesophilic and thermophilic proteins. Hopefully, such analyses will lend more insight into the obvious subtleties that must be involved in conferring intrinsic thermostability to proteins from obligate and caldoactive thermophiles. However, the picture that appears to be emerging is that the molecular mechanism(s) of thermophily may transcend this seemingly ultimate criterion in some instances, and require more detailed studies such as step-wise thermal denaturation as conducted on ribonuclease, simulation of the intracellular environment, and genetic manipulation.

Acknowledgements

We wish to express our thanks to Dr J. Kinsey in the Department of Microbiology, the University of Kansas Medical Centre for his helpful comments on the genetic aspects of thermophily. Our gratitude to Dr J. I. Harris (MRC Laboratory of Molecular Biology, Cambridge, England) for kindly furnishing us with his recent unpublished information on the thermophilic glyceraldehyde-3-phosphatedehydrogenase. For her excellent secretarial assistance, our sincere thanks to Mrs Doris Kuehn.

References*

Adman, E. T., Sieker, L. C. and Jensen, L. H. (1973). The structure of a bacterial ferredoxin. *J. Biol. Chem.* **248,** 3987–3996.

Air, G. M. and Harris, J. I. (1974). DNA-dependent RNA polymerase from the thermophilic bacterium *Thermus aquaticus*. *FEBS Lett.* **38,** 277–281.

Allen, M. B. (1953). The thermophilic aerobic sporeforming bacteria. *Bacteriol. Rev.* **17,** 125–173.

Amelunxen, R. E. (1966). Crystallization of thermostable glyceraldehyde-3-phosphate dehydrogenase from *Bacillus stearothermophilus*. *Biochim. Biophys. Acta* **122,** 175–181.

Amelunxen, R. E. (1967). Some chemical and physical properties of thermostable glyceraldehyde-3-phosphate dehydrogenase from *Bacillus stearothermophilus*. *Biochim. Biophys. Acta* **139,** 24–32.

Amelunxen, R. E. (1975). Glyceraldehyde-3-phosphate dehydrogenase from *Bacillus stearothermophilus*. *In* "Methods in Enzymology" (Ed. W. A. Wood) Vol. 41, pp. 268–273. Academic Press, New York and London.

Amelunxen, R. E. and Clark, J. (1970). Crystallization of thermostable glyceraldehyde-3-phosphate dehydrogenase after removal of coenzyme. *Biochim. Biophys. Acta* **221,** 650–652.

Amelunxen, R. E. and Grisolia, S. (1962). The mechanism of triosephosphate dehydrogenase inactivation by reduced diphosphopyridine nucleotide. *J. Biol. Chem.* **237,** 3240–3244.

Amelunxen, R. E. and Lins, M. (1968). Comparative thermostability of enzymes from *Bacillus stearothermophilus* and *Bacillus cereus*. *Arch. Biochem. Biophys.* **125,** 765–769.

*References marked with an asterisk show additional recent work dealt with in the text.

Amelunxen, R. E. and Singleton, R., Jr. (1976). Thermophilic glyceraldehyde-3-phosphate dehydrogenase. *In* "Enzymes and Proteins from Thermophilic Microorganisms", *Experientia Suppl.* 26 (Ed. H. Zuber) pp. 107–120. Birkhäuser Verlag, Basel and Stuttgart.

Amelunxen, R. E., Noelken, M. and Singleton, R., Jr. (1970). Studies on the subunit structure of thermostable glyceraldehyde-3-phosphate dehydrogenase from *Bacillus stearothermophilus*. *Arch. Biochem. Biophys.* **141,** 447–455.

Amelunxen, R. E., Sauvan, R. L. and Mira, O. J. (1974). Thermostable glyceraldehyde-3-phosphate dehydrogenase from *Bacillus stearothermophilus*. II. Kinetic and immunokinetic analysis *Physiol. Chem. Phys.* **6,** 515–526.

Andreesen, J. R., Schaupp, A., Neurauter, C., Brown, A. and Ljungdahl, L. G. (1973). Fermentation of glucose, fructose, and xylose by *Clostridium thermoaceticum*: effect of metals on growth yield, enzymes, and the synthesis of acetate from CO_2. *J. Bacteriol.* **114,** 743–751.

Babel, W., Rosenthal, H. A. and Rapoport, S. (1972). A unified hypothesis on the causes of the cardinal temperatures of microorganisms; the temperature minimum of *Bacillus stearothermophilus*. *Acta Biol. Med. Germ.* **28,** 565–576.

Balerna, M. and Zuber, H. (1974). Thermophilic aminopeptidase from *Bacillus stearothermophilus*. IV. Aminopeptidases (API_m, APIII) in mesophilic (37°C) cultures of the thermophilic bacterium *B. stearothermophilus*. *Int. J. Pept. Prot. Res.* **6,** 499–514.

Barnes, L. D. and Stellwagen, E. (1973). Enolase from the thermophile *Thermus* X-1. *Biochemistry* **12,** 1559–1565.

Bauman, A. J. and Simmonds, P. G. (1969). Fatty acids and polar lipids of extremely thermophilic filamentous bacterial masses from two Yellowstone hot springs. *J. Bacteriol.* **98,** 528–531.

Bausum, H. T. and Matney, T. S. (1965). Boundary between bacterial mesophilism and thermophilism. *J. Bacteriol.* **90,** 50–53.

Bělehrádek, S. (1931). Le mechanisme physico-chemique de l'adaptation thermique. *Protoplasma* **12,** 406–434.

*Biesecker, G., Harris, J. I., Thierry, J. C., Walker, J. E. and Wonacott, A. J. (1977). Sequence and structure of D-glyceraldehyde 3-phosphate dehydrogenase from *Bacillus stearothermophilus*. *Nature, Lond.*, **266,** 328–333.

Bigelow, C. C. (1967). On the average hydrophobicity of proteins and the relation between it and protein structure. *J. Theor. Biol.* **16,** 187–211.

Boccú, E., Veronese, F. M. and Fontana, A. (1976). Isolation and some properties of enolase from *Bacillus stearothermophilus*. *In* "Enzymes and Proteins from Thermophilic Microorganisms," *Experientia Suppl.* **26** (Ed. H. Zuber), pp. 229–236. Birkhäuser Verlag, Basel and Stuttgart.

Bradshaw, R. A., Ericsson, L. H., Walsh, K. A. and Neurath, H. (1969). The amino acid sequence of bovine carboxypeptidase A. *Proc. Natl. Acad. Sci. USA* **63,** 1389–1394.

Brandts, J. F. (1967). Heat effects on proteins and enzymes. *In* "Thermobiology" (Ed. A. H. Rose), pp. 25–72. Academic Press, New York and London.

Brewer, J. M., Ljungdahl, L., Spencer, T. E. and Neece, S. H. (1970). Physical properties of formyltetrahydrofolate synthetase from *Clostridium thermoaceticum* *J. Biol. Chem.* **245,** 4798–4803.

Bridgen, J., Harris, J. I., McDonald, P. W., Amelunxen, R. E. and Kimmel, J. R. (1972). Amino acid sequence around the catalytic site in glyceraldehyde-3-phosphate dehydrogenase from *Bacillus stearothermophilus*. *J. Bacteriol.* **111,** 797–800.

Brock, T. D. (1967). Life at high temperatures. *Science* **158,** 1012–1019.

Brock, T. D. (1970). High temperature systems. *Ann. Rev. Ecol. Systemat.* **1,** 191–220.

Brock, T. D. and Darland, G. K. (1970). Limits of microbial existence: temperature and pH. *Science* **169,** 1316–1318.

Brock, T. D. and Freeze, H. (1969). *Thermus aquaticus* gen. n. and sp. n., a non-sporulating extreme thermophile. *J. Bacteriol.* **98,** 289–297.

Bubela, B. and Holdsworth, E. S. (1966a). Amino acid uptake, protein and nucleic acid synthesis and turnover in *Bacillus stearothermophilus*. *Biochim. Biophys. Acta* **123,** 364–375.

Bubela, B. and Holdsworth, E. S. (1966b). Protein synthesis in *Bacillus stearothermophilus*. *Biochim. Biophys. Acta* **123,** 376–389.

Buchanan, B. B., Lovenberg, W. and Rabinowitz, J. C. (1963). A comparison of clostridial ferredoxins. *Proc. Natl. Acad. Sci. USA* **49,** 345–353.

Bull, H. B. and Breese, K. (1973). Thermal stability of proteins. *Arch. Biochem. Biophys.* **158,** 681–686.

Burgess, A. W. and Scheraga, H. A. (1975). A hypothesis for the pathway of the thermally-induced unfolding of bovine pancreatic ribonuclease. *J. Theor. Biol.* **53,** 403–420.

Burstein, Y., Walsh, K. A. and Neurath, H. (1974). Evidence of an essential histidine residue in thermolysin. *Biochemistry* **13,** 205–210.

Buttlaire, D. H., Hersh, R. T. and Himes, R. H. (1972). Hydrogen ion-induced reversible inactivation and dissociation of formyltetrahydrofolate synthetase. *J. Biol. Chem.* **247,** 2059–2068.

Campbell, L. L. (1954). The growth of an "obligate" thermophilic bacterium at 36°C. *J. Bacteriol.* **68,** 505–507.

Campbell, L. L. (1955). Purification and properties of an α-amylase from facultative thermophilic bacteria. *Arch. Biochem. Biophys.* **54,** 154–161.

Campbell, L. L. and Pace, B. (1968). Physiology of growth at high temperatures. *J. Appl. Bacteriol.* **31,** 24–35.

Case, K. H. and Stellwagen, E. (1975). A thermostable phosphofructokinase from the extreme thermophile, *Thermus* X-1. *Arch. Biochem. Biophys.* **171,** 682–694.

Castenholz, R. W. (1969). Thermophilic blue-green algae and the thermal environment. *Bacteriol. Rev.* **33,** 476–504.

Cavari, B. Z. and Grossowicz, N. (1973). Properties of homoserine dehydrogenase in a thermophilic bacterium. *Biochim. Biophys. Acta* **302,** 183–190.

Cavari, B. Z., Arkin-Shlank, H. and Grossowicz, N. (1972). Regulation of aspartokinase activity in a thermophilic bacterium. *Biochim. Biophys. Acta* **261,** 161–167.

Chan, M., Virmani, Y. P., Himes, R. H. and Akagi, J. M. (1973). Spin-labeling studies on the membrane of a facultative thermophilic *Bacillus*. *J. Bacteriol.* **113,** 322–328.

Chapman, D. (1967). The effect of heat on membranes and membrane constituents. *In* "Thermobiology" (Ed. A. H. Rose), pp. 123–145. Academic Press, New York and London.

Chen, M. C. and Lord, R. C. (1976). Laser Raman spectroscopic studies of the thermal unfolding of ribonuclease A. *Biochemistry* **15,** 1889–1897.

Chenoweth, D., Mitchel, R. E. J. and Smith, E. L. (1973). Aminotripeptidase of swine kidney. I. Isolation and characterization of three different forms; utility of the enzyme in sequence work. *J. Biol. Chem.* **248,** 1672–1683.

Colman, P. M., Jansonius, J. N. and Matthews, B. W. (1972). The structure of thermolysin: an electron density map at 2·3 Å resolution. *J. Mol. Biol.* **70,** 701–724.

Colowick, S. P. and Kalckar, H. M. (1943). The role of myokinase in transphosphorylations. *J. Biol. Chem.* **148,** 117–137.
*Coultate, T. P., Sundaram, T. K. and Cazzulo, J. J. (1975). Stability of protein and ribonucleic acid in *Bacillus stearothermophilus*. *J. Gen. Microbiol.* **91,** 383–390.
Crabb. J. W., Murdock, A. L. and Amelunxen, R. E. (1975). A proposed mechanism of thermophily in facultative thermophiles. *Biochem. Biophys. Res. Commun.* **62,** 627–633.
*Crabb, J. W., Murdock, A. L. and Amelunxen, R. E. (1977). Purification and characterization of thermolabile glyceraldehyde-3-phosphate dehydrogenase from the facultative thermophile *Bacillus coagulans* KU. *Biochemistry* **16,** 4840–4847.
Cronan, J. E. and Gelmann, E. P. (1975). Physical properties of membrane lipids: biological relevance and regulation. *Bacteriol. Rev.* **39,** 232–256.
Dahlquist, F. W., Long, J. W. and Bigbee, W. L. (1976). Role of calcium in the thermal stability of thermolysin. *Biochemistry* **15,** 1103–1111.
D'Alessio, G. and Josse, J. (1971). Glyceraldehyde phosphate dehydrogenase of *Escherichia coli*. Structural and catalytic properties. *J. Biol. Chem.* **246,** 4326–4333.
Damadian, R. (1973). Biological ion exchanger resins. *Ann. N.Y. Acad. Sci.* **204,** 211–244.
Daron, H. H. (1970). Fatty acid composition of lipid extracts of a thermophilic *Bacillus* species. *J. Bacteriol.* **101,** 145–151.
Date, T., Suzuki, K. and Imahori, K. (1975). Purification and some properties of DNA-dependent RNA polymerase from an extreme thermophile *Thermus thermophilus* HB8. *J. Biochem.* **78,** 845–858.
Davidson, B. E., Sajgo, M., Noller, H. F. and Harris, J. I. (1967). Amino-acid sequence of glyceraldehyde-3-phosphate dehydrogenase from lobster muscle. *Nature, Lond.* **216,** 1181–1185.
Devanathan, T., Akagi, J. M., Hersh, R. T. and Himes, R. H. (1969). Ferredoxin from two thermophilic *Clostridia*. *J. Biol. Chem.* **244,** 2846–2853.
Dick, A. J., Matheson, A. T. and Wang, J. H. (1970). A ribosomal-bound aminopeptidase in *Escherichia coli* B: purification and properties. *Can. J. Biochem.* **8,** 1181–1188.
Donovan, J. W. and Ross, K. D. (1973). Increase in the stability of avidin produced by binding of biotin. A differential scanning colorimetric study of denaturation by heat. *Biochemistry* **12,** 512–517.
Endo, S. (1962). Studies on protease produced by thermophilic bacteria. *Hakko Kogaku Zasshi* **37,** 353–410. (*Chem. Abst.* **62,** 5504, 1965.)
Esser, A. F. and Souza, K. A. (1974). Correlation between thermal death and membrane fluidity in *Bacillus stearothermophilus*. *Proc. Natl. Acad. Sci. USA* **71,** 4111–4115.
Farrell, J. and Campbell, L. L. (1969). Thermophilic bacteria and bacteriophages. *Adv. Microbial Physiol.* **3,** 83–109.
Farrell, J. and Rose, A. H. (1967a). Temperature effects on microorganisms. *Ann. Rev. Microbiol.* **21,** 101–120.
Farrell, J. and Rose, A. H. (1967b). Temperature effects on microorganisms. *In* "Thermobiology" (Ed. A. H. Rose), pp. 147–218. Academic Press, New York and London.
Feder, J., Garrett, L. R. and Wildi, B. S. (1971). Studies on the role of calcium in thermolysin. *Biochemistry* **10,** 4552–4556.
Fontana, A., Grandi, C., Boccú, E. and Veronese, F. M. (1976). Thermal properties

of glyceraldehyde-3-phosphate dehydrogenase from *Escherichia coli*. *In* "Enzymes and Proteins from Thermophilic Microorganisms," *Experientia Suppl.* **26** (Ed. H. Zuber), pp. 135–145. Birkhäuser Verlag, Basel and Stuttgart.

Frank, G., Haberstich, H., Schaer, H. P., Tratschin, J. D. and Zuber, H. (1976). Thermophilic and mesophilic enzymes from *B. caldotenax* and *B. stearothermophilus*: Properties, relationships and formation. *In* "Enzymes and Proteins from Thermophilic Microorganisms," *Experientia Suppl.* **26** (Ed. H. Zuber), pp. 375–389. Birkhäuser Verlag, Basel and Stuttgart.

Friedman, S. M. (1968). Protein-synthesizing machinery of thermophilic bacteria. *Bacteriol. Rev.* **32,** 27–38.

Fujita, S. C., Oshima, T. and Imahori, K. (1976). Purification and properties of D-glyceraldehyde-3-phosphate dehydrogenase from an extreme thermophile, *Thermus thermophilus* strain HB8. *Eur. J. Biochem.* **64,** 57–68.

Gaughran, E. R. L. (1947). The thermophilic microorganisms. *Bacteriol. Rev.* **11,** 189–225.

Greenberg, G. R., Jaenicke, L. and Silverman, M. (1955). On the occurrence of N^{10}-formyltetrahydrofolic acid by enzymic formylation of tetrahydrofolic acid on the mechanism of this reaction. *Biochim. Biophys. Acta* **17,** 589–591.

Grissom, F. E. and Kahn, J. S. (1975). Glycerladehyde-3-phosphate dehydrogenase from *Euglena gracilis*. *Arch. Biochem. Biophys.* **171,** 444–458.

Haberstich, H. V. and Zuber, H. (1974). Thermoadaptation of enzymes in thermophilic and mesophilic cultures of *Bacillus stearothermophilus* and *Bacillus caldotenax*. *Arch. Microbiol.* **98,** 275–287.

Hachimori, A., Muramatsu, N. and Nosoh, Y. (1970). Studies on an ATPase of thermophilic bacteria. I. Purification and properties. *Biochim. Biophys. Acta* **206,** 426–437.

Hachimori, A., Matsunaga, A., Shimizu, M., Samejima, T. and Nosoh, Y. (1974). Purification and properties of glutamine synthetase from *Bacillus stearothermophilus*. *Biochim. Biophys. Acta* **350,** 461–474.

Hachimori, A., Takeda, A., Kaibuchi, M., Ohkawara, R. and Samejima, T. (1975). Purification and characterization of inorganic pyrophosphatase from *Bacillus stearothermophilus*. *J. Biochem.* **77,** 1177–1183.

Han, M. H. (1972). Non-linear Arrhenius plots in temperature-dependent kinetic studies of enzyme reactions. I. Single transition processes. *J. Theor. Biol.* **35,** 543–568.

Harris, J. I. and Perham, R. N. (1968). Glyceraldehyde-3-phosphate dehydrogenase from pig muscle. *Nature, Lond.* **219,** 1025–1028.

*Hase, T., Ohmiya, N., Matsubara, H., Mullinger, R. N., Rao, K. K. and Hall, D. O. (1976). Amino acid sequence of a 4-iron-4-sulfur ferredoxin from *Bacillus stearothermophilus*. *Biochem. J.* **159,** 55–63.

Hasegawa, A., Miwa, N., Oshima, T. and Imahori, K. (1976). Studies on an α-amylase from a thermophilic bacterium. I. Purification and Characterization. *J. Biochem.* **79,** 34–42.

Heilbrunn, L. V. (1924). The colloid chemistry of protoplasm. IV. The heat coagulation of protoplasm. *Am. J. Physiol.* **69,** 190–199.

Heinen, U. J. and Heinen, W. (1972). Characteristics and properties of a caldoactive bacterium producing extracellular enzymes and two related strains. *Arch. Microbiol.* **82,** 1–23.

Heinen, W. and Lauwers, A. M. (1976). Amylase activity and stability at high and low temperatures depending on calcium and other divalent cations. *In* "Enzymes and Proteins from Thermophilic Microorganisms," *Experientia Suppl.* **26** (Ed. H. Zuber), pp. 77–89. Birkhäuser Verlag, Basel and Stuttgart.

Hengartner, H. and Zuber, H. (1973). Isolation and characterization of a thermophilic glucokinase from *Bacillus stearothermophilus*. *FEBS Lett.* **37,** 212–216.

Hengartner, H., Kolb, E. and Harris, J. I. (1976). Phosphofructokinase from thermophilic microorganisms. *In* "Enzymes and Proteins from Thermophilic Microorganisms," *Experientia Suppl.* **26** (Ed. H. Zuber), pp. 199–206. Birkhäuser Verlag, Basel and Stuttgart.

Hibino, Y., Nosoh, Y. and Samejima, T. (1974). On the conformation of NADP-dependent isocitrate dehydrogenase from *Bacillus stearothermophilus*. *J. Biochem.* **75,** 553–561.

Himes, R. H. and Harmony, J. A. K. (1973). Formyltetrahydrofolate synthetase. *CRC Crit. Rev. Biochem.* **1,** 501–535.

Himes, R. H. and Wilder, T. (1968). Formyltetrahydrofolate synthetase: effect of pH and temperature on the reaction. *Arch. Biochem. Biophys.* **124,** 230–237.

Hocking, J. D. and Harris, J. I. (1973). Purification by affinity chromatography of thermostable glyceraldehyde-3-phosphate dehydrogenase from *Thermus aquaticus*. *FEBS Lett.* **34,** 280–284.

Hocking, J. D. and Harris, J. I. (1976). Glyceraldehyde-3-phosphate dehydrogenase from an extreme thermophile, *Thermus aquaticus*. *In* "Enzymes and Proteins from Thermophilic Microorganisms," *Experientia Suppl.* **26** (Ed. H. Zuber), pp. 121–133. Birkhäuser Verlag, Basel and Stuttgart.

Hong, J. and Rabinowitz, J. C. (1967). Preparation and properties of clostridial apoferredoxins. *Biochem. Biophys. Res. Commun.* **29,** 246–252.

Howell, N., Akagi, J. M. and Himes, R. H. (1969), Thermostability of glycolytic enzymes from thermophilic *Clostridia*. *Can. J. Microbiol.* **15,** 461–464.

Hsiu, J., Fischer, E. H. and Stein, E. A. (1964). Alpha-amylases as calcium-metalloenzymes. II. Calcium and the catalytic activity. *Biochemistry* **3,** 61–66.

Iizuka, E. and Yang, J. I. (1965). Effect of salts and dioxane on the coiled conformation of poly-L-glutamic acid in aqueous solution. *Biochemistry* **4,** 1249–1257.

Irwin, C. C., Akagi, J. M. and Himes, R. H. (1973). Ribosomes, polyribosomes, and deoxyribonucleic acid from thermophilic, mesophilic and psychrophilic *Clostridia*. *J. Bacteriol.* **113,** 252–262.

Jones, G. M. T. and Harris, J. I. (1972). Glyceraldehyde 3-phosphate dehydrogenase: amino acid sequence of enzyme from baker's yeast. *FEBS Lett.* **22,** 185–189.

Jung, L., Jost, R., Stoll, E. and Zuber, H. (1974). Metabolic differences in *Bacillus stearothermophilus* grown at 55°C and 37°C. *Arch. Microbiol.* **95,** 125–138.

*Kagawa, Y., Sone, N., Yoshida, M., Hirata, H. and Okamoto, H. (1976). Proton translocating ATPase of a thermophilic bacterium. Morphology, subunits and chemical composition. *J. Biochem.* **80,** 141–151.

Kaneda, T. (1963). Biosynthesis of branched-chain fatty acids. I. Isolation and identification of fatty acids from *Bacillus subtilis* (ATCC 7059). *J. Biol. Chem.* **238,** 1222–1228.

Kobayashi, S. H., Hubbell, H. R. and Orengo, A. (1974). A homogeneous, thermostable deoxythymidine kinase from *Bacillus stearothermophilus*. *Biochemistry* **13,** 4537–4543.

Koffler, H. (1957). Protoplasmic differences between mesophiles and thermophiles. *Bacteriol. Rev.* **21,** 227–240.

Koffler, H. and Gale, G. O. (1957). The relative thermostability of cytoplasmic proteins from thermophilic bacteria. *Arch. Biochem. Biophys.* **67,** 249–251.

Kuramitsu, H. K. (1968). Concerted feedback inhibition of aspartokinase from *Bacillus stearothermophilus. Biochim. Biophys. Acta* **167,** 643–645.

Kuramitsu, H. K. (1970). Concerted feedback inhibition of aspartokinase from *Bacillus stearothermophilus*. I. Catalytic and regulatory properties. *J. Biol. Chem.* **245,** 2991–2997.

Langridge, J. (1968). Genetic and enzymatic experiments relating to the tertiary structure of β-galactosidase. *J. Bacteriol.* **96,** 1711–1717.

Lindsay, J. A. and Creaser, E. H. (1975). Enzyme thermostability is a transformable property between *Bacillus* spp. *Nature, Lond.* **255,** 650–652.

Lipscomb, W. N., Hartsuck, J. A., Quiocho, F. A. and Reeke, G. N. (1969). The structure of carboxypeptidase A. IX. The X-ray diffraction results in the light of the chemical sequence. *Proc. Natl. Acad. Sci. USA* **64,** 28–35.

Ljungdahl, L. G. and Sherod, D. (1976). Proteins from thermophilic microorganisms. *In* "Extreme Environments: Mechanisms of Microbial Adaptation" (Ed. M. R. Heinrich), pp. 147–188. Academic Press, New York and London.

Ljungdahl, L. G. and Wood, H. G. (1969). Total synthesis of acetate from CO_2 by heterotrophic bacteria. *Ann. Rev. Microbiol.* **23,** 515–538.

Ljungdahl, L., Irion, E. and Wood, H.G. (1965). Total synthesis of acetate from CO_2. I. Co-methylcobyric acid and co-(methyl)-5-methoxybenzimidazolylcobamide as intermediates with *Clostridium thermoaceticum. Biochemistry* **4,** 2771–2780.

Ljungdahl, L., Brewer, J. M., Neece, S. H. and Fairwell, T. (1970). Purification, stability, and composition of formyltetrahydrofolate synthetase from *Clostridium thermoaceticum. J. Biol. Chem.* **245,** 4791–4797.

Lodish, H. F. (1969). Species specificity of polypeptide chain initiation. *Nature, Lond.* **224,** 867–870.

Long, S. K. and Williams, O. B. (1959). Growth of obligate thermophiles at 37°C as a function of the cultural conditions employed. *J. Bacteriol.* **77,** 545–547.

Lovenberg, W., Buchanan, B. B. and Rabinowitz, J. C. (1963). Studies on the chemical nature of clostridial ferredoxin. *J. Biol. Chem.* **238,** 3899–3913.

MacKenzie, R. E. and Rabinowitz, J. C. (1971). Cation-dependent reassociation of subunits of N^{10}-formyltetrahydrofolate synthetase from *Clostridium acidiurici* and *Clostridium cylindrosporum. J. Biol. Chem.* **246,** 3731–3736.

Matsubara, H. and Feder, J. (1971). Other bacterial, mold and yeast proteases. *In* "The Enzymes" (3rd ed.) (Ed. P. D. Boyer), Vol. 3, pp. 721–795. Academic Press, New York and London.

Matsunaga, A. and Nosoh, Y. (1974). Conformational change with temperature and thermostability of glutamine synthetase from *Bacillus stearothermophilus. Biochim. Biophys. Acta* **365,** 208–211.

Matsunaga, A., Koyama, N. and Nosoh, Y. (1974). Purification and properties of esterase from *Bacillus stearothermophilus. Arch. Biochem. Biophys.* **160,** 504–513.

Matthews, B. W., Jansonius, J. N., Colman, P. M., Schoenborn, B. P. and Dupourque, D. (1972a). Three-dimensional structure of thermolysin. *Nature, Lond.* **238,** 37–41.

Matthews, B. W., Colman, P. M., Jansonius, J. N., Titani, K., Walsh, K. A. and Neurath, H. (1972b). Structure of thermolysin. *Nature, Lond.* **238,** 41–43.

Matthews, B. W., Weaver, L. H. and Kester, W. R. (1974). The conformation of thermolysin. *J. Biol. Chem.* **249,** 8030–8044.

Militzer, W., Sonderegger, T. B., Tuttle, L. C. and Georgi, C. E. (1949). Thermal enzymes. *Arch. Biochem.* **24,** 75–82.

Moore, M. R , O'Brien, W. E. and Ljungdahl, L. G. (1974). Purification and characterization of nicotinamide adenine dinucleotide-dependent methylenetetrahydrofolate dehydrogenase from *Clostridium formicoaceticum*. *J. Biol. Chem.* **249,** 5250–5253.

Moser, P., Roncari, G. and Zuber, H. (1970). Thermophilic aminopeptidases from *Bac. stearothermophilus*. II. Aminopeptidase I (API): physico-chemical properties; thermostability and activation; formation of apoenzyme and subunits. *Int. J. Prot. Res.* **11,** 191–207.

Mullinger, R. N., Cammack, R., Rao, K. K., Hall, D. O., Dickson, D. P. E., Johnson, C. E., Rush, J. D. and Simopoulos, A. (1975). Physicochemical characterization of the 4-iron-4-sulphide ferredoxin from *Bacillus stearothermophilus*. *Biochem. J.* **151,** 75–83.

Muramatsu, N. and Nosoh, Y. (1971). Purification and characterization of glucose-6-phosphate isomerase from *Bacillus stearothermophilus*. *Arch. Biochem. Biophys.* **144,** 245–252.

*Muramatsu, N. and Nosoh, Y. (1976). Some catalytic and molecular properties of threonine deaminase from *Bacillus stearothermophilus*. *J. Biochem.* **80,** 485–490.

Murdock, A. L. and Koeppe, O. J. (1964). The content and action of diphosphopyridine nucleotide in triosephosphate dehydrogenase. *J. Biol. Chem.* **239,** 1983–1988.

Murphey, W. H., Barnaby, J. F., Lin, J. F. and Kaplan, N. O. (1967). Malate dehydrogenase. II. Purification and properties of *Bacillus subtilis*, *Bacillus stearothermophilus*, and *Escherichia coli* malate dehydrogenases. *J. Biol. Chem.* **242,** 1548–1559.

Novitsky, T. J., Chan, M., Himes, R. H. and Akagi, J. M. (1974). Effect of temperature on the growth and cell wall chemistry of a facultative thermophilic *Bacillus*. *J. Bacteriol.* **117,** 858–865.

O'Brien, W. E. and Ljungdahl, L. G. (1972). Fermentation of fructose and synthesis of acetate from carbon dioxide by *Clostridium formicoaceticum*. *J. Bacteriol.* **109,** 626–632.

O'Brien, W. E., Brewer, J. M. and Ljungdahl, L. G. (1976). Chemical, physical and enzymatic comparisons of formyltetrahydrofolate synthetase from thermo- and mesophilic *Clostridia*. *In* "Enzymes and Proteins from Thermophilic Microorganisms," *Experientia Suppl.* **26** (Ed. H. Zuber), pp. 249–262. Birkhäuser Verlag, Basel and Stuttgart.

Ogasahara, K., Imanishi, A. and Isemura, T. (1970). Studies on thermophilic α-amylase from *Bacillus stearothermophilus*. II. Thermal stability of thermophilic α-amylase. *J. Biochem.* **67,** 77–82.

Ohno-Iwashita, Y., Oshima, T. and Imahori, K. (1975). Comparison of protein-synthesizing machinery of an extreme thermophile with that of *Escherichia coli*. *Z. Allg. Mikrobiol.* **15,** 131–134.

Ohta, Y. (1967). Thermostable protease from thermophilic bacteria. II. Studies on the stability of the protease. *J. Biol. Chem.* **242,** 509–515.

Ohta, Y., Ogura, Y. and Wada, A. (1966). Thermostable protease from thermophilic bacteria. I. Thermostability, physicochemical properties, and amino acid composition. *J. Biol. Chem.* **241,** 5919–5925.

Olsen, K. W., Moras, D., Rossman, M. G. and Harris, J. I. (1975). Sequence variability and structure of D-glyceraldehyde-3-phosphate dehydrogenase. *J. Biol. Chem.* **250,** 9313–9321.

Orengo, A. and Saunders, G. F. (1972). Regulation of a thermostable pyrimidine ribonucleoside kinase by cytidine triphosphate. *Biochemistry* **11,** 1761–1767.

Oshima, T. and Imahori, K. (1974). Description of *Thermus thermophilus* (Yoshida and Oshima) comb. nov., a nonsporulating thermophilic bacterium from a Japanese thermal spa. *Int. J. Syst. Bacteriol.* **24,** 102–112.
Pearse, B. M. F. and Harris, J. I. (1973). 6-Phosphogluconate dehydrogenase from *Bacillus stearothermophilus. FEBS Lett.* **38,** 49–52.
Perham, R. N. (1969). The comparative studies of mammalian glyceraldehyde-3-phosphate dehydrogenases. *Biochem. J.* **111,** 17–21.
Perutz, M. F. and Raidt, H. (1975). Stereochemical basis of host stability in bacterial ferredoxins and in haemoglobin A2. *Nature, Lond.* **255,** 256–259.
Pétra, P. H. (1970). Bovine procarboxypeptidase and carboxypeptidase A. *In* "Methods in Enzymology" (Eds G. E. Perlman and L. Lorand), Vol. 19, pp. 460–503. Academic Press, New York and London.
Pfeuller, S. L. and Elliott, W. H. (1969). The extracellular α-amylase of *Bacillus stearothermophilus. J. Biol. Chem.* **244,** 48–54.
Poston, J. M., Kuratomi, K. and Stadtman, E. R. (1966). The conversion of carbon dioxide to acetate. I. The use of cobalt-methylcobalamin as a source of methyl groups for the synthesis of acetate by cell-free extracts of *Clostridium thermoaceticum. J. Biol. Chem.* **241,** 4209–4216.
Rabinowitz, J. C. and Pricer, W. E., Jr. (1957). An enzymatic method for the determination of formic acid. *J. Biol. Chem.* **229,** 321–328.
Rabinowitz, J. C. and Pricer, W. E., Jr. (1962). Formyltetrahydrofolate synthetase. I. Isolation and crystallization of the enzyme. *J. Biol. Chem.* **237,** 2898–2902.
Rall, S. C., Bolinger, R. E. and Cole, R. D. (1969). The amino acid sequence of ferredoxin from *Clostridium acidi-urici. Biochemistry* **8,** 2486–2496.
Ramaley, R. F. and Hixson, H. (1970). Isolation of a nonpigmented, thermophilic bacterium similar to *Thermus aquaticus. J. Bacteriol.* **103,** 527–528.
Ray, P. H., White, D. C. and Brock, T. D. (1971a). Effect of temperature on the fatty acid composition of *Thermus aquaticus. J. Bacteriol.* **106,** 25–30.
Ray, P. H., White, D. C. and Brock, T. D. (1971b), Effect of growth temperature on the lipid composition of *Thermus aquaticus. J. Bacteriol.* **108,** 227–235.
Roncari, G. and Zuber, H. (1969). Thermophilic aminopeptidases from *Bacillus stearothermophilus.* I. Isolation, specificity, and general properties of the thermostable aminopeptidase I. *Int. J. Prot. Res.* **1,** 45–61.
Roncari, G. and Zuber, H. (1970). Thermophilic aminopeptidases: API from *Bacillus stearothermophilus. In* "Methods in Enzymology" (Eds G. E. Perlman and L. Lorand), Vol. 19, pp. 544–552. Academic Press, New York and London.
Roncari, G., Zuber, H. and Wyttenbach, A. (1972). Thermophilic aminopeptidases from *Bac. stearothermophilus.* III. Determination of the cobalt and zinc content in aminopeptidase I by neutron activation analysis. *Int. J. Pept. Prot. Res.* **4,** 267–271.
Rowe, J. J., Goldberg, I. D. and Amelunxen, R. E. (1973). Isolation of mutants of *Bacillus stearothermophilus* blocked in catabolic function. *Can. J. Microbiol.* **19,** 1521–1523.
Rowe, J. J., Goldberg, I. D. and Amelunxen, R. E. (1975). Development of defined and minimal media for the growth of *Bacillus stearothermophilus. J. Bacteriol.* **124,** 279–284.
Rowe, J. J., Goldberg, I. D. and Amelunxen, R. E. (1976). Characteristics of *Bacillus stearothermophilus* mutants blocked in catabolic function. *J. Bacteriol.* **126,** 520–523.
Ruwart, M. J. and Haug, A. (1974). Membrane properties of *Thermoplasma acidophila. Biochemistry* **14,** 860–866.

Saiki, T., Mahmud, I., Matsubara, N., Taya, K. and Arima, K. (1976). Purification and some properties of NADP$^+$-specific isocitrate dehydrogenase from an extreme thermophile, *Thermus flavus* AT-62. *In* "Enzymes and Proteins from Thermophilic Microorganisms," *Experientia Suppl.* **26** (Ed. H. Zuber), pp. 169–183. Birkhäuser Verlag, Basel and Stuttgart.

Sando, G. N. and Hogenkamp, H. P. C. (1973). Ribonucleotide reductase from *Thermus* X-1, a thermophilic organism. *Biochemistry* **12,** 3316–3322.

Sauvan, R. L., Mira, O. J. and Amelunxen, R. E. (1972). Thermostable glyceraldehyde-3-phosphate dehydrogenase from *Bacillus stearothermophilus.* I. Immunochemical studies. *Biochim. Biophys. Acta* **263,** 794–804.

Shing, Y. W., Akagi, J. M. and Himes, R. H. (1975). Psychrophilic, mesophilic, and thermophilic triosephosphate isomerases from three clostridial species. *J. Bacteriol.* **122,** 177–184.

Shoaf, W. T., Neece, S. H. and Ljungdahl, L. G. (1974). Effects of temperature and ammonium ions on formyltetrahydrofolate synthetase from *Clostridium thermoaceticum. Biochim. Biophys. Acta* **334,** 448–458.

Singleton, R., Jr. (1976). A comparison of the amino acid composition from thermophilic and non-thermophilic origins. *In* "Extreme Environments: Mechanisms of Microbial Adaptation" (Ed. M. R. Heinrich), pp. 189–200. Academic Press, New York and London.

Singleton, R., Jr. and Amelunxen, R. E. (1973). Proteins from thermophilic microorganisms. *Bacteriol. Rev.* **37,** 320–342.

Singleton, R., Jr., Kimmel, J. R. and Amelunxen, R. E. (1969). The amino acid composition and other properties of thermostable glyceraldehyde-3-phosphate dehydrogenase from *Bacillus stearothermophilus. J. Biol. Chem.* **244,** 1623–1630.

Smith, P. F., Langworthy, T. A., Mayberry, W. R. and Houghland, A. E. (1973). Characterization of the membranes of *Thermoplasma acidophilum. J. Bacteriol.* **116,** 1019–1028.

Stahl, S. and Ljunger, C. (1976). Calcium uptake by *Bacillus stearothermophilus*: a requirement for thermophilic growth. *FEBS Lett.* **63,** 184–187.

Stanier, R. Y., Doudoroff, M. and Adelberg, E. A. (1970). *In* "The Microbial World" (3rd ed.), pp. 315–318. Prentice-Hall, New Jersey.

Stellwagen, E. and Barnes, L. D. (1976). Analysis of the thermostability of enolases. *In* "Enzymes and Proteins from Thermophilic Microorganisms," *Experientia, Suppl.* **26,** (Ed. H. Zuber), pp. 223–227. Birkhäuser Verlag, Basel and Stuttgart.

Stoll, E. and Zuber, H. (1974). Interconversion of the different hybrids of aminopeptidase I. *FEBS Lett.* **40,** 210–212.

Stoll, E., Hermodson, M. A., Ericsson, L. H. and Zuber, H. (1972). Subunit structure of the thermophilic aminopeptidase I from *Bacillus stearothermophilus. Biochemistry* **11,** 4731–4735.

Stoll, E., Ericsson, L. H. and Zuber, H. (1973). The function of the two subunits of thermophilic aminopeptidase I. *Proc. Natl. Acad. Sci. USA* **70,** 3781–3784.

Sugimoto, S. and Nosoh, Y. (1971). Thermal properties of fructose-1,6-diphosphate aldolase from thermophilic bacteria. *Biochim. Biophys. Acta.* **235,** 210–221.

Sun, A. Y., Ljungdahl, L. G. and Wood, H. G. (1969). Total synthesis of acetate from CO_2. II. Purification and properties of formyltetrahydrofolate synthetase from *Clostridium thermoaceticum. J. Bacteriol.* **98,** 842–844.

Sundaram, T. K., Libor, S. and Chell, R. M. (1976). Anaplerotic enzymes of acetate and pyruvate metabolism: Distinctive characteristics in *Bacillus stearothermophilus. In* "Enzymes and Proteins from Thermophilic Microorganisms,"

Experientia Suppl. **26** (Ed. H. Zuber), pp. 263–275. Birkhäuser Verlag, Basel and Stuttgart.

Suzuki, K. and Harris, J. I. (1971). Glyceraldehyde-3-phosphate dehydrogenase from *Bacillus stearothermophilus. FEBS Lett.* **13,** 217–220.

Suzuki, K. and Imahori, K. (1973a). Isolation and some properties of glyceraldehyde 3-phosphate dehydrogenase from vegetative cells of *Bacillus cereus. J. Biochem.* **73,** 97–106.

Suzuki, K. and Imahori, K. (1973b). Glyceraldehyde 3-phosphate dehydrogenase of *Bacillus stearothermophilus.* Kinetics and physiocochemical studies. *J. Biochem.* **74,** 955–970.

Suzuki, K. and Imahori, K. (1974). Phosphoglycerate kinase of *Bacillus stearothermophilus. J. Biochem.* **76,** 771–782.

Tanaka, M., Nakashima, T., Benson, A. M., Mower, H. F. and Yasunobu, K. T. (1966). The amino acid sequence of *Clostridium pasteurianum* ferredoxin. *Biochemistry* **5,** 1666–1681.

Tanaka, M., Haniu, M., Matsueda, G., Yasunobu, K. T., Himes, R. H., Akagi, J. M., Barnes, E. M. and Devanathan, T. (1971). The primary structure of the *Clostridium tartarivorum* ferredoxin, a heat-stable ferredoxin. *J. Biol. Chem.* **246,** 3958–3960.

Tanaka, M., Haniu, M., Yasunobu, K. T., Himes, R. H. and Akagi, J. M. (1973). The primary structure of the *Clostridium thermosaccharolyticum* ferredoxin, a heat-stable ferredoxin. *J. Biol. Chem.* **248,** 5215–5217.

Tanford, C. (1968). Protein denaturation. *In* "Advances in Protein Chemistry" (Eds C. B. Anfinsen, Jr., M. L. Anson, J. T. Edsall, F. M. Richards), Vol. 23, pp. 122–282. Academic Press, New York and London.

Thomas, D. A. and Kuramitsu, H. K. (1971). Biosynthetic L-threonine deaminase from *Bacillus stearothermophilus* I. Catalytic and regulatory properties. *Arch. Biochem. Biophys.* **145,** 96–104.

Titani, K., Hermodson, M. A., Ericsson, L. H., Walsh, K. A. and Neurath, H. (1972). Amino-acid sequence of thermolysin. *Nature, Lond.* **238,** 35–37.

Velick, S. F. (1955). Glyceraldehyde-3-phosphate dehydrogenase from muscle. *In* "Methods in Enzymology" (Eds S. P. Colowick and N. O. Kaplan), Vol. 1, pp. 401–406. Academic Press, New York and London.

Veronese, E. M., Boccú, E., Fontana, A., Benassi, C. A. and Scoffone, E. (1974). Isolation and some properties of 6-phosphogluconate dehydrogenase from *Bacillus stearothermophilus. Biochim. Biophys. Acta* **334,** 31–44.

*Veronese, F. M., Boccú, E. and Fontana, A. (1976). Isolation and properties of 6-phosphogluconate dehydrogenase from *Escherichia coli.* Some comparisons with the thermophilic enzyme from *Bacillus stearothermophilus. Biochemistry* **15,** 4026–4033.

Von Hippel, P. H. and Schleich, T. (1969). The effects of neutral salts on the structure and conformational stability of macromolecules in solution. *In* "Biological Macromolecules" (Eds S. Timasheff and G. Fasman), Vol. 11, pp. 417–557. Marcel Dekker, New York.

*Voordouw, G. and Roche, R. S. (1975). The role of bound calcium ions in thermostable proteolytic enzymes. II. Studies on thermolysin, the thermostable protease from Bacillus *thermoproteolyticus. Biochemistry* **14,** 4667–4673.

*Voordouw, G., Milo, C., and Roche, R. S. (1976). Role of bound calcium ions in thermostable proteolytic enzymes. Separation of intrinsic and calcium ion contributions to the kinetic thermal stability. *Biochemistry* **15,** 3716–3724.

Wedler, F. C. and Hoffmann, F. M. (1974a). Glutamine synthetase of *Bacillus stearothermophilus*. I. Purification and basic properties. *Biochemistry* **13,** 3207–3214.

Wedler, F. C. and Hoffmann, F. M. (1974b). Glutamine synthetase of *Bacillus stearothermophilus*. II. Regulation and thermostability. *Biochemistry* **13,** 3215–3221.

*Wedler, F. C., Carfi, J. and Ashour, A. E. (1976). Glutamine synthetase of *Bacillus stearothermophilus*. Regulation, site interactions, and functional information. *Biochemistry* **15,** 1749–1755.

Weerkamp, A. and MacElroy, R. D. (1972). Lactate dehydrogenase from an extremely thermophilic *Bacillus*. *Arch. Microbiol.* **85,** 113–122.

Welch, W. H., Buttlaire, D. H., Hersh, R. T. and Himes, R. H. (1971). The subunit structure of formyltetrahydrofolate synthetase. *Biochim. Biophys. Acta* **236,** 599–611.

Whiteley, H. R., Osborn, M. J. and Huennekens, F. M. (1959). Purification and properties of the formate-activating enzyme from *Micrococcus aerogenes*. *J. Biol. Chem.* **234,** 1538–1543.

Williams, R. A. D. (1975). Caldoactive and thermophilic bacteria and their thermostable proteins. *Sci. Prog. Oxf.* **62,** 373–393.

Wisdom, C. and Welker, N. E. (1973). Membranes of *Bacillus stearothermophilus:* factors affecting protoplast stability and thermostability of alkaline phosphatase and reduced nicotinamide adenine dinucleotide oxidase. *J. Bacteriol.* **114,** 1336–1345.

Yasunobu, K. T. and Tanaka, M. (1973). The types, distribution and nature, structure-function, and evolutionary data of the iron-sulfur proteins. *In* "Iron-Sulfur Proteins" (Ed. W. Lovenberg), Vol. 2, pp. 27–130. Academic Press, London and New York.

*Yeh, M. F. and Trela, J. M. (1976). Purification and characterization of a repressible alkaline phosphatase from *Thermus aquaticus*. *J. Biol. Chem.* **251,** 3134–3139.

Yoshida, M. (1972). Allosteric nature of thermostable phosphofructokinase from an extreme thermophilic bacterium. *Biochemistry* **11,** 1087–1093.

Yoshida, M. and Oshima, T. (1971). The thermostable allosteric nature of fructose-1,6-diphosphatase from an extreme thermophile. *Biochem. Biophys. Res. Commun.* **45,** 495–500.

Yoshida, M., Oshima, T. and Imahori, K. (1971). The thermostable allosteric enzyme: phosphofructokinase from an extreme thermophile. *Biochem. Biophys. Res. Commun.* **43,** 36–39.

Yoshida, M., Sone, N., Hirata, H. and Kagawa, Y. (1975). A highly stable adenosine triphosphatase from thermophilic bacterium. Purification, properties and reconstitution. *J. Biol. Chem.* **250,** 7910–7916.

*Yoshida, M., Sone, N., Hirata, H. and Kagawa, Y. (1977). Reconstitution of adenosine triphosphatase of thermophilic bacterium from purified subunits. *J. Biol. Chem.* **252,** 3480–3485.

Yutani, K. (1976). Role of calcium ion in the thermostability of α-amylase produced from *Bacillus stearothermophilus*. *In* "Enzymes and Proteins from Thermophilic Microorganisms," *Experientia Suppl.* **26** (Ed. H. Zuber), pp. 91–103. Birkhäuser Verlag, Basel and Stuttgart.

Chapter 7

Microbial Life in Extreme pH Values

T. A. LANGWORTHY

University of South Dakota

I. Introduction 279
II. Occurrence of life at extremes of pH 280
A. Life at low pH 280
B. Life at high pH 285
III. Nature of the extreme pH life forms 287
A. The external environment 287
B. Internal pH 289
C. Cell surfaces and membranes 294
IV. Conclusion 305
References 306

I. Introduction

The influence of hydrogen ion concentration as a factor in defining the boundaries of living matter has long been recognized. It is probably one of the most fundamental factors affecting the growth and reproduction of an organism. The hydrogen ion concentration affects the ionic state and therefore the availability of many metabolites and inorganic ions to an organism. Its influence on the stability and function of macromolecules in biological processes cannot be overemphasized. Most natural environments usually have concentrations of hydrogen ions near pH 7 (10^{-7} M), at which most organisms thrive best. Very high concentrations of hydrogen ions (acid) or extremely low concentrations (alkaline) are normally toxic to most organisms. The generally estimated extreme limits of hydrogen ion concentration above and below which presently known organisms cease growth and reproduction approximates pH 1 (0·1 M) reported for a few bacteria and fungi and approaches pH 11 (10^{-11} M) reported for several algae, fungi and bacteria

(Souza *et al.*, 1974). Most organisms live within the limits of pH 4 to 9 and have growth optima normally near neutrality. While many organisms have the ability to grow or survive exposure outside these limits, their actual pH optimum for growth is usually found within the normal pH range. Such organisms as these are considered acid- or alkaline-tolerant. Acidophilic organisms which have an obligate requirement for extremely acid pH values of 3 or less for growth are quite rare. Even fewer are examples of alkalinophilic (alkalophilic) organisms requiring pH values in the range of 10 or higher. Many of the reports of growth at extreme pH values are based upon the ambient pH values of the environment in which the organisms are found. Carefully controlled studies required to determine whether the organisms are actually growing and reproducing at the environmental pH, however, are generally lacking.

Over the past decades, limited reports of organisms growing in extreme pH values have appeared. Many of the organisms have remained little more than interesting biological curiosities. Except for limited cases, quantitative attempts to define mechanisms of acid or alkali resistance have been all too few. Our limited knowledge is attested to by the past appearance of only a few short summaries of extreme pH life forms (Buchanan and Fulmer, 1930; Vallentyne, 1963; Kushner, 1964; Skinner, 1968; Brock, 1969).

An attempt will be made to summarize the reported extreme acidophilic and alkalinophilic organisms. Emphasis will be placed on how physiological and structural aspects may relate to acid and alkali resistance. It is intended that this chapter will be a reminder of what little is known about these organisms and their means of survival.

II. Occurrence of Life at Extremes of pH

A. Life at Low pH

1. Bacteria

Natural environments near the normal lower limits of pH 3 to 4 are relatively common. Environments more acidic than pH 3 to 4, however, are quite rare. Representative of mildly acidic habitats are many lakes, some pine soils, and acidic bogs. Such habitats support the growth of many eukaryotic algae, bacteria, plant, and animal forms (Heilbrunn, 1952). Many little-known bacteria can be isolated from these habitats. Schulz and Hirsch (1973), for example, described members of *Bactoderma*, *Caulobacter*, *Mycrocyclus*, *Planctomyces*, and *Thiovobium* found in pH 3 to 5 *Sphagnum* bogs. Members of the acetic acid bacteria grow in the range of pH 3 to 5 (Asai, 1968). An interesting example is *Acetobacter acidophilum prov.* sp. having an optimum for growth at pH 3, a lower limit of pH 2·8, and upper limit pH 4·3 (Wiame *et al.*, 1959).

More extreme acid environments having pH values of 3 or less are commonly found associated with coal mine refuse piles, drainage waters and mining effluents. Such habitats are characterized by high concentrations of dissolved sulphates, iron, and hydrogen ions. The pH values fall within the range 1·5 to 4 (Dugan *et al.*, 1970). Acidity arises from production of sulphuric acid as a result of oxidation of sulphides and pyritic materials. Abiotic oxidation occurs only slowly at pH values greater than 4. Below pH 4, oxidation is caused by the metabolic activities of the acid producing members of the thiobacilli. These bacteria have been estimated to accelerate the oxidation of pyritic materials by a factor of 10^6 (Singer and Stumm, 1970). Acid production from such mine wastes creates serious pollution problems. It has been estimated, for example, that 3×10^6 tons of sulphuric acid flow annually from such wastes into the Ohio river (Dugan and Lundgren, 1963). It is in such acid environments that are found probably the best known and most studied examples of obligate extreme acidophiles, *Thiobacillus thiooxidans* and *Thiobacillus ferrooxidans*.

Since it was first isolated and characterized by Waksman and Jåffe (1922), *T. thiooxidans* has aroused interest because of its autotrophic abilities (Vishniac and Santer, 1957). The organism is capable of producing metabolically useful energy from the oxidation of sulphur or sulphide minerals with the concomitant production of sulphuric acid. *T. thiooxidans* grows in the pH range 0·9 to 4·5 with an optimum in culture of about pH 2·5 (Rao and Berger, 1971). Occasional reports of growth at negative pH values are probably erroneous, as pointed out by Kemper (1966). There was no question however, that the organism can survive exposure to pH values close to 0, although the length of exposure was not reported (Kemper, 1966).

T. ferrooxidans, the close relative of *T. thiooxidans*, oxidizes reduced sulphur compounds, as well as ferrous iron to ferric iron (Silverman and Ehrlich, 1964; Lundgren *et al.*, 1964, 1972). The oxidation of iron is accompanied by production of acid, and in all reported instances, the hydrogen ions are associated with sulphate anions yielding sulphuric acid (Dugan and Lundgren, 1965). The pH range for growth is dependent upon the substrate utilized (McGoren *et al.*, 1969). The optimum for elemental sulphur oxidation is about pH 5. Metallic sulphide minerals, however, are oxidized at lower pH values; calcopyrite (pH 2), bornite (pH 3) and pyrite (pH 2) (Landesman *et al.*, 1966). Growth on ferrous iron is optimal at about pH 2·5 but poor above pH 4. When growth is initiated on ferrous iron above pH 4, the pH value rapidly drops to the range 1·5 to 2. Aspects of energy metabolism and interaction with minerals of the thiobacilli may be found elsewhere (Vishniac and Santer, 1957; Lundgren *et al.*, 1972, 1974; Dugan and Lundgren, 1963;

Peck, 1968; Silverman and Ehrlich, 1964; Touvinen and Kelly, 1972; Chapter 8).

Studies of acid coal mine drainage by Walsh and Mitchell (1972a, b) describe an acid-tolerant iron-oxidizing bacterium resembling a species of *Metallogenium*. Optimum for iron oxidation of this pleomorphic iron encrusted organism was found to be pH 3·5 to 5. It was proposed by these authors that a pH-dependent succession of iron-oxidizing bacteria are present in mining wastes initially involving the acid-tolerant *Metallogenium* at pH values of 3·5 to 5 followed by *T. ferrooxidans* at values below pH 4.

Other bacteria have also been isolated from acid mine wastes. Markosyan (1973) reported the isolation of a new myxotrophic sulphur bacterium from acid mine water associated with copper deposits. The organism, designated *Thiobacillus organoparus* sp. n., is morphologically and physiologically similar to *T. thiooxidans* except that it is a facultative autotroph. Belly and Brock (1974a) reported a population of heterotrophic bacteria associated with coal mine refuse piles. The predominant bacterium at pH 2·5 was a light-yellow pigmented, Gram negative, non-sporeforming rod. Dugan *et al.* (1970) isolated a slime producing bacterium from pH 2·8 acid mine waters. In the laboratory, however, the organism grew best at pH 6·9. It was concluded the production of slime protected the organism from low pH values. Joseph (in Dugan *et al.*, 1970) isolated examples of several genera of bacteria from pH 2 to 4 acid streams which included *Bacillus*, *Micrococcus*, *Sarcina*, *Crenothrix* and *Microsporium*. It was not established if these organisms were actually growing at the ambient pH values. One should caution against an immediate interpretation of active growth of organisms isolated from habitats of expreme pH values. A firmer ground can be established if the organism can be subcultured at the ambient pH value eliminating the possibility of an inactive organism or isolation of a spore (Brock, 1969; Belly and Brock, 1974a).

The most acidic naturally occurring environments are probably those associated with acid hot springs and the surrounding hot acid soils, sometimes called sulphataras. These areas contain large amounts of elemental sulphur produced by spontaneous oxidation of H_2S which is brought to the surface by steam (Fliermans and Brock, 1972; Mosser *et al.*, 1973). Acidity develops from the oxidation of the elemental sulphur. *T. thiooxidans* is found in such habitats but only in areas at temperatures below 55°C. In recent years several unique, obligate thermoacidophilic life forms have been isolated almost in pure culture from these acid habitats where temperatures are above 55°C and pH values are 1 to 3.

Sulfololus acidocaldarius has been isolated from such acid hot springs and characterized by Brock *et al.* (1972) as a facultatively autotrophic, sulphur-oxidizing bacterium. The organism is lobe-shaped and lacks the typical rigid peptidoglycan-containing bacterial cell wall. Its pH optimum is 2 to 3 and grows in the pH range 0·9 to 5·8. The optimum growth temperature is 70 to 80°C. *Sulfolobus* can be isolated from hot acid environments around the world and appears to be the major geochemical agent responsible for acid production in these high temperature areas (Fliermans and Brock, 1972; Mosser *et al.*, 1973).

An organism resembling *Sulfolobus* has been isolated from hot acid springs (Brierly and Brierly, 1973). This organism, however, is an obligate autotroph, able to oxidize both elemental sulphur and ferrous iron as energy sources.

Besides the autotrophs, the heterotrophic sporeforming rod, *Bacillus acidocaldarius*, has been isolated from hot acid environments (Darland and Brock, 1971). This species is characterized by its pH range of 2 to 6 and optimum pH 3. The temperature range is 45 to 70°C, and growth is optimal at 60°C. Unlike *Sulfolobus, B. acidocaldarius* possess a true bacterial cell wall. A similar organism has been described as *Bacillus coagulans* by Uchino and Doi (1967) but this organism has a lower limit of pH 4 and upper temperature limit of 55°C. Belly and Brock (1974b) have also isolated acidophilic strains of *B. coagulans* from acid hot spring effluents. These strains have pH optima of 3 to 4 and temperature optima of 37 to 45°C.

Probably the most unusual organism thriving in an acid environment is the obligate thermoacidophile *Thermoplasma acidophilum*, the unique member of the cell wall-less Mollicutes. The cell membrane of this organism is directly exposed to its hot acid environment. The organism has been isolated from self-heating coal refuse piles where temperatures range from 32 to 80°C and pH values approximate 2 (Darland *et al.*, 1970). *Thermoplasma* grows at pH values from 1 to 4 and optimally at pH 2. The temperature range is 45 to 65°C with an optimum 59°C. The organism has only been found in hot acid carboniferous areas but not in acid hot springs or sulphataras. It has been suggested that coal refuse piles are only a secondary habitat although a primary habitat remains unknown (Belly *et al.*, 1973). More detailed ecological aspects of the thermophilic microorganisms may be found in Chapter 5.

2. Other Microorganisms

Yeasts generally have mildly acidic pH optima of 5·5 to 6. Many species, however, are capable of growing down to pH 2 (Battley and Bartlett, 1966). *Saccharomyces ellipsoideus*, *Saccharomycopsis guttulata* and *Saccharomyces cerevisiae*, for example, can grow at pH 2·5, 2, and 1·9 respectively (Shirfine and Phaff, 1958; Hjorth-Hansen, 1939). Several yeasts reported to resemble *Rhodotorula* have been isolated from pH 2 copper mine wastes (Ehrlich, 1963).

Probably the most acid resistant organisms reported are several fungal species. Starkey and Waksman (1943) described two species as *Acontium velatium* and a *Cephalosporium* which were isolated from laboratory media. These organisms were able to grow in 2·5 N H_2SO_4 and also enjoy the presence of 4% $CuSO_4$. Growth could be initiated at pH 7 but the pH was rapidly reduced by the organism to a value of 3. A similar organism described as *Trichosporon cereberiae* was isolated by Sletten and Skinner (1948). This organism was able to grow in HCl, as well as H_2SO_4.

Many of the fungi have wide pH tolerances (Thimann, 1963). Species of *Aspergillus, Penicillium* and *Fusarium* can grow down to pH values near 2 although the upper limits for growth are near pH 10. *Phycomyces blakesleeanus* and *Marasmius foetidus* have pH optima near 3 and grow within the range pH 2 to 7. Several fungi have been isolated from acid coal wastes and acid streams. Belly and Brock (1974a) described *Acontium pullans* obtained from pH 2·5 coal refuse piles, while Ehrlich (1963) isolated a *Trichosporon* from acid copper mine drainage.

Several algae tolerate low pH values. The most extreme example is *Cyanidium caldarium*, an interesting eukaryotic alga found in acid hot springs (Allen, 1959). The organism has a pH optimum of 2 to 3 and a temperature optimum of 45°C with an upper limit near 55°C. Although growth does not occur above pH 5 the organism is able to incorporate $^{14}CO_2$ and photosynthesize equally well at pH 7 or 2. It therefore appears that the restriction to values below pH 5 is not due to the inability to photosynthesize at higher pH values (Doemel and Brock, 1971). *Chlorella pyrenoidosa* and *Chlorella ellipsoidia* can grow down to pH values near 3·5 and 2 respectively, although upper limits for growth are near pH 10 (Kessler and Kramer, 1960). *Chlamydamonas acidophila* and *Euglena mutabilis* isolated from acid (pH 1 to 2) peat waters have pH 3 optima and an upper limit near pH 5·5 (Fott and McCarthy, 1964; Cassin, 1974). Satake and Saijo (1974) also observed *C. acidophila* in addition to several other microalgae, a yeast and mould in Lake Katanuma (pH 1·8 to 2·0) while conducting studies on carbon dioxide content and metabolic activities of microorganisms in some volcanic acid lakes in Japan.

Other life forms have been reported to grow at low pH values. The flagellate *Polytomella caeca* is able to grow at pH 1·4 with an upper limit of pH 9·6 (Lwoff, 1941). Ehrlich (1963) has found flagellates resembling *Eutrepia* and several amoebae growing in pH 2 acid mine waters. Studies of acid streams in West Virginia and Indiana by Lackey (1938) revealed 11 different genera of flagellates, 12 ciliates, 7 rhizopods and several insect species including caddis fly larvae at pH 2·2 to 3·2. Several of the protozoa were found at pH 1·8. Studies to determine whether these organisms were actually growing at the ambient pH values were not done.

B. Life at High pH

1. Bacteria

Naturally occurring alkaline environments are usually found associated with soils. Typical are areas containing concentrations of alkali minerals, animal excreta or decaying proteins. Alkaline soils may result from the complete oxidation of organic matter in areas of high aeration and high temperature. Such areas are typified by some of the desert soils of the western United States where pH values may be as high as 10 (Thimann, 1963). Alkaline lakes and springs are also found with pH values ranging from 8 to 11 (Souza *et al.*, 1974; Brock and Darland, 1970). Extreme examples are the alkaline lakes, Elementia and Nakuru, in Kenya, where ambient values reach pH 10 to 11 (Jenkin, 1936).

Among the bacteria are found several examples of organisms which are resistant to values approximating pH 10. The pH optimum for growth of many of these bacteria, however, is much lower. The most alkaline-resistant bacteria are found among the nitrate reducers, suphate reducers and active ammonifiers. Meek and Lipman (1922) reported *Nitrosomonas* and *Nitrobacter* species which could survive pH values of 13. These observations, however, could not be confirmed, and the upper limit appears to be pH 10·7 (Buchanan and Fulmer, 1930). Strong alkalinity (pH 10 to 11) is tolerated by some *Rhizobium* species, of which one may be the closely related *Agrobacterium radiobacter*. This organism grows actively at pH 10 to 12 with an optimum in the range pH 6 to 9 (Thimann, 1963; Allen and Holding, 1974).

An extreme example of a bacterium which actually requires a high pH for growth is *Bacillus pasteurii*, a member of the ureaclastic bacteria. The organism hydrolizes urea in concentrations up to 10% and grows well at about pH 11 (Gibson, 1934). There is a specific requirement for ammonia and growth is poor at pH 9 or less in medium made with 1% NH_4Cl (Wiley and Stokes, 1962, 1963). Several strains of related bacilli, *B. sphaericus*, *B. pantothenticus* and *B. rotans* do not require ammonia but are capable of growth at pH 11. Their lower limits for growth however, approximate pH 5 (Bornside and Kallio, 1956). A similar organism described by Vedder (1934) as *Bacillus alkalophilous* grows actively at pH 10 but not at pH 7. The organism does not hydrolize urea or require ammonia.

Other bacilli have been isolated which are extremely resistant to alkaline pH. A strain of *Bacillus cereus* was obtained by Kushner and Lisson (1959) by multiple transfers in media at increasing pH values until the organism could grow at pH 10·3. The alkaline resistance became a stable character and was not lost on further transfers. A strain of *Bacillus circulans* was also isolated (Chislett and Kushner, 1961a) which could grow at pH 11. The spores from both strains germinated at the highest pH values giving rise to vegetative cells (Chislett and Kushner, 1961b). Growth of the *B. cereus* rapidly reduced the pH value, but growth was still active when alkali was added periodically to maintain the high pH (Kushner and Lisson, 1959). Several alkaline resistant species (pH 9 to 10) of *Bacillus* have also been isolated from soils and the alkaline stability of various extracellular enzymes examined (Horikoshi, 1971a, b, 1972; Horikoshi and Atsukawa, 1975; Kurono and Horikoshi, 1973; Boyer and Ingle, 1972). Ohta *et al.* (1975) have isolated an interesting bacillus from fermenting Indigo leaves, the source of natural indigo dye. The organism was shown to be an aerobic, sporeforming, Gram positive, motile rod which grew best at pH 10 to 10·5 while exhibiting no growth at values below pH 7 to 8.

Souza *et al.* (1974) isolated and tentatively identified a *Flavobacterium* species and an unidentified strict anerobe from a pH 11·2 to 11·6 alkaline spring, characterized as a low discharge calcium hydroxide type. The organisms grew and reproduced actively at pH 11·4 with optima between pH 9 to 10.

Many of the enteric bacteria are tolerant to pH values near 9 to 10 (Buchanan and Fulmer, 1930). Downie and Cruickshank (1928) were able to obtain pure cultures of *Streptococcus faecalis* using broth at pH 11. The ability to grow in media of about pH 10 is one of the characteristics of the enterococcus group (Chesbro and Evans, 1959).

2. Other Microorganisms

Many fungi have a wide pH tolerance. For example, *Penicillium variable, Fusarium bullatum* and *Fusarium oxysporium* have been reported to grow at pH values of 11 (Johnson, 1923), although the lower limits of pH approximate 2.

Among the green algae, *C. pyrenoidosa* and *C. ellipsoidea* are capable of growth at pH 10 (Kessler and Kramer, 1960).

The blue-green algae are usually most abundant in alkaline habitats within pH 7·5 to 10 (Holm-Hansen, 1968; Fogg, 1956). More extreme examples include *Gloeothece linaris* and *Mycrocystis aeruginosa* which have pH optima near 10 (Fogg, 1956; McLachlan and Gorham, 1962). *Arthrospira plantensis* was found to be present in highest numbers among 13 different genera of algae present in the pH 11 alkaline lakes in Kenya (Jenkin, 1936). The blue-green alga *Plectonema nostocorum* is able to grow at pH 13 which seems to be the highest pH at which life has been recorded (Geiger *et al.*, 1965).

Among other organisms, protozoa were found in the pH 11 alkaline spring examined by Souza *et al.* (1974). *Euglena gracilis* has a wide pH tolerance, growing in the range pH 2·3 to 11 and the crustacean *Chydorus* between pH 3 to 10 (Heilbrunn, 1953). Mention might also be made of the fossil-like microorganism *Kakabekia umbellata* Barghoorn (Barghoorn and Tyler, 1965) isolated from ammonia rich soils by Siegel and Giumarro (1966). The organism apparently grows, albeit rather slowly, in an atmosphere of ammonia or in 5 to 10 M NH_4OH (Siegel *et al.*, 1967).

III. Nature of the Extreme pH Life Forms

Any consideration of survival mechanisms in extremes of hydrogen ion concentration is by no means a simple task. Despite the long history of studies on the acidophilic thiobacilli and recent interest in the thermophilic acidophiles, relatively little information has accumulated on reasons for their ability to withstand low pH values. Even more meager is an understanding of pH relationships on the metabolism and stability of the alkaline-resistant organisms. Our attention, therefore, will be centred primarily on the survival and stability of the extreme obligate acidophiles.

A. The External Environment

The pH of the environment involves a tangled skein of influences all of which relate directly or indirectly to the metabolism and stability of the cell. It would therefore be an oversimplification to consider all of the effects of pH as due to the hydrogen ion alone. The hydrogen ion itself (H^+) is unique among cations. It is simply a proton with no electrons. In aqueous solutions it becomes rapidly hydrated and exists as the hydronium ion, H_3O^+. At acid pH the hydronium ion predominates and may actually exist as $(H_5O_2)^+$, $(H_7O_3)^+$, $(H_9O_4)^+$, etc., the average extent depending upon concentration and temperature (VanderWerf, 1961). At alkaline pH the hydroxyl ion (OH^-) predominates. The concentration of the hydronium or hydroxyl forms of water play a fundamental role in governing solvolysis, ionic state of nutrients, balance of electrical charges on cell surfaces and colloidal properties in the microenvironment. Very little is known of these complex relationships in naturally occurring microhabitats (Stotsky, 1972; Mclaren and Skujins, 1968). At low pH, hydrogen ions may adsorb to particulate matter and displace cations in the exchange complex which may then leach from the microenvironment. Whereas the solubility of CO_2 becomes reduced at low pH, the solubility of some ions such as Al^{3+}, Mn^{2+}, Cu^{2+} and Mo^{3+} increases to levels which may reach toxic proportions for most organisms. At acid pH values many normally non-toxic organic acids enter cells more easily in the protonated state and may become toxic. Hutner (1972) has pointed out that the amounts of trace elements required by many organisms increases at lower pH values.

At high pH, many essential elements such as Fe^{2+}, Ca^{2+}, Mg^{2+} and Mn^{2+} may become insoluble and precipitate as the carbonate, hydroxide, or phosphate salts. The pH of the environment may influence the balance of electrical charges on cell surfaces, increasing the net positive charge at low pH and increasing net negative charge at high pH. Extreme alteration likely affects cellular stability, permeability, and ability to interact with wanted or unwanted metabolites. No doubt the state of available nutrients, concentration of toxic metabolites, and cellular instability at extreme pH values limits the ability of most organisms to survive outside the normal limits of pH. Organisms that live at extreme pH values must content with all of these factors. Interestingly, many of the obligate acidophiles are resistant to high concentrations of heavy metal ions such as Cu^{2+} which are found in many of the naturally occurring acid habitats. The autotrophic, thermophilic, acidophilic isolate of Brierley and Brierley (1973) is resistant to molybdenum, as well as copper (Brierley, 1974). It would appear that increased resistance and increased requirements for certain ions at acid pH values may be due to increased surface charge. Whether resistance is due to the ability of hydrogen ion to exchange with heavy metal ions on the cell surface as suggested by Brock (1969) is not known.

The ability to grow and reproduce at low or high pH has certain survival advantages. Certainly competition from other organisms is severely limited. In the case of the acidophilic *Thiobacillus thiooxidans*, ferrous iron autooxidizes at values above pH 5 to ferric iron which would eliminate this energy producing substrate from availability to the organism (Lundgren *et al.*, 1972). At high pH values, the alkalinophile *Bacillus pasteurii* specifically requires ammonia (NH_3) for the transport and oxidation of such substrates as glutamic acid, isoleucine, threonine as well as the tricarboxylic acid cycle intermediates, acetate, α-ketoglutarate and malate (Wiley and Stokes, 1963). In this instance it would seem the advantage of high pH is the maintenance of NH_3 availability to the organism.

Alterations in environmental pH may induce compensatory enzymatic changes in many microorganisms. *Escherichia coli*, for example, responds to an increasingly acid environment by the production of amino acid decarboxylases. The resulting amines formed tend to reduce the acidity. An increase in alkalinity stimulates production of amino acid deaminases which tends to decrease the alkalinity (Gale, 1943). As viewed in the preceding section, most actively metabolizing alkaline-resistant organisms have a tendency to reduce the pH value of the medium during growth. Kushner (1964) has suggested such a response is most likely a secondary mechanism of resistance to alkaline environments. What primary mechanism(s) ensures cellular stability and growth at high pH values remains unknown. The corollary is not true for most obligate acidophiles. Active growth does not tend to increase the pH. When growth is initiated near neutrality, if growth occurs at all, the pH values are rapidly lowered. The extreme acidophiles not only survive low pH values but may actually require hydrogen ions for growth and stability. Such requirements are exemplified by *T. acidophilum* which lyses above pH 5 (Smith *et al.*, 1973). Resistance to high acidity or alkalinity would seem then to reside in structural or metabolic peculiarities of the cell.

B. Internal pH

Some mechanism for maintaining the internal pH near normal physiological values seems essential for all organisms growing in extremes of pH. Such acid labile molecules as ATP and DNA would likely not exist if the intracellular pH were that of an extracellular acid environment. However, the application of the classical concept of pH becomes a complicated affair when considering intracellular pH values. The classical concept of pH is a practical index of hydrogen ion concentration or hydrogen ion activity in aqueous solution but internally cells are colloidal and not aqueous systems. An estimate of intracellular pH values gives no information concerning undissociated protons bound to donor molecules, as pointed out by Butler *et al.* (1966). The concept and validity of the classic definition of pH and hydrogen ion concentration in relation to intracellular systems has been dealt with by Butler *et al.* (1966), Siesjö and Pontén (1966), and Waddell and Bates (1969). The notion of intracellular pH is best described by Siesjö and Pontén (1966) as a theoretically awkward concept.

> It seems as if "internal pH" is given a meaning of its own presumably because of the relation between pH in a system and enzyme kinetics. But many enzyme reactions are supposed to occur in membrane-confined tissue regions of very small sizes, and in these systems the term pH has a very debatable meaning. . . . Moreover, in such systems there is the additional complication of surface charges and internal Donnan equilibria, and many reactions studied may be occurring at surfaces where ionic activities are different from those in the bulk fluid.

Any measurement of internal pH would therefore be an average value for the contents of the cell representing the average degree of ionization of charged groups. The concept of internal pH becomes even more clouded since at normal physiological pH values the concentration of free hydrogen ions would seem to be too low to play a significant role in biological processes. For example, Thimann (1964) has calculated that a cell with a volume of $6{\cdot}1 \times 10^{-14}$ ml would contain 3·6 free hydrogen ions at pH 7 if one assumes the internal and external pH values were the same. By extrapolation, assuming internal and external pH values are the same, this would imply that an alkalinophile at pH 10 would contain $3{\cdot}6 \times 10^{-3}$ hydrogen ions, with a corresponding increase in number of hydroxyl ions. The chances of finding a hydrogen ion at all would be about 1 in 300. Conversely at pH 2 an acidophile would contain $3{\cdot}6 \times 10^{5}$ hydrogen ions. An assumption that internal and external pH values are the same in acidophiles or alkalinophiles, however, is extremely unlikely. Certainly such drastic alteration in number of internal hydrogen or hydroxyl ions from the neutral state would cause profound biological changes in the nature of these organisms.

1. Measurement

Although the concept of internal pH is not a clear one, it does have value as an indicator of the general condition within the cell. The actual measurement of intracellular pH, however, is difficult. Extensive reviews by Heilbrunn (1952), Caldwell (1956), Waddell and Bates (1969) and Rottenberg (1975) describe in detail the principles and disadvantages of the available methods of intracellular pH measurement. Several of the methods have included the use of microelectrodes, pH measurement of broken cell lysates or fluids withdrawn from cells, the use of visible indicators such as bromthymol blue, and measurement of external and internal distribution of permeant weak acids or bases. The use of the weak acid or base distribution techniques applied to the intracellular pH measurement of mitochondria and chloroplasts would appear the most applicable to microorganisms. The principle is based on the ability of weak acids such as 5,5-dimethyl-2,4-oxazolidinedione (DMO), introduced by Waddell and Butler (1959), or amines such as methylamine or NH_3^+ (Rottenberg *et al.*, 1972a, b) to permeate membranes in their neutral form. The pH is estimated from determination of the internal and external concentrations of the permeating molecule (Rottenberg, 1975). Weak acids permeate vesicles which are more alkaline internally while amines permeate vesicles which are more acidic than the surrounding medium. The use of ^{14}C-DMO has been extensively detailed by Addanki and Sotos (1969). DMO is a weak acid, metabolically inert, non-toxic, does not bind with protein or lipid and diffuses passively in the neutral form across many biological membranes (Butler *et al.*, 1966). Harold *et al.* (1970) have used DMO successfully to study transmembrane pH gradients generated in *Streptococcus faecalis*. Kashket and Wilson (1973) have used ^{14}C-methylamine to determine internal pH values while studying proton-coupled accumulation of galactosides in *Streptococcus lactis*. Based upon similar principles, Neal *et al.* (1965) utilized 2,4-dinitrophenol uptake and distribution to determine internal pH values in yeast cells. Although apparently useful in yeast, the method might not provide broad applicability since 2,4-DNP is an uncoupler of oxidative phosphorylation.

Several recent methods based upon fluorescence appear promising for the estimation of intracellular pH values. Schuldiner *et al.* (1972) have introduced the use of the fluorescent amine 9-aminoacridene. Interal pH values are estimated from the extent of fluorescence quenching in cells which have taken in the molecule. In a preliminary report Thomas *et al.* (1976) suggest fluorescein diacetate may be useful as a spectroscopic probe for the rapid estimation of intracellular pH. Fluorescein diacetate which is a nonfluorescent esterase substrate, is converted to highly fluorescent fluorescein by intracellular enzymes when taken into the cell. Advantage is taken of the initial absorption and excitation spectra of fluorescein which is highly pH sensitive. The method appears useful over the range pH 2 to 7. Application of the technique to *Bacillus acidocaldarius* grown at pH 3 and 55°C indicated an internal pH value of 5·5 which compared favourably to the value of pH 5·2 to 5·8 when estimated by the DMO method.

2. *Acidophiles*

The internal pH values of acidophiles or alkalinophiles is almost certainly not that of the external environment. Many reports of enzyme systems isolated from these organisms indicate pH optima far different from the external environment. Such evidence has generally been considered to imply that internal pH values are not the same as external pH values. Until quite recently, few direct estimates of internal pH values have been reported. Crude estimates have usually been made by disrupting cells in small volumes of water and measuring the resulting pH of the cell "juice" without any consideration for the buffering capacity of the lysate in water. Even more meagre is information relating to the actual mechanism(s) by which these organisms maintain such pH gradients.

a. T. acidophilum. An extensive analysis of the intracellular pH of *Thermoplasma* (optimum growth, pH 2 and 59°C) has been reported by Hsung and Haug (1975). The internal pH was found to lie close to neutrality. The pH of cell lysates was 6·3 to 6·8 which agreed with an internal pH value of 6·5 determined by the DMO method. Additionally, malate dehydrogenase was isolated which had a pH optimum of 8·5 to 10 at 56°C but elicited less than 25% of maximal activity at values less than pH 6·5. It was also demonstrated that alteration of temperature (56°C, 24°C) or extracellular pH (2, 4, 6) did not affect the intracellular pH value. Neither was internal pH altered in heat killed cells nor by exposure of cells to the metabolic inhibitors, 2,4-dinitrophenol, iodoacetate or sodium azide. It was concluded that *Thermoplasma* likely maintains its pH gradient of 4·5 units by passive or impermeable properties of the cell rather than by active metabolic extrusion of hydrogen ions.

b. S. acidocaldarius. De Rosa *et al.* (1975) have estimated a near neutral internal pH value of 6·3 for *Sulfolobus* strains (growth optimum, pH 3 and 70°C) measured on cell lysates prepared by exposure to 0·5% sodium luaryl sulphate. It was also found that viability was quickly lost when cells at pH 3 were held for unspecified times at temperatures below which active growth occurred (45°C). Microscopic examination indicated cells became vacoulated and the cytoplasm congealed. Viability could be maintained at sub-optimum temperatures if the culture medium was adjusted to pH 6. The authors suggested that these findings were evidence that the organism requires active metabolism for maintenance of the pH gradient across the membrane.

c. B. acidocaldarius. The intracellular pH of *B. acidocaldarius* (growth optimum pH 3 and 55°C) appears to be in the range 5·6 to 5·8 measured both by the fluorescent probe, fluorescein diacetate and by the DMO method (Thomas *et al.*, 1976). Studies by Yamazaki *et al.* (1973) on a bacillus, presumably related to *B. acidocaldarius*, give similar implications. The organism has a pH range of 2 to 5·5 (optimum pH 4) and temperature range of 35 to 60°C (optimum 55°C). Using an oxygen electrode, it was found that intact cells or prepared protoplasts showed maximal respiratory activity at both pH 4 and 7. Cell free extracts, however, showed respiratory activity at pH 7 but no activity at pH 4. It was further demonstrated that a resting cell suspension at pH 4 consumed H^+ but there was no consumption of H^+ when glucose was added as the respiratory substrate. It was therefore suggested that the internal pH value was near neutrality and that the Δ pH may depend on a membrane function as well as an energy dependent H^+ exclusion mechanism.

d. Thiobacilli. Blaylock and Nason (1963) reported an internal pH value of 4·8 to 5 for *T. ferrooxidans* (growth optimum pH 2·5). The intracellular pH of related *T. thiooxidans* would also appear to be independent of the extracellular pH (Rao and Berger, 1971; Dewey and Beecher, 1966). Estimates of internal pH values near neutrality are further fortified by reports of enzyme systems purified from both *T. thiooxidans* and *T. ferrooxidans* which have optimal pH values ranging from 5 to 9. Several examples include: inorganic phosphatase, optimum pH 7·5 to 8·5 (Adele and Lundgren, 1970); a sulphur-oxidizing enzyme system optimum pH 7·8 (Silver and Lundgren, 1968a); the thiosulphate splitting enzyme rhodanase, optimum pH 7·5 to 9 (Tabita *et al.*, 1969); the thiosulphate oxidizing enzyme tetrathionase, optimum pH 5 (Silver and Lundgren, 1968b); iron-cytochrome oxidase, optimum pH 7 (Din *et al.*, 1967); adenosine triphosphatase, optimum pH 9 to 10 (Adapoe and Silver, 1975). The only reported enzymatic reaction having a pH optimum 2·5 to 3 is a crude iron oxidase system associated with the cell envelope (Bodo and Lundgren, 1972). Similar findings of acidic pH optima have been reported for several sulphur-oxidizing systems from *T. thiooxidans* (Suzuki, 1965; Adair, 1966).

Amemiya and Umbreit (1974) studied a cell free protein-synthesizing system from *T. thiooxidans* in which amino acid incorporation was optimal at pH 7·2. No incorporation of amino acids occurred below pH 5. The *in vitro* system was dependent upon exogenous message (poly-U), a ribosomal and supernatant fraction and an energy source (GTP). The ribosomal and supernatant fractions from *T. thiooxidans* were found to be functional when exchanged for the same fractions obtained from *E. coli* or *B. thuringiensis*. Interestingly the temperature optimum for amino acid incorporation was 37 to 45°C, exceeded only by thermophilic organisms. This may be important in considering the thermophilic acidophiles. Unfortunately no information is yet available concerning enzymes or protein synthesis in these organisms.

e. Other acidophiles. A few recent studies imply intracellular pH values near neutrality in several acidophilic organisms other than bacteria. Cassin (1974) examined the flagellated green alga *C. acidophila* (growth range pH 2 to 4) and observed that the organism remained green at acid pH values in which other acid non-tolerant *Chlamydomonas* species lost green colour (pheophytinization) suggesting some type of H^+ exclusion mechanism.

Enami and Fukuda (1975) and Enami *et al.* (1975) initiated studies on the physiology and photosynthesis of the eukaryotic alga *C. caldarium* (optimum growth pH 2 to 3 and 45°C). Partially disintigrated cells were prepared which completely lacked the outer cell wall but still maintained subcellular structure. Hill activity was measured using *p*-benzoquinone as the Hill oxidant. Hill activity was maximal in outer wall-less cells at pH 7 and 35°C. Activity was completely lost at pH 3 and 50°C, conditions under which intact cells showed maximal activity, suggesting the cell wall plays an important role in the acid resistance of the organism.

3. *Alkalinophiles*

As with the acidophiles, it appears that the alkalinophiles have intracellular pH values near neutrality. Although reports are all too few, the following examples support the view of differential pH values in these organisms.

Larson and Kallio (1954) isolated the enzyme urease from *B. pasteurii* (optimum growth pH 9·2). The purified enzyme exhibited a pH optimum at 6·5 to 7. Urease activity was five times more active in cell lysates than in whole cells. Wiley and Stokes (1962, 1963) found that disrupted cell suspensions of *B. pasteurii* had pH values of 6·8. It was further demonstrated that a resting cell suspension required pH 8·5 to 9 and the specific presence of NH_3 for the oxidation of low concentrations of glycine, glutamic acid, alanine, serine, fumaric acid and other substrates. No oxidation occurred in whole cell suspensions in the absence of NH_3. Disrupted cells, however, oxidized these substrates at pH 7·2 in the absence of NH_3 and in fact oxidation was inhibited by raising the pH or an addition of NH_3. It was concluded from these studies that the internal pH lies near neutrality and that the alkaline pH affects cells externally and not internally.

More recently, several alkaline-requiring bacilli have been isolated, with studies directed toward the alkaline stability of various extracellular enzymes. As would be expected these enzymes have alkaline pH optima. Several examples include: an alkaline protease, pH 11·5 to 12 and pectinase, pH 10 to 10·5 (Horikoshi, 1971a, 1972); alkaline amylase, pH 9·2 to 10·5 (Horikoshi, 1971b; Boyer and Ingle, 1972); alkaline catalase, pH 10 (Kurono and Horikoshi, 1973) and a β-1,3-gluconase, pH 8·5 (Horikoshi and Atsukawa, 1975). As these studies give no information on the internal condition of these cells, Ohta *et al.* (1975) initiated an investigation on the obligate alkaline requirements of a bacillus (growth optimum pH 10 to 10·5) isolated from fermenting Indigo leaves. The organism demonstrated maximal protein synthesis, nucleic acid synthesis, amino acid uptake and oxygen consumption at pH 9 to 10·5. Membrane preparations oxidized NADH maximally at pH 7·5 and membrane bound ATPase gave maximal activity at pH 7. The soluble cell fraction gave maximal L-lactate, L-alanine and malate dehydrogenase activity at pH 7·4 to 8·9. In neither case did these membrane-associated or soluble enzymes demonstrate any activity at pH 10, the optimum for cell growth. The suggestion was made that the internal pH approaches neutrality and a membrane associated OH^- exclusion mechanism may be responsible for the alkaline resistant properties of the organism.

Other examples could be cited indicating that organisms requiring extreme pH values do not have peculiarities in the functioning of their enzyme systems and have internal pH values unrelated to the external environment. The foregoing examples suggest the differential between external and internal pH values is probably accomplished by exclusion of H^+ or OH^- from the cell by the cell wall or membrane and/or expulsion by a metabolically active pumping mechanism. Complete exclusion of ions by the latter mechanism would very likely place a stringent demand on cellular energy. It would seem more likely that both the nature of the cell wall and membrane, and cellular metabolism serve important functions in the maintenance of cellular pH gradients. As pointed out by Harold and Altendorf (1974), studies on active ion regulation mechanisms in these organisms remain unexplored. Such studies should greatly extend an understanding of survival in extreme pH values. In view of the lack of available information pertaining to these organisms, a consideration of transmembrane pH gradients and active ion regulation will not be pursued here. These concepts have been well reviewed elsewhere (Henderson, 1971; Harold, 1969, 1972; Harold and Altendorf, 1974).

C. Cell Surfaces and Membranes

The ability of extreme acidophiles or alkalinophiles to exclude H^+ on OH^- and maintain structural stability implies a relationship between the nature of the cell wall and cytoplasmic membrane, and the external environment. Most available information has been derived from studies on the acidophilic thiobacilli. More recently investigations have been initiated on several of the extremely thermophilic acidophiles, *Bacillus acidocaldarius*, the cell wall-deficient *S. acidocaldarius*, and the cell wall-less *T. acidophilum*. Features of the latter organisms reflect the ability to survive both low pH and high temperature. In view of the lack of meaningful information regarding cell walls and membranes of extreme alkalinophiles, only the acidophilic organisms ranging from the cell wall-containing thiobacilli to the cell wall-less *Thermoplasma* will be considered here.

1. The Thiobacilli

Several investigations have been reported on the cell wall structure and membrane lipids of the acidophilic thiobacilli in search of structural peculiarities which may be related to acid stability. Thus far there appears to be no obvious structures connected with acid stability. The fine structure is that of a typical Gram negative bacterium for both *T. thiooxidans* (Mahoney and Edwards, 1966) and *T. ferrooxidans* (Remsen and Lundgren, 1966). Electron micrographs of thin-sectioned and freeze-etched cells reveal a typical outer layer, lipoprotein–lipopolysaccharide (LPS) layer, globular protein attached to peptidoglycan and an inner cytoplasmic membrane. In either case no extramural structures are apparent which might relate to acid resistance.

Chemical analysis showed the *T. ferrooxidans* peptidoglycan, the macromolecular complex responsible for cell rigidity, contained normal proportions of glutamic acid, α, ε-diaminopimelic acid, alanine, glucosamine and muramic acid (Wang and Lundgren, 1968). Crum and Siehr (1967) found no unusual amino acid composition in cell walls of *T. thiooxidans*. The composition of LPS from *T. ferrooxidans* varies slightly depending upon the substrate utilized for growth. It contains the typical LPS carbohydrates: 2-keto-3-deoxyoctulosonate, glucose, and galactose (Wang *et al.*, 1970; Vestal *et al.*, 1973). Calcium, magnesium and high amounts of associated iron were present which led to speculation that the LPS may play a role in interaction with various substrates. There was a noticeable absense of significant amounts of phosphorous when compared to LPS from a *Salmonella*. The low phosphorous content suggested a significantly different composition of the lipid A moiety, although this aspect was not pursued further. The significance of low phosphorous content is not known. Its absence, and therefore absence of a potential negative charge (and potential for protonation by H^+), might render the lipid A portion much more hydrophobic and therefore possibly more stable in its surface exposure to the aqueous acid environment.

Doetsch *et al.* (1967) isolated flagella from *T. thiooxidans* which were resistant to both high acidity and high temperature. Although acid resistance would be expected since the flagella are exposed directly to the acid environment, it demonstrates that the thiobacilli can synthesize an acid-stable protein.

The phospholipid metabolism of *T. ferrooxidans* was investigated by Short *et al.* (1969) to determine if phospholipids play an active role in maintaining the internal pH. The diacyl phospholipids were composed of phosphatidyl monomethylethanolamine (42%), phosphatidyl glycerol (23%), phosphatidyl serine (20%), cardiolipin (13%), with traces of phosphatidyl choline and phosphatidyl dimethylethanolamine. Similar phospholipids have been found in *T. thiooxidans* (Jones and Benson, 1965; Shively and Benson, 1967; Barridge and Shively, 1968). No difference was observed in the turnover of either ^{14}C- or ^{32}P-labelled phospholipids in cells grown at pH 1·5 or 3·5. In fact, the phospholipid metabolism was relatively inactive when compared to that of other Gram negative organisms growing at pH 7. Levin (1971) has shown that these thiobacilli do contain a preponderance of cyclopropane fatty acids. An ornithine-containing lipid which had ester linked *cis*-11,12-methylene-2-hydroxyoctadecanoic acid was reported in *T. thiooxidans* (Knoche and Shively, 1969; Shively and Knoche, 1969).

At present it appears difficult to relate any features of cell wall structure or membrane composition to acid stability in the thiobacilli. A preponderance of amine containing phospholipids might initially suggest exclusion of H^+ through protonation of the amine groups, thus acting in a buffering capacity in an attempt to maintain membrane integrity. Evidence that amine groups inhibit protons in acidic media by increased protonation of the amine function has been reported by Haest *et al.* (1972). It was demonstrated that *Staphylococcus aureus* synthesized large amounts of lysylphosphatidyl glycerol when the pH of cultures was adjusted from pH 7 to pH 4·7. Adjustment back to pH 7 resulted in a rapid decrease in lysylphosphatidyl glycerol and a corresponding increase in phosphatidyl glycerol. There was also a strong reduction in Rb^+ uptake at lower pH values. Similarly, acidified (pH 5) cultures of *Acholeplasma laidlawii* strain B produce D- or L-alanylphosphatidyl glycerols but not at pH 8, the growth optimum for this organism (Koostra and Smith, 1969). Such a consideration may imply a minor role at best in acid resistance in the acidophilic thiobacilli, since phosphatidylethanolamine and related derivatives are found ubiquitously in many acid non-tolerant Gram negative members of the *Eubacteriales* and *Pseudomonadales* (Goldfine, 1972; Shaw, 1974). The significance of the abundance of cyclopropane fatty acids is also uncertain since they are found in many acid non-tolerant bacteria such as the lactobacilli, streptococci, clostridia and *Brucellaceae* (Goldfine, 1972).

2. *B. acidocaldarius*

This organism grows optimally at 60°C and pH 3. It is an aerobic sporeforming rod and possesses a peptidoglycan containing cell wall (Darland and Brock, 1971). Detailed information relating to cell wall structure has not been reported. Recent studies have centred primarily on membrane lipid composition. De Rosa *et al.* (1971) examined the fatty acid composition and first described the occurrence of the unusual ω-cyclohexyl C_{17} and C_{19} fatty acids, 11-cyclohexyl undecanoic and 13-cyclohexyl tridecanoic acid. Depending upon growth conditions these ω-cyclohexyl acids represent 60 to 90% of the fatty acid composition with lesser amounts of the C_{17} branched chain, 15-methyl hexadecanoic and 14-methyl hexadecanoic acids. The structural identifications have been further confirmed by Oshima and Ariga (1975) and the fatty acids shown to be in esterified and amide-linked form in the glyceride type complex lipids of the organism (Langworthy *et al.*, 1976). A series of studies (De Rosa *et al.*, 1971a, 1972, 1974a, b, c; De Rosa and Gambacorta, 1975) have demonstrated that the ω-cyclohexyl fatty acids originate biosynthetically from the precursor shikimic acid. Shikimic acid is deoxygenated to cyclohexene-1-carboxylic acid which is subsequently reduced to cyclohexane carboxylic acid. Fatty acids arise by C-2 elongation from the CoA derivative of cyclohexane carboxylic acid. The biosynthetic route was confirmed and extended by Oshima and Ariga (1975) to show that shikimate is formed from glucose. It was noted that many of the branched and cyclopropane fatty acids found in acid non-tolerant bacteria are attuned to amino acid metabolism and not glucose metabolism. It was further shown that varying the conditions of temperature or pH did not influence the relative proportions of ω-cycohexyl and acyclic fatty acids. This was in contrast to a report by De Rosa *et al.* (1974a) who discussed a relationship between decreased ω-cyclohexyl acid formation at higher pH or lower temperatures and increased formation at lower pH or higher temperatures.

Several lipids of the neutral class have been structurally identified (De Rosa *et al.*, 1971b, 1973) including C_{50} and C_{55} polyprenols, menaquinones, squalene and pentacyclic triterpene hydrocarbons belonging to the hopane class, the major one being hop-22(29)-ene. A detailed analysis of the complex lipids (Langworthy *et al.*, 1976; Langworthy and Mayberry, 1976) indicated a composition of 15·7% neutral lipids, 64% glycolipids and 20·3% acidic lipids. The glycolipids were comprised primarily of glucosyl(1 $\xrightarrow{\beta}$ 4)N-acylglucosaminyl(1 $\xrightarrow{\beta}$ 1)diacyl glycerol (24·9%), glucosyl(1 $\xrightarrow{\beta}$ 4)N-acylglucosaminyl-(1 $\xrightarrow{\beta}$ 1)monoacyl glycerol (41·3%) and a unique pentacyclic triterpene derived tetrol N-acylglucosaminoside (26%). The latter glycolipid was identified as an N-acylglucosamine β-linked to the primary hydroxyl of a fully saturated 1,2,3,4-tetrahydroxy pentane-substituted triterpene ($C_{35}H_{62}O_4$, mol. wt. 546). Nearly all of the NH_2 groups in the glycolipids were N-acylated with fatty acids having a distribution similar to the esterified fatty acids in the glyceride portion of the glycolipids. The acidic lipids were comprised of the phospholipids, diphosphatidyl glycerol (32·3%), lysodiphosphatidyl glycerol (5·3%), phosphatidic acid (5·8%) and phosphatidyl glycerol (13·4%), typical of many acid non-tolerant bacteria. Also present was a large amount (43·2%) of a sulphonolipid very similar to, if not the same as, the plant sulphonolipid, 6-sulphoquinovosyl diacyl glycerol. The sulphonolipid contained the acid resistant C–S bond.

Features of *B. acidocaldarius* membrane lipids reveal many unique and unusual structures. One might speculate on the physiological relationship of the membrane lipids to stability at low pH and high temperature but too little is yet known to establish very significant correlations. It is likely that the occurrence of fully saturated ω-cyclohexyl and branched chain fatty acids aid in maintenance of the appropriate membrane fluidity at the high temperatures. Branched-chain and saturated fatty acids are thought to serve this function and increase membrane rigidity in the thermophiles (Singleton and Amelunxen, 1973). What the specific role of the ring-containing structures, e.g. ω-cyclohexyl fatty acids, and pentacyclic triterpenes and derivatives might play in H^+ exclusion is not clear. Perhaps these structures influence the packing and therefore physical state of the membrane in such a way as to make it even more rigid and therefore less permeable to H^+ at high temperature. The significance of the triterpenes and derivatives in the organism is not known. The natural occurrence of triterpenes is quite rare, having only been reported in several eukaryotic mosses and ferns (Marsili and Morelli, 1968, 1970; Marsili *et al.*, 1971) and in only one other prokaryote, *Methylococcus capsulatus* (Bird *et al.*, 1971). Amino sugar glycosides are considerably more resistant to acid hydrolysis than neutral sugar glycosides. Amino sugar glycosides, however, are hydrolysed at about the same rate as neutral sugar glycosides when the amino group is N-acylated, as are nearly all of the glucosamine residues in *B. acidocaldarius*. The organism does contain a sulphonolipid similar to the plant sulphonolipid, 6-sulphoquinovosyl diacyl glycerol, which contains the acid resistant C–S bond. At present, 6-sulphoquinovosyl diacyl glycerol has only been found in photosynthetic plants, bacteria and algae. It is thought to have some function in photosynthesis and respiration (Haines, 1971). *B. acidocaldarius*, however, is a non-photosynthetic organism. Sulphonolipids are also the most acidic of the polar lipids being ionized at all pH values in aqueous solution. It might be conjectured that the sulphonolipid in *B. acidocaldarius* may function in stimulation of respiration at the high temperatures for growth and/or function in some H^+ exclusion capacity (Langworthy *et al.*, 1976).

3. S. acidocaldarius

The organism, a facultative autotroph, is characterized by its multi-lobe shape and its ability to grow optimally at pH 3 and 70°C. The unusual cell surface which is in direct contact with the harsh environment lacks a morphologically and chemically defined peptidoglycan layer characteristic of other bacteria but does contain an overlying ultrastructural element (Brock *et al.*, 1972; Millong *et al.*, 1975). The organism possess irregular shaped pili, apparently for adhesion to surfaces such as elemental sulphur. These pili are extremely acid and heat stable (Weiss, 1973). In a detailed study, Weiss (1974) has shown the cell wall consists of a regular polyhexagonal array of protein subunits 13 to 15 nm diameter on the cell surface. The subunits are in close association with the cytoplasmic membrane. Freeze-fracturing did not expose typical interior wall layers or the inner hydrophobic cytoplasmic membrane. Instead cells cross-fractured exposing a cross section of wall-membrane and cytoplasm indicative of the unusual nature of the envelope. The subunit cell wall did not disaggregate when separated from the membrane, but did lose any resemblance of a lobe shape. It seems likely that subunit cell wall-membrane interaction is necessary for maintaining cellular integrity.

Trypsin, pronase or lysozyme-EDTA had no effect on the intact organisms. Extraction of cells with dimethylsulphoxide, which presumably removes LPS, had no effect on cell wall stability or appearance. The cell wall contained only traces of hexosamine and neutral sugars. Amino acid analysis of the cell wall showed about a 10% excess of acidic over basic amino acids, similar to organisms living in neutral habitats. It was speculated "that in low pH, high temperature habitats, peptidoglycan is unstable and the non-peptidoglycan wall of *S. acidocaldarius* may be a factor in permitting adaption to the extreme conditions." The speculation might not be valid however, since peptidoglycan-containing *B. acidocaldarius* and the cell wall-less *T. acidophilum* are both found in similar environments.

The unusual nature of the complex lipids of *S. acidocaldarius* have been investigated by Langworthy *et al.* (1974). Cells grown heterotrophically at 70°C, pH 3 had a total lipid distribution of 10·5% neutral lipids, 21·7% phospholipids and a very high 67·6% glycolipids. All of the complex lipid structures contain very long chain C_{40} fully saturated isopranol glycerol diethers rather than typical glycerol fatty acid ester-linked diglycerides. In further detailed critical analyses, De Rosa *et al.* (1974, 1976a) have shown the glycerol diether moiety to be composed of equimolar proportions of glycerol and a C_{40} isopranoid methyl branched alkyl side chain which is either acyclic ($C_{40}H_{82}$), monocyclic ($C_{40}H_{80}$) or bicyclic ($C_{40}H_{78}$). Degradation of the glycerol diether to give the alkyl side chain as the alcohol revealed C_{40} diols which contained an hydroxyl group substituted on each terminal end of the hydrocarbon chain. Because glycerol and C_{40} diols were found in equimolar amounts, it was proposed that each of the terminal hydroxyls of the C_{40} diols were in ether linkages to glycerol, thus forming a glycerol diether containing a 40-carbon macrocyclic loop. It was thought that this gives rise to a glycerol diether that is dimensionally similar to the glycerol diethers from the extreme halophiles which contain two C_{20} dihydrophytanyl hydrocarbon chains (Kates, 1972). The *Sulfolobus* glycerol diether was also shown to be based on the *sn*-2,3-glycerol configuration as is the *sn*-2,3-di-0-phytanyl glycerol from the extreme halophiles. In this author's opinion, however, the exact structure of the *Sulfolobus* glycerol diethers is still in doubt. The molecular weight appears to be much higher than would be expected for the proposed cyclic glycerol diether and it may in fact contain more than one free hydroxyl group. (The ethers are in fact diglycerol tetraethers, Langworthy, 1977a.)

Analysis of the glycolipids (Langworthy *et al.*, 1974) showed they were composed of glucosyl galactosyl glycerol diether and a glucosyl polyol glycerol diether. The latter glycolipid contained an unidentified polyalcohol attached to the glycerol diether through an ether bond. Approximately 40% of the phospholipids were present as the diether analogue of phosphatidyl inositol. Two unique phosphoglycolipids were detected. These were the inositol monophosphate derivatives of the two glycolipids found in the organism: inositolphosphoryl glucosyl galactosyl glycerol diether and inositolphosphoryl glucosyl polyol glycerol diether. A small amount of the monosulphate derivative of glucosyl galactosyl glycerol diether also was found. No acid resistant carbon–phosphorous or carbon–sulphur bonds were detected in the lipids as might be expected to be found at such high temperature and low pH. The lipid composition of *Sulfolobus* grown autotrophically on sulphur is essentially identical to heterotrophically grown cells except for the occurrence of a new as yet unidentified phosphoglycolipid. The lipid composition of the obligate autotrophic *Sulfolobus*-like isolate of Brierley and Brierley (1973) is identical to *Sulfolobus* lipids when both cells are grown on sulphur suggesting a close taxonomic relationship (Langworthy, 1977b).

The survival of *S. acidocaldarius* seems to depend on an intimate relationship between the cell wall and membrane. The long C_{40} isopranoid alkyl chain of the glycerol diether lipids would seem to be related to the maintenance of the proper liquid–crystalline state of the membrane at high temperature. The occurrence of ether bonds, which are resistant to most chemical degradations, apparently relate to the acid stability of the organism. The significance of the high glycolipid content and the exclusive presence of carbohydrate or carbohydrate-derived phospholipids is not clear. Perhaps their presence may be related to the nature of the deficient cell wall. It has been suggested that the accumulation of phosphoglycolipids in certain mycoplasmas may be related to the inability of these organisms to synthesize a normal cell wall structure (Smith *et al.*, 1973).

4. T. acidophilum

The existence of this cell wall-less organism which grows optimally at pH 2 and 59°C provides a unique model for the study of membrane stability in a harsh acid environment. Where other reported acidophiles possess some manner of cell wall, the membrane of *Thermoplasma* is in direct contact with its hot acid environment. General physiological and morphological features have been described (Belly *et al.*, 1973; Mayberry-Carson *et al.*, 1974). The unusual nature of the membrane is indicated by freeze-fracturing. As in the case of *Sulfolobus*, cells are cross-fractured revealing only membrane bilayer and cytoplasm (Ververgaert, personal communication). The unusual stability of *T. acidophilum* to several physical agents has been reported by Belly and Brock (1972). Cells are resistant to lysis by heating to temperatures of 100°C, non-ionic detergents, and the enzymes pronase, trypsin, or lysozyme. Cells are lysed by cationic detergents and more rapidly by anionic detergents. The extreme rigidity of the membrane is further exemplified in electron spin resonance studies reported by Smith *et al.* (1974). Lipid phase transitions were found at 45 and 60°C which are the temperature limits for growth. The mobility of spin-lables in the lipid domain of the membrane was even less than that reported for *Halobacterium cutirubrum* suggesting that the *Thermoplasma* membrane is probably the most rigid yet known.

A more detailed investigation of membrane properties and cellular stability was reported by Smith *et al.* (1973) and confirmed by Ruwart and Haug (1975). Varying both the pH and ionic strength of the suspending medium produced effects on the stability and viability of the organism. Apparent cell lysis occurred at values below pH 2 and above pH 6. Maximum stability based on viability measurements was seen at pH 3 to 4 in spite of the growth optimum at pH 2. Above pH 8 lysis was essentially complete. An increase in the ionic strength of the suspending medium brought about lysis of organisms at pH values of normally maximum stability. Lysis resulting from high ionic strength or alkaline pH gave no sediment upon centrifugation at 100,000 × *g* indicating the enveloping membranes are solubilized under these conditions. Interestingly, dialysis of the solubilized membranes against distilled water (pH 5·5) resulted in a re-aggregation of membrane vesicles. Cells were also found to be extremely resistant to sonic oscillation. Membranes could be prepared, however, by sonic oscillation at moderate ionic strength (0·05) and pH 5 which permitted recovery of about 50% membranous material. Polyacrylamide gel electrophoresis of isolated membranes produced a multiplicity of bands indicating a heterogeneous mixture of proteins. The amino acid profile of the membrane proteins did not reveal an unusual composition. The ratio of free carboxyl groups to free amino groups in *Thermoplasma* membranes was approximately 4 to 1. A similar proportion was found in *Acholeplasma laidlawii* B, a normal mycoplasma grown at pH 7. The total number of charged groups in *Thermoplasma* membrane, however, was only half that found in the usual mycoplasmal membrane, indicative of the former's hydrophobic nature.

The role of electrical charges in maintaining membrane stability was examined (Smith *et al.*, 1973) by effecting alterations to the free carboxyl groups in the membrane. Normally, membranes or whole cells of *Thermoplasma* are solubilized at pH values greater than 5 to 6. When glycine methyl ester was attached to carboxyl groups in the membrane, which would remove one potential negative charge, the membranes were stable over the entire pH range. By attaching ethylene diamine to the membrane carboxyl groups, which removes one potential negative charge and adds one potential positive charge, the membranes became stable at alkaline pH but solubilized at acid pH. Such evidence would indicate that hydrogen ions are necessary for a reduction in the ionization of the available carboxyl groups in the membrane. Apparently, at increasing pH values the resulting repulsion of negative charges causes a disruption of membrane proteins resulting in cellular lysis. The requirement for hydrogen ion is specific. Various monovalent or divalent cations or sucrose do not stabilize the cell at higher pH values. The hydrogen ion concentration has a stabilizing effect on *Thermoplasma* analogous to that of the high salt concentration on the extreme halophiles. Raising the pH for *Thermoplasma* or lowering the salt concentration for the halophiles results in membrane disaggregation.

Gross chemical analysis of the membrane reveals about 10% carbohydrate, 25% lipid and 60% protein. Most of the carbohydrate is accounted for by the unusual lipopolysaccharide, $(mannose)_{24}$-glucose-glycerol diether (Mayberry-Carson *et al.*, 1974a). The LPS exhibits a significant antigenic activity (Sugiyama *et al.*, 1974), is a stable high molecular weight aggregate at 59°C, pH 2, and has physical properties typical of Gram-negative LPS (Mayberry-Carson *et al.*, 1975).

The lipids of the organism have been investigated by Langworthy *et al.*, (1972). *Thermoplasma* complex lipids contain C_{40} isopranol glycerol diethers which are structurally the same as those in *Sulfolobus* (De Rosa *et al.*, 1976a). The phospholipids are not typical phosphatidyl forms but rather phosphoglycolipids. Although the exact lipid structures remain elusive, nearly 50% of the lipids is accounted for by a glycerolphosphoryl glycosyl glycerol diether. Of the total lipids, phospholipids represented about 57%, glycolipids comprised of monoglycosyl and diglycosyl glycerol diethers 25% and neutral lipids 17%. The presence of the naphthoquinone, K_2-7 among the neutral lipids indicated the existence of a respiratory chain in the organism.

Several interesting aspects of *Thermoplasma* metabolism have also been reported which may relate to its ability to exist in a hot acid environment. Searcy (1975) has reported the existence internally of an histone-like nucleoprotein associated with the DNA. The molecule was characterized as rich in basic amino acids and amides of acidic amino acids, soluble in acid, positively charged at neutral pH and showing the same mobility as histone IV(F2al) from calf thymus by polyacrylamide gel electrophoresis. Its function is uncertain, although it was suggested it may increase the thermal stability of the DNA or perhaps protect it from depurination by acid. Smith *et al.* (1975) have also demonstrated the requirement for a partially characterized polypeptide growth factor isolated from yeast extract. The availability of the polypeptide was shown to be an absolute requirement for growth of *Thermoplasma*. Amino acid supplementation did not override this specific a requirement. Analysis of the polypeptide showed that it was comprised largely of arginine, glutamic acid, aspartic acid and ammonia and that it had a molecular weight of approximately 1000. It was speculated that the specific polypeptide requirement may actually reflect its role as either an ion scavenger for some trace element, involvement in ion transport, protection at the cell surface from H^+ or possibly as an essential amino acid supply in a permeable form in the acidic environment.

The emerging picture in *T. acidophilum* suggests that obligate acidophily is related to the hydrophobic nature of the membrane due to a drastic reduction in the number of polar groups on the membrane proteins and to the presence of ether lipids. A balance of ionic interactions with the membrane proteins also appears necessary for cellular stability. These conditions are near optimum in laboratory culture medium which has a calculated ionic strength of 0·25 and a pH of 2. The long isopranol chains in the lipids would seem to relate to the thermophily of the organism. The invariable nature of the lipids and proteins comprising the major portion of the membrane very likely reflect the obligatory requirement for a hot acid environment.

5. *Acidophiles: An Overview*

In comparison to the thiobacilli which contain a rather typical wall and membrane composition, the thermophilic acidophiles are replete with unusual and unique chemical structures. In terms of composition, the extreme thermoacidophiles appear to have ensured their survival by synthesis of novel chemical structures which apparently take advantage of the dual stress of high temperature and low pH. *Thermoplasma* and *Sulfolobus* both contain almost exclusively C_{40} alkyl glycerol diether based complex lipids. The long hydrocarbon chains would appear relevant to maintenance of appropriate membrane fluidity at high temperature while the ether linkages may impart stability of the lipids to acid hydrolysis. In this respect these two organisms share structural features in common with the extreme halophiles which contain exclusively dihydrophytanyl glycerol diether derived complex lipids. In each case the cell membrane is nearly completely exposed to its extreme ionic environment. In addition, *Thermoplasma* has a specific H^+ requirement for the maintenance of cellular stability just as the extreme halophiles require salt for stability (Larsen, 1967; Kushner, 1968; Chapter 8). Perhaps other properties and interactions with H^+, besides its requirement for cellular stability, will be found analagous to the role of salt in the extreme halophiles. *Thermoplasma* would thus appear to possess properties relevant to the three different extremes of thermophily, halophily and acidophily. Both *Thermoplasma* and *Sulfolobus* contain phospholipids which are present exclusively as carbohydrate derived phosphoglycolipids, for reasons which are not yet clear. Their occurrence might be related to the absence of a true cell wall in these two organisms since the cell wall-containing thermoacidophile *B. acidocaldarius* does not contain phosphoglycolipids but rather typical phosphatides. *B. acidocaldarius* does not contain long chain alkyl glycerol diethers but rather diglyceride based complex lipids. Its only present claim to a truely acid resistant bond is found in the presence of a sulphonolipid. The resistance of *B. acidocaldarius* to its hot acid environment would therefore appear to include the cell wall which may prove to have unusual features. Its presence apparently allows synthesis of membrane lipid structures which are intermediate in conventionality between *Thermoplasma* and *Sulfolobus* and less extreme organisms.

A common parameter of the thermoacidophiles appears to be the occurrence of large amounts of carbohydrate derived membrane lipid structures. All of the complex lipids in *Thermoplasma* and *Sulfolobus* contain carbohydrate derived residues and even *B. acidocaldarius* lipids contain 75% glycolipid derived structures. Similarly, the tetraglycosyl diacyl glycerol found in the thermophile *Flavobacterium thermophilum* represents almost 70% of the total lipids (Oshima and Yamakawa, 1974). Ray *et al.* (1971) have also shown the glycolipids in *Thermus aquaticus* increase four-fold with a temperature increase from 50 to 75°C. It was suggested that these structures may be involved in increasing membrane stability at high temperatures. The preponderance of carbohydrate derived lipids in the thermoacidophiles may also be a response to high temperature and function in a similar capacity.

As an overall consideration of the known acidophiles, it appears that as temperature stress is decreased the obligate acidophiles become morphologically and chemically similar to organisms which grow at more neutral pH values. At this time no single unifying structural feature is outstandingly apparent as a specific requirement for acidophily. Perhaps a common denominator which might be relative to all of the acidophiles studied thus far is the occurrence of cyclization within the hydrocarbon portions of the membrane lipids. *Thermoplasma* and *Sulfolobus* lipids possess cyclohexyl rings in the alkyl chains and *B. acidocaldarius* possesses cyclohexyl fatty acids and many pentacyclic triterpene derivatives. The thiobacilli contain large amounts of cyclopropane fatty acids. Cyclization affects the fluidity within the lipid domain of membranes. It is thought that cholesterol affects membrane packing thereby increasing membrane rigidity and decreasing membrane permeability (De Kruyff *et al.*, 1972; Rottem *et al.*, 1973). Cyclized hydrocarbons in acidophiles may function by in a similar manner. De Rosa *et al.* (1974d) have suggested that with the cyclized hydrocarbons the "cholesterol effect" is in fact built into the hydrophobic portion of the membrane. The actual physiological role this cyclization may play in relation to acidophily, however, awaits experimental data. The mechanisms of acidophily are most certainly a complex of interwoven structural and metabolic subtilties which await further investigation in order to detail the acidophilic mode of life.

As an interesting final note, the occurrence of many of the unusual lipid structures in the thermoacidophiles may be of potentially important evolutionary significance. *B. acidocaldarius* for example contains triterpene lipid derivatives found in the primitive eukaryotic ferns and mosses and contains the sulphonolipid which has been found only in photosynthetic cells. *Thermoplasma* contains a histone-like nucleoprotein, while histones are normally associated only with eukaryotic cells. The previously unexplained presence of cyclohexane hydrocarbon derivatives associated with many geochemical deposits may have arisen from such organisms as the thermoacidophiles if they existed in prehistoric times. The occurrence of hot acid environments associated with increased geological activity long ago would likely have been quite common. It may therefore not be unreasonable to suppose that the thermophilic acidophiles may be rather primitive life forms related to evolutionary precursors of modern day cells.

IV. Conclusion

As demonstrated by the brevity of this chapter, our knowledge of life at extremes of pH is quite limited. The temptation to speculate is always present, but considering the dearth of knowledge concerning these organisms such attempts may not have much meaning. For example, most alkaline-resistant organisms tend to reduce pH values toward neutrality during active growth. There are few examples of obligate alkalinophiles which grow optimally at pH 10 or higher. At the other extreme, obligate acidophiles do not increase pH values during growth and relatively more examples have been reported. It would appear that the presence of excess hydrogen ions is less stringent on organisms than the absence of hydrogen ions. Such an observation, however, is simply a reflection of the limited attempts to thoroughly search alkaline environments for truly extreme obligate alkalinophiles. It is probably safe to conclude that hydrogen ions are excluded by the acidophilic organisms. By analogy, the alkalinophiles most likely exclude hydroxyl ions or inversely, retain hydrogen ions. It should be fairly certain by now that these organisms maintain internal pH values different from the external environment. At the present time it appears likely that both metabolic activities and structural features of the cell wall and/or membrane function in maintaining internal–external pH gradients. Although the concept of hydrogen ions pumps and presumably anion pumps has often been invoked as an active mechanism for ion exclusion, direct experimental evidence remains unavailable. Such information would be extremely valuable for an understanding of survival in extreme pH values. Further work on acidophiles and alkalinophiles in terms of physiology and envelope structure is also necessary before clear mechanisms of resistance to extreme pH values may emerge. Perhaps the recent interest generated by the discovery of the thermophilic acidophiles will restimulate interest in the mesophilic counterparts. Further studies on the nature and occurrence of the extreme pH life forms is clearly warranted. Certainly these organisms should gain an increasing importance in view of the increasing number of extreme habitats associated with the rise in environmental pollution.

References

Adair, R. W. (1966). Membrane-associated sulfur oxidation by the autotroph *Thiobacillus thiooxidans*. *J. Bacteriol.* **92,** 899–904.

Adapoe, C. and Silver, M. (1975). The soluble adenosine triphosphatase of *Thiobacillus ferooxidans*. *Can. J. Microbiol.* **21,** 1–7.

Addanki, S. and Sotos, J. F. (1969). Observations on intramitochondrial pH and ion transport by the 5,5-dimethyl-2,4-oxazolidinedione (DMO) method. *Ann. N.Y. Acad. Sci.* **147,** 756–804.

Adele, H. and Lundgren, D. G. (1970). Inorganic pyrophosphatase from *Ferrobacillus ferrooxidans* (*Thiobacillus ferrooxidans*). *Can. J. Biochem.* **48,** 1302–1307.

Allen, M. B. (1959). Studies with *Cyanidium caldarium* an anomously pigmented chlorophyte. *Arch. Mikrobiol.* **32,** 270–277.

Allen, O. N. and Holding, A. J. (1974). *In* "Bergy's Manual of Determinative Bacteriology" (8th ed.) (Eds R. E. Buchanan and N. E. Gibbons), pp. 264–267. Williams and Wilkins, Baltimore.

Amemiya, K. and Umbreit, W. W. (1974). Heterotrophic nature of the cell-free protein-synthesizing system from the strict chemolithotroph, *Thiobacillus thiooxidans*. *J. Bacteriol.* **117,** 834–839.

Asai, T. (1968). *In* "Acetic Acid Bacteria." University Park Press, Baltimore.

Barghoorn, E. S. and Tyler, S. A. (1965). Microorganisms from the Gunflint chert. *Science* **147,** 563.

Barridge, J. K. and Shively, J. M. (1968). Phospholipids of the thiobacilli. *J. Bacteriol.* **95,** 2182–2185.

Battley, E. M. and Bartlett, E. J. (1966), A convenient pH-gradient method for the determination of the maximum and minumum pH for microbial growth. *Antonie van Leeuwenhoek* **32,** 245–255.

Belly, R. T. and Brock, T. D. (1972). Cellular stability of a thermophilic, acidophilic mycoplasma. *J. Gen. Microbiol.* **73,** 465–469.

Belly, R. T. and Brock, T. D. (1974a). Ecology of iron-oxidizing bacteria in pyritic materials associated with coal. *J. Bacteriol.* **117,** 726–732.

Belly, R. T. and Brock, T. D. (1974b). Wide spread occurrence of acidophilic strains of *Bacillus coagulans* in hot springs. *J. Appl. Bacteriol.* **37,** 175–177.

Belly, R. T., Bohlool, B. B. and Brock, T. D. (1973). The genus *Thermoplasma*. *Ann. N.Y. Acad. Sci.* **225,** 94–107.

Bird, C. W., Lynch, J. M., Pirt, S. J. and Reid, W. W. (1971). Identification of hop-22(29)-ene in procaryotic organisms. *Tetrahedron Lett.* **34,** 3189–3190.

Blaylock, B. A. and Nason, A. (1963). Electron transport systems of the chemo-autotroph *Ferrobacillus ferrooxidans*. *J. Biol. Chem.* **238,** 3453–3462.

Bodo, C. A. and Lundgren, D. G. (1972). Oxidation of iron by cell free envelopes of *Thiobacillus ferrooxidans*. *Abst. Am. Soc. Microbiol.* **72,** 174.

Bornside, G. H. and Kallio, R. E. (1956). Urea-hydrolyzing bacilli. I. A physiological approach to identification. *J. Bacteriol.* **28,** 627–634.

Boyer, E. W. and Ingle, M. B. (1972). Extracellular alkaline amylase from a *Bacillus* species. *J. Bacteriol.* **110,** 992–1000.

Brierley, C. L. (1974). Molybdenite-leaching: use of a high-temperature microbe. *J. Less-Common Metals* **36,** 237–247.

Brierley, C. L. and Brierley, J. A. (1973). A chemoautotrophic and thermophilic microorganism isolated from an acid hot spring. *Can. J. Microbiol.* **19,** 183–188.

Brock, T. D. (1969). Microbial growth under extreme conditions. *Symp. Soc. Gen. Microbiol.* **19,** 15–41.

Brock, T. D. and Darland, G. (1970). The limits of microbial existence: temperature and pH. *Science* **169,** 1316–1318.

Brock, T. D. and Gustafson, J. (1976). Ferric iron reduction by sulfur- and iron-Oxidizing bacteria. *Appl. Environ. Microbiol.* **32,** 567–571.

Brock, T. D., Brock, K. M., Belly, R. T. and Weiss, R. L. (1972). *Sulfolobus*: a new genus of sulfur-oxidizing bacteria living at low pH and high temperature. *Arch.Mikrobiol.* **84,** 54–68.

Buchanan, R. E. and Fulmer, E. I. (1930). *In* "Physiology and Biochemistry of Bacteria," Vol. 2. Williams and Wilkins, Baltimore.

Butler, T. C., Waddell, W. J. and Poole, D. T. (1966). The pH of intracellular water. *Ann. N.Y. Acad. Sci.* **133,** 73–77.

Caldwell, P. C. (1956). Intracellular pH. *Int. Rev. Cytol.* **5,** 229–277.

Cassin, P. E. (1974). Isolation growth and physiology of acidophilic chlamydo-monads. *J. Phycol.* **10,** 439–447.

Chesbro, W. R. and Evans, J. B. (1959). Factors affecting the growth of enterococci in highly alkaline media. *J. Bacteriol.* **78,** 858–862.

Chislett, M. E. and Kushner, D. J. (1961a). A strain of *Bacillus circulans* capable of growing under highly alkaline conditions. *J. Gen. Microbiol.* **24,** 187–190.

Chislett, M. E. and Kushner, D. J. (1961b). Germination under alkaline conditions and transmission of alkali resistance by endospores of certain strains of *Bacillus cerus* and *Bacillus circulans*. *J. Gen. Microbiol.* **25,** 151–156.

Crum, E. H. and Siehr, D. J. (1967). *Thiobacillus thiooxidans* cell wall amino acids and monosaccharides. *J. Bacteriol.* **94,** 2069–2070.

Darland, G. and Brock, T. D. (1971). *Bacillus acidocaldarius* sp. nov., an acidophilic thermophilic spore-forming bacterium. *J. Gen. Microbiol.* **67,** 9–15.

Darland, G., Brock, T. D., Samsonoff, W. and Conti, S. F. (1970). A thermophilic, acidophilic mycoplasma isolated from a coal refuse pile. *Science* **170,** 1416–1418.

De Kruyff, B., Demel, R. A. and van Deenen, L. L. M. (1972). The effect of cholesterol and epicholesterol incorporation on the permeability and on the phase transition of intact *Acholeplasma laidlawii* cell membranes and derived liposomes. *Biochim. Biophys. Acta* **255,** 331–347.

DeRosa, M. and Gambacorta, A. (1975). Identification of natural and semisynthetic ω-cyclohexyl fatty acids. *Phytochemistry* **14,** 209–210.

DeRosa, M., Gamacorta, A., Minale, L. and Bu'Lock, J. D. (1971a). Cyclohexane fatty acids from a thermophilic bacterium. *J. Chem. Soc. Chem. Commun.* **1971,** 1334.

DeRosa, M., Gambacorta, A., Minale, L. and Bu'Lock, J. D. (1971b). Bacterial triterpenes. *J. Chem. Soc. Chem. Commun.* **1971,** 618–620.

DeRosa, M., Gambacorta, A., Minale, L. and Bu'Lock, J. D. (1972). A new biosynthetic pathway: the formation of ω-cyclohexyl fatty acids from shikimate in an acidophilic thermophilic bacillus. *Biochem. J.* **128,** 751–754.

DeRosa, M., Gambacorta, A., Minale, L. and Bu'Lock, J. D. (1973). Isoprenoids of *Bacillus acidocaldarius. Phytochemistry* **12,** 1117–1123.

DeRosa, M., Gambacorta, A., and Bu'Lock, J. D. (1974a). Effects of pH and temperature on the fatty acid composition of *Bacillus acidocaldarius. J. Bacteriol.* **117,** 212–214.

DeRosa, M., Gambacorta, A. and Bu'Lock, J. D. (1974b). Origin of cyclohexane carboxylic acid in *Bacillus acidocaldarius. Phytochemistry* **13,** 1793–1794.

DeRosa, M., Gambacorta, A. and Bu'Lock, J. D. (1974c). Specificity effects in the biosynthesis of fatty acids in *Bacillus acidocaldarius. Phytochemistry* **13,** 905–910.

DeRosa, M., Gambacorta, A., Minale, L. and Bu'Lock, J. D. (1974d). Cyclic diether lipids from very thermophilic acidophilic bacteria. *J. Chem. Soc. Chem. Comm.* **1974,** 543–544.

DeRosa, M., Gambacorta, A. and Bu'Lock, J. D. (1975). Extremely thermophilic acidophilic bacteria convergent with *Sulfolobus acidocaldarius. J. Gen. Microbiol.* **86,** 156–164.

DeRosa, M., Gambacorta, A. and Bu'Lock, J. D. (1976a). The Caldariella group of extreme thermoacidophile bacteria: Direct comparison of lipids in Sulfolobus Thermoplasma, and the MT strains. *Phytochemistry* **15,** 143–145.

DeRosa, M., DeRosa, S., Gambacorta, A., Carteni-Farina, M. and Zappia, V. (1976b). Occurrence and characterization of new polyamines in the extreme thermophile *Caldarella acidophila. Biochem. Biophys. Res. Commun.* **69,** 253–261.

DeRosa, M., DeRosa, S., Gambacorta, A. and Bu'Lock, J. D. (1976c). Isoprenoid triether lipids from Caldariella. *Phytochemistry* **15,** 1995–1996.

Dewey, D. L. and Beecher, J. (1966). The internal hydrogen ion concentration of *Thiobacillus thiooxidans* and survival after irradiation. *Radiat. Res.* **28,** 289–295.

Din, G. A., Suzuki, I. and Lees, H. (1967). Ferrous iron oxidation by *Ferrobacillus ferrooxidans*: Purification and properties of Fe^{++}-cytochrome c reductase. *Can. J. Biochem.* **45,** 1523–1546.

Doemel, W. N. and Brock, T. D. (1971). The physiological ecology of *Cyanidium caldarium. J. Gen. Microbiol.* **67,** 17–32.

Doetsch, R. N., Cook, T. M. and Vaituzis, Z. (1967). On the uniqueness of the flagellum of *Thiobacillus thiooxidans. Antonie van Leeuwenhoek* **33,** 196–202.

Downie, A. W. and Cruickshank, J. (1928). The resistance of *Streptococcus faecalis* to acid and alkaline media. *Brit. J. Exp. Pathol.* **9,** 171–173.

Dugan, P. R. (1975). Bacterial ecology of strip mine areas and its relationship to the production of acidic mine drainage. *Ohio J. Sci.* **75,** 266–279.

Dugan, P. and Lundgren, D. G. (1963). Acid production by *Ferrobacillus ferrooxidans* and its relation to water pollution. *Dev. Ind. Microbiol.* **5,** 250–257.

Dugan, P. R. and Lundgren, D. G. (1965). Energy supply for the chemolithotroph *Ferrobacillus ferrooxidans, J. Bacteriol.* **89,** 825–834.

Dugan, P. R., MacMillan, C. D. and Pfister, R. M. (1970). Aerobic heterotrophic bacteria indigenous to pH 2·8 acid mine water: microscopic enumeration of acid streams. *J. Bacteriol.* **101,** 973–981.

Ehrlich, H. L. (1963). Microorganisms in acid drainage from a copper mine. *J. Bacteriol.* **86,** 350–352.

Enami, I. and Fukuda, I. (1975). Mechanisms of the acido- and thermo-phily of *Cyanidium caldarium* Geitler. I. Effects of temperature, pH and light intensity on photosynthetic oxygen evolution of intact and treated cells. *Plant Cell Physiol.* **16,** 211–220.

Enami, I., Nagashina, H. and Fukuda. I. (1975). Mechanisms of the acido- and thermo-phily of *Cyanidium caldarium* Geither. II. Physiological role of the cell wall. *Plant Cell Physiol.* **16,** 221–232.

Fliermans, C. B. and Brock, T. D. (1972). Ecology of sulfur-oxidizing bacteria in hot acid soils. *J. Bacteriol.* **111,** 343–350.

Fogg, G. E. (1956). The comparative physiology and biochemistry of the blue-green algae. *Bacteriol. Rev.* **20,** 148–165.

Fott, B. and McCarthy, S. J. (1964). Three acidophilic volvocine flagellates in pure culture. *J. Protozol.* **11,** 116–120.

Gale, E. F. 1943). Factors influencing the enzymatic activities of bacteria. *Bact. Rev.* **7,** 139–173.

Geiger, P. J., Jaffe, L. D. and Mamikunian, G. (1965). Biological contamination of the planets. *In* "Current Aspects of Exobiology," (Eds G. Mamikunian and M. H. Briggs), pp. 283–322. Pergamon Press, New York.

Gibson, T. (1934). An investigation of the *Bacillus pasteurii* group. II. Special physiology of the organisms. *J. Bacteriol.* **28,** 313–322.

Goldfine, H. (1972). Comparative aspects of bacteriol lipids. *Adv. Microbiol. Physiol.* **8,** 1–58.

Haddock, B. and Cobley, J. G. (1976). Electron transport in the alkalophile *Bacillus pasteurii* (NCIB 8841). *Biochem Soc. Trans.* **4,** 709–711.

Haest, C. W. M., De Gier, J., op Den Kamp, J. A. F., Bartels, P. and van Deenen, L. L. M. (1972). Changes in permeability of *Staphylococcus aureus* and derived liposomes with varying lipid compositions. *Biochim. Biophys. Acta* **255,** 720–733.

Haines, T. H. (1971). The chemistry of the sulfolipids. *In* "Progress in the Chemistry of Fats and other Lipids" (Ed. R. T. Holman), Vol. 2, pp. 297–345. Pergamon Press, New York.

Handley, P. S. and Knight, D. G. (1975). Ultrastructural changes occurring during germination and outgrowth of spores of the thermophile *Bacillus acidocaldarius. Arch. Microbiol.* **102,** 155–161.

Harold, F. M. (1969). Antimicrobial agents and membrane function. *Adv. Microbiol. Physiol.* **4,** 46–104.

Harold, F. M. (1972). Conservation and transformation of energy by bacterial membranes. *Bacteriol. Rev.* **36,** 172–230.

Harold, F. M. and Altendorf, K. (1974). Cation transport in bacteria: K^+, Na^+,

H^+. *In* "Current Topics in Membranes and Transport" (Eds F. Bonner and A. Kleinzeller), Vol. 5, pp. 1–50. Academic Press, New York and London.

Harold, F. M., Pavlasová, E. and Baarda, J. R. (1970). A transmembrane pH gradient in *Streptococcus faecalis*. Origin and dissipation by proton conductors and *N'*, *N'*-Dicyclohexylcarbodiimide. *Biochem. Biophys. Acta* **196,** 235–244.

Heilbrunn, L. V. (1952). "An Outline of General Physiology" (3rd ed.). Saunders, Philadelphia.

Henderson, P. J. (1971). Ion transport by energy-conserving biological membranes. *Ann. Rev. Microbiol.* **25,** 393–428.

Hjorth-Hansen, S. (1939). Über das Wachstum der Hefe in synthetischer Nährlösung bei konstantem pH. *Biochem. Z.* **301,** 292–300.

Holm-Hansen, O. (1968). Ecology, physiology and biochemistry of blue-green algae. *Ann. Rev. Microbiol.* **22,** 47–70.

Horikoshi, K. (1971a). Production of alkaline enzymes by alkalophilic microorganisms I. Alkaline protease produced by bacillus No. 221. *Agric. Biol. Chem.* **35,** 1407–1414.

Horikoshi, K. (1971b). Production of alkaline enzymes by alkalophilic microorganisms. II. Alkaline amylase produced by bacillus No. A-40-2. *Agric. Biol. Chem.* **35,** 1783–1791.

Horikoshi, K. (1972). Production of alkaline enzymes by alkalophilic microorganisms. III. Alkaline pectinase of bacillus No. P-4-N. *Agric. Biol. Chem.* **36,** 285–293.

Horikoshi, K. and Atsukawa, Y. (1975). β-1,3-Gluconase produced by alkalophilic bacteria, bacillus No. K-12-5. *Agric. Biol. Chem.* **37,** 1449–1456.

Hsung, J. C. and Haug, A. (1975). Intracellular pH of *Thermoplasma acidophila. Biochim. Biophys. Acta* **389,** 477–482.

Hsung, J. C. and Haug, A. (1977). Membrane potential of *Thermoplasma acidophila. FEBS Lett.* **73,** 47–50.

Hutner, S. H. (1972). Inorganic nutrition. *Ann. Rev. Microbiol.* **26,** 313–346.

Jenkin, P. M. (1936). Reports on the Percy Sloden expedition to some rift valley lakes in Kenya in 1929. VII. Summary of the ecological results, with special reference to the alkaline lakes. *Ann. Mag. Nat. Hist.* **18,** 133–181.

Johnson, H. W. (1923). Relationships between hydrogen ion, hydroxyl ion and salt concentrations and the growth of seven soil molds. *Iowa Agr. Exp. Stn. Res. Bull.* **76,** 307–344.

Jones, G. E. and Benson, A. A. (1965). Phosphatidyl glycerol in *Thiobacillus thiooxidans. J. Bacteriol.* **89,** 260–261.

Kashket, E. R. and Wilson, T. H. (1973). Proton-coupled accumulation of galactoside in *Sheptococcus lactis* 7962, *Proc. Natl. Acad. Sci. USA* **70,** 2866–2869.

Kates, M. (1972). Ether-linked lipids in extremely halophilic bacteria. *In* "Ether lipids: Chemistry and Biology" (Ed. F. Snyder), pp. 351–398. Academic Press, New York and London.

Kemper, E. S. (1966). Acid production by *Thiobacillus thiooxidans. J. Bacteriol.* **92,** 1842–1843.

Kessler, E. and Kramer, H. (1960). Physiologische untersuchungen an einer ungewöhnlich saureresistenten *Chlorella. Arch. Mikrobiol.* **37,** 245–255.

Knoche, N. W. and Shively, J. M. (1969). The identification of *cis*-11, 12-methylene-2-hydroxyoctadecanoic acid from *Thiobacillus thiooxidans. J. Biol. Chem.* **244,** 4773–4778.

Koostra, W. L. and Smith, P. F. (1969). D- and L- Alanylphosphatidylglycerols from *Mycoplasma laidlawii*, strain B. *Biochemistry* **8,** 4794–4806.

Kurono, Y. and Horikoshi, K. (1973). Alkaline catalase produced by bacillus No. KU-1. *Agric. Biol. Chem.* **37,** 2565–2570.

Kushner, D. J. (1964). Microbial resistance to harsh and destructive environmental conditions. *In* "Experimental Chemotherapy" (Eds R. J. Schnitzer and F. Hawking), Vol. 2, pp. 113–168. Academic Press, New York and London.

Kushner, D. J. (1968). Halophilic bacteria. *Adv. Appl. Microbiol.* **10,** 73–99.

Kushner, D. J. and Lisson, T. A. (1959). Alkali resistance in *Bacillus cereus*. *J. Gen. Microbiol.* **21,** 96–108.

Lackey, J. B. (1938). The flora and fauna of surface waters polluted by acid mine drainage. *Publ. Health Rep., Wash.* **53,** 1499–1507.

Landesman, T., Duncan, D. W. and Walden, C. C. (1966). Oxidation of inorganic sulfur compounds by washed cell suspensions of *Thiobacillus ferrooxidans*. *Can. J. Microbiol.* **12,** 957–964.

Langworthy, T. A. (1977a). Long-chain diglycerol tetraethers from *Thermoplasma acidophilum*. *Biochim. Biophys. Acta* **487,** 37–50.

Langworthy, T. A. (1977b). Comparative lipid composition of heterotrophically and autotrophically grown *Sulfolobus acidocaldarius*. *J. Bacteriol.* **130,** 1326–1332.

Langworthy, T. A. and Mayberry, W. R. (1976a). A 1,2,3,4-Tetrahydroxy pentane-substituted pentacyclic triterpene from *Bacillus acidocaldarius*. *Biochim. Biophys. Acta* **431,** 570–577.

Langworthy, T. A., Smith, P. F. and Mayberry, W. R. (1972). Lipids of *Thermoplasma acidophilum*. *J. Bacteriol.* **112,** 1193–1200.

Langworthy, T. A., Mayberry, W. R. and Smith, P. F. (1974). Long-chain glycerol diether and polyol dialkyl glycerol triether lipids of *Sulfolobus acidocaldarius*. *J. Bacteriol.* **119,** 106–116.

Langworthy, T. A., Mayberry, W. R. and Smith, P. F. (1976). A sulfonolipid and novel glucosamidyl glycolipids from the extreme thermoacidophile *Bacillus acidocaldarius*. *Biochim. Biophys. Acta* **431,** 550–569.

Larsen, H. (1967). Biochemical aspects of extreme halophilism. *Adv. Microbiol Physiol.* **1,** 97–132.

Larson, A. D. and Kallio, R. E. (1954). Purification and properties of a bacterial urease. *J. Bacteriol.* **68,** 67–73.

Levin, R. A. (1971). Fatty acids of *Thiobacillus thiooxidans*. *J. Bacteriol.* **108,** 992–995.

Lundgren, D. G., Andersen, K. J., Remsen, C. C., and Mahoney, R. P. (1964). Culture, structure, and physiology of the chemoautotrophic *Ferrobacillus ferrooxidans*. *Dev. Ind. Microbiol.* **6,** 250–259.

Lundgren, D. G., Vestal, J. R. and Tabita, F. R. (1972). The microbiology of mine drainage pollution. *In* "Water Pollution Microbiology" (Ed R. Mitchell), pp. 69–88. Wiley, New York.

Lundgren, D. G., Vestal, J. R. and Tabita, F. R. (1974). The iron-oxidizing bacteria. *In* "Microbial Iron Metabolism, a Comprehensive Treatise" (Ed. J. B. Neilands), pp. 457–473. Academic Press, New York and London.

Lwoff, A. (1941). Limites de concentrations en ions H et OH compatibles avec le dévelopment *in vitro* du flagellé *Polytomella caeca*. *Ann. Inst. Pasteur.* **66,** 407–416.

Mahoney, R. P. and Edwards, M. R. (1966). Fine structure of *Thiobacillus thiooxidans*. *J. Bacteriol.* **92,** 487–495.

Markosyan, G. E. (1973). A new myxotrophic sulfur bacterium developing in acidic media, *Thiobacillus organoparus* sp. n. *Dokl. Akad. Nauk. SSSR. Ser. Biol.* **211,** 1205–1208.

Marsili, A. and Morelli, I. (1968). Triterpenes from mosses-I. The occurrence of 22(29)-hopene in *Thamnium alopecurum* (L.) Br. Eur. ssp. Eu-*alopecurum* Giac. *Phytochemistry* **7,** 1705–1706.

Marsili, A. and Morelli, I. (1970). Triterpenes from *Thuidium tamariscifolium. Phytochemistry* **9,** 651–653.

Marsili, A., Morelli, I. and Iori, A. M. (1971). 21-Hopene and some other constituents of *Pseudoscleropodium purum. Phytochemistry* **10,** 432–433.

Mayberry-Carson, K. J., Langworthy, T. A., Mayberry, W. R. and Smith, P. F. (1974a). A new class of lipopolysaccharide from *Thermoplasma acidophilum. Biochim. Biophys. Acta* **360,** 217–229.

Mayberry-Carson, K. J., Roth, I. L., Harris, J. L. and Smith, P. F. (1974b). Scanning electron microscopy of *Thermoplasma acidophilum. J. Bacteriol.* **120,** 1472–1475.

Mayberry-Carson, K. J., Roth, I. L. and Smith, P. F. (1975). Ultrastructure of lipopolysaccharide isolated from *Thermoplasma acidophilum. J. Bacteriol.* **121,** 700–703.

Meek, C. S. and Lipman, C. B. (1922). The relation of the reactions of the salt concentration of the medium to nitrifying bacteria. *J. Gen. Physiol.* **5,** 195–204.

McGoran, C. J. M., Duncan, D. W. and Walden, C. C. (1969). Growth of *Thiobacillus ferrooxidans* on various substrates. *Can. J. Microbiol.* **15,** 135–138.

McLachlan, J. and Gorham, P. R. (1962). Effects of pH and nitrogen sources on growth of *Microcystis aeurginosa* Kütz. *Can. J. Microbiol.* **8,** 1–11.

Mclaren, A. D. and Skujins, J. (1968). The physical environments of microorganisms in soil. *In* "The Ecology of Soil Bacteria" (Eds T. R. G. Gray and D. Parkinson), Liverpool University Press, Liverpool.

Millong, G., DeRosa, M., Gambacorta, A. and Bu'Lock, J. D. (1975). Ultrastructure of an extremely thermophilic acidophilic microorganism. *J. Gen. Microbiol.* **86,** 165–173.

Mosser, J. L., Mosser, A. G. and Brock, T. D. (1973). Bacterial origin of sulfuric acid in geothermal habitats. *Science.* **179,** 1323–1324.

Neal, A. L., Weinstock, J. O. and Lampen, J. O. (1965). Mechanisms of fatty acid toxicity for yeast. *J. Bacteriol* **90,** 126–131.

Ohta, K., Kiyomiya, A., Koyama, N. and Nosoh, Y. (1975). The basis of the alkalophilic property of a species of *Bacillus. J. Gen. Microbiol.* **86,** 259–266.

Oshima, M. and Ariga, T. (1975). *ω*-Cyclohexyl fatty acids in acidophilic thermophilic bacteria. *J. Biol. Chem.* **250,** 6963–6968.

Oshima, T. and Kawasaki, Y. (1976). Bacteria in hot springs. Introduction to studies on acidophilic and thermophilic bacteria. *Gendai Kagaku* **66,** 12–17.

Oshima, M. and Yamakawa, T. (1974). Chemical structure of a novel glycolipid from an extreme thermophile, *Flavobacterium thermophilum. Biochemistry* **13,** 1140–1146.

Peck, H. D. (1968). Energy-coupling mechanisms in chemolithotrophic bacteria. *Ann. Rev. Microbiol.* **22,** 489.

Rao, G. S. and Berger, L. R. (1971). The requirement of low pH for growth of *Thiobacillus thiooxidans. Arch. Mikrobiol.* **79,** 338–344.

Ray, P. H., White, D. C. and Brock, T. D. (1971). Effect of growth temperature on the lipid composition of *Thermus aquaticus. J. Bacteriol.* **108,** 227–235.

Remsen, C. and Lundgren, D. G. (1966). Electron microscopy of the cell envelope of *Ferrobacillus ferrooxidans* prepared by freeze-etching and chemical fractionation techniques. *J. Bacteriol.* **92,** 1765–1771.

Rottem, S., Yashomv, J., Ne'eman, Z. and Razin, S. (1973). Cholesterol in Mycoplasma membranes: composition, ultrastructure and biological properties from *Mycoplasma mycoides* var. Capri cells adapted to grow with low cholesterol concentrations. *Biochim. Biophys, Acta* **323,** 495–508.

Rottenberg, H. (1975). The measurement of transmembrane electrochemical proton gradients. *Bioenergetics* **7,** 61–74.

Rottenberg, H. and Grunwald, T. (1972a). Determination of Δ pH in chloroplasts. 3. Ammonium uptake as a measure of Δ pH in chloroplasts and sub-chloroplast particles. *Eur. J. Biochem.* **25,** 71–74.

Rottenberg, H., Grunwald, T. and Avron, M. (1972b). Determination of Δ pH in chloroplasts. 1. Distribution of [^{14}C] methylamine. *Eur. J. Biochem.* **25,** 54–63.

Ruwart, M. J. and Haug, A. (1975). Membrane properties of *Thermoplasma acidophila. Biochemistry* **14,** 860–866.

Sakaki, Y. and Oshima, T. (1976). A new lipid-containing phage infecting acidophilic, thermophilic bacteria. *Virology* **76,** 256–259.

Satake, J. and Saijo, Y. (1974). Carbon dioxide content and metabolic activity of microorganisms in some acid lakes in Japan. *Limnol. Oceanogr.* **19,** 331–338.

Schuldiner, S., Rottenberg, H. and Avron, M. (1972). Determination of Δ pH in chloroplasts. 2. Fluorescent amines as a probe for the determination of Δ pH in chloroplasts. *Eur. J. Biochem.* **25,** 64–70.

Schulz, E. and Hirsch, P. (1973). Morphologically unusual bacteria in acid bog water habitats. *Abst. Am. Soc. Microbiol.* **73,** 60.

Searcy, D. G. (1975). Histone-like protein in the prokaryote *Thermoplasma acidophilum. Biochim. Biophys. Acta* **395,** 535–547.

Searcy, D. G. (1976). *Thermoplasma acidophilum*: Intracellular pH and potassium concentration. *Biochim. Biophys. Acta* **451,** 278–286.

Searcy, D. G. (1976). *Thermoplasma acidophilum*: Studies on a procaryote that contains a histone-like protein. *Symp. Mol. Cell. Biol.* **5,** 51–57.

Shaw, N. (1974). Lipid composition as a guide to the classification of bacteria. *Adv. Appl. Microbiol.* **17,** 63–108.

Shirfine, M. and Phaff, H. J. (1958). On the isolation, ecology and taxonomy of *Saccharomycopsis guttulatta. Antonie van Leeuwenhoek* **24,** 193–209.

Shively, J. M. and Benson, A. A. (1967). Phospholipids of *Thiobacillus thiooxidans. J. Bacteriol.* **94,** 1679–1683.

Shively, J. M. and Knoche, H. W. (1969). Isolation of an ornithine-containing lipid from *Thiobacillus thiooxidans. J. Bacteriol.* **98,** 829.

Short, S. A., White, D. C. and Aleem, M. I. H. (1969). Phospholipid metabolism in *Ferrobacillus ferrooxidans. J. Bacteriol.* **99,** 142–150.

Siegel, S. M. and Giumarro, C. (1966). On the culture of a microorganism similar to the Precambrian microfossil *Kakabekia umbellata* Barghoorn in ammonia-rich atmospheres. *Proc. Natl. Acad. Sci. USA* **55,** 349–353.

Siegal, S. M., Roberts, K., Nathan, H. and Daly, O. (1967). Living relative of the microfossil *Kakabekia. Science.* **156,** 1231–1234.

Siesjö, B. K. and Pontén, U. (1966). Intracellular pH-true parameter or misnomer? *Ann. N.Y. Acad. Sci.* **133,** 78–86.

Silver, M. and Lundgren, D. G. (1968a). Sulfur-oxidizing enzyme of *Ferrobacillus ferrooxidans* (*Thiobacillus ferrooxidans*). *Can. J. Biochem.* **46,** 457–461.

Silver, M. and Lundgren, D. G. (1968b). The thiosulfate-oxidizing enzyme of

Ferrobacillus ferrooxidans (*Thiobacillus ferrooxidans*). *Can. J. Biochem.* **46,** 1215–1219.

Silverman, M. P. and Ehrlich, H. L. (1964). Microbial formation and degradation of minerals. *Adv. Appl. Microbiol.* **6,** 153–206.

Singer, P. C. and Stumm, W. (1970). Acidic mine drainage: the rate-determining step. *Science* **167,** 1121–1123.

Singleton, R. and Amelunxen, R. E. (1973). Proteins from thermophilic micro-organisms. *Bacteriol. Rev.* **37,** 320–342.

Skinner, F. A. (1968). Limits of microbial existence. *Proc. R. Soc. Ser. B. Biol. Sci.* **171,** 77–89.

Sletten, O. and Skinner, C. E. (1948). Fungi capable of growing in strongly acid media and in concentrated copper sulfate solutions. *J. Bacteriol.* **56,** 679–681.

Smith, P. F., Langworthy, T. A. and Mayberry, W. R. (1973). Lipids of mycoplasmas. *Ann. N.Y. Acad. Sci.* **225,** 22–27.

Smith, P. F., Langworthy, T. A., Mayberry, W. R. and Houghland, A. E. (1973). Characterization of the membranes of *Thermoplasma acidophilum J. Bacteriol.* **116,** 1019–1028.

Smith, P. F., Langworthy, T. A. and Smith, M. R. (1975). Polypeptide nature of growth requirement in yeast extract for *Thermoplasma acidophilum. J. Bacteriol.* **124,** 884–892.

Smith, G. G., Ruwart, M. J. and Haug, A. (1974). Lipid phase transitions in membrane vesicles from *Thermoplasmaacidophila. FEBS Lett.* **45,** 96–98.

Souza, K. A., Deal, P. H., Mack, H. M. and Turnbill, E. E. (1974). Growth and reproduction of microorganisms under extremely alkaline conditions. *Appl. Microbiol.* **28,** 1066–1068.

Starkey, R. L. and Waksman, S. A. (1943). Fungi tolerant to extreme acidity and high concentrations of copper sulfate. *J. Bacteriol.* **45,** 509–519.

Stotsky, G. (1972). Activity, ecology and population dynamics of microorganisms in soil. *In* "Critical Reviews in Microbiology" (Eds A. I. Laskin and H. Lechevalier), CRC Press, Cleveland.

Sugiyama, T., Smith, P. F., Langworthy, T. A. and Mayberry, W. R. (1974). Immunological analysis of glycolipids and lipopolysaccharides derived from various mycoplasmas. *Infect. Immun.* **10,** 1273–1279.

Suzuki, I. (1965). Oxidation of elemental sulfur by an enzyme system of *Thiobacillus thiooxidans. Biochim. Biophys. Acta* **104,** 359–371.

Tabita, R., Silver, M. and Lundgren, D. G. (1969). The rhodenase enzyme of *Ferrobacillus ferrooxidans* (*Thiobacillus ferrooxidans*). *Can. J. Biochem.* **47,** 1141–1145.

Thimann, K. V. (1963). "The Life of Bacteria" (2nd ed.), Macmillan Co., New York.

Thomas, J. A., Cole, R. E. and Langworthy, T. A. (1976). Intracellular pH measurements with a spectroscopic probe generated *in situ. Fed. Proc.* **35,** 1455.

Touvinen, O. H. and Kelly, D. P. (1972). Biology of *Thiobacillus ferrooxidans* in relation to the microbiological leaching of sulfide ores. *Z. Allg. Microbiol.* **12,** 311–346.

Uchino, F. and Doi, S. (1967). Acido-thermophilic bacteria from thermal waters. *Agric. Biol. Chem.* **31,** 817–822.

Vallentyne, J. R. (1963). Environmental biophysics and microbial ubiquity. *Ann. N.Y. Acad. Sci.* **108,** 342–352.

VanderWerf, C. A. (1961). Acids, Bases, and the Chemistry of the Covalent Bond.

In "Selected Topics in Modern Chemistry" (Eds H. H. Sisler and C. A. Vander-Werf), pp. 33–39. Reinhold, New York.

Vedder, A. (1934). *Bacillus alcalophilus* n. sp., benevens enkele ervaringen met sterk alcalische voedingsbodems. *Antonie van Leeuwenhoek* **1,** 141–147.

Vestal, J. R., Lundgren, D. G. and Milner, K. C. (1973). Toxic and immunological differences among lipopolysaccharides from *Thiobacillus ferrooxidans* grown autotrophically and heterotrophically. *Can. J. Microbiol.* **19,** 1335–1339.

Vishniac, W. and Santer, M. (1957). The Thiobacilli. *Bacteriol. Rev.* **21,** 195–213.

Waddell, W. J. and Bates, R. G. (1969). Intracellular pH. *Physiol. Rev.* **49,** 285–329.

Waddell, W. J. and Butler, T. C. (1959). Calculation of intracellular pH from the distribution of 5,5-dimethyl-2,4-oxazolidinedione (DMO). Application to skeletal muscle of the dog. *J. Clin. Invest.* **38,** 720–729.

Waksman, S. A. and Jåffe, J. S. (1922). Microorganisms concerned in the soil. II. *Thiobacillus thiooxidans*, a new sulfur-oxidizing organism isolated from the soil. *J. Bacteriol.* **7,** 239–256.

Walsh, F. and Mitchell, R. (1972a). An acid-tolerant iron-oxidizing *Metallogenium*. *J. Gen. Microbiol.* **72,** 369–376.

Walsh, F. and Mitchell, R. (1972b). A pH-dependent succession of iron bacteria. *Environ. Sci. Technol.* **6,** 809–812.

Wang, W. S. and Lundgren, D. G. (1968). Peptidoglycan of a chemolithotrophic bacterium, *Ferrobacillus ferrooxidans*. *J. Bacteriol.* **95,** 1851–1856.

Wang, W. S., Korczynski, M. S. and Lundgren, D. G. (1970). Cell envelope of an iron-oxidizing bacterium: structure of lipopolysaccharide and peptidoglycan. *J. Bacteriol.* **104,** 556–565.

Weiss, R. L. (1973). Attachment of bacteria to sulfur in extreme environments. *J. Gen. Microbiol.* **77,** 501–507.

Weiss, R. L. (1974). Subunit cell wall of *Sulfolobus acidocaldarius*. *J. Bacteriol.* **118,** 275–284.

Wiame, J. M., Harpigny, R. and Dothey, R. G. (1959). A new type of Acetobacter: *Acetobacter acidophilum* prov. sp. *J. Gen. Microbiol.* **20,** 165–172.

Wiley, W. R. and Stokes, J. C. (1962). Requirement of an alkaline pH and ammonia for substrate oxidation by *Bacillus pasteurii*. *J. Bacteriol.* **84,** 730–734.

Wiley, W. R. and Stokes, J. L. (1963). Effect of pH and ammonium ions on the permeability of *Bacillus pasteurii*. *J. Bacteriol.* **86,** 1152–1156.

Yamazaki, Y., Koyama, N. and Nosoh, Y. (1973). On the acidostability of an acidophilic thermophilic bacterium. *Biochim. Biophys. Acta* **314,** 257–260.

Chapter 8

Life in High Salt and Solute Concentrations: Halophilic Bacteria*

D. J. KUSHNER

University of Ottawa

I. Introduction 318
II. Natural zones of high salt and solute concentrations 319
III. Biological effects of high solute concentrations 320
IV. Taxonomic distribution of solute-tolerant and halophilic microorganisms 323
A. Isolation of more and less solute-requiring mutants 325
V. Effects of extracellular solutes on internal solute concentrations . . 325
A. "Compatible solutes" 329
VI. Physiology of extreme halophiles 330
A. Nutrition and metabolism 330
B. Enzyme activity 331
C. *In vitro* protein synthesis 334
D. Mechanisms of the effects of salts on enzymes and other salt-dependent proteins of extremely halophilic bacteria 336
E. Peculiarities of the DNA and bacteriophages of extreme halophiles 340
F. External layers of extremely halophilic bacteria 340
VII. Special problems posed by moderately halophilic bacteria and salt-tolerant microorganisms 347
A. *In vitro* protein synthesis 351
VIII. Concluding remarks 354
A. Water and solutes 354
B. What next with the halophiles? 355
Addendum 356
Acknowledgements 357
References 357

* Dedicated to the memory of Dr Norman Gibbons (1906–1977), a pioneer student of halophilic bacteria. I am personally indebted to him for much kindness and for introducing me to the study of microorganisms that live in extreme environments.

I. Introduction

In living cells water serves as a medium of interactions of small and large molecules between themselves and with each other. The solutes dissolved in cell water and the structure of this water control all the vital processes: enzyme action and regulation, assembly and disassembly of organelles, membrane structure and function. Small changes in solute concentration and water activity can result in such large physiological changes that it is hardly surprising that metazoans have evolved physiological mechanisms for keeping constant the composition of their body fluids and, even more, of their intracellular fluids. In mammals, to take only one example, sodium and potassium ions in the blood are kept in balance by elaborate hormonal controls acting at the level of the kidneys and on the exchange between blood and tissues.

Each microbial cell, however, must come to its own terms with its external aqueous environment. We may consider it a kind of "extreme condition" (though a very common one) when cells grow in solutes much more dilute than their internal milieu, as do all freshwater microorganisms. Animals protect their internal cells from osmotic lysis by maintaining solute concentrations in extracellular tissue fluids corresponding to the intracellular concentration. Often their outer covering is quite impermeable to water.* Most microorganisms (as well as plant cells) protect themselves by a rigid cell wall that permits high internal osmotic pressure to build up without causing lysis. Protozoa, with more flexible walls, have dealt with their osmotic problems by concentrating the water that enters in contractile vacuoles and excreting the vacuoles.

In contrast, cells that grow in high solute concentrations do not seem able to keep a much more dilute cytoplasm. Indeed, as pointed out by Brown (1964a, 1976) it would be very difficult for them to do so since this would involve impermeability to water or the continued active excretion of solutes. Though microorganisms can make their internal composition very different from the external one, none are known to maintain overall more dilute conditions within their cells.

This chapter discusses the natural occurrence, classification, and physiology of microorganism which can or must live in the presence of high salt or other solutes, with special reference to the mechanisms they use to adapt to such environments. This topic has been reviewed several times in the past, including the recent past (Flannery, 1956; Brown, 1964a, 1976; MacLeod 1965; Larsen, 1962, 1967, 1973; Ingram, 1957; Kushner, 1968, 1971; Ohnishi, 1963; Lanyi, 1974a; Gibbons, 1970). The reviews cited should be consulted for a detailed survey of the earlier literature. Justifiably, any treatment of this

* And to solutes: I have been informed that many arthropods can survive in fixing fluids for a long time if their outer integument remains intact.

subject must pay special attention to those stars of the show, the extremely halophilic bacteria, on whom the most extensive work has been done. More recently it has been realized that microorganisms that can grow over a wide range of solute concentration can, in their own way, pose just as difficult and fascinating problems as the extreme halophiles.

II. Natural Zones of High Salt and Solute Concentrations

High concentrations of salt and sugar have long been used as food preservatives. Much of the early interest in salt and solute tolerant microorganisms arose from the observations that they could spoil such foods (Ingram, 1957; Scott, 1957). Interest in the red halophiles arose because of their ability to spoil salted fish or hides—although their presence was obvious centuries earlier due to the red colour they imparted to salterns, used for preparing solar salt from sea water (Baas-Becking, 1931). These salterns have long been rich sources for red halophilic bacteria; some of those currently being studied were obtained from salterns; one such is conveniently located near the NASA laboratories at Moffett Field near San Francisco Bay. Natural salt lakes provide other good sources of halophilic bacteria; the Dead Sea has served for some strains used by Israeli scientists, since the pioneering studies of the microflora of this body of water by Volcani (1940). The ionic contents of some salt lakes are presented in Table 1. It should be appreciated that NaCl

Table 1

Ionic composition of some natural salt lakes

Lake	Cations				Anions				Total salts	References
	Na^+	K^+	Mg^{2+}	Ca^{2+}	Cl^-	SO_4^{2-}	HCO_3^-	Br^-		
Great Salt Lake	8·99	0·47	0·8	—	15·0	0·20	—	—	27·5	Brock (1969)[b]
Dead Sea[a]	3·49	0·75	4·19	1·58	20·8	0·54	0·24	—	31·5	Brock (1969)
Lake Assal										
Surface	7·78	0·54	0·80	1·46	16·4	0·23	—	0·47		Brisou *et al.* (1974)
20 m deep	9·1	0·63	2·92	1·58	24·6	0·15	—	0·51		

Concentration in g/100 ml

[a] Brisou *et al.* (1974) give figures for the Dead Sea showing that concentrations at a depth of 50 m are substantially greater than those on the surface. Their figures suggest that those cited by Brock (1969) are surface values

[b] Other lakes cited by Brock (1969) contain more than 7·5% $NaHCO_3$ as practically the only solute, or contain mainly $MgSO_4$ (total salinity >25%). His article also contains references to hypersaline lagoons

is not the major salt in all such lakes; the Mg^{2+} content of the Dead Sea is higher than its Na^+ content, and there are other fascinating salt lakes (Table 1) on which few if any microbiological studies have been carried out. The lakes seem devoid of life above the microbial level, but many bacteria and other microorganisms have been found in them. As pointed out by Brock (1969) this does not mean that such microorganisms normally live in these lakes. In the study by Brisou *et al.* (1974) of Lake Assal, in a remote almost inaccessible part of French Somaliland, a number of bacteria were found with fresh-water and euryhaline species (growing in 1 to 5% NaCl) predominating; moderately and extremely halophilic forms were also found. Most of the species isolated could not grow in the salt concentrations found in the lake and it seems most likely that only the halophiles are native to the lake, the others being carried in by the few streams that flow into it. The high proportion of bacilli found suggests that these species survive well in the very salt water. For a time it was thought that bacteria could grow in the Don Juan pond in the Antarctic, where water contains saturated calcium chloride water activity (a_w) about 0·45, which never freezes. More recent results strongly suggested that bacteria found in the pond were carried there by outside streams (Horowitz, 1976). An a_w of 0·62 may still remain the limit for microbial growth.

In addition to these natural or man-made lakes, certain food products provide rich media for the growth of salt and solute tolerant microorganisms. Soy sauce and miso paste (which contain approximately 18% and 7 to 20% NaCl respectively (Onishi, 1963) have provided many salt and solute tolerant bacteria and yeasts studied by Japanese workers. Fortunately for the consumers, these food products contain salt tolerant microorganisms, but not any extreme halophiles. The fermentations needed to produce soy and mizo sauce take place in the dark. The red halophiles seem to have a selective advantage in bright sunlight because of their characteristic pigments which both protect them from visible light and help in photoreactivation after ultraviolet irradiation (Dundas and Larsen, 1962; Hescox and Carlberg, 1972); but they grow slowly, and we would not expect them to predominate in a dark environment, where the salt concentration is below their optimum and the conditions anaerobic.

III. Biological Effects of High Solute Concentrations

Effects of high solute concentrations may be due to the solutes themselves, or to the effect the solutes have on water activity, a_w. Everything that dissolves in a solvent attracted is to the molecules of that solvent and thereby reduces their freedom. The water activity (a_w) may be defined by the equation

$$a_w \equiv \frac{P}{Po} \equiv \frac{n_2}{n_1 - n_2},$$

where p = the vapour pressure of the solution and Po that of the solvent, pure water; n_1 and n_2 are the number of moles of solvent and solute respectively. More detailed treatments of the concept of "water activity" and ways of calculating it for solutions of known concentrations are found in the reviews of Scott (1957), Kushner (1971) and Brown (1976). The a_w's of solutions in which microbial growth is possible, are shown in Fig. I. Effects of a_w on microbial growth are determined by adding such solutes to growth media or

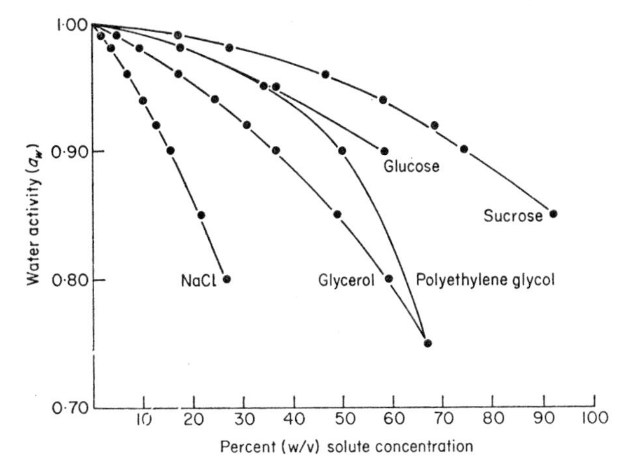

Fig. 1. Relation between the concentrations of different solutes (expressed as g per 100 ml solution) and the water activity (a_w) of these solutions.

(especially when studying the water relations of fungi that can grow under very dry conditions) by equilibrating a small volume of growth medium against a much larger volume of salt or H_2SO_4 solution of the desired water activity (Scott, 1957). Table 2 presents a survey of the limiting a_w for growth of different microorganisms over the whole a_w range in which life is possible.

For a few microorganisms it has been found that the lowest a_w in which growth is possible is the same whether produced by ionic or non-ionic solutes. For example, the fungus *Aspergillus amstelodamii* could not grow above an a_w of 0·99 and grew best at an a_w of 0·96, whether the latter was adjusted by adding sucrose, glucose, glycerol, $MgCl_2$ or NaCl (Scott, 1957). Similarly, the limiting a_w for growth of food-poisoning strains of *Staphylococcus aureus* and of several salmonella strains was the same whether the solute used was sugars, salts, or dried soups, meat or milk powders (Scott, 1957). The nutritional richness of the medium could extend the minimum a_w in which *Salmonella oranienburg* grows (Christian, 1955) just as it could extend the minimum and maximum a_w values (upper and lower NaCl concentrations) in which the moderately halophilic bacteria, *Vibiro costicola* and *Paracoccus* (formerly *Micrococcus*) *halodenitrificans* could grow (Forsyth and Kushner,

Table 2
Limiting water activities (a_w) *for microbial growth*

Organism	Lower limit	Range for best growth
Fungi		
Rhizopus nigricans	0·93	
Penicillium (8 sp.)	0·80–0·90	
Aspergillus flavus	0·90	
A. niger	0·84	0·96–0·98
A. amstelodamii		0·70–0·95
Sporendonema sebi	0·77	
Xeromyces bisporus[a]	0·65	0·90–0·97
Yeasts		
Saccharomyces cerevisiae	0·94 (in NaCl)	
	0·92 (in glucose)	
	0·90 (in agar, equilibrated against salt solution)	
Candida pseudotropicalis	0·93 (in NaCl)	
Saccharomyces rouxii[b]	0·860 (in NaCl)	
	0·857 (in sucrose)	
	0·845 (in glucose)	
Bacteria[c]		
Bacillus subtilis	0·949	
Pseudomonas aeruginosa	0·945	
Escherichia coli	0·932	
Vibrio metchnikovii	0·995–0·96	
Lactobacillus[b]	0·90	
Staphylococcus aureus (14 sp.)	0·86–0·88	
Moderate halophiles		
Vibrio costicola	0·86–0·98 (in NaCl)	
Paracoccus halodenitrificans	0·86–0·98 (in NaCl)	
Extreme halophiles		
Halobacteria and halococci	0·75–0·88 (in NaCl)	

[a] *X. bisporus* will not grow at an a_w greater than 0·97
[b] From Brown (1976). Brown assigns *Saccharomyces rouxii* a much lower limiting a_w value (0·60) in sugar
[c] Except for the halophiles, most of these bacteria grow best in media such as nutrient broth ($a_w = 0{\cdot}999$) without added solutes. *V. metchni Kovii* and *V. cholerae* will not grow in nutrient broth unless salt is added

1970). In other cases, the nature of the external solute certainly makes a difference in the lower limit of a_w permitting growth. For a striking example, *Saccharomyces rouxii* can grow at a much lower a_w in sugars than in salts (Table 2). We cannot always distinguish in advance between specific solute effects and effects on availability of water.

IV. Taxonomic Distribution of Solute-Tolerant and Halophilic Microorganisms

Salt and solute-tolerant organisms are widely-distributed among the bacteria, fungi, yeasts, algae and protozoa and among the viruses for some of these organisms (Tables 2 and 3); other examples are given in the text. The most

Table 3

Salt response of different microorganisms[a]

Category	Reaction	Examples
Non-halophile	Grows best in media containing less than 0·2 M salt[b]	Most normal eubacteria and most freshwater microorganisms
Slight-halophile	Grows best in media containing 0·2–0·5 M salt	Many marine microorganisms
Moderate-halophile	Grows best in media containing 0·5–2·5 M salt. Organisms able to grow in less than 0·1 M salt are considered facultative halophiles	Bacteria and some algae (see Table 6 for more complete list)
Borderline extreme halophile[c]	Grows best in media containing 1·5–4·0 M salt	*Ectothiorhodospira halophila* *Actinopolyspora halophila*
Extreme halophile	Grows best in media containing 2·5–5·2 M salt (saturated)	The "red halophiles", halobacteria and halococci
Halotolerant	Non-halophile which can tolerate salt. If the growth range extends above 2·5 M salt, it may be considered extremely halotolerant	*Staphylococcus aureus* and other staphylococci; solute-tolerant yeasts and fungi[d]

[a] Classification extended from those of Larsen (1962) and Shindler (1976)

[b] "Salt" is usually NaCl, but can be other salts in addition to a minimum amount of NaCl

[c] A halophilic blue green alga, *Aphanothece halophytica*, unable to grow in less than 1·5 M NaCl was described by Miller *et al.* (1976), and Brock (1976)

[d] A halotolerant blue green alga, *Cocchochloris elabens*, can grow in 25% NaCl (4·3 M) Kao *et al.*, (1973). *Staphylococcus epidermidis* can grow in 4·0 M NaCl (Komaratat and Kates, 1975)

distinctive organisms in this group are the extremely halophilic bacteria of the genera *Halobacterium* and *Halococcus* all of which can grow in saturated NaCl (*c.* 32% or 5·2 M). They are critically characterized by the lower limit of NaCl required for growth of 12 to 15% (Buchanan and Gibbons, 1974), since quite a number of organisms that can grow in saturated NaCl require much lower amounts than the extreme halophiles. The halobacteria and halococci have a number of distinctive biochemical properties which set them apart from other forms of microbial life. A recently discovered actinomycete *Actinopolyspora halophila* (Gochnauer *et al.*, 1975) whose minimum NaCl requirement for growth is 10% on solid medium and 12% on liquid medium may also be considered an extreme halophile, if a borderline one. As will be seen, it lacks some of the distinctive biochemical "markers" of other extreme halophiles. A borderline status should also be given the photosynthetic bacterium, *Ectothiorhodospira halophila* which can grow at slightly lower NaCl concentrations than the extreme halophiles.

Moderate halophiles were first described by Baxter and Gibbons (1956). Several food-spoilage organisms have been classed as moderate halophiles, being able to grow in the range 0·50 to 3·5 M NaCl (about 3 to 20%). This definition also fits many marine bacteria, many of which require about 0·5 M (3%) NaCl for growth (MacLeod, 1965) and have been found on further examination to withstand, 20, 25 or even 30% NaCl (Forsyth *et al.*, 1971).

Recent findings have shown that the definition of the lower salt concentration for growth of moderately halophilic bacteria is not complete unless the temperature is also specified. At 20°C the newly described species, *Planococcus halophilus* (Novitsky and Kushner, 1976) grows in the virtual absence of Na^+ ions (in 0·01 M, the amount found in the growth medium). If the temperature is raised to 25°C of higher, at least 0·5 M NaCl is needed for growth and NaCl cannot be replaced by KCl or by non-ionic solutes (Novitsky and Kushner, 1975). If growth experiments were carried out at 20°C, this organism (which can grow in 4·0 M NaCl) would be classified as a highly salt tolerant organism, whereas at 25°C it must be considered a moderate halophile.

These experiments raise the question: how many true moderate halophiles exist? Will others become merely salt tolerant at lower temperatures? *Vibrio costicola*, probably the best studied moderate halophile, has been re-examined in this respect. This organism grows in the NaCl range 0·5 to 4·0 M at 30°C. At 20°C it can grow down to 0·2 M NaCl, but no lower (Shindler, unpublished observations). In terms of non-halophilic bacteria, this might be considered a substantial salt requirement, but it is certainly less than the requirement at the higher temperature. Other "moderate halophiles" should certainly be re-examined in this way. Our experiments suggest that their salt requirement might be lowered or abolished at lower temperatures.

The reasons for the findings with *P. halophilus* and *V. costicola* are not known, but they are by no means the only examples of temperature changing a solute requirement—or rather of solutes changing temperature response. A number of temperature-sensitive (T_S) mutants of *Escherichia coli* were protected at the non-permissive temperature by both ionic and non-ionic solutes (Bilsky and Armstrong, 1973). There was no special ionic requirement for protection in these experiments. Rather, it seemed another example of the general phenomenon of "osmotic protection" of certain mutants which, as already noted for several species, can be protected by solutes from the lethal consequences of a mutation.

The so-called *rod* mutants of *Bacillus subtilis* and *B. licheniformis* provide interesting examples of organisms that depend on salts or solutes for normal morphology. These mutants grow as normal cells in the presence of 0·8 M NaCl, but in the absence of NaCl grow in quite abnormal ways, as strings of coccoid bodies with distorted walls—though apparently many of these are still viable. At high temperatures (45°C) the *rod* mutants but not the normal cells required NaCl to grow (Rogers *et al.*, 1970.)

A. Isolation of More and Less Solute-Requiring Mutants

It will be clear from subsequent descriptions that there are many genetic differences between extremely halophilic bacteria and all other micro-organisms. It is hardly surprising that past efforts to convert extreme halophiles into moderate ones, and vice versa, have failed (Larsen, 1967; Kushner, 1968). Those working with some eukaryotic organisms have been more lucky, and it may soon be possible to study the genetic basis of solute tolerance and requirement. The salt range of more and less halophilic *Dunaliella* species has been extended in both directions (Brown, 1976). Koh (1975a, b) was able to isolate a number of obligately osmophilic mutants of *Saccharomyces rouxii*. Many of the mutations apparently involved only a single gene change. Most osmophilic mutants seemed to have weakend cell envelopes; one, studied in more detail (Koh, 1975b) proved to have significant changes in the chemical composition of its envelope. The osmolarity-temperature relationships of Koh's mutants were not studied. However, a number of temperature-conditioned obligate osmophiles have been isolated earlier (Onishi, 1963).

V. Effects of Extracellular Solutes on Internal Solute Concentrations

Adaptation of most microorganisms to high external solute concentrations is probably never achieved by maintaining a low internal solute concentration. The global water activity inside cells is never higher than that outside, and is frequently lower. This conclusion (Ingram, 1957; Larsen, 1962 and 1967; Brown 1964a and 1976; Kushner, 1964a, and 1968) follows from measurements of the freezing point of internal solutes of cells grown at different salt concentrations, and from a number of direct chemical analyses to be described. The simple fact that cells can be centrifuged down from high solute concentrations is good evidence that they contain high solute concentrations: otherwise they would be less dense than the external medium (Ingram, 1957). It will be seen, however, that whatever the overall solute content, the water in a cell can by no means be considered a "simple" solution of the ions and compounds present.

The known internal solute concentrations of a number of microorganisms that are able to grow at high solute concentrations are listed in Table 4.

Table 4

Internal ionic concentrations of bacteria growing in different NaCl concentrations

Bacterium	Ion concentration (M) in medium Na^+	in medium K^+	cell-associated Na^+	cell-associated K^+	References
Vibrio costicola[a]	1·0	0·004	0·684	0·221	Christian and Waltho (1962); Shindler *et al.* (1977)
	0·6	0·008	0·505	0·524	
	1·0	0·008	0·584	0·661	
	1·6	0·008	1·09	0·594	
	2·0	0·008	0·898	0·567	
Paracoccus halodinitrificans	1·0	0·004	0·311	0·474	Christian and Waltho (1962)
Pseudomonas 101	1·0	0·0055	0·90	0·71	Masui and Wada (1973)
	2·0	0·0055	1·15	0·89	
	3·0	0·0055	1·04	0·67	
Marine pseudomonad B-16	0·3		0·123	0·374	Thompson and Macleod (1973)
Unidentified salt-tolerant rod[b]	0·6	0·04	0·05	0·34	Matheson *et al.* (1976)
	4·4	0·04	0·62	0·58	
Halobacterium cutirubrum[c]	3·33	0·05	0·80	5·32	Matheson *et al.* (1973)

Values for Mg^{2+} contents (which vary little with external Na^+ concentration) and for Cl^- concentration are included in some of these papers, but complete ionic balances are not known. Harold and Altendorf (1974) have pointed out that a complete ionic balance has been made only for *E. coli*

[a] The difference between the K^+ concentration found by the two laboratories may be due to the fact that Christian and Waltho (1962) measured ions in stationary-phase cells and Shindler *et al.* (1977) those in exponentially growing cells, re-aerated after centrifuging.

[b] G + C analyses suggest this is a pseudomonad (Matheson, personal communication)

[c] Several workers have reported similar or higher results for K^+ in *H. cutirubrum* or *H. halobium*

Perhaps the most striking and intriguing example of adaptation by maintenance of high internal solute concentrations is provided by the extremely halophilic bacteria which can grow in saturated NaCl but contain concentrations of K^+ which, as KCl, would more than saturate the available water. It has been calculated that K^+ may constitute 30 to 40% of the dry weight of *Halobacterium halobium* (Gochnauer and Kushner, 1971). The contents can differ with the physiological state of the cell, being highest in the early stages of growth (Gochnauer and Kushner, 1971).

In the halobacteria the gradient of intracellular/extracellular K^+ can be as high as 1000 : 1 (Ginzburg *et al.*, 1971a). The demand for K^+ ions can be so high that this ion can easily become growth-limiting, and high growth yields depend on substantial K^+ ion concentrations, at least 1 mg/ml (26 m M) in the medium (Gochnauer and Kushner, 1969). However, there is no evidence that large energy expenditures are needed to maintain such gradients. They can be maintained in the cold, in cells apparently carrying out very little metabolism (Ginzburg *et al.*, 1971a). Cells of *H. halobium* could live for days in a 25% NaCl solution without K^+ added to the external medium, or any added source of energy and maintain their K^+ content unchanged. They lived longer thus if Mg^{2+} and Ca^{2+} ions were also present. Only when the cells died did they lose K to the external medium (Gochnauer and Kushner, 1971). Halobacteria can also lose K^+ on treatment with trinitrocresolate or other ions (Ginzburg *et al.*, 1971b).

Such experiments have suggested that K^+ might be bound in some way inside the cell. Nuclear magnetic resonance studies suggested that cell-bound K^+ was complexed by fixed charges or was solvated in semi-crystalline cell water (Cope and Damadian, 1970). However, if such binding does exist it seems very easily broken. Lanyi and Silverman (1972) found that the K^+ activity associated with lysed *H. cutirubrum* was almost identical to that of equimolar solutions of KCl, though the lysed cells still kept their ability to bind Mg^{2+}. Thus, any substantial binding ability existed only in intact cells.

Interpreting of the "freedom" of K^+ ions in the cytoplasm of extreme halophiles and other bacteria is complicated by the probability that cytoplasmic water has a more ordered structure than external water (Harold and Altendorf, 1974); cytoplasmic ions such as K^+ and Na^+ may be less free than in external solutions, but still free enough to be osmotically and physiologically active. As we will see, the behaviour of intracellular enzymes and protein-synthesizing systems of extreme halophiles is consistent with the total ionic concentrations measured being those that are active in the cell; this conclusion does not necessarily hold for moderate halophiles.

Though the internal salt concentrations within moderately halophilic bacteria might be expected to be as high as those outside, and though the first measurements by Christian and Waltho (1962) tended to confirm this for *V. costicola* and *P. halodenotrificans* (Table 4) more recent measurements show that in other bacteria the internal salt concentrations can be substantially lower than the external one.

Masui and Wada (1973) found that in the moderate halophile, "*Pseudomonas* 101", growing in the range 1·0 to 3·0 M NaCl, the intracellular Na^+, K^+, and Cl^- concentrations were almost independent of the NaCl concentration of the medium. They also found that isolated cell envelope preparations could concentrate both Na^+ and K^+ ions. It is not certain from their results if this represented binding to the envelopes or uptake into envelope vesicles.

In the unidentified Gram negative rod studied by Matheson *et al.* (1976) (Table 4) the internal ionic concentration was usually much lower than that outside. As will be seen, other microorganisms can maintain lower internal ionic concentrations than those of the growth medium, using non-ionic solutes to lower the intracellular water activity. This is probably true for the above moderate halophiles, though so far no measurements have been made of their non-ionic solutes.

Regardless of the total ionic concentrations in moderate halophiles, the concentration of K^+ ions inside the cells was substantially higher than that outside, and the Na^+ concentrations somewhat lower (Table 4). Such figures must be interpreted with caution, since the phase of growth at which cells are harvested and the treatment after harvesting can greatly affect the measured intracellular ionic composition. We found (Shindler *et al.*, 1977) that cells of *V. costicola* harvested in mid-logarithmic phase had higher K^+ and lower Na^+ contents than those harvested in the stationary phase. Furthermore, cells could apparently lose substantial amounts of K^+ and gain substantial amounts of Na^+ during harvesting and washing (a necessarily long process, since for accurate determinations of intercellular and intracellular water relatively large amounts of pelleted cells are needed). If concentrated cell suspensions of *V. costicola* were aerated with an energy source, such as glucose, K^+ ions were taken up and Na^+ ions released. Under anaerobic conditions, or in cells poisoned by metabolic inhibitors, the process was reversed. These experiments strongly suggest that the anaerobic conditions which can very easily accompany centrifuging and washing, could substantially change the ionic contents of those found in growing cells. The values we have obtained for intracellular contents of *V. costicola* are substantially greater than those given earlier by Christian and Waltho (1962), probably because of the differences in time of harvesting or of handling. Our experiences have convinced us that any measured values for internal salt concentrations of this organism, and probably other moderate halophiles, are at the best close approximations of those in actively growing cells.

The concentration of Mg^{2+}, the major divalent cation of *V. costicola*, and as far as is known of other halophilic bacteria, changed relatively little with times of growth and conditions of handling (Shindler *et al.*, 1977).

V. costicola differs from the other moderate halophiles in that the total concentrations of monovalent cations vary with the external salt concentration, being highest in cells grown at the highest NaCl concentrations. However, the former do not change as much as the latter: in cells growing in a three-fold range of external salt concentration the internal total salt concentration changes only about 1·5 times.

A. "Compatible Solutes"

Although it is virtually impossible for most microorganisms growing in high solute concentrations to maintain intracellular contents much more dilute (that is, of much higher water activity) than the external one (Brown, 1964a, 1976), some organisms do grow in concentrations of salts that can inhibit many of their essential intracellular enzymes. This is true for *Saccharomyces rouxii* (Onishi, 1963) and for the halophilic green alga *Dunaliella viridis* which can grow over a wide range of NaCl concentrations, up to saturation, but whose enzymes are as sensitive to high salt concentrations as those of freshwater algae (Johnson *et al.*, 1968). Different *Dunaliella* species contain very large amounts of intracellular glycerol, the amount varying with the external salt concentrations (Ben-Amotz and Avron, 1973; Borowitzka and Brown, 1974). The variation in intracellular glycerol concentration appears to depend on active synthesis and destruction of glycerol in response to different external salt concentrations. Possible mechanisms are discussed by Brown (1976). Borowitzka and Brown (1974) also compared the internal glycerol concentrations of the marine alga *Dunaliella terticolecta*, which can grow in 1·36 M NaCl, and the more halophilic *D. viridis* which can grow in 1·7 M NaCl and higher. The internal glycerol concentrations of the halophilic algae could reach 4·4 mol kg^{-1} those of the marine one, 1·4 mol kg^{-1}. The internal Na^+ content of these algae was difficult to measure accurately (because of the large amounts of intercellular Na^+ in the cell pellets used for analysis) but was assumed to be low on indirect evidence (Borowitzka and Brown, 1974). Internal Na^+ and Cl^- concentrations of a halophilic *Chlamydomonas* were also found to be much lower than the external ones (0·2 M for cells growing in 1·7 M NaCl; Okamoto and Suzuki, 1964).

Borowitzka and Brown (1974) pointed out that the glucose-6 phosphate dehydrogenase and a NADP-specific glycerol dehydrogenase of *D. tertiolecta* and *D. viridis* were inhibited by high NaCl, but not by high glycerol, concentrations. In fact, glycerol partially protected these enzymes from inhibition by high salt concentrations. These authors suggested that glycerol plays the role of a "compatible solute," one which maintains low a_w value of cells grown in high NaCl concentrations, while permitting enzyme activity to continue.

Polyols, including glycerol and arabitol, may act as compatible solutes for salt and sucrose-tolerant yeasts (Brown and Simpson, 1972; Brown, 1974). In the tolerant yeast, *Saccharomyces rouxii*, a high intracellular polyol concentration was associated with growth in high salt or sucrose concentrations. Non-tolerant yeasts such as *S. cerevisiae* did not contain high concentrations of polyols. High concentrations of salt and sucrose inhibited the isocitrate dehydrogenase of both *S. rouxii* and *S. cerevisiae*, but glycerol did not (Brown, 1976). These results suggest that polyols may permit enzyme activity in low a_w in salt or sucrose tolerant yeasts.

Brown's group (Brown, 1976) is examining the effects of non-ionic solutes on enzyme activity, a subject on which only a few previous studies exist. The effects of such solutes are clearly not due only to a lowering of a_w nor to hydrophobicity of the inhibitor molecules, but rather to their specific structures. For one example, sucrose is a much stronger enzyme inhibitor than maltose. Brown's (1976) review should be consulted for a detailed treatment of this subject and of the physiology of organisms that grow in high concentrations of non-ionic solutes.

VI. Physiology of Extreme Halophiles

A. Nutrition and Metabolism

Extremely halophilic rods and cocci have complex nutritional requirements. In contrast, *A. halophila* can grow on much simpler media (Gochnauer *et al.*, 1975). Defined growth media for *Halobacterium* all contain a number of amino acids, and growth may be stimulated by the presence of vitamins as well (Dundas *et al.*, 1963; Onishi *et al.*, 1965; Gochnauer and Kushner, 1969). Because of the very high K^+ concentration of these bacteria, this ion itself may be growth limiting unless it is present in substantial concentrations; otherwise, growth stops when all the K^+ in the medium becomes cell-bound (Gochnauer and Kushner, 1971). Quite recently Grey and Fitt (1976a) found that growth of *H. cutirubrum* is much better if only L-amino acids are used (rather than twice the quantity of DL-amino acids), suggesting that certain D-amino acids may inhibit growth.

It was thought for some years that most of the halobacteria did not metabolize carbohydrates. This was based on their inability to produce acid from glucose and other carbohydrates in standard laboratory tests. Only one species, *H. marismortui* had ealier been reported to produce acid from glucose in Bergeys Manual (7th ed.) (Breed *et al.*, 1957); but this species was listed as no longer in existence in culture collections in the 8th ed. (Buchanan and Gibbons, 1974).

Gochnauer and Kushner (1969) pointed out that carbohydrates stimulated growth of *H. halobium*, suggesting that these substances might be used, even though the pH did not fall. More recently Tomlinson and Hochstein (1972a, b, 1976) have isolated a number of *Halobacteria* that produce acid from carbohydrates. One of them has been designated *H. saccharovorum*. Tomlinson *et al.* (1974) showed that this organism metabolizes glucose by the Entner-Doudoroff pathway, and it can also produce galactonic acid from galactose lactobionic acid from lactose (Hochstein *et al.*, 1976). Labelled carbon was incorporated into lipids of *H. cutirubrum* containing ^{14}C-glucose (Deroo, 1974).

B. Enzyme Activity

The internal machinery of the extreme halophiles seems very well adapted to the high salt concentrations, and in some cases to the high concentrations of K^+ ions prevailing in the cytoplasm. This is shown first by the fact that most of the many enzymes from these organisms that have been studied require high salt concentrations for activity, stability, or both. This subject has been reviewed by Lanyi (1974) and Cazzulo (1975).

There is not just one pattern of behaviour of enzymes of extreme halophiles towards salts. Rather, one can distinguish three basic types of salt response: (1) those needing at least 1 M monovalent salt for activity and showing maximal activity in from 2 to 4 M salt. Examples include aspartate transcarbamylase (ATCase) many amino acid-activating enzymes, and enzymes associated with the cell membranes; (2) those such as NADP-specific isocitrate dehydrogenase, which show maximal activity at 0·5 to 1·5 M salt and are inhibited by higher salt concentrations, and (3) those which show the highest activity in the absence of salt and which are strongly inhibited by high salt concentrations. These last include fatty acid synthetase (Pugh *et al.*, 1971) and under certain conditions the DNA-dependent RNA polymerase of *H. cutirubrum* (Louis and Fitt, 1972a, b). Fatty acid synthetase is present in relatively small amounts, and its inhibition by high salt concentrations may further explain why only trace amounts of fatty acids are formed by these organisms (Kates *et al.*, 1966). The RNA polymerase isolated by Louis and Fitt (1972a, b) is a curious enzyme. It needs high salt concentrations for its activity if *H. cutirubrum* DNA is the template, but if calf thymus DNA is the template it is most active in the absence of salt and is inhibited 90% in the presence of 0·5 M or higher NaCl or KCl. The molecular weight of this enzyme, which is composed of two subunits, is only about 36,000, which is much smaller than other known DNA-dependent RNA polymerases. Chazan and Bayley (1973), however, partially purified from a DNA-protein-membrane complex a polymerase of much higher molecular weight (300,000 to 400,000) and different salt response. A more recent study (Fitt *et al.*, 1975) suggests that when highly purified both enzymes are identical. Neither catalyses extensive RNA chain elongation. Apparently the RNA-synthesizing system in these bacteria is still incompletely understood.

1. Specific Cation Effects on Enzymes

The salts most studied in relation to extremely halophilic enzymes are NaCl and KCl, the major salts outside and inside the cell, respectively. More than half the enzymes so far studied respond about equally to the two salts, while others show substantially more activity in the presence of KCl than of NaCl (Lanyi, 1974a). Li^+ and NH_4^+ salts can also support the activity of some enzymes of extreme halophiles. The malic enzyme of *H. cutirubrum* is especially interesting in being optimally active in 1 M NH_4Cl, having about 60% maximal activity in 3 M KCl and none in any NaCl concentration up to 5 M (Cazzulo and Vidal, 1972).

Some of the enzymes that require high concentrations of monovalent salts for activity can be activated by much lower concentrations of divalent salts, such as those of Mg^{2+} or Ca^{2+} or of polyvalent ions such as spermidine or spermine (Lanyi, 1974a). Examples include citrate synthase (Cazzulo, 1973), and aspartate transcarbamylase (Norberg *et al.*, 1973).

Past studies have not always distinguished clearly between salt effects on activity and stability; it is necessary to do so to interpret the effects of salt on enzyme activity, especially when the assay period is long. Sometimes stability is much more dependent than activity on high salt concentrations. For one example, the aspartate transcarbamylase (ATCase) of *H. cutirubrum* is about four times as active in 3·5 M NaCl as in 2 M NaCl but almost 500 times as stable, judged by the half-life (8 h and 1 min) of the enzyme at the two salt concentrations (Norberg *et al.*, 1973). For the malic enzyme of this organism, NH_4Cl is the best activator, but is not nearly as good as NaCl (which does not activate the enzyme at all) in maintaining enzyme stability (Cazzulo and Vidal, 1972). Hubbard and Miller (1969) also pointed out that different salts had contrasting effects on the activity and stability of isocitric dehydrogenase of *H. cutirubrum*.

2. Regulatory Properties of Enzymes of Extremely Halophilic Bacteria

Regulation by allosteric effectors and inhibitors depends on interactions between enzyme subunits. Only a few studies have been carried out thus far with the regulatory enzymes of extreme (or other) halophiles and never with the highly purified enzymes.

The ATCase of *H. cutirubrum* was one of the first such enzymes studied (Liebl *et al.*, 1969; Norberg *et al.*, 1973). This enzyme depends on high concentrations of salt for both activity and regulation. Maximum activity and regulation (as measured by feedback inhibition by CTP) occur at 3 M salt.

Both NaCl and KCl are approximately equally effective in maintaining activity, but feedback inhibition is slightly stronger in KCl than in NaCl (Liebl, unpublished observations). As the salt concentration is lowered below 2 M the activity falls but the feedback inhibition falls more, so that the remaining enzyme activity is hardly inhibited by CTP. The behaviour of ATCase from yeast is practically a "mirror image" of that from *H. cutirubrum*, maximal activity and feedback inhibition (by UTP) are obtained in the absence of salt. As the salt concentration rises, the activity is lost but feedback inhibition is lost more rapidly. No activity remains at salt concentrations higher than 1·0 M (Liebl *et al.*, 1969).

The ATCase of *H. cutirubrum* has only been partly purified. It is very unstable, even in 2·0 M NaCl, and some of its kinetic properties change with time (Norberg *et al.*, 1973). Its instability interfered with attempts to determine if the loss of regulatory properties at lower salt concentrations were due to dissociation into catalytic subunits. Catalytic and regulatory activity could not readily be separated in *H. cutirubrum* ATCase. Heating at 50°C, which can separate catalytic from regulatory subunits in yeast ATCase (Kaplan and Messmer, 1969) so that feedback inhibition disappears before enzyme activity does, had an opposite effect on the *H. cutirubrum* enzyme. Other treatments, successful with mesophilic enzymes, failed to separate activity from regulation in the halophilic ATCase. However, treatment of the enzyme in a high salt concentration with polyethylene glycol lowered the molecular weight from about 160,000 to 34,000 (as determined by chromatography on agarose columns) apparently splitting it into four subunits. These low molecular weight forms were no longer inhibited by CTP. The results suggest that in this enzyme the same dependence on a multisubunit structure for regulation exists as in the enzyme from more "normal" microorganisms. Other examples of dissociation of multi-subunit enzymes by non-ionic solutes are discussed by Brown (1976).

Lieberman and Lanyi (1972) found that ADP, an allosteric effector of the catabolic threonine deaminase (TDase) of *H. cutirubrum*, greatly changed the salt response of the enzyme. In the absence of ADP the enzyme worked well over a wide range of NaCl concentrations, with an optimum at 2 to 3 M; in the presence of ATP, NaCl inhibited the enzyme. The malic enzyme of *H. cutirubrum* which plays a central role in metabolism, is inhibited by acetyl CoA, NADH, oxalacetate and glyoxylate. These inhibitors, especially NADH (Vidal and Cazzulo, 1976), were more effective in 3 M KCl than in 1 M NH_4Cl, even though activity was twice as great in the latter salt (Cazzulo and Vidal, 1972). Vidal and Cazzulo (1976) pointed out that the halophilic malice enzyme was, except for its salt requirement, similar to the corresponding enzymes from non-halophilic organisms in both kinetic and regulatory properties.

Citrate synthases of Gram negative bacteria (molecular weight about 250,000) are inhibited by low concentration of NADH and reactivated by AMP. The same enzyme from Gram positive bacteria and eucaryotes (molecular weight about 80,000) are not under such regulatory control. The citrate synthase from *H. cutirubrum* was only inhibited by very high NADH concentrations, and inhibition was greater in 45 mM $MgCl_2$ than in 2·6 M KCl, in both of which maximal activity was expressed (Cazzulo, 1973). Thus, it seems unlikely that regulation by NADH is physiologically significant in *H. cutirubrum*. The molecular weight of the enzyme is about 75,000; in this, it resembles citrate synthase of eukaryotes and Gram positive bacteria rather than Gram negative ones. This enzyme has since been highly purified (400 times), though not quite to homogeneity (Higa and Cazzulo, 1975).

C. *In Vitro* Protein Synthesis

The protein synthesizing system of *H. cutirubrum* shows how far adaptation to a unique internal environment can go. Work on this subject, which began with a study of the ribosomes of *H. cutirubrum* (Bayley and Kushner, 1964) has recently been summarized by Bayley (1976). The following account is largely drawn from his review. The ribosomes of *H. cutirubrum* (and probably of other halobacteria and halococci) are unique in their high salt requirement for stability, and in their possession of large numbers of acidic proteins. To maintain stable structures of the monomers (70S) and subunits (30S and 50S forms) high concentrations of both KCl (3 M) and Mg^{2+} ions (*c*. 0·1 M) are needed. KCl cannot be replaced by NaCl; at high NaCl concentrations an irregular aggregation of ribosomes occurs. At low salt concentrations, e.g. in 0·001 M $MgCl_2$ which is able to maintain the structure of *E. coli* ribosomes, most of the proteins are lost from the 30S and 50S subunits, leaving rRNA with a few basic protein attached.

Protein synthesis, as measured by incorporation of radioactive amino acids by cell free systems, can take place optimally in a system containing 3·8 M KCl, 1·0 M NaCl, 0·4 M NH_4Cl and 0·04 M Mg acetate. Different experiments have shown that ionic dependence is due to the protein components of the protein-synthesizing system. As noted above, ribosomes specifically require high K^+ and Mg^{2+} concentrations for stability. A number of amino acyl tRNA synthetases require 3·8 M KCl for optimal activity. The sites of action of the other monovalent cations are still uncertain.

In contrast, the tRNA's from extreme halophiles and from mesophiles do not seem to control the salt response of protein synthesis in these organisms. The amino acyl tRNA synthetases from *H. cutirubrum* could use the tRNA's from *E. coli*, and vice versa. Once the tRNA's of each organism had been charged by the homologous synthetase, the amino acid could be transferred to polypeptides using ribosomes and transfer factors from the other micro-organism.

Bayley and his co-workers considered the attractive possibility that high salt concentrations caused a misreading of the genetic code, specifically one leading to an increased production of acidic amino acids and a decreased production of basic ones. However, experiments with a number of synthetic messengers showed that the codon assignments were essentially the same as those found in non-halophilic organisms. The special properties of proteins of the extreme halophiles seem due rather to the special genetic information of these bacteria that to a misreading of more "ordinary" genetic information.

The protein synthesizing systems of *H. cutirubrum* may, in addition to its high salt requirement, be unusual in that protein synthesis is initiated with a non-formylated methionyl-tRNA Met; in this it resembles eukaryotic rather than prokaryotic protein-synthesizing systems.

A further unusual property of protein synthesis in extreme halophiles is its insensitivity to antibiotics that affect ribosomes. Moore and McCarthy (1969a) found that "standard antibiotics" did not affect the growth of halophiles. Tomlinson and Hochstein (1976) found that *H. saccharovorum* was insensitive to chloramphenicol, streptomycin, tetracycline, and erythromycin. These cells were also insensitive to penicillin (as might be expected considering the lack of mucopeptide and the insensitivity of other extreme halophiles (Larsen, 1967)) and to bacitracin, but sensitive to polymyxin.

1. Ribosomal Proteins of H. cutirubrum

The major N-terminal residue of the 50S ribosomal protein of *H. cutirubrum* is serine; in contrast, such proteins from *E. coli* and *Bacillus stearothermophilus* contain methionine $>$ alanine $>$ serine as the predominating N-terminal residues. In all three bacteria, the predominating N-terminal residues on proteins from the 30S subunit are: alanine $\gg$ methionine $>$ serine. Distribution of N-terminal residues of the total soluble proteins of the cytoplasm is generally similar for the halophile, thermophile and mesophile (Matheson *et al.*, 1975).

In low salt concentrations 75 to 80% of the proteins of 30S subunits and 60 to 65% of those of the 50S subunit are released, as well as the 5S RNA from these ribosomes. The proteins released are the most acidic ones. The more basic proteins, which remain attached to the 16S and 23S RNA can be removed by a Li^+—EDTA extraction (Visentin *et al.*, 1972). A much more detailed study of the effects of lowering K^+ and Mg^{2+} concentrations on the loss of 5S RNA and of specific proteins from the ribosomal proteins has since been carried out. Some 21 different proteins have been separated from the 30S subunit and 32 from the 50S subunit (Strøm and Visentin, 1973; Strøm *et al.*, 1975a) and the properties of some of the proteins studied.

Two of the 50S proteins, HL20 and HL21 were distinguished by high alanine content as well as high acidity. Experiments with cells grown at different temperatures suggest that protein HL21 is a cleavage product of protein HL20 (Strøm *et al.*, 1975b). These proteins are chemically very similar to the L7-L12 proteins of *E. coli* and the "A" proteins of *B. stearothermophilus* which are found in the 50S subunits of the ribosomes of these bacteria and which are thought to be involved in polypeptide translocation (Strøm *et al.*, 1975b; Visentin *et al.*, 1974). Some regions of homologous amino acid sequences have been found in the proteins from the extreme halophile and the mesophiles (Oda *et al.*, 1974; Strøm *et al.*, 1975b).

More recent work (Matheson *et al.*, personal communication) comparing the L7-L12 proteins of two moderately halophilic bacteria (*V. costicola* and an unidentified halophile) suggests that on the basis of the number of mutations required to transform one protein to another, the moderate halophiles are much more closely related to each other and to *E. coli* than to the extreme halophile. Indeed, it may well be that the "A" proteins of *H. cutirubrum* are more closely related to the corresponding ones from certain eukaryotic ribosomes than to any known prokaryotic ribosomal protein of corresponding function (Matheson, personal communication).

D. Mechanisms of the Effects of Salts on Enzymes and Other Salt-Dependent Proteins of Extremely Halophilic Bacteria

Lanyi's (1974a) review should be consulted for a detailed consideration of the effects of salts on protein conformation. Our concepts of chemical properties of the enzymes of extreme halophiles rely a good deal on analyses of non-enzymatic proteins. Very few enzymes of extreme (or other) halophiles have been purified to homogeneity. Many methods of protein purification depend on the separation of charged molecules by column chromatography; charges are effectively blocked by the high salt concentrations on which most enzymes of extreme halophiles depend for stability. Amino acid analyses seem available only for the DNA dependent and the RNA dependent RNA polymerases of *H. cutirubrum* (Louis and Fitt, 1972a, b, c).

We know that proteins of extreme halophiles as a whole are highly acidic. This has been determined for the bulk cytoplasmic proteins of several extremely halophilic bacteria, for the envelope proteins of others, for the ribosomal proteins of *H. cutirubrum* (Table 5). Composition has been determined also for the protein of the gas vacuoles and "purple membrane" of *H. halobium*, neither of which are especially salt-dependent. Indeed, isolation of the purple membrane depends on the fact that it is stable under conditions of low ionic strength in which most other cell structures disintegrate. Except for these last two structures, there is a clear pattern of acidity of proteins (as measured by excess of acidic over basic amino acids) much greater than in the corresponding structures from non-halophilic bacteria.

Lanyi (1974a) pointed out that in these determinations amide contents were not given or were estimated indirectly from the NH_3 release. He himself

Table 5

Properties of amino acids in proteins of extremely halophilic and non-halophilic bacteria[a]

Material	Bacteria	Amino acids mol %			Polar/non polar[b]
		acidic	basic	excess acidic	
Cytoplasmic proteins	*Halobacterium salinarium*	26·8	9·7	17·1	1·089
	Halococcus No. 24	27·8	9·9	17·9	1·107
	Pseudomonas fluorescens	20·9	13·8	7·1	0·953
	Sarcina lutea	21·2	12·6	8·6	0·917
Ribosomes, 70S	*H. cutirubrum*	26·6	13·1	13·5	1·321
	E. coli[e]	18·4	18·2	0·2	1·003
Cell envelope, total	*H. cutirubrium*	27·6	6·9	20·7	1·204
Cell envelope	*H. halobium*[c]	28·0	7·3	20·7	1·423
outer layer membrane fraction		25·7	8·2	17·5	1·124
Red pellet		24·0	7·2	16·8	1·022
Purple membrane		13·6	7·4	6·2	0·610
Supernatant		29·2	7·1	22·1	1·371
Gas vacuoles[d]	*H. halobium*	21·2	10·9	10·3	0·997

[a] Condensed from Lanyi (1974), Table 1.
[b] Ratio of volumes of residues.
[c] See text for procedures of separating out different parts of the envelopes. "Red pellets" and "purple membranes" were obtained after dialysis against distilled water; the parts of the envelopes remaining soluble are the supernatants.
[d] Mean values from two separate laboratories.
[e] Falkenberg *et al.* (1976).

showed that at least 80% of the total acidic groups in *H. cutirubrum* envelopes were in the free, rather than the amide form. Such determinations seem essential to estimate accurately the acidity of halophilic proteins.

Because the general acidity of proteins of extreme halophiles has been recognized for some years, it has usually been assumed that the enzymes also have a preponderance of acidic groups and to consider that salts act on these as they would on other polyanions. Baxter (1959) suggested some years ago that salts supported the activity of halophilic enzymes because the cations screened negatively-charged groups and prevented their mutual repulsion (by these charges) from distorting protein conformation. If this were so, we might expect that di- and polyvalent cations, e.g. Mg^{2+} Ca^{2+}, spermine which have much higher charge density than monovalent cations might be much more effective than the latter in supporting enzyme activity and/or stability. This has been found true for a number of enzymes of extreme halophiles (Lanyi, 1974a).

It has also been found that much lower concentrations of di- than of monovalent cations can protect envelopes of extreme halophiles (Brown, 1964a; Kushner, 1964b; Kushner and Onishi, 1966).

Though shielding of negative charges by cations undoubtedly plays an important part in the effects of salts on the enzymes and other proteins of extremely halophilic bacteria, Lanyi (1974a) and Lanyi and Stevenson (1970) have argued in some detail that all the effects of salts cannot be due to charge-shielding action. The concentrations of monovalent salts required for activity or stability are simply too high. It is known that cations of monovalent salts can also shield the charges of highly charged polyanions such as DNA or polyglutamic acid, but 0·2 M NaCl, is sufficient to neutralize these charges, much less than is needed for the protection or activation of extremely halophilic enzymes. Lanyi (1974a) points out that though DNA and polyglutamic acid are freely permeable to the solvent (water) and to dissolved ions, the same is not necessarily true of proteins. At the highest salt concentrations, collapse of the polypeptide chain might allow for less charge shielding. Indeed, polyglutamic acid becomes unstable at very high NaCl or KCl concentrations. Such considerations show that it is neither realistic or fruitful to think of the effects of salts on enzymes or other proteins of halophiles as due solely to a charge-shielding effect on polyanions.

Lanyi (1974a) and Lanyi and Stevenson (1970) considered the possibility that very high concentrations of NaCl or certain other salts might be needed to maintain hydrophobic interactions in proteins of extreme halophiles because such interactions, normally very important in maintaining protein structure, are weaker than usual in halophilic proteins.

It is known that salts can affect hydrophobic interactions, largely by influencing the structure of water. When hydrophobic compounds, such as toluene, or the non-polar groups of amino acids, dissolve in water there is a decrease in entropy due to the increased ordering of water structure around these groups. Entropy increases when non-polar groups turn away from the water phase and interact with each other, to form "hydrophobic bonds". The bond seems driven more by an avoidance of water than by an active attraction between the non-polar molecules. The role of hydrophobic bonds in protein structure and in the formation of biological structures made of many protein subunits is reviewed in detail elsewhere (Lanyi, 1974a; Kushner, 1969).

"Salting out", i.e. decreasing the solubility, of proteins by salts such as NaCl or Na_2SO_4 and "salting in" or increasing their solubility by salts such as NaSCN or $NaClO_4$ may be considered primarily due to anionic effects on the structure of water. The "salting out" anions decrease interactions of non-polar groups with water and increase hydrophobic interactions within the non-aqueous part of the molecule. "Salting in" anions make it easier for non-polar groups to interact with water and hence decrease the internal hydrophobic interactions.

High concentrations of salting-in salts can denature many enzymes, including halophilic ones; that is, such salts do not maintain the stability of the latter enzymes. "Salting out" type salts seem specifically required for the stability of menadione reductase, cytochrome oxidase, and threonine deaminase of *H. cutirubrum* (Lanyi, 1974a). Inhibition of menadione reductase (in high NaCl concentrations) by a number of protein denaturants, including alkyl derivatives of urea and formamide, seemed correlated with the latters' ability to increase the solubility of toluene in water (Lanyi and Stevenson, 1970). This suggests that these compounds act by interfering with the hydrophobic bonds needed to maintain the enzyme in an active configuration.

Another indication of the importance of hydrophobic bonds in enzymes of extreme halophiles is that several of these enzymes (Lanyi, 1974a), including ATCase (Norberg *et al.*, 1973) and citrate synthase (in 3 M KCl but not 5 M KCl—that is in the physiological solution; Higa and Cazzulo, 1975) show "cold lability". That is, they have maximal stability at temperatures greater than 0°C and lower stability at colder temperatures. This is also a characteristic of hydrophobic bonds; the effect may be considered most simply in terms of water structure. At lower temperatures the size of the cluster of water molecules is increased and hydrophobic groups can interact more easily with them, breaking hydrophobic bonds (Lanyi, 1974a). Cold sensitivity has been discussed in terms of stability of self-assembling structures (Kushner, 1969).

The above experiments support an important role of hydrophobic bonds, maintained by high salt concentrations, in determining the active configuration of proteins of extremely halophilic bacteria. Lanyi (1974a) points out that individual proteins of extreme halophiles (ribosomal proteins; the DNA- and RNA- dependent RNA polymerases; Louis and Fitt, 1972a, b; two purified enzymes) as well as bulk cytoplasmic and envelope proteins contain substantially lower amounts of non-polar amino acids than similar proteins from mesophilic bacteria (Table 5). In some cases, the lowered amount of non-polar residues is counterbalanced by higher amounts of the "borderline hydrophobic" amino acids, serine and threonine.

Lower amounts of the non-polar amino acids might well make for weaker hydrophobic interactions that could only exist in the presence of very high salt concentrations. Of course, we know little about the sequence and nothing about the internal conformation of any of these proteins. It is to be hoped that the very interesting indirect evidence for very weak hydrophobic interactions as a cause for a high salt requirement will stimulate further searches for more direct evidence in other, purified halophilic proteins.

E. Peculiarities of the DNA and Bacteriophages of Extreme Halophiles

A number of halobacteria and halococci (in fact, all studied so far) contain a satelite band of DNA which makes up 11 to 36% of the total DNA (Joshi *et al.*, 1963; Moore and McCarthy, 1969a). All satellite bands have substantially lower G + C contents (57 to 60%) than those of the major DNA (66 to 68%). These bands do not appear to be episomes (Moore and McCarthy, 1969a) but integral parts of the genome. Lou (in Bayley, 1976) found that the satellite DNA of *H. salinarium* consisted of close circular molecules about 37 μm in circumference. It was calculated that each cell contained seven to eight such circles and suggested that the difference in proportions of satellite DNA between different species of extreme halophiles might reflect different numbers of circles in each species.

DNA/DNA and DNA/RNA hybridization studies (Moon and McCarthy, 1969b) suggested that different halobacteria were related to each other, but not to the halococci; similar findings have been made by techniques of numerical taxonomy (Colwell *et al.*, unpublished) even though the biochemical peculiarities of the halobacteria and halococci make one suspect that these genera are more closely related to each other than they are to anything else.

The high salt concentrations surrounding extreme halophiles do not prevent phage infection. Phages with isometric heads and with tails have been described for several species (Torsvic and Dundas, 1974; Wais *et al.*, 1975). The most recent work characterized their phages as containing double-stranded DNA. A phage against one halobacterium species could infect several others. Phages required high concentrations of monovalent salts or lower concentrations of Mg^{2+} ions for stability. KCl was much more effective than NaCl in conferring stability.

Recently Grey and Fitt (1976b) discovered that *H. cutirubrum* cannot carry out dark repair after ultraviolet irradiation, though it can carry out repair by photoreactivation.

F. External Layers of Extremely Halophilic Bacteria

1. *Walls and Membranes*

The outer layers of the extremely halophilic rods and cocci are quite different from those of most other prokaryotes. The designations Gram positive (halococci) and Gram negative (halobacteria) applied to these groups (Buchanan and Gibbons, 1974) are chemically, and probably taxonomically, meaningless. The layers of halobacteria bear some (though little) resemblance to those of the mycoplasma and of *Sulfolobus*, a bacterium growing in acid hot springs (Chapter 7). It has been known for some years that the halobacteria lack muramic acid, diaminopimelic acid, and D-amino acids and hence lack a mucopeptide layer (Brown, 1964a; Larson, 1967; Kushner, 1968). Furthermore, no lipoprotein layer appears to be present (Steensland and Larsen, 1966). Because of the absence of mucopeptides halobacteria have been considered to lack a proper cell wall entirely and to consist of a cytoplasmic membrane covered by an external layer of hexagonally-arranged subunits which may be composed of one species of protein. This hexagonally arranged outer layer has also been observed in such non-halophilic bacterial species as *Spirillum, Acinetobacter* and others (Beveridge and Murray, 1976). Thin sections of halobacteria examined under the electron microscope show no definite wall layer. However, the absence of muramic acid and other mucopeptide components is not in itself enough to show that no wall exists, but only that the "standard" wall does not. This is dramatically illustrated by the extremely halophilic cocci, whose walls are thick and strong but which also contain no muramic acid (Brown and Cho, 1970; Forsyth, 1971; Reistad, 1975). These cells obviously possess a strong shape-determining structure, but its chemical nature is still unknown. Forsyth (1971) found that one-third to half of the walls of different halococci were composed of carbohydrates, mainly hexoses with glucose and galactose predominating. Hexosamines were also present. These were studied in detail by Reistad (1972, 1975) who found glucosamine, galactosamine and 2-amino-2-deoxyguluronic acid. The last compound could conceivably serve a function similar to that of muramic acid, by forming bonds with other sugars and with amino acids. The walls of the extreme halophiles present a real challenge to the biochemist concerned with bacterial walls—one that may well be left unanswered, through the absence of medical interest that stimulated so many studies of wall structure in human pathogens.

The walls of halococci retain their shape in distilled water, but halobacterial cells require high concentrations of salt to maintain their rod shape and stability of their envelopes. They normally grow and are stable in about 4 M NaCl. If the NaCl concentration is lowered much below 3 M, cells change into irregular forms and finally become spherical. At yet lower concentrations (from 1 to 2 M NaCl depending on the strain) the outer layers of the cells disintegrate. Of the monovalent salts, NaCl is much more effective than KCl or NH_4Cl for maintaining cellular integrity. In 2 M NaCl, for example, cells of *Halobacterium cutirubrum* have become spherical but do not leak, whereas in 2 M KCl or NH_4Cl most of the intracellular contents are lost, though envelopes can still be seen (Kushner, 1964b; Stoeckenius and Rowen, 1967).

Outer layers ("cell envelopes") isolated mechanically in 4 M NaCl disintegrate when suspended in lower salt concentrations. High concentrations of non-ionic solutes do not stabilize envelopes. Disintegration, which takes place very rapidly in the cold, is not due to enzymic reactions. Disintegration involves the formation of a rather ill-defined mixture of macromolecules of S value about 5 (Kushner *et al.*, 1964) except in these species such as *H. halobium* whose envelopes contain substantial amounts of the "purple membrane". Envelopes can also be partly stabilized in low concentrations (0·02 M) of Mg^{2+} salts (though the outer layer is lost). If cells are placed in such solutions, osmotic lysis occurs but most of the envelope remains; this method has been used as a convenient way to prepare envelopes for further study (Brown *et al.*, 1965).

In contrast to cells, envelopes are maintained almost as well by KCl as by NH_4Cl or by NaCl (Kushner, 1964b). Different explanations for this have been put forward. An early one (Kushner, 1964b) that Na^+ acted on the outside and K^+ or NH_4^+ on the inside of the cell envelope is probably too simple-minded—though to be sure the outside of the envelope is apparently exposed to high Na^+ and the inside to high K^+ concentrations. Soo-Hoo and Brown (1967) suggested that cells lysed in intermediate K^+ concentrations because K^+ acts as a permeable solute. Since such cells would already contain saturating amounts of K^+ ions, this argument probably does not hold. It seems more likely that Na^+ is not specifically needed to maintain the stability of the whole envelope, but rather to prevent certain parts from leaking, and that it can do this better than the other ions. Certainly, high Na^+ concentrations are needed for one aspect of membrane function: active transport. Stevenson (1966) showed that NaCl was needed for active transport of L-glutamate by *H. salinarium*, and that concentrations of KCl or $MgCl_2$ which were able to maintain cell structure could not substitute for NaCl. Na^+ is also needed for the light-induced accumulation of leucine by membrane vesicles of *H. halobium* (Macdonald and Lanyi, 1975). A specific Na^+ requirement for active transport has also been demonstrated in marine bacteria (MacLeod, 1965).

The requirement for a high monovalent salt, or a lower divalent salt, concentration for envelope stability appears due, at least in part, to the acidity of the envelope protein (Table 5), which leads to mutual repulsion by the anionic groups unless they are shielded by cations. The contribution of negatively charged groups to the halophilic (salt-dependent) character of envelopes was shown dramatically by Brown's (1964b) experiment: membranes of a marine pseudomonad (NCMB 845) were treated with succinic anhydride, which acted on free NH_3^+ groups to increase the number of carboxylic acid groups on the membrane. The treated membranes required much higher salt concentrations for stability than the untreated membranes.

The discussion on the effects of very high salt concentrations on enzyme action shows that effects on hydrophobic bonds must also be considered. Brown (1976) suggested that high salt concentrations may also affect envelope stability by supporting hydrophobic bonds (which certainly exist in these membranes, as in others). Membrane proteins, like other proteins (except those of the "purple membrane"), of extreme halophiles contain low amounts of non-polar amino acids (Table 5). It is reasonable to expect that hydrophobic bonds will be involved in membrane stabilization. Lanyi (1971) presented evidence for such an involvement during a detailed study of the components (proteins, lipids, flavoproteins, and cytochrome oxidase) released as *H. cutirubrum* envelopes were suspended in decreasing salt concentrations. At the highest NaCl concentration (3·4 M) a low Mg^{2+} concentration ($\geqslant$ 0·2 mM) seemed necessary for complete stability, suggesting that there was a Mg^{2+}-bound envelope fraction. As the salt concentration was lowered, the outer layer of the envelope was lost first (at 1·0 to 1·2 M NaCl). Then, at concentrations down to 0·6 M NaCl other envelope components were lost. The fact that NaCl was more effective than $NaNO_3$ or $NaClO_4$ in stabilizing envelopes in this range, as well as the effects of hydrophobic bond-breaking agents, suggested that the envelope components lost had been bound mainly by hydrophobic bonds. At lower salt concentrations than 0·6 M stability of the residue of the envelope showed little anionic specificity, suggesting that the effects of salts were now primarily ionic (Lanyi, 1971).

The envelopes of extreme halophiles have other curious properties, some of which may well be related to their salt requirement. A detailed study of the cell envelope of *H. salinarium* strain 1 (Mescher *et al.*, 1974; Mescher and Strominger, 1976a, b) showed that 40 to 50% of its protein was accounted for by a single protein of molecular weight about 194,000, containing covalently bound carbohydrate. The rest of the envelope was accounted for by 15 to 20 proteins of lower molecular weight. If the envelopes were suspended in 5 mM $MgSO_4$, 0·14 M NaCl and 0·01 M KCl (a concentration of Mg^{2+} that should maintain a good part of envelope structure), most of the envelope proteins were solubilized, but the high molecular weight protein remained in the fraction sedimenting after 1 h at 100,000 $\times$ *g*.

The envelopes of *E. coli* and other Gram-negative bacteria contain proteins complexed with lipopolysaccharides (lacking in the extreme halophiles). True glycoproteins are characteristic of eukaryotic cells, but are not known in other prokaryotic cells than *H. salinarium* and *H. halobium* (Mescher and Strominger, 1976a, b; Koncewicz, 1972.) The glycoprotein is the major surface component of *H. salinarium*. If it is absent, or its glycosylation blocked by treatment with bacitracin, the cells lose their rod shape and become spheres

(Mescher and Strominger, 1976b.) It is possible that the complete glycoprotein is essential in determining shape of these cells.

The envelope lipids of the extreme halophiles are unique in the biological world and are of special interest, both in pointing out the very unusual biochemical characteristics of these microorganisms and possibly in explaining some of their reactions to high salt environments. Almost all the lipids of *H. cutirubrum* are found in the cell envelope (Kushner *et al.*, 1964). The main components of these lipids are the diphytanyl ether analog of phosphatidyl glycerophosphate (63% of the total lipid) and of a glycolipid sulphate (23% of the total lipid; Kates, 1972). In addition to these polar lipids, envelopes of these and other extremely halophilic rods and cocci contain non-ionic lipids such as the red C_{50} carotenoid pigments, bacterioruberins, and squalenes, as well as phytoene, lycopene, β carotene, vitamin MK8 and the visual pigment retinal (Kushwaha *et al.*, 1972, 1974, 1975a; Kates, 1972; Kates and Kushwaha, 1976; Plachy *et al.*, 1974). Most of the lipids of extremely halophilic rods and cocci, are isoprenoid compounds, but these organisms contain only trace amounts of fatty acids (Kates *et al.*, 1966) a fact that may be due to the very low levels of fatty acid synthase in these organisms, and to the inhibition of this enzyme by high salt concentrations (Pugh *et al.*, 1971).

Pathways of lipid biosynthesis in *H. cutirubrum* are currently being investigated. The formation of C_{40} carotenes follows a pathway similar, but not identical, to that found in higher plants (Kushwaha *et al.*, 1976a). Changes in the growth medium of halobacteria can change the pattern of lipid (especially pigmented lipid) formation (Gochnauer *et al.*, 1972; Kushwaha *et al.*, 1974). Quite recently Kushwaha and Kates (1976) found that nicotine inhibited the formation of C_{50} bacterioruberins; the pattern of bacterioruberin after removal of the inhibitor suggested that these pigments are formed from a C_{40} carotene, lycopene.

This unusual pattern of lipid composition has until recently served as a biochemical "key" for detecting extreme halophiles. Extremely halophilic rods and cocci are characterized by possession of a diether lipid and the absence of fatty acids and mucopeptide constituents. Conversely, the absence of diether lipids and the presence of fatty acids and mucopeptide components has served to distinguish microorganisms in culture collections, thought to be extreme halophiles, but actually moderately halophilic or very salt tolerant organisms (Kates *et al.*, 1966; Novitsky and Kushner, 1975; Komaratat and Kates, 1975). The extremely halophilic actinomycete, *Actinopolyspora halophila* (Gochnauer *et al.*, 1975), however, does not follow this rule. It requires at least 10% NaCl for growth (admittedly a low concentration for extreme halophiles) but has lipids similar to mesophiles (ester-linked phospholipids, and fatty acids) and mucopeptide components (as shown by the possession of diaminopimelic acid and sensitivity to lysozyme). The biochemical peculiarities noted above may be confined to the halobacteria and halococci alone.

The lipids of extremely halophilic rods and cocci are unusually acidic, containing few or no basic groups. Brown's (1965) titration of envelope proteins suggested that their acidic groups were at or near the surface, while the basic groups were buried beneath the surface, in which case they could well be interacting with the acidic groups of phospholipids (Brown, 1976). Thus, the acidity of the phospholipids could effectively increase that of the envelope proteins and increase a tendency towards mutual repulsion between the latter's negatively charged groups.

Extensive studies have been made of the physical state of lipids in the envelopes of *H. cutirubrum*, and in artificial dispersions of lipids from these envelopes, using fluorescent or spin-labelled probes. Esser and Lanyi (1973) found (using stearic acid with a nitroxide label attached to different carbon atoms so that the label could penetrate to different depths within the envelope or lipid bilayer) the environment of the artificial bilayer was much more flexible than that of the envelope. In the latter, only in a narrow central portion of the lipid bilayer did the environment become flexible, so that is fluidity changed with changes in temperature. These workers pointed out that the lipid protein ratio in these envelopes is less than 0·2, unusually low for bacterial envelopes. That is, their protein content is higher than that of many other membranes. They suggested that protein molecules caused an immobilization of all but the central portion of the lipids. For most of the depth of the bilayer the structure was highly ordered. Indeed, the restriction on movement of spin labels was the highest reported so far for any biological membrane. There are very few other biological membranes known (sarcoplasmic vesicles being another) in which strong protein-lipid interactions within the bilayer seem to take place. A comparison of the behaviour of the fluorescent probes, perylene and 8-anilino naphthalene sulphonic acid, in lipid dispersions and envelope vesicles also supported the idea that membrane proteins of *H. cutirubrum* immobilize most of all of the lipid phase (Lanyi, 1974b).

Recent studies with dispersions of the total lipids and of the isolated polar lipids from *H. cutirubrum* envelopes provide clues on the membrane architecture needed for such cation binding. (Lanyi *et al.*, 1974; Plachy *et al.*, 1974) The polar lipids consist of an approximately 2:1 mixture of a diphytanylether analog of phosphatidyl glycerophosphate and of a diphytanyl ether analog of a glycolipid sulphate. In dispersions containing only polar lipid, spin labelling and dilatometric studies suggest that a good deal of kinking of the packed lipid hydrocarbon chains occurs. If squalene, one of the major non-polar lipids, was added, changes in the behaviour of probes suggested that this compound is accommodated in the hydrocarbon region of the bilayer perpendicular to the plane of the membrane, in the interstitial space between hydrocarbon chains. The polar lipids from *H. cutirubrum* were only aggregated by Mg^{2+} or Ca^{2+} if squalene was present. Thus, squalene may be required to space the polar lipid molecules apart sufficiently to permit the entry of divalent cations to the charged phosphate groups. These effects were observed at temperatures and salt (NaCl and KCl) concentrations in which these organisms live, which suggested an important physiological role for squalene.

That the polar lipids of extreme halophiles may serve as important binding sites for divalent cations has also been suggested by earlier work (Kushner and Onishi, 1966; Rayman *et al.*, 1967; McClare, 1967; Brown, 1976).

2. *The Purple Membrane of the Halobacteria*

Much of the recent work on extreme halophiles has been inspired by this curious structure, found in strains of *H. halobium* and *H. cutirubrum.* Stoeckenius and his co-workers (Stoeckenius, 1976) first observed that *H. halobium* envelopes exposed to distilled water did not completely dissociate, but left small amounts of purple material that could be collected on ultracentrifuge density gradients as bands of purple membranes. A separate red band could also be isolated. The purple material represented differentiated regions of the envelope which, in cells grown in the light under limited aeration, could occupy up to 50% of the membrane area. Purple membranes contain 20 to 25% lipid, both polar and non-polar (Kushwaha *et al.*, 1975b). These included the visual pigment, retinal and the sulphated glycolipid mentioned earlier, but none of the characteristic red pigments of halobacteria, the bacterioruberins. The latter were found in the red membranes, which contained no retinal or sulphated lipids (Kushwaha *et al.*, 1975). The purple membrane had only one protein, bacteriorhodopsin, so called because like the rhodopsin of animal eyes it is associated with retinal (Oesterhelt and Stoeckenius, 1971; Kushwaha *et al.*, 1975). Although the molecular weight of bacteriorhodopsin from *H. cutirubrum* and *H. halobium* were first reported to be different by different laboratories (Oesterhelt and Stoeckenius, 1971; Kushwaha *et al.*, 1975), a more recnt joint survey (Kushwaha *et al.*, 1976b) has shown that they are the same (19,300 ± 200*); as near as can be determined the purple membranes of the two bacterial species are identical. Although the purple membrane has only been studied in *Halobacterium* retinal is present in a number of pigmented extreme halophiles (though not in moderate ones) which suggests that they all may contain purple membranes (Kushwaha *et al.*, 1974).

The great interest in the purple membrane arises because this structure is responsible for a unique kind of photophosphorylation. Oesterhelt and

* As determined by SDS-gel electrophoresis which gives low values; the correct value is now accepted to be 25,000 ±100 (Bridgen and Walker, 1976).

Stoeckenius (1973) showed that light led to an excretion of H^+ ions from cells, Danon and Stoeckenius (1974) found that light could catalyse the intracellular formation of ATP under anaerobic conditions. The effects of light were correlated with shifts in the absorption of bacteriorhodopsin from 560 to 415 nm. This shift could be transient, with a quick return to 560 nm in the dark, or else could be an oscillation between the two wavelengths under continuous illumination (Oesterhelt and Stoeckenius 1973). The net result was an outward translocation of protons from the cell.

The involvement of purple membranes in cell energetics is also shown by the fact that light can induce the transport of leucine into membrane vesicles of *H. halobium* (MacDonald and Lanyi, 1975).

Isolated purple membranes could be incorporated into vesicles containing animal or plant phospholipids, where they also caused H^+ translocation in the light (in this case, into the vesicles, since the membranes were apparently turned around in these artificial vesicles). An ATPase from bovine heart mitochondria could be incorporated into the vesicles, and under the influence of light ATP was produced (Racker and Stoeckenius, 1974).

The arrangement of bacteriorhodopsin in the purple membrane has been studied by X-ray diffraction, as well as by electron microscopic techniques (Blaurock and Stoeckenius, 1971; Blaurock, 1975; Henderson, 1975). The available evidence suggests that molecules of bacteriorhodopsin are centred in one plane in groups of three, with a three-fold axis of rotation at the centre of each cluster. These groups are arranged in a hexagonal array on the membrane surface. Each protein, which contains a considerable portion of α-helical arrangements, spans the entire thickness of the membrane. The proposed pumping function is possible without requiring additional membrane or cytoplasmic agents; the lipids that are present seem to form patches between the protein groups.

VII. Special Problems Posed by Moderately Halophilic Bacteria and Salt-Tolerant Microorganisms

Despite the attractiveness and excitement of the extreme halophiles, they are a special breed of bacteria. Most microorganisms that can grow in high salt concentrations can also grow in much lower ones. This was brought out by a survey of a number of marine bacteria isolated (on agar containing 3% NaCl) off the coast of New Brunswick (Forsyth *et al.*, 1971). Most of these bacteria could grow in up to 20% NaCl, and some could grow in 30% NaCl. These studies imply that many or most marine bacteria can grow in much higher salt concentrations than those of the ocean, and such bacteria may be considered moderately halophilic or at least very salt-tolerant microorganisms. Perhaps the most interesting aspect of the lives of such microorganisms is the range of salt concentration in which they can grow.

The adaptation of extreme halophiles (at least, the halobacteria and halococci) to high salt concentrations is associated with changes in their proteins, and with other biochemical changes reflected in the chemistry of the cell envelope and cellular lipids. Will such changes be apparent in all bacteria growing in high salt concentrations? When these cells grow in much lower salt concentrations, will they be adapted to lower salt concentrations? To take the most intriguing possibility, will be composition of proteins change with salt concentrations at which they are formed?

Some of the moderate halophiles presently being studied are listed in Table 6. *V. costicola* seems an especially desirable organism for physiological

Table 6

Moderately halophilic bacteria under recent or current investigation[a]

Organism	M NaCl Growth range	Best growth	Source	References
Vibrio costicola NRC 37001, NCMB 701	0·2–4·0	0·7–2·0	bacon curing brines	Smith (1938); Robinson (1950)
V. alginolyticus 138-2	0·1–1·7	0·5	marine fish	Unemoto and Hayashi (1969)
Paracoccus halodenitrificans[b]	0·4–4·0	0·5–2·0	bacon curing brines	Robinson (1950); Robinson and Gibbons (1952)
Micrococcus 203	0·5–3·4	1·9–2·1	salted whale meat	Hiwatashi *et al.* (1958)
Pseudomonas 101[c]	0·5–4·3	2·1	crude salt imported into Japan	Hiwatashi *et al.* (1958); Hiramatsu *et al.* (1976); Ohno *et al.* (1976)
Pseudomonas Ba_1	0–4·0	0·5–1·0	unrefined Dead Sea solar salt	Rafaeli-Eshkol (1968)
Micrococcus halobius	0·5–4·0	1·0–2·0	unrefined solar salt	Onishi (1972); Onishi and Kamekura (1972)
M. varians	0–4·0	1·0–3·0	soy sauce mash	Kamekura and Onishi (1974a)
Planococcus halophilus[d] NRC 14033	0–5·0	0·7–1·5	possible contaminant of halophile stock culture	Novitski and Kushner (1975, 1976)
Bacillus 21-1	0–4·0	1·0–2·0	unrefined solar salt	Kamekura and Onishi (1974b)

[a] Not an exhaustive listing

[b] Originally called *Micrococcus halodenitrificans* but reclassified on basis of new taxonomic evidence

[c] In reference given it is referred to as possibly an *Achromobacter*, but in most papers it is referred to as *Pseudomonas*

[d] Novitski and Kushner (1976) use the term "facultative halophile" to describe this strain

studies because of its simple nutritional requirements (it grows on glucose minimal medium) and also because its internal ionic concentration is usually as high or higher than the external one, a relation that does not seem to hold for all other moderate halophiles (Table 4).

Forsyth and Kushner (1970) examined the pattern of growth at different salt concentrations of *Vibrio costicola* and *Paracoccus halodenitrificans*. For both bacteria the pattern was the same, with the highest rate of growth at 0·5 to 1·0 M NaCl whether or not the inocula were precultured on the highest (3·5 M) or lowest (0·5 M) NaCl concentrations permitting growth. These experiments, and others in which colony counts were carried out at different salt concentrations, showed that the particular salt concentration did not select populations of cells having different salt responses; rather, each cell could grow over the entire range of NaCl concentration. It seems likely that such an ability holds for other organisms capable of growing over a wide range of solute concentrations (though so far it does not seem to have been tested for any of them). In any case, before comparing physiological properties of cells grown at different solute concentrations it is essential to confirm that we are dealing with a genetically homogeneous population.

We must also know how the intracellular ionic content varies with different external salt concentrations, to compare the behaviour of intracellular components of cells grown at different concentrations. Information on the amounts of cell-associated ions (expressed as intracellular concentrations in terms of cell water) over a wide range of external salt concentrations exists for only a few moderately halophilic bacteria (Table 4). As already discussed, certain yeasts and algae that grow in high external salt concentrations have high internal concentrations of non-ionic solutes. Thus, no *a priori* assumptions can be made about the internal solute composition of any organism growing over a wide range of salt (or other solute) concentrations.

In the few cases so far examined, changing the salt concentration of the growth medium does not seem to change the salt response of enzymes of moderate halophiles. This is true for the extracellular amylase of a moderate halophile (Onishi, 1972), and for the $NADH_2$ oxidase of *V. costicola* (Hochstein, personal communication).

The biosynthetic threonine deaminase (TDase) of *V. costicola* has also been studied in this respect. This enzyme is formed when cells are grown in a glucose mineral medium. In a medium with several amino acids, formation of this enzyme is repressed; furthermore, no catabolic TDase is formed. The salt response of both activity and feedback inhibition of this enzyme present an interesting contrast to those of the biosynthetic TDase of *E. coli* (Fig. 2). Activity of both enzymes is highest in the absence of salt (through salts stabilize *V. costicola* TDase). As the salt concentration increases, activity of the halophilic TDase decreases, eventually levelling off at about 30% of the maximal value. Feedback inhibition by isoleucine is high and remains so over the whole range of salt concentration. Thus, this enzyme functions pretty well over a wide range of salt concentrations, both as regards activity and regulation. The salt response of both activity and feedback inhibition of TDase were the same regardless of the salt concentration in which *V. costicola* were grown (Shindler, 1976).

The *E. coli* enzyme also decreases with increased salt concentrations, but loses almost all activity at concentrations of 2 M and higher. In 3 M salt, isoleucine is ineffective as a feedback inhibitor.

Few other regulatory enzymes of moderate halophiles have been studied. The ATCase of *V. costicola* (molecular weight 160,000) has maximal activity in 2 M salt. However, this enzyme is not subject to feedback inhibition by CTP or UTP, nor is it affected by a number of other nucleotides. It does not appear subject to allosteric regulation (Shindler, 1976).

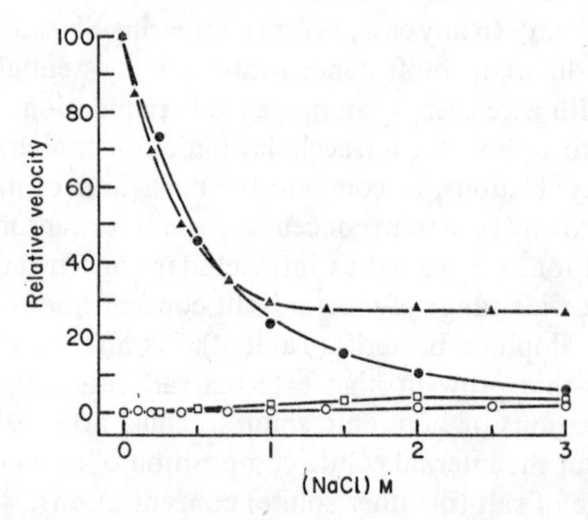

Fig. 2. Effect of NaCl concentration on activity and isoleucine inhibition of threonine deaminases from *V. costicola* and *E. coli* (Shindler, 1976) ▲ *V. costicola* activity; ○ *V. costicola* + 10^3 M Ile; ● *E. coli* activity; □ *E. coli* activity + 10^{-3} M Ile.

The salt response of transport systems of moderate halophiles might be expected to change after growth at different salt concentrations, but this intriguing possibility has never been studied.

Ribosomes of *V. costicola* maintain the same sedimentation properties over a wide range of salt concentrations, and growth in media of higher or lower salt concentration does not seem to affect this behaviour (Wydro *et al.*, 1975).

Growth at different salt concentrations does not seem to cause major changes in chemical composition of the cell walls or cellular lipids of moderate halophiles. The lipid pattern characteristic of halophiles (di-hydrophytyl ether lipids and a virtual absence of fatty acids) does not appear in moderate halophiles when they are grown in high salt concentrations (Kates *et al.*,1966; Kushner and Forsyth, unpublished work with *V. costicola* and *P. halodenitrificans*). These last two moderate halophiles, and also the salt tolerant "facultative halophile" *Planococcus halophilus* also retain muramic acid (and presumably a mucopeptide structure) when grown in high salt concentrations (Forsyth, 1971; Novitsky and Kushner, 1975). Such experiments, and similar findings with the borderline extreme halophile *Actinopolyspora halophila* (Gochnauer *et al.*, 1975) suggest that the unusual lipids and walls of the halobacteria and halococci are not necessarily connected with growth at high salt concentrations, or even the requirement for such concentrations. Indeed, we still do not know why the red halophiles have such unusual walls and lipids. Enzymes that act on these structures seem very rare in nature, and this may protect the extreme halophiles. An isoprenoid basis for all lipids might produce more pigmented compounds, which do have survival value in the bright sunlight in which the extreme halophiles grow (Dundas and Larsen, 1962).

A. *In Vitro* Protein Synthesis

One fundamental problem with moderate, as with extreme, halophiles is to know the true concentrations of ions free in the cytoplasm. The amounts of cell-associated ions can fairly easily be determined, but not their intracellular location. Masui and Wada (1973) found that substantial amounts of Na^+ and K^+ ions could be bound to the envelopes of a moderately halophilic rod. Though this indicates that some of these cell-bound ions are not free in the cytoplasm, it is probably impossible to extrapolate from the amounts bound by envelope layers in broken cell preparations to those bound in the living cell. First, the binding equilibrium of monovalent ions would be shifted by the dilution of intracellular material necessary to break cells and isolate envelopes. Furthermore, it seems very likely that the binding properties of envelopes would be changed when these are disrupted and removed from the environment of the living cell.

The true internal environment may also be estimated by the effects of salts or other solutes on physiological functions. Such an approach can involve the assumption that the solute concentration that best supports such functions is the one actually existing in the living cell. A great deal of attention to this approach was given in Brown's (1976) review, as part of his treatment of "compatible solutes", those that permit a cell to carry out its internal functions while maintaining an appropriate internal a_w. He points out that for one very well-studied enzyme, the isocitrate dehydrogenase (ICDH) of *H. salinarium*, KCl is less inhibitory at high concentrations than NaCl and seems to follow a simple non-competitive inhibition pattern. NaCl gives complex non-linear inhibition pattern. Detailed kinetic studies suggested that at higher salt concentrations NaCl is much more tightly bound to the enzyme than KCl. Even though NaCl is more inhibitory than KCl it is not completely inhibitory, and substantial amounts of enzyme activity remain at the highest NaCl concentrations (Aitken and Brown, 1972; Brown, 1976).

Other experiments (Brown, 1976) have shown that the NADP-specific ICDH of a salt tolerant yeast, *Saccharomyces rouxii* is more severely inhibited by sucrose than by glycerol, which may well be consistent with the fact that these organisms exclude sucrose but accumulate polyols.

Interesting as these experiments are, one can obviously not assume that the actual solute conditions inside the cell are those that give the greatest activity of one important enzyme. It is not certain that any enzyme will be working at its full capacity within the cell, where substrate concentrations will necessarily be low. Changes that lead to adaptation in response to temperature often involve changes in regulatory properties that permit enzymes to function well under cellular conditions, rather than to changes in V_{max} (Hochachka and Somero, 1974). It may also be doubted that it is best for the cell to have most enzymes working at full capacity. Regulatory enzymes may provide a better physiological index of the internal solute conditions. The striking example of the malic enzyme of *H. cutirubrum* (Vidal and Cazzulo, 1976) which is most active in 1 M NH_4 Cl but best regulated in 3 M KCl strongly supports the idea that 3 M KCl is near the physiological concentration of this salt within the cell.

Perhaps the best physiological index of all, which should be further explored in other microorganisms, is *in vitro* protein synthesis. To date, the only detailed studies with halophilic bacteria are those of Bayley and his collaborators (Bayley, 1976) described above, in the extreme halophile *H. cutirubrum*. Here, the stability of ribosomes, the activity of amino acid activating enzymes and the incorporation of amino acids into protein take place at ionic conditions close to those calculated to exist within the cell by measurements of cell-associated ions. One obvious implication of these studies is that the cell-associated ions are indeed "free" in the cytoplasm (though from the discussion in section V it is doubtful that such freedom is absolute).

Recent experiments with the moderate halophile, *V. costicola* suggest that most of the cell-associated ions in this organism may not be free in the cytoplasm. A cell-free protein synthesizing system was prepared and tested in the presence of mixtures of the composition and concentration judged likely to be found in the cytoplasm, on the basis of analyses of cell-associated ions (Shindler *et al.*, 1977). No activity was found, until the system was tested in lower salt concentrations. In fact, only in 0·2 M or lower salt was substantial activity found. Protein synthetic activity (either endogenous, as measured by incorporation of a mixture of amino acids without added messenger or poly-U-directed phenylalanine incorporation) was highest in about 0·1 to 0·2 M salt, KCl and NH_4Cl being most active. Mg^{2+} ions were needed, with a

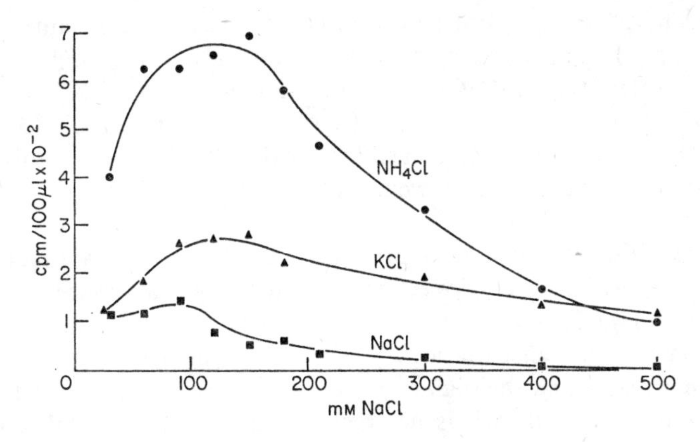

Fig. 3a. Effect of monovalent salts on protein synthesis (poly-U directed phenylalanine incorporation) by cell free extracts of *V. costicola.*

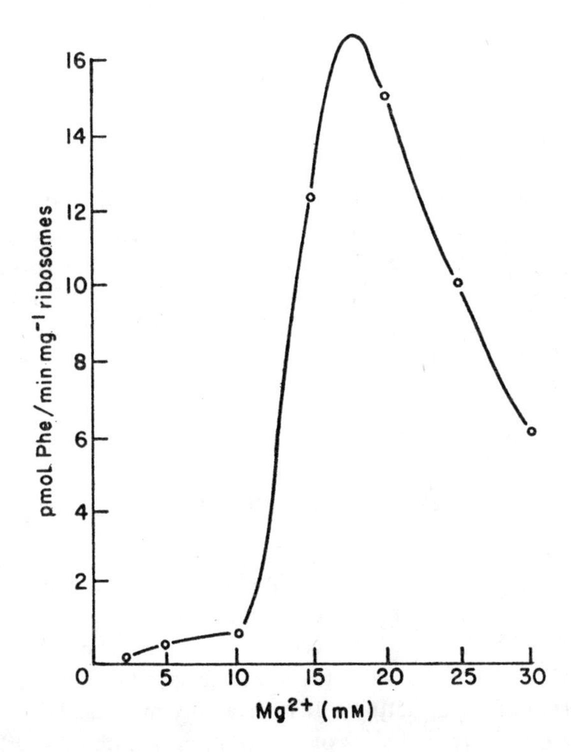

Fig. 3b. Effect of $MgCl_2$ on protein synthesis (poly-U directed phenylalanine incorporation) by cell free extracts of *V. costicola.*

peak of activity at about 20 mM ,substantially less than the total amount in the cell (Figs 3a, b). Higher concentrations of monovalent salts inhibited protein synthesis. Interestingly, NaCl was more inhibitory than KCl or NH_4Cl. These salts also affected stability, but concentrations that inhibited activity of the protein-synthesizing system still maintained the stability of this system. High NaCl concentrations, however, quickly inactivated it (Wydro *et al.*, 1977).

Since growing cells must make proteins, these results suggest that there are some regions within *V. costicola*, secluded from high salt concentrations, in which protein synthesis takes place. If this is so, there must be other regions of the cell in which the concentrations of salts are very high, say the cell envelope. It is even possible that the high regions of salt serve to stabilize the protein-synthesizing system when it is not being used, even though other ways of stabilizing this system, for example by polyamines, have not been excluded.

These suggestions could not arise from studies of the ribosomes of these organisms (which are stable over a very wide range of salt concentration) or from studies of individual enzymes, which, as discussed above, have different salt optima. It is not possible to predict a physiological ionic concentration within the cytoplasm from studies of any one enzyme. Protein synthesis, an essential process that involves many enzymes and cell structures, seems a much more suitable physiological indicator.

It should be profitable to extend such studies to other more, or less, salt and solute-tolerant microorganisms. One might even find that there are solute-protected regions within the cytoplasm of mesophilic organisms. The ribosome of *E. coli* are aggregated in 0·5 M NaCl, but cells can grow in this salt concentration (Wydro *et al.*, 1975). Possibly this organism can also keep salts away from its very sensitive parts, though we would not expect it to do so by keeping its total internal a_w higher than the external solution.

VIII. Concluding Remarks

A. Water and Solutes

The role of water stress, that is low water activity, in limiting growth of microorganisms in nature is discussed in Chapter 9. In the laboratory, many examples show that it is usually not possible to distinguish the concept of water stress from that of high concentrations of specific solutes. Certainly, at the enzyme level, effects seem to be solute-specific rather than due to a general lowering of water activity. It is probably most useful to consider a_w as limiting growth during studies of microorganisms on dry surfaces, where growth can be measured as a function of the relative humidity of the atmosphere. The fact, however, that organisms on a surface stop growing after the external a_w reaches a certain minimum value does not prove that the a_w alone is responsible. A low external a_w will draw water from the cell and result in a concentration of intracellular solutes. It may not be possible to differentiate between the effects of water loss and those of such solutes.

The actions of certain solutes, especially at high concentrations, may be attributable to effects on water structure, that is on formation of water aggregates, which in turn influence the formation of hydrophobic bonds. Again, as the discussion in this chapter (and in more detail in Lanyi's (1974) review) shows, it is not possible to distinguish effects of solutes on protein molecules from those on water structure.

B. What Next With the Halophiles?

From being once considered as out-of-the-way biological curiosities, the halobacteria and halococci have now developed into a considerable focus of research. This interest is justified by the unique biochemical properties of these organisms. They make up for some of the difficulties of working with them, which include: complex nutrition, so that nutritional mutants are probably very hard to find (and none have as yet been sought seriously); insensitivity to most antibiotics, and hence loss of the possibility of selecting other useful mutants; the facts that much of the fine structure is lost when salts are removed, and that salts greatly interfere with detailed studies of cellular organelles. This last may explain why apparently nothing is known of the flagella of extreme halophiles except that they apparently break up in the absence of salt (as do the flagella of the marine bacterium, *Vibrio alginolyticus* (Ulitzur and Kessel, 1973), since no trace of flagella have been reported in water-lysed halobacteria. The subcellular structures most studied by electron microscopy are those stable in the absence of salts, such as the purple membrane discussed above and gas vacuoles of *H. halobium* (Krantz and Ballou 1973).

Instability of proteins in the absence of salts and difficulties of purifying then in the presence of salts has also limited work on enzymes of extreme halophiles. I have met a good half-dozen biochemists who, having made significant contributions to the enzymology of extreme halophiles have abandoned them for more accessible microorganisms whose genetics can be studied, and whose enzymes are stable without salt. It may be no accident that the most detailed work now being carried out on extreme halophiles, that on the purple membrane, is concerned with a structure that does not require salt for its physiological functions. (This is not to deny the great intrinsic interest of this structure, but old halophilologists may wonder if purple membranes would be studied so ardently if high salt concentrations had to be present all the time.)

These and other unusual properties of the halobacteria and halococci, including lipids, walls and satellite DNA are fully worth study in themselves, but there is no clear indication that they are related to life in high salt concentrations. The fact that a microorganism living under extreme conditions possess a certain unusual property does not prove that one is related to the other.

Despite this warning, the unusual biochemical properties of the red halophiles should be prized as possible clues to the taxonomic and evolutionary status of these organisms, a sufficiently fascinating subject in its own right. The halobacteria were once thought related to the pseudomonads, and possible ways in which they were derived from this group have been discussed (Larsen, 1967). I think it a mistake to consider the red halophiles as very closely related to any other known bacterial genus. We may hope for further revelations of their taxonomic status by comparisons of specific ribosomal proteins, of RNA's, and of other macromolecules of known function.

Though they are less exciting at first glance than the extreme halophiles the moderately halophilic bacteria, and solute-tolerant microorganisms in general, pose quite sufficiently interesting questions, especially those implied by their ability to grow over wide ranges of solute concentrations. Further work on these relatively little-studied microorganisms may be expected to bring dividends in the form of insight on the relation of internal and external solute concentrations, and on the state of cell-associated ions within the cytoplasm. If the last decade has been that of the extreme halophiles, we can hope that the next one will see their more modest, moderate cousins (in the spiritual sense only) take their proper place in the scientific canon.

Addendum

In the last year, work has proceeded very rapidly on the molecular biology of extreme halophiles, especially the purple membrane of halobacteria (Bayley and Morton, in press; Heinrich *et al.*, 1977; Henderson, 1977). Most recent literature can be found in these reviews. More recent references are indicated by asterisks in the References.

Reviews of the microbial ecology of the Dead Sea (Nissenbaum, 1975; 1977a, b) and of the Great Salt Lake (Post, 1977) have been presented. Studies have also been made of microbial populations in natural waters that show a fluctuating salt concentration, including estuaries, lagoons, and salterns (Blevins *et al.*, 1976; Jeske, 1975; Vreeland and Litchfield, 1976). A number of other salt-tolerant or halophilic microorganisms, including those found in the sea, in blood, sausage, cheese, and lobsters, are listed below.

Other halophiles are being studied at a more moderate rate than the halobacteria. Those whose physiology has been most studied include the "classical" moderate halophiles, *Vibrio costicola* (de Médicis and Rossignol, 1977) and *Paracoccus halodentrificans* (Alico and Hochstein, 1976); the interesting photosynthetic halophile, *Ectothiorhodospira halophila* (Rinehart and Hubbard, 1976; Tabita and McFadden, 1976) and the halophilic blue-green alga, *Aphanothece halophitica* (Koelsch *et al.*, 1976; Miller *et al.*, 1976; Yopp *et al.*, 1976).

Acknowledgements

I would like to thank Drs M. Kate and A. D. Matheson for the invaluable advice they gave during the preparation of this chapter.

References*

Abdulaev, N. G., Kiselev, A. V. and Ovchinnikov, Y. A. (1976). Effects of proteolytic enzymes on *Halobacterium halobium* purple membranes. *Bioorg. Khim.* **2,** 1148–1150.

Aitken, D. M. and Brown, A. D. (1972). Properties of halophil nicotinamide-adenine dinucleotide phosphate-specific isocitrate dehydrogenase. *Biochem. J.* **130,** 645–662.

*Aldova, E. and Ryc, M. (1976). Halophilic vibrio isolated from presswurst. *Folia Microbiol.* **21,** 327.

*Alico, R. K. and Hochstein, L. I. (1976). Internal cation concentrations in the moderate halophile *Paracoccus halodenitrificans. Abst. Ann. Meet. Am. Soc. Microbiol.* **76,** 182.

Baas-Becking, L. G. M. (1931). Historical notes on salt and salt manufacture. *Sci. Month.* **32,** 434–446.

*Balashov, S. P. and Litvin, F. F. (1976). Initial photochemical transformations of bacteriorhodopsin in *Halobacterium* purple membranes at 4° K. *Bioorg. Khim.* **2,** 565–566.

Barskii, E. L., Drachev, L. A., Kaulen, A. D. and Kondrashin, A. A. (1975). Direct measurement of the electric current generation by lipoprotein complexes. *Bioorg. Khim.* **1,** 113–126.

Baxter, R. M. (1959). An interpretation of the effects of salts on the lactic dehydrogenase of *Halobacterium salinarium. Can. J. Microbiol.* **5,** 47–57.

Baxter, R. M. and Gibbons, N. E. (1956). Effects of sodium and potassium chloride on certain enzymes of *Micrococcus halodenitrificans* and *Pseudomonas salinaria. Can. J. Microbiol.* **2,** 599–606.

Bayley, S. T. (1976). Information transfer: salt effects, *In* "Extreme Environments: mechanisms of microbial adaptation" (Ed. M. R. Heinrich), pp. 119–136. Academic Press, New York and London.

Bayley, S. T. and Kushner, D. J. (1964). The ribosomes of the extremely halophilic bacterium. *Halobacterium cutirubrum. J. Mol. Biol.* **9,** 654–669.

*Bayley, S. T. and Morton, R. A. (1978). Recent developments in the molecular biology of extremely halophilic bacteria. *CRC Crit. Rev. Microbiol.* (in press).

*References marked with an asterisk are those given or referred to iu the Addendnm

Ben-Amotz, A. (1974). Osmoregulation mechanism in the halophilic alga *Dunaliella parva*. *In* "Membrane Transport in Plants" (Eds U. Zimmerman and J. Dainty), pp. 95–100. Springer Verlag, New York.

Ben-Amotz, A. and Avron, M. (1973). The role of glycerol in the osmotic regulation of the halophilic alga *Dunaliella parva*. *Plant Physiol.* **51,** 875–878.

Beveridge, T. J. and Murray, R. G. E. (1976). Dependence of the superficial layers of *Spirillum putridiconchylium* on Ca^{2+} or Sr^{2+}. *Can. J. Microbiol.* **22,** 1233–1244.

Bilsky, A. Z. and Armstrong, J. B. (1973). Osmotic reversal of temperature sensitivity in *Escherichia coli*. *J. Bacteriol.* **113,** 76–81.

Blaurock, A. E. (1975). Bacteriorhodopsin: A trans-membrane pump containing α-helix. *J. Mol. Biol.* **93,** 139–158.

Blaurock, A. E. and Stoeckenius, W. (1971). Structure of the purple membrane. *Nature New Biol.* **233,** 152–155.

*Blevins, W. L., Nevin, T. A. and Noble, C. L. (1976). Thiol synthesis by halophilic bacteria indigenous in a coastal lagoon. *Bull. Environ. Contam. Toxicol.* **15,** 330–334.

*Borovyagin, V. L., Plakunova, V. G. and Sherman, M. B. (1976). Study of ultra-structure of membranes of *Halobacterium halobium*. *Dokl. Akad. Nauk. SSSR.* **228,** 726–728.

Borowitzka, L. J. and Brown, A. D. (1974). The salt relations of marine and halophilic species of the unicellular green alga, *Dunaliella*. *Arch. Microbiol.* **96,** 37–52.

Breed, R. S., Murray, E. G. D. and Smith, N. R. (1957). *In* Bergey's Manual of Determinative Bacteriology" (7th ed.). Williams and Wilkins, Baltimore.

Bridgen, J. and Walker, I. D. (1976). Photoreceptor protein from the purple membrane of *Halobacterium halobium*. Molecular weight and retinal binding site. *Biochemistry* **15,** 792–798.

*Brinkley, A. W., Rommel, F. A. and Huber, T. W. (1976). The isolation of *Vibrio parahemolyticus* from moribund aquarium lobsters. *Can. J. Microbiol.* **22,** 315–317.

Brisou, J., Courtois, D. and Denis, F. (1974). Microbiological study of a hypersaline lake in French Somaliland. *Appl. Microbiol.* **27,** 819–822.

Brock, T. (1969). Microbial growth under extreme conditions. *Symp. Soc. Gen. Microbiol.* **19,** 15–42.

Brock, T. (1976). Halophilic blue-green algae. *Arch. Microbiol.* **107,** 109–111.

Brown, A. D. (1964a). Aspects of bacterial response to the ionic environment. *Bacteriol. Rev.* **28,** 296–329.

Brown, A. D. (1964b). The development of halophilic properties in bacterial membranes by acylation. *Biochem. Biophys. Acta* **93,** 136–142.

Brown, A. D. (1965). Hydrogen ion titrations of intact and dissolved lipoprotein membranes. *J. Mol. Biol.* **12,** 491–508.

Brown, A. D. (1974). Microbial water relations: features of the intracellular composition of sugar-tolerant yeasts. *J. Bacteriol.* **118,** 769–777.

Brown, A. D. (1976). Microbial water stress. *Bacteriol. Rev.* **40,** 803–846.

Brown, A. D. and Simpson, J. R. (1972). Water relations of sugar-tolerant yeasts: the role of intracellular polyols. *J. Gen. Microbiol.* **72,** 589–591.

Brown, A. D., Shorey, C. D. and Turner, H. P. (1965). An alternative method of isolating the membrane of a halophilic bacterium. *J. Gen. Microbiol.* **41,** 225–231.

Buchanan, R. E. and Gibbons, N. E. (1974). *In* "Bergey's Manual of Determinative Bacteriology." Williams and Wilkins, Baltimore.

Cazzulo, J. J. (1973). On the regulatory properties of a halophilic citrate synthase. *FEBS Lett.* **30,** 339–342.

Cazzulo, J. J. (1975). Las bacterias halophilas extremas. I. Generalidades, composicion quimica y estructura. II. Enzimologia y metabolismo. *Rev. Asoc. Arg. Microbiol.* **7,** 28–37; 68–80.

Cazzulo, J. J. and Vidal, M. C. (1972). Effect of monovalent cations on the malic enzyme from the extreme halophile, *Halobacterium cutirubrum. J. Bacteriol.*, **109,** 437–439.

Chazen, L. L. and Bayley, S. T. (1973). Some properties of a DNA-dependent RNA poymerase from *Halobacterium cutirubrum. Can. J. Biochem.* **51,** 1297–1304.

*Chekulaeva, L. N., Korolev, Yu, N., Telegin, N. L. and Rikhireva, G. T. (1975). Study of the formation of purple membranes during the cultivation of halophilic bacteria. *Biofizika* **20,** 839–843.

*Cherry, R. J., Heyn, M. P. and Oesterhelt. D. (1977). Rotational diffusion and exciton coupling of bacteriorhodopsin in cell membrane on *Halobacterium halobium. FEBS Lett.* **78,** 25–30.

*Chignell, C. F. and Chignell, D. A. (1975). A spin label study of purple membranes from *Halobacterium halobium. Biochem. Biophys. Res. Commun.* **62,** 136–143.

Christian, J. H. B. (1955). The influence of nutrition on the water relations of *Salmonella oranienburg. Aust. J. Biol. Sci.* **8,** 75–82.

Christian, J. H. B. and Waltho, J. (1962). Solute concentrations within cells of halophilic and non-halophilic bacteria. *Biochim. Biophys. Acta* **65,** 506–508.

*Cinco, M., Tamaro, M. and Cociancich, L. (1975). Taxonomical, cultural and metabolic characteristics of halophilic leptospirae. *Zent. Bakteriol. Parasit. Infekt. Hyg. Erste.* **233,** 400–405.

Cope, F. W. and Damadian, R. (1970). Cell potassium by 39K spin echo nuclear magnetic resonance. *Nature, Lond.* **228,** 76–77.

*Daiku, F., Fujita, Y., Ezura, Y. and Sakai, M. (1976). Physiological studies on the inorganic salt requirements of marine bacteria. 1. Minimum salt concentration in the cell suspension required to prevent lysis. *Bull. Jap. Soc. Sci. Fish.* **42,** 307–313.

*Daiku, F., Fujita, Y., Ezura, Y. and Saka, M. (1976). Physiological Studies on the inorganic salt requirements of marine bacteria. 2. Effects of inorganic salts on the oxidations of succinic acid and of fumaric acids. *Bull. Jap. Soc. Sci. Fish.* **42,** 315–322.

Danon, A. and Stoeckenius, W. (1974). Photophosphorylation in *Halobacterium halobium. Proc. Natl. Acad. Sci. USA* **71,** 1234–1238.

*de Médicis, E. and Rossignol, B. (1977). Pyruvate kinase from the moderate halophile, *Vibrio costicola. Can. J. Biochem.* **55,** 825–833.

Deo, P. W. (1974). Studies on the structure and metabolism of the glycolipids and glycolytic sulphate in the extreme halophile *Halobacterium cutirubrum.* Ph.D. thesis, Univ. Ottawa.

Dundas, I. D. and Larsen, H. (1962). The physiological role of the carotenoid pigments. of *Halobacterium salinarium. Arch. Mikrobiol.* **44,** 233–239.

Dundas, I. D., Srinivasan, V. R. and Halvorson, H. O. (1963). A chemically defined medium for *Halobacterium salinarium* strain 1. *Can. J. Microbiol.* **9,** 619–624.

*Dynev, T., Rajkov, P., Stamenova, C., Nackova, V. and Atanasova, V. (1976). Isolation of non-agglutinating and halophile vibrios in man. *Folia. Microbiol.* **21,** 326–327.

*Edzes, H. T., Gzinburg, M., Ginzburg, B. Z. pan Berendsen, H. J. C. (1977). Physical state of alkali ions in a halobacterium: some NMR results. Experientia **33**, 732–734.

Esser, A. F. and Lanyi, J. K. (1973). Structure of the lipid phase in cell envelope vesicles from *Halobacterium cutirubrum*. *Biochemistry* **12**, 1933–1939.

Falkenberg, P., Matheson, A. T. and Rollin, C. F. (1976). The properties of ribosomal proteins from a moderate halophile. *Biochim. Biophys. Acta.* **434**, 474–482.

*Fitt, P. S. and Peterkin, P. I. (1976). Isolation and properties of a small manganese ion stimulated bacterial alkaline phosphatase. *Biochem. J.* **157**, 161–167.

Fitt, P. S., Peterkin, P. I. and Barua, N. N. (1975). The relationship between the deoxyribonucleic acid-bound and low-molecular-weight soluble forms of *Halobacterium cutirubrum* deoxyribonucleic acid-dependent ribonucleic acid polymerase. *Biochem. J.* **149**, 719–724.

Flannery, W. L. (1956), Current status of knowledge of halophilic bacteria. *Bacteriol. Rev.* **20**, 49–66.

Forsyth, M. P. (1971). Physiological studies on halophilic bacteria. Ph.D. thesis, Univ. Ottawa.

Forsyth, M. P. and Kushner, D. J. (1970). Nutrition and distribution of salt response in populations of moderately halophilic bacteria. *Can. J. Microbiol.* **16**, 253–261.

Forsyth, M. P., Shindler, D. B., Gochnauer, M. B. and Kushner, D. J. (1971). Salt tolerance of intertidal marine bacteria. *Can. J. Microbiol.* **17**, 825–828.

*Garty, H. and Caplan, S. R. (1977). Light dependent rubidium transport in intact *Halobacterium halobium* cells. *Biochim. Biophys. Acta* **459**, 532–545.

Gibbons, N. E. (1970). Isolation, growth and requirements of halophilic bacteria (Eds J. R. Norris and D. W. Ribbons), *In* "Methods in Microbiology", Vol. 3B, pp. 169–182. Academic Press, New York and London.

*Gillbro, T., Kriebel, A. N. and Wild, U. P. (1977). Origin of red emission of light-adapted purple membrane of *Halobacterium halobium*. *FEBS Lett.* **78**, 57–60.

Ginzburg, M., Sachs, L. and Ginzburg, B. Z. (1971a) Ion metabolism in a *Halobacterium*. II. Ion concentration in cells at different levels of metabolism. *J. Memb. Biol.* **5**, 78–101.

Ginzburg, M., Ginzburg, B. Z. and Tosteson, D. C. (1971b). The effect of anions on K^+-binding in a *Halobacterium* species. *J. Memb. Biol.* **6**, 259–268.

Gochnauer, M. B. and Kushner, D. J. (1969). Growth and nutrition of extremely halophilic bacteria. *Can. J. Microbiol.* **15**, 1157–1165.

Gochnauer, M. B. and Kushner, D. J. (1971). Potassium binding, growth and survival of an extremely halophilic bacterium. *Can. J. Microbiol.* **17**, 17–23.

Gochnauer, M. B., Kushwaha, S. C., Kates, M. and Kushner, D. J. (1972). Nutritional control of pigment and isoprenoid compound formation in extremely halophilic bacteria. *Arch. Mikrobiol.* **84**, 339–349.

Gochnauer, M. B., Leppard, G. G., Komaratat, P., Kates, M., Novitsky, T. and Kushner, D. J. (1975). Isolation and characterization of *Actinopolyspora halophila*, gen. et sp. nov., an extremely halophilic actinomycete. *Can. J. Microbiol.* **21**, 1500–1511.

Grey, V. L. and Fitt, P. S. (1976a). An improved synthetic growth medium for *Halobacterium cutirubrum*. *Can. J. Microbiol.* **22**, 440–442.

Grey, V. L. and Fitt, P. S. (1976b). Evidence for the lack of deoxyribonucleic acid dark-repair in *Halobacterium cutirubrum*. *Biochem. J.* **156**, 569–575.

*Griffiths, D. E., Hyams, R. L. and Partis, M. D. (1977). Studies of energy linked reactions: role for lipoic acid in purple membrane of *Halobacterium halobium. FEBS Lett.* **78,** 155–160.

*Gupte, S. S., Haug, A. and Elbavoumi, M. A. (1976). Phase transitions in purple membrane and cell membrane vesicles in *Halobacterium halobium. Biophys. J.* **16,** A103.

Harold, F. M. and Altendorf, K. (1974). Cation transport in bacteria: K^+, Na^+, and H^+. *In* "Current Topics in Membranes and Transport" (Eds F. Bronner and A. Kleinzeller), Vol. 5, pp. 1–50. Academic Press, London and New York.

*Heinrich, M. R., Lanyi, J. K., Stoeckenius, W., and Oesterhelt D. (1977). Light energy transduction by the purple membrane of halophilic bacteria. *Fed. Proc.* **36,** 1797–1839.

*Helgerson, S. L. and Lanyi, J. K. (1977). Effects of cysteine on methionine transport in *Halobacterium halobium* vesicles. *J. Supramol. Struct.* **1977,** 139.

Henderson, R. (1975). The structure of the purple membrane from *Halobacterium halobium*: analysis of the X-ray diffraction pattern. *J. Mol. Biol.* **93,** 123–138.

*Henderson, R. (1977). The purple membrane from *Halobacterium halobium. Ann. Rev. Biophys. Bioeng.* **6,** 87–109.

Hescox, M. A. and Carlberg, D. M. (1972). Photoreactivation in *Halobacterium cutirubrum. Can. J. Microbiol.* **18,** 981–985.

*Hess, B. and Kuschmitz, D. (1977). The photochemical reaction of the 412 nm chromophore of bacteriorhodopsin. *FEBS Lett.* **74,** 20–24.

Higa, A. and Cazzulo, J. J. (1975). Some properties of the citrate synthase from the extreme halophile, *Halobacterium cutirubrum. Biochem. J.* **147,** 267–274.

Hiramatsu, T., Yokoyama, T., Ohno, Y., Yano, I., Masui, M. and Iwamoto, T. (1976). Preparation and chemical properties of the outer membrane of a moderately halophilic gram-negative bacterium. *Can. J. Microbiol.* **22,** 731–740.

Hiwatashi, T., Hara, M. and Yamada, A. (1958). Biological properties of halophilic bacteria no. 101 and no. 203. *J. Osaka City Med. Cent.* **7,** 550–552.

Hochachka, P. W. and Somero, G. N. (1973). "Strategies of Biochemical Adaptation," 358 pp. W. B. Saunders, London and Philadelphia.

*Hochstein, L. I. (1975). Studies of a halophilic NADH dehydrogenase. 2. Kinetic properties of the enzyme in relation to salt activation. *Biochem. Biophys. Acta* **403,** 58–66.

Hochstein, L. I., Dalton, B. P. and Pollock, G. (1976). The metabolism of carbohydrates by extremely halophilic bacteria: identification of galactonic acid as a product of galactose metabolism. *Can. J. Microbiol.* **22,** 1191–1196.

*Hollis, D. G., Weaver, R. E., Baker, C. N. and Thornsberry, C. (1976). Halophilic *Vibrio* species isolated from blood cultures. *J. Clin. Microbiol.* **3,** 425–431.

*Hong, F. T. (1977). Photoelectric and magneto orientation effects in pigmented biological membranes. *J. Colloid Interface Sci.* **58,** 471–497.

Horowitz, N. H. (1976). Life in extreme environments: Biological water requirements. *In* "Chemical Evolution of the Giant Planets" (Ed. C. Ponnam peruma), pp. 121–128. Academic Press, New York and London.

Hubbard, J. S. and Miller, A. B. (1969). Purification and reversible inactivation of the isocitrate dehydrogenase from an obligate halophile. *J. Bacteriol.* **99,** 161–168.

Ingram, M. (1957). Microorganisms resisting high concentrations of sugars or salt. *Symp. Soc. Gen. Microbiol.* **7,** 90–133.

*Jeske, R. (1975). A study on the influence of salinity fluctuations upon the bacterial flora of the lagoon Cienaga Grande de Santa Marta, Colombia and the adjacent coastal region. *Kiel Meeresforsch.* **31,** 7–16.

Johnson, M. K., Johnson, E. J., MacElroy, R. D., Speer, H. L. and Bruff, B. J. (1968). Effects of salts on the halophilic alga *Dundaliella viridis. J. Bacteriol.* **95,** 1461–1468.

Joshi, J. G., Guild, W. R. and Handler, P. (1963). The presence of two species of DNA in some halobacteria. *J. Mol. Biol.* **6,** 34–38.

Kamekura, M. and Onishi, H. (1974a). Halophilic nuclease from a moderately halophilic *Micrococcus varians. J. Bacteriol.* **119,** 339–344.

Kamekura, M. and Onishi, H. (1974b). Protease formation by a moderately halophilic *Bacillus* strain. *Appl. Microbiol.* **27,** 809–810.

Kao, O. H. W., Berns, D. S. and Town, W. R. (1973). The characterization of C-phycocyanin from an extremely halotolerant blue-green alga, *Coccochloris elabens. Biochem. J.* **131,** 39–50.

Kaplan, J. G, and Messmer, I. (1969). The combined effect of temperature and dilution on the activity and feedback inhibition of yeast aspartate transcarbamylase. *Can. J. Biochem.* **47,** 477–479.

Kates, M. (1972). Ether linked lipids in extremely halophilic bacteria. *In* "Ether Lipids, Chemistry and Biology" (Ed. F. Snyder), pp. 351–398. Academic Press, New York and London.

Kates, M. and Kushwaha, S. C. (1976). The diphytanyl glycerol ether analogues of phospholipids and glycolipids in membranes of *Halobacterium cutirubrum. In* "Lipids, Vol. 1, Biochemistry" (Eds R. Paoletti, G. Procellati and G. Jacini), pp. 267–275. Raven Press, New York.

Kates, M., Palameta, B., Joo, C. N., Kushner, D. J. and Gibbons, N. E. (1966). Allphatic diether analogs of glyceride-derived lipids. IV. The occurrence of *di*-o-dihydrophytyl-glycerol ether containing lipids in extremely halophilic bacteria. *Biochemistry* **5,** 4092–4099.

*Kayamura, Y. and Takada, H. (1975). Establishment of salt hypertonic medium which allows halophilic *Streptomyces* species to grow. *Trans. Mycol. Soc. Jap.* **16,** 282–288.

*Keradjopolos, D. and Holldorf, A. W. (1977). Thermophilic character of enzymes from extreme halophilic bacteria. *FEMS Microb.* **1,** 179–182.

*Koelsch, J. C., Tindall, D. R. and Yopp, J. H. (1976). Utilization of inorganic and organic nitrogen by the halophilic blue-green alga *Aphanothece halophytica. Plant. Physiol.* **57,** 96.

Koh, T. Y. (1975a). The isolation of obligate osmophilic mutants of the yeast *Saccharomyces rouxii. J. Gen. Microbiol.* **88,** 184–188.

Koh, T. Y. (1975b). Studies on the 'osmophilic' yeast *Saccharomyces rouxii* and an obligate osmophilic mutant. *J. Gen. Microbiol.* **88,** 101–114.

Komaratat, P. and Kates, M. (1975). The lipid composition of a halotolerant species of *Staphylococcus epidermidis. Biochim. Biophys. Acta* **398,** 464–484.

Koncewicz, M. A. (1972). Glycoproteins in the cell envelope of *Halobacterium halobium. Biochem. J.* **128,** 124P.

*Koops, H. P., Harms, H. and Wehrmann, H. (1976). Isolation of a moderate halophilic ammonia oxidizing bacterium, *Nitrosococcus mobilis* new species. *Arch. Microbiol.* **107,** 277–282.

Krantz, M. J. and Ballou, C. E. (1973). Analysis of *Halobacterium halobium* gas vesicles. *J. Bacteriol.* **114,** 1058–1067.

Kushner, D. J. (1964a). Microbial resistance to harsh and destructive environmental conditions. *Exp. Chemother.* **2,** 113–168.

Kushner, D. J. (1964b). Lysis and dissolution of cells and envelopes of an extremely halophilic bacterium. *J. Bacteriol.* **87,** 1147–1156.

Kushner, D. J. (1968). Halophilic bacteria. *Adv. Appl. Microbiol.* **10,** 73–99.

Kushner, D. J. (1971). Influence of solutes and ions on microorganism. *In* "Inhibition and Destruction of the Microbial Cell" (Ed. W. B. Hugo), pp. 259–283. Academic Press, New York and London.

Kushner, D. J. and Onishi, H. (1966). Contributions of protein and lipid components to the salt response of envelopes of an extremely halophilic bacterium. *J. Bacteriol.* **91,** 653–660.

Kushner, D. J., Bayley, S. T., Boring, J., Kates, M. and Gibbons, N. E. (1964). Morphological and chemical properties of cell envelopes of the extreme halophile, *Halobacterium cutirubrum. Can. J. Microbiol.* **10,** 483–497.

Kushwaha, S. C. and Kates, M. (1976). Effect of nicotine on biosynthesis of C_{50} carotenoids in *Halobacterium cutirubrum. Can. J. Biochem.* **54,** 824–829.

Kushwaha, S. C., Gochnauer, M. B., Kushner, D. J. and Kates, M. (1974). Pigments and isoprenoid compounds in extremely and moderately halophilic bacteria. *Can. J. Microbiol.* **20,** 241–245.

Kushwaha, S. C., Kramer, J. and Kates, M. (1975a). Isolation and characterization of C_{50}-carotenoid pigments and other polar isoprenoids from *Halobacterium cutirubrum. Biochim. Biophys. Acta* **398,** 303–314.

Kushwaha, S. C., Kates, M. and Martin, W. G. (1975b). Characterization and composition of the red and purple membrane from *Halobacterium cutirubrum. Can. J. Biochem.* **53,** 284–292.

Kushwaha, S C., Kates, M. and Porter, J. W. (1976a). Enzymatic synthesis of C_{40} carotenes by cell-free preparation from *Halobacterium cutirubrum. Can. J. Biochem.* **54,** 816–823.

Kushwaha, S. C., Kates, M. and Stoeckenius, W. (1976b). Comparison of purple membrane from *Halobacterium cutirubrum* and *Halobacterium halobium. Biochim. Biophys. Acta.* **426,** 703–710.

*Kushwaha, S. C., Kates, M. and Kramer, J. K. G. (1977). Occurrence of indole in cells of extremely halophilic bacteria. *Can. J. Microbiol.* **23,** 826–828.

Lanyi, J. K. (1971). Studies of the electron transport chain of extremely halophilic bacteria. VI. Salt-dependent dissolution of the cell envelope. *J. Biol. Chem.* **246,** 4552–4559.

Lanyi, J. K. (1974a). Salt dependent properties of proteins from extremely halophilic bacteria. *Bacteriol. Rev.* **38,** 272–290.

Lanyi, J. K. (1974b). Irregular bilayer syructure in vesicles prepared from *Halobacterium cutirubrum* lipids. *Biochim. Biophys. Acta* **356,** 245–256.

*Lanyi, J. K. (1977). Transport in *Halobacterium halobium*: Light-induced cation gradients, amino acid uptake kinetics and reconstitution. *J. Supramol. Struct.* **1977,** 133.

Lanyi, J. K. and Silverman, M. P. (1972). The state of binding of intracellular K^+ in *Halobacterium cutirubrum. Can. J. Microbiol.* **18,** 993–995.

Lanyi, J. K. and Stevenson, J. (1970). Studies of the electron transport chain of extremely halophilic bacteria. IV. Role of hydrophobic forces in the structure of menadione reductase. *J. Biol. Chem.* **245,** 4074–4080.

Lanyi, J. K., Plachy, W. Z. and Kates, M. (1974). Lipid interactions in membranes of extremely halophilic bacteria. II. Modification of the bilayer structure by squalene. *Biochemistry* **13,** 4914–4920.

Larsen, H. (1962). Halophilism. *In* "The Bacteria" (Eds I. C. Gunsalus and R. Y. Stanier), Vol. 4, pp. 297–342. Academic Press, New York and London.

Larsen, H. (1967). Biochemical aspects of extreme halophilism. *Adv. Microbiol. Physiol.* **1,** 97–132.

Larsen, H. (1973). The halobacteria's confusion to biology. *Antonie van Leeuwenhoek* **39,** 383–396.

*Libinzon, A. E., Demina, A. I., Kulov, G. I. and Shestialtynova, I. S. (1977). Halophilic vibrios of the Azov Sea. *Zhur. Mikrobiol. Epidemiol. Immunobiol.* **1977,** 77–80.

Lieberman, M. M. and Lanyi, J. K. (1972). Threonine deaminase from extremely halophilic bacteria. Cooperative substrate kinetics and salt dependence. *Biochemistry* **11,** 211–216.

Liebl, V., Kaplan, J. G. and Kushner, D. J. (1969). Regulation of a salt-dependent enzyme: the aspartate transcarbamylase of an extreme halophile. *Can. J. Biochem.* **47,** 1095–1097.

Louis, B. G. and Fitt, P. S. (1972a). Isolation and properties of highly purified *Halobacterium cutirubrum* deoxyribonucleic acid-dependent ribonucleic acid polymerase. *Biochem. J.* **127,** 69–80.

Louis, B. G. and Fitt, P. S. (1972b). The role of *Halobacterium cutirubrum* deoxyribonucleic acid-dependent ribonucleic acid polymerase subunits in initiation and polymerization. *Biochem. J.* **127,** 81–86.

Louis, B. G. and Fitt, P. S. (1972c). Purification and properties of the ribonucleic acid-dependent ribonucleic acid polymerase from *Halobacterium cutirubrum. Biochem. J.* **128,** 755–762.

MacDonald, R. E. and Lanyi, J. K. (1975). Light-induced leucine transport in *Halobacterium halobium* envelope vesicles: a chemiosmotic system. *Biochemistry* **14,** 2882–2889.

MacLeod, R. A. (1965). The question of the existence of specific marine bacteria. *Bacteriol. Rev.* **29,** 9–23.

Masui, M. and Wada, S. (1973). Intracellular concentrations of Na^+, K^+, and Cl^- of a moderately halophilic bacterium. *Can. J. Microbiol.* **19,** 1181–1186.

Matheson, A. T., Yaguchi, M. and Visentin, L. P. (1975). The conservation of amino acids in the N-terminal position of ribosomal and cytosol proteins from *Escherichia coli, Bacillus stearothermophilus,* and *Halobacterium cutirubrum. Can. J. Biochem.* **53,** 1323–1327.

Matheson, A. T., Sprott, G. D., McDonald, I. J. and Tessier, H. (1976). Some properties of an unidentified halophile: growth characteristics, internal salt concentration, and morphology. *Can. J. Microbiol.* **22,** 780–786.

*McCarthy, D. H. (1975). *Aeromonas proteolytica,* a halophilic aeromonad. *Can. J. Microbiol.* **21,** 902–904.

McClare, C. W. F. (1967). Bonding between proteins and lipids in the envelopes of *Halobacterium halobium. Nature, Lond.* **216,** 766–771.

Mescher, M. F. and Strominger, J. L. (1976a). Purification and characterization of a prokaryotic glycoprotein from the cell envelope of *Halobacterium salinarium. J. Biol. Chem.* **251,** 2005–2014.

Mescher, M. F. and Strominger, J. L. (1976b). Structural (shape-maintaining) role of the cell surface glycoprotein of *Halobacterium salinarium. Proc. Natl. Acad. Sci. USA* **73,** 2687–2691.

Mescher, M. F., Strominger, J. L. and Watson, S. W. (1974). Protein and carbohydrate composition of the cell envelope of *Halobacterium salinarium*. *J. Bacteriol.* **120,** 945–954.

Miller, D. M., Jones, J. H., Yopp, J. H., Tindall, D. R. and Schmid, W. E. (1976a). Ion metabolism in a halophilic blue-green alga, *Aphanothece halophytica*. *Arch. Microbiol.* **111,** 145–149.

Miller, D. M., Yopp, J. H., Schmid, W. E. and Tindall, D. R. (1976b). Intracellular ions and water in response to environmental changes in an obligately halophilic blue-green alga. *Plant Physiol.* **57,** 96.

Moore, R. L. and McCarthy, B. J. (1969a). Characterization of the deoxyribonucleic acid of various strains of halophilic bacteria. *J. Bacteriol.* **99,** 248–254.

Moore, R. L. and McCarthy, B. J. (1969b). Base sequence homology and renaturation studies of the deoxyribonucleic acid of extremely halophilic bacteria. *J. Bacteriol.* **99,** 255–262.

*Mullakhanbhai, M. F. and Larsen, H. (1975). *Halobacterium volcanii* new species, a Dead Sea halobacterium with a moderate salt requirement. *Arch. Microbiol.* **104,** 207–214.

*Nissenbaum, A. (1975). The microbiology and biogeochemistry of the Dead Sea. *Microb. Ecol.* **2,** 139–161.

*Nissenbaum, A. (1977a). Minor and trace elements in Dead Sea Water. *Chem. Geol.* **19,** 99–111.

*Nissenbaum, A. (1977b). The physicochemical basis of legends of the Dead Sea. Proc. Conference on Terminal Lakes, Utah.

*Nissenbaum, A. and Kaplan, I. R. (1976). Sulfur and carbon isotopic evidence for biogeochemical processes in the Dead Sea ecosystem. Environmental Biogeochemistry (Ed. J. O. Nriagu), pp. 309–325. Ann Arbor Science Publishers, Inc., Ann Arbor, Michigan.

Norberg, P., Kaplan, J. G. and Kushner, D. J. (1973). Kinetics and regulation of the salt-dependent aspartate transcarbamylase of *Halobacterium cutirubrum*. *J. Bacteriol.* **113,** 680–686.

Novitsky, T. J. and Kushner, D. J. (1975). Influences of temperature and salt concentration on the growth of a facultatively halophilic "Micrococcus" sp. *Can. J. Microbiol.* **21,** 107–110.

Novitsky, T. J. and Kushner, D. J. (1976). A facultatively halophilic coccus, *Planococcus halophilus* sp. nov. *Int. J. Syst. Bacteriol.* **26,** 53–57.

Oda, G., Strøm, A. R., Visentin, L. P. and Yaguchi, M. (1974). An acidic, alanine-rich 50S ribosomal protein from *Halobacterium cutirubrum*: amino acid sequence homology with *Escherichia coli* protein L7 and L12. *FEBS Lett.* **43,** 127–130.

Oesterhelt, D. and Stoeckenius, W. (1971). Rhodopsin-like protein from the purple membrane of *Halobacterium halobium*. *Nature New Biol.* **233,** 149–152.

Oesterhelt, D. and Stoeckenius, W. (1973). Functions of a new photoreceptor membrane. *Proc. Natl. Acad. Sci. USA* **70,** 2853–2857.

Ohno, Y., Yano, I., Hiramatsu, T. and Masui, M. (1976). Lipids and fatty acids of a moderately halophilic bacterium, No. 101. *Biochim. Biophys. Acta* **424,** 337–350.

Okamoto, H. and Suzuki, Y. (1964). Intracellular concentration of ions in a halophilic strain of *Chlamydomonas* I. Concentrations of Na^+, K^+, and Cl^- in the cell. *Z. Allg. Mikrobiol.* **4,** 350–357.

Onishi, H. (1963). Osmophilic yeasts. *Adv. Food Res.* **12,** 53–94.

Onishi, H. (1972). Salt response of amylase produced in media of different NaCl or

KCl concentrations by a moderately halophilic Micrococcus. *Can. J. Microbiol.* **18,** 1617–1620.

Onishi, H. and Kamekura, M. (1972). *Micrococcus halobius* sp. n. *Int. J. Syst. Bacteriol.* **22,** 193–210.

Onishi, H., McCance, M. E. and Gibbons, N. E. (1965). A synthetic medium for extremely halophilic bacteria. *Can. J. Microbiol.* **11,** 365–373.

Plachy, W. Z., Lanyi, J. K. and Kates, M. (1974), Lipid interactions in membranes of extremely halophilic bacteria. I. Electron spin resonance and dilatometric studies of bilayer structure. *Biochemistry* **13,** 4906–4913.

*Plakunova, V. G. (1976). Biphasic character of photodependent transport of alanine C^{14} into *Halobacterium halobium* cells. *Biofizika* **21,** 661–664.

*Post, F. J. (1977). The microbial ecology of the Great Salt Lake. *Microb. Ecol.* **3,** 143–165.

Pugh, E. L., Wassef, M. K. and Kates, M. (1971). Inhibition of fatty acid synthetase in *Halobacterium cutirubrum* and *Escherichia coli* by high salt concentrations. *Can. J. Biochem.* **49,** 953–958.

Racker, E. and Stoeckenius, W. (1974). Reconstitution of purple membrane vesicules catalyzing light-driven proton uptake and adenosine triphosphate formation. *J. Biol. Chem.* **249,** 662–663.

Rafeali-Eshkol, D. (1968). Studies on halotolerance in a moderately halophilic bacterium. Effect of growth conditions on salt resistance of the respiratory system. *Biochem. J.* **109,** 679–685.

Rayman, M. K., Gordon, R. C. and MacLeod, R. A. (1967). Isolation of a Mg^{++} phospholipid from *Halobacterium cutirubrum. J. Bacteriol.* **93,** 1465–1466.

Reistad, R. (1972). Cell wall of an extremely halophilic coccus: investigation of ninhydrin-positive compound. *Arch. Mikrobiol.* **82,** 24–30.

*Rinehart, C. A. and Hubbard, J. S. (1976). Energy coupling in active transport of proline and glutamate by the photosynthetic halophile. *Ectothiorhodospira halophila. J. Bacteriol.* **127,** 1255–1264.

Robinson, J. (1952). The effects of salts on the nitratase and lactic acid dehydrogenase activity of *Micrococcus halodenitrificans. Can. J. Bot.* **30,** 155–163.

Robinson, J. and Gibbons, N. E. (1952). The effect of salts on the growth of *Micrococcus halodenitrificans* n. sp. *Can. J. Bot.* **30,** 147–154.

Rogers, H. J., McConnell, M. and Burdett, I. D. (1970). The isolation and characterization of mutants of *Bacillus subtilis* and *Bacillus lichemformis* with disturbed morphology and cell division. *J. Gen Microbiol.* **61,** 155–171.

*Rose, A. H. (1976). Osmotic stress and microbial surviv al. *Symp. Soc. Gen. Microbiol.* **26,** 155–182.

Scott, W. J. (1957). Water relations of food spoilage microorganisms. *Adv. Food Res.* **7,** 83–127.

Shindler, D. B. (1976). Physiology and enzymatic aspects of moderately halophilic microorganisms. Ph.D. thesis, University of Ottawa.

Shindler, D. B., Wydro, R. W. and Kushner, D. J. (1977). Cell-bound cations of the moderately halophilic bacterium, *Vibrio costicola. J. Bacteriol.* **130,** 698–703.

*Sineshchekov, V. A. and Litvin, F. F. (1976). Luminescence of bacteriorhodopsin in purple membranes from *Halobacterium halobium* cells. *Biofizika* **21,** 313–320.

Smith, F. B. (1938). An investigation of a taint in rib bones of bacon. The determination of halophilic *Vibrios* (n. spp.) *Proc. R. Soc. Queensland* **49,** 29–52.

Soo-Hoo, T. S. and Brown, A. D. (1967). A basis of the specific sodium requirement

for morphological integrity of *Halobacterium halobium*. *Biochim. Biophys. Acta* **135,** 164–166.
*Steber, J. and Schleifer, K. H. (1975). *Halococcus morrhuae*: A sulfated heteropolysaccharide as the structural component of the bacterial cell wall. *Arch. Microbiol.* **105,** 173–177.
Steensland, H. and Larsen, H. (1969). A study of the cell envelope of the halobacteria. *J. Gen. Microbiol.* **55,** 325–336.
Stevenson, J. (1966). The specific requirement for sodium chloride for the active uptake of L-glutamate by *Halobacterium salinarium*. *Biochem. J.* **99,** 257–260.
Stoeckenius, W. (1976). The purple membrane of salt-loving bacteria. *Sci. Am.* **234,** 38–46.
Stoeckenius, W. and Rowen, R. (1967). A morphological study of *Halobacterium halobium* and its lysis in media of low salt concentration. *J. Cell Biol.* **34,** 365–393.
Strøm, A. R. and Visentin, L. P. (1973). Acidic ribosomal proteins from the extreme halophile, *Halobacterium cutirubrum*. The simultaneous separation, identification and molecular weight determination. *FEBS Lett.* **37,** 274–280.
Strøm, A. R., Hasnain, S., Smith, N., Matheson, A. T. and Visentin, L. P. (1975a). Ion effects on protein-nucleic acid interactions: the disassembly of the 50S ribosomal subunit from the halophilic bacterium, *Halobacterium cutirubrum*. *Biochim. Biophys. Acta* **383,** 325–337.
Strøm, A. R., Oda, G., Hasnain, S., Yaguchi, M. and Visentin, L. P. (1975b). Temperature related alterations in the acidic alanine-rich "A" protein from the 50S ribosomal particle of the extreme halophile, *Halobacterium cutirubrum*. *Mol. Gen. Genet.* **140,** 15–27.
*Tabita, F. R. and McFadden, B. A. (1976). Molecular and catalytic properties of ribulose 1,5 bis phosphate carboxylase from the photosynthetic extreme halophile *Ectothiorhodospira halophila*. *J. Bacteriol.* **126,** 1271–1277.
*Takeuchi, T. (1976). Growth factors for *Pediococcus halophilus* in polypeptone. *J. Ferment. Technol.* **54,** 302–307.
Thompson, J. and MacLeod, R. A. (1973). Na^+ and K^+ gradients and α-aminoisobutyric acid transport in a marine pseudomonad. *J. Biol. Chem.* **248,** 7106–7111.
Tomlinson, G. A. and Hochstein, L. I. (1972a). Isolation of carbohydrate-metabolizing extremely halophilic bacteria. *Can. J. Microbiol.* **18,** 698–701.
Tomlinson, G. A. and Hochstein, L. I. (1972b) Studies on acid production during carbohydrate metabolism by extremely halophilic bacteria. *Can. J. Microbiol.* **18,** 1973–1976.
Tomlinson, G. A. and Hochstein, L. I. (1976). *Halobacterium saccharavorum* sp. nov., a carbohydrate-metabolizing, extremely halophilic bacterium. *Can. J. Microbiol.* **22,** 587–591.
Tomlinson, G. A., Koch, T. K. and Hochstein, L. I. (1974). The metabolism of carbohydrates by extremely halophilic bacteria: glucose metabolism via a modified Entner-Doudoroff pathway. *Can. J. Microbiol.* **20,** 1085–1091.
Torsvik, T. and Dundas, I. (1974). Bacteriophage of *Halobacterium salinarium*. *Nature, Lond.* **248,** 680–681.
Ulitzur, S. and Kessel, M. (1973). Giant flagellar bundles of *Vibro alginolyticus*. *Arch. Mikrobiol.* **94,** 331–339.
Unemoto, T. and Hayashi, M. (1969). Chloride ion as a modifier of 2′3′-cyclic phosphodiesterase purified from halophilic *Vibrio alginolyticus*. *Biochim. Biophys. Acta* **171,** 89–102.

Vidal, M. C. and Cazzulo, J. J. (1976). On the regulatory properties of a halophilic malic enzyme from *Halobacterium cutirubrum. Experientia* **32,** 441–442.

Visentin, L. P., Chow, C., Matheson, A. T., Yaguchi, M. and Rollin, F. (1972). *Halobacterium cutirubrum* ribosomes. Properties of the ribosomal proteins and ribonucleic acids. *Biochem. J.* **130,** 103–110.

Visentin, L. P., Matheson, A. T. and Yaguchi, M. (1974). Homologies in procaryotic ribosomal proteins: alanine rich acidic proteins associated with polypeptide translocation. *FEBS Lett.* **41,** 310–314.

Volcani, B. E. (1940). Studies of the microflora of the Dead Sea. Ph.D. thesis., Hebrew University, Jerusalem.

*Vreeland, R. H. and Litchfield, C. D. (1976). Identification of bacteria from a solar salt facility. *Abst. Ann. Meeting Am. Soc. Microbiol.* **76,** 181.

Wais, A. C., Kon, M., MacDonald, R. E. and Stollar, B. D. (1975). Salt-dependent bacteriophage infecting *Halobacterium cutirubrum* and *Halobacterium halobium. Nature, Lond.* **256,** 314–315.

Wydro, R. W., Madira, W., Hiramatsu, T., Kogut, M. and Kushner, D. J. (1977). Salt-sensitive *in vitro* protein synthesis by moderately halophilic bacterium. *Nature, Lond.* **269,** 824–825.

*Yamanaka, K. and Okada, T. (1975). Production of arginase by a halophilic bacterium, especially on effect of manganese salt. *J. Agric. Chem. Soc. Japan* **49,** 363–369.

*Yanai, Y., Rosen, B., Pinsky, A. and Sklan, D. (1977). The microbiology of pickled cheese during manufacture and maturation. *J. Dairy Res.* **44,** 149–153.

*Yopp, J. H., Miller, D. M., Tindall, D. R. and Schmid, W. E. (1976). Physiochemical parameters of solutes differentially affecting glucose-6-phosphate dehydrogenase from a halophilic blue-green alga. *Plant Physiol.* **57,** 100.

Chapter 9

Water Relations of Microorganisms in Nature

D. W. SMITH

University of Delaware

I. Introduction 369
II. Fungi 370
III. Lichens 371
IV. Bacteria 373
V. Algae 374
VI. Mosses 376
VII. Mechanisms of damage by water stress 376
References 377

I. Introduction

There are several recent works on various aspects of survival and growth of microorganisms in nature (Holding *et al.*, 1974; Gray and Postgate, 1976; Brown, 1974; Heinrich, 1976). There is, however, relatively little mention of the importance of water in microbial ecological studies. Friedmann and Galun (1974) give a good review of microorganisms in desert habitats, but even their report does not emphasize water as a controlling factor *per se*. Similarly Gray (1976) considers water as a factor in the survival of soil microorganisms, but does not discuss general effects of soil water status.

The focus of this review will include the importance of water to microorganisms in dry environments as well as the general significance of water to microorganisms in nature. There are special cases such as halophilic bacteria and airborne microorganisms which will not be considered here. Kushner discusses halophiles in Chapter 8 and information concerning airborne microorganisms is reviewed by Gregory (1971).

The water status of microorganisms and their environment has been expressed in several ways, the simplest of which is moisture or water content. Although useful, water content measurements reveal nothing about the availability of water to a given organism. Two different soils may have the same water content, but very different water availabilities, due to adsorption and solution phenomena. Therefore it is becoming more common for water status to be presented in terms of the energy relations or potential of the water in the soil. This concept describes the amount of thermodynamic work which must be done by an organism to obtain water. The units for water potential are usually bars. 1 bar = 10^6 dyne cm^{-2} = 0·987 atm = 75 cm mercury pressure = 1022 cm water pressure. Pure water has a potential of 0. All solutions have negative potentials with respect to pure water, and their potentials are therefore expressed as —bars. Water availability may also be expressed in terms of water activity (a_w) or relative humidity (RH). These terms, most commonly used by food microbiologists, are related to each other and to water potential (ψ) as follows (Lang, 1967):

$$RH = \frac{a_w}{100}$$

$$\psi = \frac{1000\,RT}{W_A} \ln (a_w)$$

where R is the gas constant, T the absolute temperature and W_A the molecular weight of water. Water potential may be altered in two ways, which are termed matric and osmotic (Griffin, 1969). Matric water potential results from the adsorption of water molecules to surfaces at interfaces in solid substrates, such as soil. Osmotic contributions to water potential occur in solutions as the result of interactions between solute molecules and water molecules.

Detailed studies of the effects of water stress have been conducted on fungi, lichens, bacteria, algae, and mosses. The results for each of these groups will be summarized separately.

II. Fungi

Fungi have consistently been shown to be the most desiccation resistant members of the soil community (Griffin, 1963, 1969). This desiccation tolerance of fungi has implications in many areas: (1) fungal diseases of plants; (2) value of mycorrhizal fungi to their hosts; (3) role of fungi as decomposers; (4) fungal production of antibiotics and their possible role in nature; and (5) contribution to the physiology and ecology of lichens.

Griffin (1969) has concluded that many fungal diseases increase in incidence with increasing matric water potential (i.e. greater water stress). This disease pattern is related to the ability of the fungus to withstand the gradients of water potential which are found around plant roots (Schippers *et al.*, 1967). These fungi are also capable of withstanding periodic dry periods which occur in soil.

The improved growth of many plants when they are associated with mycorrhizal fungi is well known. It has recently been suggested (Mexal and Reid, 1973) that the contribution of the fungus in the association may be more than increased mineral absorption; the fungi may also offer the plants protection against desiccation, since they are so efficient at absorbing water. This concept of protection, or water-gathering, has also been proposed in connection with lichens (Brock, 1975b).

The importance of fungi in the decomposition of dead plant material is obvious, but worthy of mention. The recycling of plant material between growing seasons is essential to the balance of elements in nature. The largest part of this recycling occurs when conditions are not at their optimal for biological activity. Temperature is often considered, but also of great importance is the water potential of the soil where the decomposition is taking place (Shameemullah *et al.*, 1971). Growing plants contribute water to the soil (Schippers *et al.*, 1967); therefore the cessation of active plant growth may lower the water potential of the soil. The ability of the fungi to survive this water stress and function at the low water potential is essential for continuity in nutrient cycling.

The production and function of antibiotics in nature has been a question of considerable interest and debate for some time (Brian, 1957). Bruehl *et al.* (1972) have found that *Cephalosporium gramineum* has its peak production of antibiotics at water potentials of −30 to −40 bars, while reduction of growth of the fungus begins to occur at −10 bars. They claim that antagonists of *Cephalosporium* in its natural straw environment grow best at water potentials from −10 to −67 bars. The survival of the fungus under these conditions where water potential limits its growth rate would seem to depend on its water stress-stimulated antibiotic production.

III. Lichens

Lichens are usually the predominant macroscopic biota found in dry regions. It may be argued in fact that these alga-fungus symbiotic associations exist solely because of harsh environmental conditions. This view is supported by the frequent observation that the lichen association is extremely difficult to maintain in the laboratory. Removal of the stressful environmental conditions leads to the outgrowth of one or the other member of the association with the resultant breakdown of the lichen as an entity. Reassociation of the phycobiont (algal partner) and mycobiont (fungal partner) into a complete lichen is possible, but only under carefully controlled, stressful conditions. The key feature which must be controlled is the water regime. Only when the lichen is subjected to the stress of alternate wetting and drying does the reassociation take place (Ahmadjian, 1967).

These observations have led to a great deal of research into the water relations of the intact lichen thallus as well as the individual phycobionts and mycobionts (Kappen, 1973; Blum, 1973). Distribution of lichens near aquatic environments (Rouse and Kershaw, 1973), in deserts (Rogers, 1971; Lange *et al.*, 1970; in Rogers, 1971), and in forests (Jesberger and Sheard, 1973; Lechowicz and Adams, 1974) has been shown to relate to water availability. Laboratory studies with intact thalli have shown that the net CO_2 assimilation capabilities of lichens are related to water content of the thallus. There appears to be an optimum assimilation when the lichen water content is near 50% of the saturation value for the thallus (Kershaw, 1972; Adams, 1971). It has also been suggested that fluxes in moisture, i.e. alternate wet and dry cycles, may be more important to lichen physiology than the water content at any given moment (Harris and Kershaw, 1971). These facts combine to limit many lichens to being photosynthetically active only for a few hours in the morning when they are still wet from night time dew (Lange *et al.*, 1970). Farrar (1976a, b) has recently investigated in some detail the effect of cyclical wet–dry regimes on lichens. He has found that continued saturation is harmful to the metabolic activity of his specimens. His laboratory wet–dry cycles (attempting to simulate natural conditions) have led him to conclude that the lichens actually require the fluctuation.

As summarized above, much work with fungi has been done in terms of water potential, or energy relations, of the environment. Unfortunately this is not the case for lichens. Almost all results are expressed as water content (moisture) or percentage of thallus saturation. The recent work of Brock (1975b) is a welcome incursion into the study of the relation of water energetics to lichen ecology and physiology.

Why are lichens able to survive in dry habitats? Fungi as a group are tolerant to water stress while algae are quite susceptible to desiccation. The logical inference therefore is that the fungus somehow confers desiccation resistance upon the lichen association. This interpretation has recently received some experimental support (Brock, 1975b). The photosynthetic capabilities of intact thalli and of algae liberated from the thalli by grinding were measured in response to water potential stress. In all cases the minimum water potential which allowed photosynthesis was lower in the intact thalli. In one case this difference was −162 bars, a considerable water potential change. There is a difficulty with the explanation of fungus protection, however. If it is true that the fungi are able to absorb water from dry environments by doing the requisite thermodynamic work, then how can the desiccation-susceptible algae obtain this water from the fungus? As Brock suggests, it would seem to be necessary to postulate active water secretion by the fungus. In contrast Ahmadjian (1967) claims that water absorption by lichens is almost completely a physical process and does not involve active absorption of water. In support of this view he cites evidence that water uptake by dead and live lichens is similar. These classic works (Bachmann, 1922; Smyth, 1934, in Ahmadjian, 1967) are concerned only with water contents of thalli and not with water potential relations.

IV. Bacteria

The water relations of bacterial populations have been investigated under a wide variety of conditions. Food microbiology (Scott, 1957), microbial survival studies (Bateman *et al.*, 1962; Goepfert *et al.*, 1970) and, of course, soil microbiology have all made major contributions to our understanding of microbial water relations. This presentation will concentrate on soil bacteria.

The simplest type of study is the measurement of bacterial survival as a function of time in dried soil. Such studies have indicated a wide range of desiccation susceptibilities by different types of bacteria. The number of viable *Pseudomonas* was shown to decline by a factor of 100 within a month of inoculation and incubation in air-dried soil (Robinson *et al.*, 1965). On the other hand, *Arthrobacter* cells incubated under similar conditions showed a 50% initial decrease and then remained viable for at least six months (Boylen, 1973). Examination of native soil samples which have been stored for various periods of time shows that *Azotobacter* may remain viable for as long as 13 years (Vela, 1974). This latter report does not discuss any decreases in numbers of viable cells nor are the dry conditions quantified. It had previously been demonstrated that it is probably the cysts of *Azotobacter* which are responsible for the survival rather than vegetative cells (Socolofsky and Wyss, 1962).

Williams *et al.* (1972) investigated survival of actinomycetes in both pure culture and natural soil systems. Their work is the best example of soil bacterial survival studies because they conducted their experiments in terms of water potential. An extremely cogent point of their findings is that as the water stress on natural soil samples increases the percentage of actinomycetes among microbial survivors increases at the expense of fungi and eubacteria in the soil. It should be noted that the numbers of all three types of organisms decreased at the lower water potentials. Pure culture studies presented in the same report demonstrate that the desiccation resistance of the actinomycetes reflects the tolerance of spores. Vegetative actinomycete hyphae were considerably more easily damaged by water stress.

It appears therefore that the survival of bacteria in soil is greatly increased if the organisms possess some type of resistant structure. The vegetative cells of *Pseudomonas* are very susceptible while the *Azotobacter* cysts and actinomycete (*Streptomyces*) spores are considerably more resistant. *Arthrobacter* presents a more confusing picture. This bacterium does not possess any obvious resting or protective structure. Perhaps the peculiar rod–coccus conversions which *Arthrobacter* undergoes are relevant. It would be quite interesting to learn if one form were significantly more desiccation resistant than the other. If so perhaps a correlation could be made between this resistance and the frequency of occurrence of the two types in natural soil.

Natural nutrient cycles, especially the nitrogen cycle, are directly dependent on specific bacterial activities. It is therefore of considerable interest to know what effect water stress has on these organisms. Both nitrification (Dubey, 1968) and symbiotic nitrogen-fixation (Sprent, 1971) are reduced by lowered water potentials. The importance of these effects should be considered when evaluating the overall impact of dry conditions on productivity, either agricultural or natural. The water relations of the plants are of course important, but the effect of water stress on potential pathogens and beneficial microorganisms should not be overlooked.

V. Algae

Algae are the most experimentally neglected organisms in the soil. This situation is made worse by the common acknowledgement that these organisms are important in terrestrial productivity and revegetation of denuded areas (Shields and Durrell, 1964; Lund, 1967; Bold, 1970).

It was to fill this gap that studies were initiated with natural populations of *Cyanidium caldarium* in hot, acid soils of Yellowstone National Park, Wyoming, USA (Smith and Brock, 1973a, b). We found that natural populations of *Cyanidium* never occurred at water potentials below −4 bars. The minimum water potential which allowed growth in the laboratory was about −80 bars. This discrepancy between laboratory capabilities and natural reality is at least partially related to the sharp spatial water potential gradients found in the Yellowstone soils where *Cyanidium* flourishes. This gradient is from −50 bars at the surface to −2 to −4 bars at 3 to 5 mm (Smith and Brock, 1973b). All natural soil populations of *Cyanidium* occur in a band 3 to 5 mm below the soil surface. It therefore appears that there is sufficient ecological pressure exerted by water stress to force the phototrophic *Cyanidium* to live in a shaded environment. It should be noted that the soil in these geothermal areas is translucent opaline silica (Walter, personal communication; Smith and Brock, 1973b).

These studies with Yellowstone populations of *Cyanidium* were made easier because the extreme conditions of pH and temperature simplify the community structure so that virtually pure cultures of *Cyanidium* exist in most areas. It was also necessary to develop new quantitative procedures to measure the activity of soil populations. The technique which was developed (Smith *et al.*, 1972, 1973) allows sensitive determination of CO_2 fixation. A key feature of the process is that the radioactive tracer, $^{14}CO_2$ gas, can be added without alteration of the water potential of the sample. This procedure has been modified and used in studying the activity of bacteria (Belly and Brock, 1974), lichens (Brock, 1975b), and other algae (Brock, 1975a).

Brock's (1975a) study with desert mats of a blue-green alga made a most interesting distinction between survival and the ability of microorganisms to function at a low water potential. The material he used is normally inactive in nature because of the dry conditions. It was in fact necessary to first incubate the samples under light and wet conditions in order to obtain active populations. However, these revived populations were inactivated by relatively slight water potential stress (−18 to −28 bars). It therefore appears that these algae are only active intermittently throughout the year, when rainfall or dew have sufficiently wetted them. The mucilaginous sheath of these organisms is very likely an important factor in allowing desiccation survival (Shields *et al.*, 1957). Brock also found that stressing the algae matrically rather than osmotically was more harmful. This latter finding is significant since the main stress experienced by soil organisms is matric. The ecological consequences of osmotic versus matric stress will be considered below.

The aerosol study of Ehresmann and Hatch (1975) showed that prokaryotic algae are more desiccation resistant than eukaryotic algae. They found the blue-green alga, *Synechococcus*, behaved much like airborne bacteria, thus extending the investigation of the water relations of algae into a new area.

Aquatic algae are stressed only osmotically unless they are attached to rocks or some other solid substrate. These osmotic effects are most easily seen in natural saline environments. The germination of akinetes of the Chaetophoralean alga *Ctenocladus circinnatus* was impaired at −40 bars and eliminated at −70 bars (Blinn, 1971). The halophilic alga *Dunaliella* is readily found in Great Salt Lake which has salinities more than four times that of sea water. Brock (1975c) has demonstrated in the laboratory that the optimum water potential of this alga for growth and photosynthesis is much higher (−100 to −150 bars; 12 to 17% NaCl) than that of its natural environment (−300 bars; 25% NaCl). The survival of this alga under these conditions appears to depend at least partially on escape from competition and predation at the lower salinities. The existence and growth, however slow, of *Dunaliella* in saturated NaCl establish it as the alga most resistant to low water potential (Brock, 1975c).

The last algal type to be considered is the green alga *Trebouxia*, which is a common phycobiont in many lichens. Brock (1975b) has shown that this alga is able to photosynthesize at water potentials of up to −200 bars when in symbiotic association. Lichens which contain *Trebouxia* as the phycobiont appear to be the most desiccation resistant lichens (Ahmadjian, 1967).

VI. Mosses

Mosses have ecological determinants similar to those of several lichens. Therefore these simple plants are of interest to briefly consider with the Protists already discussed. Lee and Stewart (1971) measured the effect of desiccation on the rate of photosynthesis in mosses from different habitats. They found that mosses which were collected from drier areas were more resistant to desiccation than were those from wetter areas. Dilks and Proctor (1974) showed that the ability to survive in a dry environment was not necessarily related to the ability to be active in that environment. This finding is similar to that of Brock (1975a) with desert crusts of blue-green algae. Bewley (1973, 1974) has discovered considerable desiccation resistance among mosses with respect to protein synthesis capability. He found that an aquatic moss was more desiccation sensitive than a terrestrial one. He has extended his investigations in a more molecular direction in an effort to explain the mechanism of desiccation damage in these organisms (Dhindsa and Bewley, 1976). It is somewhat paradoxical that such detailed molecular information should exist for the mosses, about whose ecology we know relatively little.

VII. Mechanisms of Damage by Water Stress

Many theories have been put forth to explain why organisms need a certain amount of water, or rather why they are harmed when this water is absent. Among these suggestions is that under certain conditions of water stress energy is diverted from growth to osmoregulation (Bernstein, 1963). It has also been hypothesized that intracellular biochemical activities and enzyme reactions are inhibited by low internal water potentials (Christian and Waltho, 1962).

It is interesting that, where the effects of both types of stress have been measured, organisms are more sensitive to matric water stress than to osmotic (Sommers *et al.*, 1970; Adebayo and Harris, 1971; Brock, 1975b). As pointed out by Adebayo *et al.* (1971), this greater resistance to matric stress may be related to the requirement of most microorganisms for a positive turgor pressure. In order for a cell to be turgid, it must possess an internal water potential lower than that of its environment or else the cell will collapse. If a cell (or organism) is stressed osmotically, then the possibility exists that it can incorporate some of the solute and thus achieve a low internal water potential. In some instances this requirement is met by the synthesis of low molecular weight solutes. *Dunaliella* lives in saturated NaCl, but its enzymes are sensitive to high concentrations of salt. This alga produces high concentrations of glycerol in order to maintain a low internal water potential (Ben-Amotz and Avron, 1973). An organism which is being stressed matrically in an environment with low solute concentrations (e.g. many soils) does not have the opportunity to assimilate osmotically active substances. The alternatives for such an organism are either to synthesize solutes *de novo* as *Dunaliella* does or perhaps to degrade some intracellular macromolecules such as nucleic acids or proteins to obtain larger numbers of solute molecules. Such degradation would have to be quite selective in order for the cell to survive. It therefore seems that organisms faced with matric stress may have more difficulty in responding satisfactorily; it is thus not surprising that matric water potential stress is seen to be more damaging than osmotic.

This presentation has emphasized water relations as an important ecological determinant. It must be remembered, however, that organisms in a natural environment are subject to many physical and chemical factors. The effect of any one of these factors is quite likely to influence, or be influenced by, the others. Water potential considered alone will not answer all questions concerning the ecology of microorganisms. But the advances in technique and experimental design in the past 20 years have contributed greatly to advancing water relations as a significant consideration in studies of microorganisms in nature.

References

Adams, M. S. (1971). Effects of drying at three temperatures on carbon dioxide exchange of *Cladonia rangiferina* (L.) Wigg. *Photosynthetica* **5,** 124–127.

Adebayo, A. A. and Harris, R. F. (1971). Fungal growth responses to osmotic as compared to matric water potential. *Soil Sci. Soc. Am. Proc.* **35,** 465–469.

Adebayo, A. A., Harris, R. F. and Gardner, W. R. (1971). Turgor pressure of fungal mycelia. *Trans. Br. Mycol. Soc.* **57,** 145–151.

Ahmadjian, V. (1967). "The Lichen Symbiosis," 152 pp. Blaisdell Publishing Co., Walthan, Massachusetts.

Bachmann, E. (1922). Zur Physiologie der Krustenflechten. *Z. Bot.* **14,** 193–233.

Bateman, J. B., Stevens, C. L., Mercer, W. B. and Carstensen, E. L. (1962). Relative humidity and the killing of bacteria: the variation of cellular water content with external relative humidity of osmolality. *J. Gen. Microbiol.* **29,** 207–219.

Belly, R. T. and Brock, T. D. (1974). Ecology of iron-oxidizing bacteria in pyritic materials associated with coal. *J. Bacteriol.* **117,** 726–732.

Ben-Amotz, A. and Avron, M. (1973). The role of glycerol in the osmotic regulation of the halophilic alga *Dunaliella parva*. *Plant Physiol.* **51,** 875–878.

Bernstein, L. (1963). Osmotic adjustments of plants to saline media. II. Dynamic phase. *Am. J. Bot.* **50,** 360–370.

Bewley, J. D. (1973). Desiccation and protein synthesis in the moss *Tortula ruralis*. *Can. J. Bot.* **51,** 203–206.

Bewley, J. D. (1974). Protein synthesis and polyribosome stability upon desiccation of the aquatic moss *Hygrohyphum luridum*. *Can. J. Bot.* **52,** 423–427.

Blinn, D. W. (1971). Autecology of a filamentous alga, *Ctenocladus circinnatus* (Chlorophyceae), in saline environments. *Can. J. Bot.* **49,** 735–743.

Blum, O. B. (1973). Water relations. *In* "The Lichens" (Eds V. Ahmadjian and M. E. Hale), pp. 381–400. Academic Press, New York and London.

Bold, H. C. (1970). Some aspects of the taxonomy of soil algae. *Ann. N.Y. Acad. Sci.* **175,** 601–616.

Boylen, C. W. (1973). Survival of *Arthrobacter crystallopoites* during prolonged periods of extreme desiccation. *J. Bacteriol.* **113,** 33–57.

Brian, P. W. (1957). The ecological significance of antibiotic production. *Symp. Soc. Gen. Microbiol.*, **7,** 168–188.

Brock, T. D. (1975a). Effect of water potential on a *Microcoleus* (Cyanophyceae) from a desert crust. *J. Phycol.* **11,** 316–320.

Brock, T. D. (1975b). The effect of water potential on photosynthesis in whole lichens and in their liberated algal components. *Planta* **124,** 13–23.

Brock, T. D. (1975c). Salinity and the ecology of *Dunaliella* from Great Salt Lake. *J. Gen. Microbiol.* **89,** 285–292.

Brown, G. W., Jr. (Ed.) (1974). "Desert Biology" Vol. 2, 601 pp. Academic Press, New York and London.

Bruehl, G. W., Cunfer, B. and Toivinanen, M. (1972). Influence of water potential on growth, antibiotic production, and survival of *Cephalosporium gramineum*. *Can. J. Plant Sci.* **52,** 417–423.

Christian, J. H. B. and Waltho, J. A. (1962). The water relations of *Staphylococci* and *Micrococci*. *J. Appl. Bacteriol.* **25,** 369–377.

Dhindsa, R. S. and Bewley, J. D. (1976). Plant desiccation: polysome loss not due to ribonuclease. *Science* **191,** 181–182.

Dilks, T. J. K. and Proctor, M. C. F. (1974). The pattern of recovery of bryophytes after desiccation. *J. Bryol.* **8,** 97–115.

Dubey, H. D. (1968). Effect of soil moisture levels on nitrification. *Can. J. Microbiol.* **14,** 1348–1350.

Ehresmann, D. W. and Hatch, M. T. (1975). Effect of relative humidity on the survival of airborne unicellular algae. *Appl. Microbiol.* **29,** 352–357.

Farrar, J. F. (1976a). Ecological physiology of the lichen *Hypogymnia physodes*. I. Some effects of constant water saturation. *New Phytol.* **77,** 93–103.

Farrar, J. F. (1976b). Ecological physiology of the lichen *Hypogymnia physodes*. II. Effects of wetting and drying cycles and the concept of 'physiological buffering.' *New Phytol.* **77,** 105–113.

Friedman, E.I. and Galun, M. (1974). Desert algae, lichens, and fungi. *In* "Desert Biology" (Ed. G. W. Brown, Jr.), pp. 165–212. Academic Press, New York and London.

Goepfert, J. M., Iskander, I. K. and Amundson, C. H. (1970). Relation of the heat resistance of salmonellae to the water activity of the environment. *Appl. Microbiol.* **19,** 429–433.

Gray, T. R. G. (1976). Survival of vegetative microbes in soil. *Symp. Soc. Gen. Microbiol.* **26,** 327–364.

Gray, T. R. G. and Postgate, J. R. (Eds) (1976). "The Survival of Vegetative Microbes." *Symp. Soc. Gen. Microbiol.* **26,** 432 pp. Cambridge University Press, Cambridge.

Gregory, P. H. (1971). Airborne microbes: their significance and distribution. The Leeuwenhoek Lecture, 1970. *Proc. R. Soc. Lond. B.* **177,** 469–483.

Griffin, D. M. (1963). Soil moisture and the ecology of soil fungi. *Biol. Rev.* **38,** 141–166.

Griffin, D. M. (1969). Soil water in the ecology of fungi. *Ann. Rev. Phytopathol.* **7,** 289–310.

Harris, G. P. and Kershaw, K. A. (1971). Thallus growth and the distribution of stored metabolites in the phycobiont of the lichens *Parmelia sulcata* and *Parmelia physodes. Can. J. Bot.* **49,** 1367.

Heinrich, M. R. (Ed.) (1976). "Extreme Environments," 362 pp. Academic Press, New York and London.

Holding, A. J., Heal, O. W., MacLean, S. F., Jr. and Flanagan, P. W. (Eds) (1974). "Soil Organisms and Decomposition in Tundra," 398 pp. Swedish IBP Committee, Stockholm.

Jesberger, J. A. and Sheard, J. W. (1973). A quantitative study and multivariate analysis of corticolous lichen communities in the southern boreal forest of Saskatchewan. *Can. J. Bot.* **51,** 185–201.

Kappen, L. (1973). Response to extreme environments. *In* "The Lichens." (Eds V. Ahmadjian and M. E. Hale), pp. 311–380. Academic Press, New York and London.

Kershaw, K. A. (1972). The relationship between moisture content and net assimilation rate of lichen thalli and its ecological significance. *Can. J. Bot.* **50,** 543–555.

Lang, A. R. G. (1967). Osmotic coefficients and water potentials of sodium chloride solutions from 0 to 40°C. *Aust. J. Chem.* **20,** 2017–2023.

Lange, O. L., Schulze, E. D. and Koch, W. (1970). Experimentellökologische Untersuchungen an Flechten der Negev-Wuste. II. CO_2-Gaswechsel und Wasserhaushalt von *Ramalina maciformis* (Del.) Bory am natürlichen Standort während der sommerlichen Trochen-periode. *Flora, Jena* **149,** 345.

Lechowicz. M. J. and Adams, M. S. (1974). Ecology of *Cladonia* lichens. II. Comparative physiological ecology of *C. mitis, C. rangiferina,* and *C. uncialis. Can. J. Bot.* **52,** 411–422.

Lee, J. A. and Stewart, G. R. (1971). Desiccation injury in mosses. I. Intraspecific differences in the effect of moisture stress on photosynthesis. *New Phytol.* **70,** 1061–1068.

Lund, J. W. G. (1967). Soil algae. *In* "Soil Biology" (Eds A. Burges and F. Raw), pp. 129–147. Academic Press, New York and London.

Mexal, J. and Reid, C. P. P. (1973). The growth of selected mycorrhizal fungi in response to induced water stress. *Can. J. Bot.* **51,** 1579.

Robinson, J. B., Salonius, P. O. and Chase, F. E. (1965). A note on the differential response of *Arthrobacter* spp. and *Pseudomonas* spp. to drying soil. *Can. J. Microbiol.* **11,** 746–748.

Rogers, R. W. (1971). Distribution of the lichen *Chondropsis semiviridis* in relation to its heat and drought resistance. *New Phytol.* **70,** 1069–1077.

Rouse, W. R. and Kershaw, K. A. (1973). Studies of the lichen-dominated systems.

VI. Interrelations of vegetation and soil moisture in the Hudson Bay Lowlands. *Can. J. Bot.* **51,** 1309–1316.

Schippers, B., Schroth, M. N. and Hildebrand, D. C. (1967). Emanation of water from underground plant parts. *Plant Soil.* **27,** 81–91.

Scott, W. J. (1957). Water relations of food spoilage microorganisms. *Food Res.* **7,** 83–127.

Shameemullah, M., Parkinson, D. and Burges, A. (1971). The influence of soil moisture tension on the fungal population of a pinewood soil. *Can. J. Microbiol.* **17,** 975–986.

Shields, L. M. and Durrell, L. W. (1964). Algae in relation to soil fertility. *Bot. Rev.* **30,** 92–128.

Shields, L. M., Mitchell, C. and Drouet, F. (1957). Alga- and lichen-stabilized surface crust as a soil nitrogen source. *Am. J. Bot.* **44,** 489–498.

Smith, D. W. and Brock, T. D. (1973a). The water relations of the alga *Cyanidium caldarium* in soil. *J. Gen. Microbiol.* **79,** 219–231.

Smith, D. W. and Brock, T. D. (1973b). Water status and the distribution of *Cyanidium caldarium* in soil. *J. Phycol.* **9,** 330–332.

Smith, D. W., Fliermans, C. B. and Brock, T. D. (1972). Technique for measuring $^{14}CO_2$ uptake by soil microorganisms *in situ. Appl. Microbiol.* **23,** 595–600.

Smith, D. W., Fliermans, C. B. and Brock, T. D. (1973). An isotopic technique for measuring the autotrophic activity of soil microorganisms *in situ. Bull. Ecol. Res. Comm. (Stockholm)* **17,** 243–246.

Smyth, E. S. (1934). A contribution to the physiology and ecology of *Peltigera canina* and *P. polydactyla. Ann. Bot.* **48,** 781–818.

Socolofsky, M. D. and Wyss, O. (1962). Resistance of the *Azotobacter* cyst. *J. Bacteriol.* **84,** 119–124.

Sommers, L. E., Harris, R. F., Dalton, F. N. and Gardner, W. R. (1970). Water potential relations of three root-infecting *Phytophthora* species. *Phytopathology* **60,** 932–934.

Sprent, J. I. (1971). The effects of water stress on nitrogen-fixing root nodules. *New Phytol.* **70,** 9–17.

Vela, G. R. (1974), Survival of *Azotobacter* in dry soil. *Appl. Microbiol.* **28,** 77–79.

Williams, S. T., Shameemullah, M., Watson, E. T. and Mayfield, C. I. (1972). Studies on the ecology of actinomycetes in soil. VI. The influence of moisture tension on growth and survival. *Soil. Biol Biochem.* **4,** 215–225.

Chapter 10

How Microbes Cope with Heavy Metals, Arsenic and Antimony in their Environment

H. L. EHRLICH

Rensselaer Polytechnic Institute

I. Introduction 381
II. The sources of heavy metals and of arsenic and antimony in the environment 382
III. Microbial responses to heavy metals, arsenic and antimony in the environment 385
IV. The bacteriology of acid mine-drainage 389
V. Microbial interaction with manganese and iron compounds . . 392
VI. Microbial interactions with mercury 395
VII. Conclusion 397
Addendum 398
References 398

I. Introduction

Although not common, some natural contemporary environments, unmodified by man, may contain unusually high concentrations of metals. For instance, the 56°C-brine of the Atlantis Deep in the Red Sea has been reported to contain (in grams of metal per kilogram of brine): $8{\cdot}1 \times 10^{-2}$ Fe, $8{\cdot}2 \times 10^{-2}$ Mn, $5{\cdot}4 \times 10^{-3}$ Zn, $2{\cdot}6 \times 10^{-4}$ Cu, $1{\cdot}6 \times 10^{-4}$ Co, and $6{\cdot}3 \times 10^{-4}$ Pb (Brewer and Spencer, 1969). The peat of Black Bay Bog, Ontario, has been found to contain a maximum of 100 ppm of Zn, 60 ppm of Cu, 40 ppm of Ni, and 10 ppm of Pb (Gleeson and Coope, 1967). The peat of Sawyer Bog, Quebec, has been found to contain a maximum of 140 ppm or Zn, 30 ppm of Cu, and 60 ppm of Ni (Gleeson and Coope, 1967). Kendrick (1962) studied soil from a swamp with cupriferous peat with copper concentrations as high as 68,000 ppm, in which fungi were growing. De Grys (1964) reported copper concentrations as high as 100 ppb in the waters of some Andean streams and 100 ppm in their sediments. The copper derived from copper mineralization in the drainage basins of these streams. Kovalskii and Letunova (1974) studied lakes and soils with high amounts of Co, Cu, Mo, U, and V in which the microorganisms were generally resistant to these elements. Of the foregoing examples, only the brine pools of the Red Sea have been directly shown not to harbour living microorganisms (Watson and Waterbury, 1969).

Man's modification of parts of his environment has initiated or accentuated localized heavy metal concentrations in some water bodies, soils and even the atmosphere. The development of acid mine-drainage from the oxidation of iron pyrites associated with bituminous coal seams upon exposure to air and moisture during coal mining is a dramatic example of this (Lundgren *et al.*, 1972). This drainage has become an environmental nuisance. The production of acid drainage from mine tailings and low-grade ores in leach dumps of copper mines is another example (Beck, 1967). In this case, the drainage has economic value.

Very high concentrations of heavy metals, arsenic and antimony, i.e. in some cases in excess of 1 g $litre^{-1}$ and in others in excess of 1, 10 or 100 mg $litre^{-1}$, are often detrimental to microbes. Lesser concentrations, if not toxic, may be chemically modified by some microbes and thereby removed. Certain groups of microorganisms may use some heavy metal-, arsenic- or antimony-compounds at non-toxic concentrations either as sources of energy or as terminal electron acceptors. Low concentrations of some heavy metals may stimulate growth (Andreeson and Ljungdahl, 1973; Andreesen *et al.*, 1974). Indeed, some heavy metals generally are required in microbial nutrition. In this chapter the response of microbes to different concentrations of heavy metals, arsenic and antimony will be considered.

II. The Sources of Heavy Metals and of Arsenic and Antimony in the Environment

The concentrations of various metals and of arsenic and antimony in the environment vary widely. Table 1 lists the average concentration or the concentration ranges in soil, fresh water and sea water. Table 2 lists toxic levels of some heavy metals, arsenic and antimony as recorded in the literature. It must be understood that the criteria for toxicity are not the same for each item in the table, and that the toxic levels listed for each element in the table were determined under one specific set of test conditions and may not be the same for different test conditions. The degree of toxicity of any heavy metal-, arsenic-, or antimony-compound depends on its exact state; i.e. it depends on

Table 1

Concentration of heavy metals, arsenic and antimony in the environment

Element (mg litre^{-1})	Soil	Fresh water[c] (mg litre^{-1})	Sea water[c] (mg litre^{-1a})
As	6 mg g^{-1a}	4×10^{-4}	$2{\cdot}6 \times 10^{-3}$
Cd	6×10^{-2} mg g^{-1a}	<8	1×10^{-4}
Co	1–300 mg g^{-1b} usual concentration is 10–15 mg g^{-1b}	9×10^{-4}	4×10^{-4}
Cu	5–5000 mg g^{-1b}; in U.S. soils 6–67 mg g^{-1b}	1×10^{-2}	3×10^{-3}
Fe	200 mg g^{-1} to 100 g kg^{-1}; in U.S. soils 2·3 to 112 g kg^{-1b}	0·67	3×10^{-3}
Hg	0·03–0·8 mg g^{-1a}	8×10^{-5}	2×10^{-4}
Mn	trace to 150 g kg^{-1b}	$1{\cdot}2 \times 10^{-2}$	2×10^{-3}
Ni	40 mg g^{-1a}	1×10^{-2}	7×10^{-3}
Pb	10 mg g^{-1a}	5×10^{-3}	3×10^{-5}
Sb	2–10 mg litre^{-1a}	$(0{\cdot}27\text{–}4{\cdot}9) \times 10^{-3d}$	3×10^{-4}
U	1 mg g^{-1a}	1×10^{-3}	3×10^{-3}
V	100 mg g^{-1a}	1×10^{-3}	2×10^{-3}
Zn	2–50 mg g^{-1b}	1×10^{-2}	1×10^{-2}

[a] Bowen (1966)
[b] Sauchelli (1969)
[c] Anonymous (1971)
[d] Kharkar *et al.* (1968)

whether in solution it is a free ion or a salt which is largely undissociated, or whether it is complexed organically or inorganically. Ramamoorthy and Kushner (1975a, b) have studied the extent of chelation of Hg^{2+}, Pb^{2+}, Cu^{2+}, and Cd^{2+} by organic and inorganic ligands occurring naturally in river water and in microbial growth media. They showed that the degree of complexation may be extensive. Jerneloev and Martin (1975) have reviewed theories of complex formation and metal binding. The degree of toxicity of heavy metal-, arsenic- or antimony-compounds is also influenced by the presence of competing cations or anions, whether they be toxic or non-toxic (Adiga *et al.*, 1962; Bowen, 1966; Da Costa, 1972; De Turk and Bernheim, 1960; Gonye and Jones, 1973; Jones, 1973; Laborey and Lavollay, 1967; Sadler and Trudinger, 1967; Sastry *et al.*, 1962). Undissociated salts and complexed ions usually tend to be less toxic, mole for mole, than free ions (e.g. Temple, 1964; Temple and LeRoux, 1964).

Table 2

Toxic levels of some heavy metals, arsenic and antimony for microorganisms

Element	Compound tested	Toxic level	Organism tested	References
As	Na_2HAsO_4	10^{-4} M	*Azotobacter*	Den Dooren De Jong and Roman (1971)
	arsenate	2×10^{-2} M	yeast	Mosin *et al.* (1974)
Ag	$AgNO_3$	6 mg litre^{-1}[b]	*Rhodotorula, Torula, Hansenula, Serratia, Azotobacter, Pseudomonas, Escherichia coli*	Avakyan (1967)
Cd	$CdCl_2$	10^{-4} M	*Bacterium communis*	Winslow and Hotchkiss (1921/22)
	$CdCl_2$	74–83 mg litre^{-1}[a]	*Azotobacter, Pseudomonas*	Avakyan (1967)
		4×10^{-6} M	*Tetrahymena*	Yamaguchi *et al.* (1973)
Co	$CoCl_2.6H_2O$	23–26 mg litre^{-1}[a]	*Torula, Serratia Azotobacter, Pseudomonas, Escherichia coli*	Avakyan (1967)
	$CoCl_2.6H_2O$	1·2 mg litre^{-1}	*Aspergillus niger*	Adiga (1961)
	$CoCl_2.6H_2O$	10^{-4} M	*Azotobacter*	Den Dooren De Jong and Roman (1971)
	$CoCl_2$	0·7 mg litre^{-1}	*Aspergillus niger*	Nowosielski *et al.* (1971)
Cu	$CuSO_4$	10^{-3}–10^{-4} M	*Pseudomonas* C-1	Sadler and Trudinger (1967)
	$CuSO_4.5H_2O$	10^{-4} M	*Azotobacter*	Den Dooren De Jong and Roman (1971)
Fe	$FeCl_3$	10^{-3} M	*Bacterium communis*	Winslow and Hotchkiss (1921/22)
Hg	$HgCl_2$	10^{-5} M	*Bacteriun communis*	Winslow and Hotchkiss (1921/22)
	$HgCl_2$	7·4 mg litre^{-1}[b]	*Rhodotorula, Torula, Hansenula, Serratia, Azotobacter, Pseudomonas, Escherichia coli*	Avakyan (1967)
		$7{\cdot}6 \times 10^{-6}$ M	*Tetrahymena*	Yamaguchi *et al.* (1973)
	Hg metal	0·1 mg ml^{-1}	*Aspergillus niger*	Nowosielski *et al.* (1971)
Mn	$MnCl_2$	5×10^{-2} M	*Bacterium communis*	Winslow and Hotchkiss (1921/22)
	$MnSO_4.H_2O$	>0·05%	*Sphaerotilus discophorus*	Ali and Stokes (1971)
	$MnSO_4.4H_2O$	0·1%	*Corynebacterium* and *Chromobacterium* (Mixed culture)	Bromfield (1956)
Ni	$NiCl_2.6H_2O$	23–27 mg litre^{-1}[a]	*Rhodotorula, Hansenula, Pseudomonas, Escherichia coli*	Avakyan (1967)
	$NiSO_4.6H_2O$	0·4 mg 10 ml^{-1}	*Aspergillus niger*	Adiga *et al.* (1961)
	$Ni(NO_3)_2.6H_2O$	10^{-4} M	*Azotobacter*	Den Dooren De Jong and Roman (1971)
	$NiCl_2$	0·9 mg ml^{-1}	*Aspergillus niger*	Nowosielski *et al.* (1971)
Pb	$Pb(NO_3)_2$	0·1 g litre^{-1}[b]	*Pseudomonas pyocyanea* strains 8 and 10	Avakyan (1967)
	$PbCl_2$	67 mg ml^{-1}	*Aspergillus niger*	Nowosielski *et al.* (1971)
	$Pb(NO_3)_2$	0·007 M	*Aspergillus niger*	Zlochevskaya and Rabotnova (1966)
	$Pb(NO_3)_2$	5 mg litre^{-1}	*Cosmarium*	Malanchuk and Gruendling (1973)
	$Pb(NO_3)_2$	15–18 mg litre^{-1}	*Anaboena, Navicula, Chlamydomonas*	Malanchuk and Genendling (1973)
Sb	Sb_2O_3SbOCl	10^{-4} M	*Azotobacter*	Den Dooren De Jong and Roman (1971)
U	$UO_2(NO_3)_2.6H_2O$	10^{-4} M	*Azotobacter*	Den Dooren De Jong and Roman (1971)
V	$VO(SO_4)$	10^{-4} M	*Azotobacter*	Den Dooren De Jong and Roman (1971)
Zn	$ZnSO_4.7H_2O$	1 mg litre^{-1}	*Aspergillus niger*	Adiga *et al.* (1961)

[a] Actual ionic concentration in medium determined polarographically

[b] Amount of salt added to medium

The sources of heavy metals, arsenic and antimony in nature can be divided into two major categories, natural and artificial. The artificial source is usually associated with man-made pollution. The primary natural sources are magnetic and hydrothermal activity, and igneous rocks. Sediments, and metamorphic and sedimentary rocks are secondary sources. The weathering of rocks may release heavy metals, arsenic and antimony into solution which may then be transported to new sites where they may be reprecipitated or adsorbed. Sometimes, heavy metal-, arsenic-, or antimony-compounds may occur in locally high concentrations in host rock. Such rock is then called an ore.

Man has been responsible for dispersing some heavy metals, arsenic and antimony in the environment by exposing ore bodies to air and moisture. The ore bodies thus become subject to extensive weathering. Man has also introduced heavy metals, arsenic and antimony into the environment in the disposal of certain solid, liquid and gaseous wastes. The metals, arsenic or antimony from these sources may reach locally high concentrations, at times achieving toxic levels, or they may be dispersed over wide areas as they are carried away by water or by air currents.

Microbes may contribute to solubilization of insoluble metal-, arsenic-, and antimony-compounds through oxidation, as in the case of metal-, arsenic-, and antimony-sulphides (Ehrlich, 1971; Silverman and Ehrlich, 1964); through the reduction of oxides, as in the case of iron- and manganese oxides (Ehrlich, 1971; Silverman and Ehrlich, 1964; Ottow, 1968); through the action of inorganic or organic acids which the microorganisms may produce, as in the weathering of granites (Silverman and Munoz, 1970); or through the production of chelating agents (Kee and Bloomfield, 1961). One organism, *Stibiobacter*, has been discovered which oxidizes insoluble Sb_2O_3 to Sb_2O_5 (Lyalikova, 1972; Lyalikova *et al.*, 1974).

III. Microbial Responses to Heavy Metals, Arsenic and Antimony in the Environment

Microbes respond to heavy metals in different ways, depending on the kind of microorganisms and depending on the concentration of the heavy metals in the environment. This is also true for arsenic and antimony. All microbes require certain heavy metals, including Co, Cu, Fe, Mn and Zn, in their nutrition (Brock, 1974; Enoch and Lester, 1972). Some also require Mo, V, and Ni (Bartha and Ordal, 1965; Bertrand, 1974; Bertrand and De Wolff, 1973; Ehrlich, 1971; Esposito and Wilson, 1956). All these metals are mainly involved in enzyme function, and they are needed only in very low concentrations in the nutrient medium, usually in the range of micrograms $litre^{-1}$. Some microorganisms are able to take some of these elements into the cell by active transport (Bhattacharyya, 1970; Eisenstadt *et al.*, 1973; Khovrychev, 1973; Silver *et al.*, 1970; Silver and Kralovic, 1969; Wang and Newton, 1969). Some bacteria and fungi produce special chelating agents to facilitate the uptake of iron at neutral pH (Arceneaux *et al.*, 1973; Corbin and Bulin, 1969; Cox *et al.*, 1970; Davis and Byers, 1971; Emery, 1971; Haydon *et al.*, 1973; Luckey *et al.*, 1972). This uptake involves active transport of the chelated iron with breakdown of the chelate after penetration through the plasma membrane. Even the toxic arsenate ion may be taken up by active transport, as in the case of *Saccharomyces cerevisiae* (Cerbon, 1969).

Sufficiently high concentrations of any of the metals or of arsenic or antimony become toxic to microorganisms (Table 2). This toxicity may manifest itself in various ways, e.g. altered cell morphology (Bubela, 1970; Cobet *et al.*, 1970; Cobet *et al.*, 1971; O'Callaghan *et al.*, 1973; Renshaw *et al.*, 1966; Sadler and Trudinger, 1967; Weed and Longfellow, 1954), altered cell metabolism (Albright and Wilson, 1974; De Turk and Bernheim, 1960; Leahy, 1969; Ou and Anderson, 1972; Sadler and Trudinger, 1967; Tandon and Mishra, 1969; Weinberg, 1970), bacteriostasis (Oster and Golden, 1954; Sadler and Trudinger, 1965), or lethality (Romans, 1954; Sadler and Trudinger, 1965). Resistant strains may arise that are more tolerant to metal, arsenic or antimony, i.e. they may require higher concentrations of the toxicant than the parent culture before being affected. This resistance is usually due to a genetic modification, frequently plasmid-linked (Hedges and Baumberg, 1973; Kondo *et al.*, 1974; Novick and Roth, 1968; Smith, 1967; Smith and Novick, 1972; Summers and Silver, 1972), but occasionally sex-factor linked (Loutit, 1970) or chromosomally linked (Dyke *et al.*, 1970). The increased resistance may be due to a lessened permeability of the cell to the toxicant or to a biochemically mediated detoxification of the toxicant. Starkey (1973) related extreme copper tolerance (exposure to about 1 M $CuSO_4$) of *Scytalidium* to low pH of the medium (pH 2·0 to 0·3) and an inability of copper ions to penetrate the cells in that pH range, since near neutrality, the fungus was sensitive to 4×10^{-5} M $CuSO_4$. Some microbes can detoxify heavy metals, arsenic or antimony by the elaboration of detoxifying agents which react with the metals, arsenic and antimony inside the cell (e.g. in the methylation of mercury or arsenic; Wood, 1974), or are reacted with them outside the cell, thus rendering them harmless, i.e. unassimilable by the microbe (e.g. the precipitation of arsenate and arsenite by ferric iron in arsenopyrite oxidation by *Thiobacillus ferrooxidans*, Ehrlich, 1964). Others can detoxify by enzymatically modifying the toxicant to a less harmful compound (e.g. the reduction of $HgCl_2$ to HgO; Komura *et al.*, 1970). Although not well documented, the physiological state of an organism can also determine its susceptibility to heavy metal-, arsenic- or antimony-intoxication as suggested by the work of Sherman and Albus (1923).

The mechanism by which heavy metals, arsenic and antimony exert their toxicity depends on the nature of the compound and the organism involved. Some elements, like Cu, may be bound mainly on the cell surface and do their damage in that location (Sadler and Trudinger, 1967; Yang, 1974). Other elements, like Hg, may be taken into the cell where they may bind to certain functional groups, especially —SH groups, thus rendering vital molecules like enzyme proteins inactive, or where they may be deposited in metallic form (Brunker and Bott, 1974). Bowen (1966) cites additional mechanisms as follows: (1) acting as antimetabolites, (2) producing stable precipitates or chelates with essential metabolites or catalysing the decomposition of essential metabolites and thereby making the metabolites unavailable to the cells, or (3) replacing structurally or electrochemically important elements and thereby interfering with enzymatic or cellular function.

Some microbes can oxidize reduced forms of heavy metal, arsenic or antimony compounds, while others can reduce oxidized forms of these elements on a large scale (Ehrlich, 1971; Silverman and Ehrlich, 1964). When oxidizing reduced metal compounds, at least some organisms can derive useful energy and reducing power from the process. When reducing oxidized metal compounds, some microbes carry out a process that may be a form of respiration in which the oxidized metal compounds, arsenic or antimony serve partially or exclusively as terminal electron acceptors. Such oxidation and reduction reactions may be fundamental in the redistribution of the elements in the environment. Table 3 lists some minerals, many of which are associated with ores, that are acted upon by microorganisms.

Microbes may concentrate some heavy metals intracellularly or on their surface. Bowen (1966), for instance, mentions that plankton concentrates cadmium from seawater 910 times, cobalt 4600 times, copper 7000 times, iron 87,000 times, lead 41,000 times, manganese 9400 times, titanium 20,000 times, and zinc 65,000 times. Except for cadmium, copper, manganese and zinc, the metals may, at least in part, be in particulate form in seawater. Engel and Owen (1970) in a study of metal accumulation by fuel-utilizing bacteria found that the organisms accumulated more than 50% of added zirconium, titanium and zinc from media containing 10 ppm or each metal in volumes from 50 ml to 2 litres. Friedman and Dugan (1968) observed that 0·1 g (dry weight) of zoogloea-producing bacterial strain 115 took up 99·3% of 500 ppm of Co^{2+}, 84·9% of 800 ppm of Cu^{2+}, 98·8% of 350 ppm of Fe^{3+}, and 50% of 500 ppm of Ni^{2+} in 18 h in 51 ml of medium. The metals taken up were probably tied, at least in part, to the slime of the zoogloea. Zajic and Chiu (1972) found that two strains of *Penicillium* resistant to uranium, strontium, titanium silver and platinum, and isolated from sewage, took up 70% of the uranium in 100 ml of a medium containing 100 ppm of UO_2SO_4. In general, the final cellular concentration of a metal

Table 3

Some natural metal-containing minerals acted upon by microorganisms

Mineral	Formula	Active microbes	References[a]
Arsenopyrite	$FeS_2 \cdot FeAs_2$	*T. ferrooxidans*	Ehrlich (1964)
Birnessite	MnO_2	*Bacillus* 29 and other marine isolates	Ehrlich (1963c)
Bornite	Cu_5FeS_4	*T. ferrooxidans*	Bryner *et al.* (1954)
Chalcocite	Cu_2S	*T. ferrooxidans*	Bryner *et al.* (1954)
Chalcopyrite	$CuFeS_2$	*T. ferrooxidans*	Bryner *et al.* (1954)
Covellite	CuS	*T. ferrooxidans*	Bryner *et al.* (1954)
Enargite	$3Cu_2S \cdot As_2S_5$	*T. ferrooxidans*	Ehrlich (1964)
Galena	PbS	*T. ferrooxidans*	Torma and Subramanian (1974)
Goethite	$Fe_2O_3 \cdot 2H_2O$	*Bacillus* 29	De Castro and Ehrlich (1970)
Hematite	Fe_2O_3	*Bacillus* 29	De Castro and Ehrlich (1970)
Limonite	$Fe_2O_3 \cdot nH_2O$	*Bacillus* 29	De Castro and Ehrlich (1970)
Millerite	NiS	*T. ferrooxidans*	Razzell and Trussell (1963)
Molybdenite	MoS_2	*Sulfolobus* sp.	Brierley and Murr (1973)
Orpiment	AsS_2	*T. ferrooxidans*	Ehrlich (1963b)
Pyrite, marcasite	FeS_2	*T. ferrooxidans*	Leathen *et al.* (1953); Temple and Delchamps (1953)
Pyrolusite	MnO_2	*Bacillus* 29 and other marine isolates	Ehrlich (1966)[b]
Sphalerite	ZnS	*T. ferrooxidans*	Malouf and Prater (1961) Ivanov *et al.* (1961)
Todorokite	$Mn_2Mn_5O_{12} \cdot 2H_2O$[c]	*Bacillus* 29 and other marine isolates	Ehrlich (1963c)

[a] First reports
[b] Earlier reports exist, but mineral form was not defined
[c] Mg, Ca, Na, K, B, Zn and Ag may substitute for the divalent Mn of todorokite

may be several orders of magnitude over its concentration in the surrounding environment. In some instances, the accumulation of these substances may be lethal, in others not. The physiological state of the cells and the environmental conditions may affect the uptake of the metal ions (e.g. Fisher *et al.*, 1973).

IV. The Bacteriology of Acid Mine-Drainage

Acid mine-drainage is an example of an environment with extreme heavy metal- and, possibly, arsenic- and antimony-concentrations which are toxic to many microorganisms. Yet, such drainage has been found to harbour a mixed microbial flora that contains algal, fungal, protozoan and bacterial forms which seem to be specially adapted to cope with the conditions (Lackey, 1938; Ehrlich, 1963a; Marchlewitz and Schwartz, 1961; Tuttle *et al.*, 1968). Acid mine-drainage derives from two kinds of sources, the oxidation of iron pyrites in bituminous coal seams and the oxidation of metal sulphide deposits (e.g. copper sulphides, iron sulphides) in ore bodies and leach dumps. Its pH may range from 1·4 to 5·5 (Colmer and Hinkle, 1947; Corbett and Gorwitz, 1967; Lundgren *et al.*, 1972); its iron content may range from 0·5 to 10 g $litre^{-1}$; its copper content may range from traces to more than 10 g $litre^{-1}$; its aluminum content may range from 0 to 2 g $litre^{-1}$ (Lundgren *et al.*, 1972). Organisms which have been implicated in the formation of this drainage include a *Metallogenium*-like organism (Walsh and Mitchell, 1972); *Thiobacillus ferrooxidans* (Colmer *et al.*, 1950); *T. thiooxidans* (Colmer and Hinkle, 1947) and, perhaps, *Sulfolobus* (Brierley and Brierley, 1973). Only *T. ferrooxidans* and *T. thiooxidans* have been studied extensively for their role in acid mine-drainage formation.

Acid coal mine-drainage forms when FeS_2 (pyrite, marcasite, sulphur ball) in coal seams becomes exposed to moisture and air. Initially, FeS_2 oxidizes according to the following reaction.

$$FeS_2 + 3\tfrac{1}{2}\,O_2 + H_2O \rightarrow FeSO_4 + H_2SO_4 \qquad (1)$$

Although this reaction can occur under sterile conditions, it is accelerated by the direct catalysis at the mineral surface by *T. ferrooxidans* (Beck and Brown, 1968; Silverman, 1967). The oxidation of the pyrite is further accelerated by chemical interaction with ferric iron (Lundgren *et al.*, 1972) as follows.

$$FeS_2 + 14Fe^{3+} + 8H_2O \rightarrow 15Fe^{2+} + 2SO_4{}^{2-} + 16H^+ \qquad (2)$$

The ferric iron needed for this reaction derives initially from ferrous iron oxidation catalysed, possibly, by *Metallogenium* until the pH has dropped to 3·5 and the ferrous iron concentration has risen to 0·1 g $litre^{-1}$ (Walsh and Mitchell, 1972), and below pH 4 catalysed by *T. ferrooxidans* (Temple and Colmer, 1951; Silverman and Lundgren, 1959a). The reaction can be written as follows.

$$2Fe^{2+} + \tfrac{1}{2}O_2 + 2H^+ \rightarrow 2Fe^{3+} + H_2O \qquad (3)$$

The source of the ferrous iron in this reaction is initially reaction 1, and then later also reaction 2. Although reaction 3 will proceed uncatalysed by bacteria, it is extremely slow without them at pH values below 4 (Silverman and Lundgren, 1959b). According to Singer and Stumm (1970), reaction 3 is the rate determining reaction in FeS_2 oxidation, and through it the bacteria help in regulating the rate of FeS_2 oxidation. The acid in the drainage results from the hydrolysis of much of the ferric iron produced in the oxidation in reaction 3.

$$Fe^{3+} + 3H_2O \rightarrow Fe(OH)_3 + 3H^+ \qquad (4)$$

The ferric hydroxide thus formed reacts further with some of the H_2SO_4 in the drainage, resulting in insoluble ferric sulphate.

$$2Fe(OH)_3 + 3H_2SO_4 \rightarrow Fe_2(SO_4)_3 + 6H_2O \qquad (5)$$

Considering reactions 1 to 5, it can be shown that for every 2 mol of FeS_2 transformed to $Fe_2(SO_4)_3$, 1 mol of sulphuric acid is formed. Actually somewhat more sulphuric acid may form since ferric iron from reaction 3 is more frequently converted to basic ferric sulphate ($Fe(OH)SO_4$) or to jarosite ($AFe_3(SO_4)_2(OH)_6$), where A may represent K or NH_4 (Duncan and Walden, 1972). Any aluminum occurring in the acid drainage results mainly from the corrosive action of H_2SO_4 on alumino-silicate minerals in rocks, sands and sediments with which it comes in contact.

Acid mine-drainage from copper sulphide-containing ore deposits results, in part, from reactions similar to those responsible for such drainage from bituminous coal mines (Beck, 1967; Kuznetsov *et al.*, 1963; Silverman and Ehrlich, 1964; Zajic, 1969). Iron pyrite (FeS_2) ,which generally accompanies the copper sulphides such as chalcopyrite ($CuFeS_2$), chalcocite (Cu_2S), covellite (CuS), bornite (Cu_5FeS_4), etc., is oxidized as already discussed (reactions 1 to 5) and may be a major source of acid and iron in the drainage. In addition, chalcopyrite crystals may be oxidized on exposure to air and water according to the following reaction, which assumes $Fe(OH)SO_4$ to be one of the products.

$$2CuFeS_2 + H_2O + 8{\cdot}5\,O_2 \rightarrow 2Cu^{2+} + 2Fe(OH)SO_4 + 2SO_4{}^{2-} \qquad (6)$$

Its rate of oxidation is significantly accelerated by *T. ferrooxidans* since as a mineral it is relatively stable under sterile conditions (Razzell and Trussell, 1963). The bacterial action is further improved by the presence of surface-active agents (Duncan *et al.*, 1964). Acid ferric sulphate is not a very effective oxidant for chalcopyrite unless the ferric iron concentration is low, i.e. below 1 g litre^{-1} (Ehrlich, unpublished data).

Chalcocite (Cu_2S) crystals in an ore body on exposure to air and moisture may be oxidized according to the following series of reactions.

$$Cu_2S + \tfrac{1}{2}O_2 + 2H^+ \rightarrow Cu^{2+} + CuS + H_2O \qquad (7)$$

Digenite (Cu_9S_5) appears to be an intermediate in reaction 7 (Nielsen and Beck, 1972). The CuS is then further oxidized to $CuSO_4$ as follows.

$$CuS + \tfrac{1}{2}O_2 + 2H^+ \rightarrow Cu^{2+} + S^0 + H_2O \quad (8)$$

$$S^0 + \tfrac{3}{2}O_2 + H_2O \rightarrow H_2SO_4 \quad (9)$$

These reactions are catalysed by *T. ferrooxidans,* although they also proceed slowly in the absence of the organism (Fox, 1967; Nielsen and Beck, 1972). Reaction 9 may be catalysed by *T. thiooxidans* besides *T. ferrooxidans.* Nielsen and Beck (1972) have shown that *T. ferrooxidans* can derive energy from reaction 7 and assimilate CO_2 with its help. The organism, of course, can derive additional energy from reactions 8 and 9. Imai *et al.* (1973) have demonstrated Cu_2S oxidation with a cell free extract of *T. ferrooxidans.*

Covellite (CuS) crystals in an ore body on exposure to air and moisture may be oxidized according to reactions 8 and 9 (Fox, 1967). Although this oxidation can proceed slowly under sterile conditions, it, like chalcocite oxidation, is significantly accelerated by *T. ferrooxidans.* It should be noted that while covellite oxidation (reactions 8 and 9) does not involve a net consumption of acid, chalcocite oxidation (reactions 7 to 9) consumes 2 mol of H^+ per mol of chalcocite. In the field, the consumed acid may be replaced by acid from iron pyrite oxidation (reactions 1 to 5).

Chalcocite and covellite may be oxidized by ferric iron instead of oxygen in acid solution. Sullivan (1930) showed that under these conditions chalcocite reacts as follows.

$$Cu_2S + 2Fe^{3+} \rightarrow Cu^{2+} + CuS + 2Fe^{2+} \quad (10)$$

$$CuS + 2Fe^{3+} \rightarrow Cu^{2+} + S^0 + 2Fe^{2+} \quad (11)$$

Covellite was shown by him to be oxidized according to reaction 11. Although bacteria do not directly catalyse reactions 10 and 11, in the field they help in regenerating the oxidant, Fe^{3+}, and the Fe^{2+} produced.

It has been claimed that significant amounts of heat are generated in the oxidation of copper sulphide ore bodies and leach dumps, which may raise the interior temperature to 80°C (Beck, 1967). This is too high for *T. ferrooxidans,* whose temperature optimum is between 25 and 45°C. However, a thermophilic and acidophilic *Sulfolobus* has been isolated recently from a hot spring, which is capable of oxidizing iron at 70°C (Brierley and Brierley 1973). This organism is also capable of promoting the leaching of molybdenite (MoS_2) and chalcopyrite at 60°C (Brierley and Murr, 1973). Although *Sulfolobus* has not yet been reported from ore bodies and leach dumps, it may dominate the bacterial oxidizing activity on the interior of these structures.

Because of their physiological requirements, it is not surprising that *T. ferrooxidans* and *T. thiooxidans* are found associated with iron pyrites and acid mine-drainage of bituminous coal mines. Similarly, it is not surprising that these organisms are found associated with copper sulphide ore bodies and acid drainage from them. *T. ferrooxidans* lives at the expense of ferrous iron in acid mine-drainage. *T. thiooxidans* lives at the expense of any sulphur compounds of intermediate oxidation state formed in the oxidation of metal sulphides. In all instances, the mineral substances that are oxidized serve as energy sources to the autotrophic organisms, which are specially adapted to grow at the very acid pH that prevails in their milieux and which, moreover, may become specially adapted to cope with the high metal concentrations they encounter in these situations, which may be in the order of g litre^{-1} for Cu, Ni and Zn (Bryner *et al.*, 1954; Tuovinen *et al.*, 1971).

T. ferrooxidans has also been reported to oxidize cobalt sulphide (Torma, 1971); nickel sulphide (Torma, 1971; Silver and Torma, 1974); galena (PbS) (Silver and Torma, 1974); zinc sulphide (Ivanov *et al.*, 1961; Ivanov, 1962; Malouf and Prater, 1961; Torma, 1971); arsenic minerals such as arsenopyrite ($Fe_2As_2S_2$), enargite ($3Cu_2S \cdot As_2S_5$), and orpiment (As_2S_3) (Ehrlich, 1963b, 1964); and red antimony trisulphide (Sb_2S_3) (Lyalikova, 1961; Silver and Torma, 1974). All are attacked in insoluble form. Silver and Torma (1974) observed CO_2 uptake by the organism at the expense of lead sulphide and antimony trisulphide oxidation, but not at the expense of nickel sulphide oxidation. Ehrlich (1963b and 1964) obtained growth of *T. ferrooxidans*, and, by implication, CO_2 fixation, on arsenopyrite, enargite and orpiment. In the oxidation of arsenopyrite, arsenite and arsenate were produced which precipitated a significant portion of the solubilized iron.

It has been reported that molybdenite (MoS_2) is oxidized by *T. ferrooxidans* (Bryner and Anderson, 1957). This finding has been questioned because more recent work suggests that the molybdate ion formed in the oxidation is very toxic to the organism, 5 mg Mo as molybdate being lethal (Tuovinen *et al.*, 1971). It is possible, however, that in the earlier experiments the molybdate was detoxified by reacting with iron from concurrent pyrite oxidation to form insoluble iron molybdate compounds.

V. Microbial Interaction with Manganese and Iron Compounds

In nature, at neutral pH, Mn and Fe are not known to occur at levels toxic to microorganisms (Tables 1 and 2). At that pH, appropriate compounds of the elements may, however, be microbially oxidized or reduced, depending on their valence state. Manganese oxidation by bacteria and fungi has been reported by various investigators, starting with Beijerinck (Silverman and Ehrlich, 1964; Ehrlich, 1971; Schweisfurth, 1968 and 1971). Except for observations of enzymatic oxidation of manganese by certain soil bacteria studied by Bromfield (1956), the mechanism of microbial manganese oxidation in soil by different microorganisms is not clearly established. In some cases, it may be enzymatic, and in others it may result from alkalinization of microenvironments as a result of microbial metabolism, e.g. ammonia production, which promotes autoxidation of Mn(II). In fresh water, *Sphaerotilus discophorus* appears to be able to oxidize Mn(II) with the derivation of useful energy, since it can grow quasi-autotrophically with Mn(II) as sole energy source and CO_2 as sole carbon source except for the trace requirements of cyanocobalamin, thiamin, and biotin (Ali and Stokes, 1971). It may also grow mixotrophically in a casamino acids-mineral salts medium containing 0·05% $MnSO_4 . H_2O$ (Ali and Stokes, 1971).

Some Hyphomicrobia have been reported to oxidize Mn(II) in fresh water (Tyler, 1970), and *Metallogenium* has been reported to oxidize Mn(II) in soil and fresh water (Perfil'ev and Gabe, 1965; Zavarzin, 1961). Whether these microorganisms oxidize manganese enzymatically has not been clearly established. The work of Kossaya (1967) suggests that Mn(II) oxidation by *Metallogenium* is due partly to a heat-stable oxidase and partly to autoxidation. The genus has been implicated in manganese deposition in Lake Punnus-Yarvi on the Karelian Peninsula (Sokolova-Dubinina and Deryugina, 1967 and 1968).

In marine environments, a number of different bacterial types have been found to oxidize Mn(II) in the formation of ferromanganese concretions, also known as manganese nodules (Ehrlich, 1963c; Ehrlich *et al.*, 1972). These concretions are a potentially important mineral resource, especially because of the Cu (0·53%, ave.), Ni (0·99%, ave.) and Co (0·35%, ave.) they contain, which may be commercially exploited in the not so distant future. The bacterial Mn(II) oxidation in nodule growth has been shown to be enzymatic in all instances tested (Ehrlich, 1968; Ehrlich *et al.*, 1972). At least some of these organisms seem to be able to derive useful energy from Mn(II) oxidation (Ehrlich, 1976 and unpublished data). Other reports of manganese-oxidizing bacteria and fungi from marine environments have appeared (Krumbein, 1971; Thiel, 1925).

Reduction of Mn(IV), as in MnO_2, by bacteria and fungi has been observed in soil (Mann and Quastel, 1946); in freshwater environments (Troshanov, 1968; Schweisfurth, 1968; Tortoriello, 1971), and in marine environments (Ehrlich, 1963c; Ehrlich *et al.*, 1972). This reduction may be enzymatic or non-enzymatic. In the latter case, the microbes may produce metabolic products which can reduce Mn(IV) chemically to Mn(II) (Tortoriello, 1971). Examples of such metabolic products are formate and oxalate. It is noteworthy that bacterial isolates from the marine environment capable of reducing MnO_2 have so far always been found to promote the reduction enzymatically (Ehrlich *et al.*, 1972). Although reduced oxygen tension has been said to favour Mn(IV) reduction (Alexander, 1961; Troshanov, 1969), specific instances have been found in which anaerobiosis did not especially stimulate this process (Trimble and Ehrlich, 1968; Troshanov, 1969).

A primary environmental effect of Mn(II) oxidation is the removal of manganese from solution since the products of oxidation, i.e. Mn(IV) oxides, are generally insoluble. A secondary effect is the removal of other cations, especially heavy metals, from solution through adsorption to or ion exchanges with hydrous oxides of Mn(IV) (Ehrlich *et al.*, 1973; Jenne, 1968; McKenzie, 1967). The environmental effect of Mn(IV) reduction is the opposite of Mn(II) oxidation, namely the release of manganese and of many of the scavenged heavy metal cations into solution (Ehrlich *et al.*, 1973). Manganese oxidation may make manganese limiting in the environment while manganese reduction may raise the manganese concentration to toxic levels, for higher plants at least (Sauchelli, 1969).

Questions exist in regard to enzymatic iron oxidation by microbes at neutral pH. Since iron is very susceptible to autoxidation at pH values above 4, demonstration of enzyme-catalysed oxidation is very difficult. Although older reports implicated such organisms as *Leptothrix ochracea* and *Gallionella ferruginea* in ferrous iron oxidation (Hanert, 1968; Kucera and Wolfe, 1957; Lieske, 1911 and 1919; Praeve, 1957; Sartory and Meyer, 1947; Winogradsky, 1888), this has been questioned by some investigators. If these organisms do oxidize ferrous iron enzymatically, they may require it in the form of ferrous carbonate or ferrous sulphide at reduced oxygen tension (Aristovskaya and Zavarzin, 1971).

Many cases of so-called iron oxidation may really involve the decomposition of the organic portion of soluble organic iron complexes (ferrous or ferric) resulting in the freeing of the iron and its subsequent precipitation as iron hydroxides or oxides. In these instances, any oxidation of freed ferrous iron under aerobic conditions and at neutral pH may be non-enzymatic. Any ferric iron liberated in the digestion would very likely precipitate as a result of hydrolysis to ferric hydroxide. However, Aristovskaya and Zavarzin (1971) believe that the complex-digesting organisms can also catalyse oxidation of ferrous iron released by them from the complex. Experimental proof is needed here.

Iron oxide reduction by bacteria has been studied by a number of investigators (Silverman and Ehrlich, 1964), but especially by Ottow (1968, 1969, 1970, 1971). He found this process to be enzymatic in those cases which he studied. One or both of two possible enzymatic systems may be involved: nitrate reductase and an as yet unidentified enzymatic redox system. Ottow made his observations with soil bacteria. De Castro and Ehrlich (1970) obtained iron reduction of limonite ($Fe_2O_3 . nH_2O$), goethite ($Fe_2O_3 . H_2O$), and hematite (Fe_2O_3) with a marine *Bacillus*, but neither De Castro (1969) nor Ehrlich *et al.* (1973) observed very significant bacterial reduction of ferric iron in ferromanganese nodules. Troshanov (1968), on the other hand, observed iron reducing bacteria which could simultaneously reduce iron and manganese oxides of ores in certain lakes on the Karelian Isthmus. As in the case of manganese, iron oxidation results in iron removal from solution while iron reduction results in iron solution.

VI. Microbial Interactions with Mercury

Mercury appears to be ubiquitous in the environment, although it occurs at low concentrations in most places. Representative concentrations in air range from 1 to 50 ng m^{-3}; in soil from 0·02 to 0·92 $\mu g\ g^{-1}$ (average, 0·07 mg g^{-1}); in fresh waters from 0·02 to 0·07 ng ml^{-1}; and in freshwater sediments from less than 0·01 to 0·15 $\mu g\ g^{-1}$ (Klein, 1972; Konrad, 1972). The ubiquity of mercury can be attributed, in part, to certain chemical and physical properties of the metal and some of its compounds. Mercury metal at room temperature is a slightly volatile liquid, which allows for its ready dissemination. The chloride, sulphate and nitrate salts of mercuric mercury are very soluble as are many of its organic salts and derivatives, which also contribute to its dissemination. Dimethylmercury is volatile and readily enters the atmosphere (Bratt, 1967). Humic acids and fulvic acids also contribute to the dissemination of mercury by their ability to complex mercury ions (Ramamoorthy and Kushner, 1975b). The natural foci of mercury dissemination are usually considered to be ore deposits of mercury, mainly in the form of cinnabar (HgS), and non-mercury ore deposits like those of lead, arsenic and antimony, which are of igneous origin and contain traces of mercury (Anonymous, 1965). Moiseyev (1971) believes that a significant amount of mercury in the environment is released from sediments through the action of volcanic heat and superficial hydrothermal solutions. Natural weathering of mercury-containing deposits and man's exploitation of them is considered to be a major source of the mercury disseminated throughout the world. Industrial activities of man which help to spread mercury through the environment include chlorine and caustic soda production, paper production, tanning, manufacture of mercury batteries and control instruments, manufacture and application of fungicides, manufacture and application of mercurials in medicine, and mining activities (D'Itri, 1972).

The mercuric ion of mercury salts tends to be bound by cells, especially by their proteins. Sulphydryl groups are especially susceptible. Hence mercury salts are strong enzyme inhibitors. Methylmercury ion is also very poisonous. In vertebrate animals it seems to exert its greatest effect on the nervous system, and, in particular, the brain owing to its lipid solubility. According to Storm and Gunsalus (1974), methylmercury is also a very strong inhibitor of adenyl cyclase in rat liver plasma-membrane. Its mode of action on microbes is as yet unclear.

A number of microbes, including bacteria (Jensen and Jerneloev, 1969; Wood, 1974; Yamada and Tonomura, 1972a, b) and fungi (Landner, 1971; Vonk and Sijpesteijn, 1973) are able to modify mercury compounds biochemically. Jerneloev and Martin (1975) recently summarized the state of knowledge of this subject. Bacteria which are capable of it, transform uncomplexed mercury salts to methylmercury and dimethylmercury by methylation with methylcobalamine (CH_3B_{12}) using the following reaction sequence (Wood *et al.*, 1968),

$$Hg^{2+} \xrightarrow{CH_3B_{12}} CH_3Hg^+ \xrightarrow{CH_3B_{12}} (CH_3)_2Hg \qquad (12)$$

On the other hand, *Neurospora crassa*, a fungus, methylates mercuric salts after first complexing mercuric ion with homocysteine or cysteine as follows (Landner, 1971).

$$\underset{\text{mercuryl homocysteine}}{\begin{array}{l} \text{S-Hg}^+ \\ CH_2 \\ CH_2 \\ CHNH_2 \\ COOH \end{array}} \xrightarrow[\text{transmethylase}]{\text{methly donor}} \underset{\text{homocysteine}}{\begin{array}{l} SH \\ CH_2 \\ CH_2 \\ CHNH_2 \\ COOH \end{array}} + CH_3Hg^+ \qquad (13)$$

According to Wood (1974), the bacterial methylation of methylmercury is about 6000 times slower than the bacterial methylation of Hg^{2+}. The methylation of inorganic mercury with methylcobalamin (CH_3B_{12}) in bacteria seems to occur in the absence of any enzyme (Bertilsson and Neujahr, 1971; Imura *et al.*, 1971; Schrauzer *et al.*, 1971). In contrast, the methylation of mercury complexed by homocysteine or cysteine in a fungus like *N. crassa* requires a methyl donor such as choline or betaine, but not CH_3B_{12}, and a transmethylating enzyme (Landner, 1971). The formation of dimethylmercury can be viewed as a natural mechanism of detoxification of soil and water environments since the compound is volatile. However, in air, dimethylmercury is decomposed by the ultraviolet component of sunlight to elementary mercury plus methane and ethane (Wood, 1974).

Methylmercury can be formed by abiological means, e.g. by transalkylation with methyl tin compounds (Jerneloev and Martin, 1975), or photochemically with ultraviolet or visible light from mercuric chloride and acetic acid (Akagi and Sakagami, 1972; Akagi *et al.*, 1972). This raises a question about the relative contribution of microbes and chemical, abiological processes to the formation of methylmercury in the environment.

Phenyl- and methylmercury compounds can also be attacked microbiologically, being converted to volatile Hg^0 and benzene or methane, respectively. This activity has been observed in lake and estuarine sediments and in soil (Nelson *et al.*, 1973; Spangler *et al.*, 1973a; Tonomura *et al.*, 1968). The organisms involved were members of the genus *Pseudomonas*. They are generally mercury-resistant strains (e.g. Furukawa *et al.*, 1969; Nelson *et al.*, 1973). Some phenylmercuric acetate can also be converted microbiologically to diphenylmercury (Matsumura *et al.*, 1971).

Mercuric ion can be reduced to metallic mercury (Hg^0) by *Pseudomonas*, enteric bacteria, *Staphylococcus aureus*, and *Cryptococcus* (Brunker and Bott, 1974; Komura *et al.*, 1970; Nelson *et al.*, 1973; Summers and Lewis, 1973). The production of Hg^0 is generally associated with aerobic conditions (e.g. Nelson *et al.*, 1973; Spangler *et al.*, 1973a, b). While the formation of CH_3Hg^+ has been generally associated with anaerobic conditions in the laboratory (e.g. Wood *et al.*, 1968), it, too, is favoured by at least partially aerobic conditions in nature, owing to the fact that H_2S, which is produced in natural anaerobic environments, converts Hg^{2+} to HgS (Fagerstroem and Jerneloev, 1971). The HgS is not convertible to CH_3Hg^+ without conversion to a soluble salt or to HgO (Yamada and Tonomura, 1972c). The enzymatics of HgO production from phenylmercuric acetate by mercury resistant *Pseudomonas* has been studied by Japanese workers (Furukawa and Tonomura, 1971; 1972a, b), who showed an involvement of a reduced nicotinamide adenine dinucleotide phosphate (NADP) generating system, glucose- or arabinose-dyhydrogenase, cytochrome c I, and a metallic mercury-releasing enzyme. The metallic mercury-releasing enzyme contains flavin adenine dinucleotide (FAD) as a prosthetic group and is inducible.

Observation of methylmercury formation in sediments has been reported by Jerneloev (1970) and by Fagerstroem and Jerneloev (1971), and in soil by Beckert *et al.* (1974). The degradation of methylmercury in sediments has been observed by Spangler *et al.* (1973a, b). This latter reaction probably prevents major build-up of methylmercury in many environments.

VII. Conclusion

The foregoing discussion has shown that some microbes have developed special ways of interacting with heavy metals, arsenic and antimony in their environment, sometimes at levels toxic to many other microbes and higher forms of life. They may use these substances as energy sources, or as electron acceptors in their respiratory process. In some cases, microbes have developed ways of eliminating such substances from their environment through precipitation, adsorption or volatilization. These reactions may amount to detoxification which may not only render the environment more favourable for the microbes promoting them but also for other organisms which could not develop without this help.

Addendum

Recent work by Lyalikova and co-workers has provided additional information on *Stibiobacter senarmontii* (Lyalikova *et al.*, 1974, 1976). Lewis and Miller (1977) have published evidence for cytochrome reduction in *Thiobacillus ferrooxidans* by cuprous ions, thus providing strong supporting evidence that cuprous copper can serve as energy source for this organism. Hajj and Makemson (1976) have questioned growth of *Sphaerotilus discophorus* under mixotrophic conditions on the basis of new experiments. The toxic effects of copper, nickel, cobalt and manganese ions on some marine bacteria have been demonstrated on cultures growing at 1 atm (Yang and Ehrlich, 1976), and the effect of hydrostatic pressure on the interaction of these metal ions with the marine cultures has been demonstrated by Arcuri and Ehrlich (1977).

References

Adiga, P. R., Sastry, K. S., Venkatasubramanyam, V. and Sarma, P. S. (1961). Interrelationships in trace-element metabolism in *Aspergillus niger*. *Biochem. J.* **81,** 545–550.

Adiga, P. R., Sastry, K. S. and Sarma, P. S. (1962). The influence of iron and magnesium on the uptake of heavy metals in metal toxicities in *Aspergillus niger*. *Biochim. Biophys. Acta* **64,** 546–548.

Akagi, H. and Sakagami, Y. (1972). Studies on photochemical alkylation of inorganic mercury. II. Alkylation of inorganic mercury in water by irradiation with sunlight or blacklight. *J. Hyg. Chem.* **18,** 358–362.

Akagi, H., Fujita, M. and Sakagami, Y. (1972). Studies on photochemical alkylation of inorganic mercury. I. Alkylation of inorganic mercurial in water by ultraviolet irradiation. *J. Hyg. Chem.* **18,** 309–314.

Albright, L. J. and Wilson, E. M. (1974). Sublethal effects of several metallic salts-organic compound combinations upon the heterotrophic microflora of a natural water. *Water Res.* **8,** 101–105.

Alexander, M. (1961). "Introduction to Soil Microbiology." Wiley, New York and London.

Ali, S. H. and Stokes, J. L. (1971). Stimulation of heterotrophic and autotrophic growth of *Sphaerotilus discophorus* by manganous ions. *Antonie van Leeuwenhoek* **37,** 519–528.

Andreesen, J. R. and Ljungdahl, L. G. (1973). Formate dehydrogenase of *Clostridium thermoaceticum:* Incorporation of selenium-75, and the effects of selenite, molybdate, and tungstate on the enzyme. *J. Bacteriol.* **116,** 867–873.

Andreesen, J. R., El Ghazzawi, E. and Gottschalk, G. (1974). The effect of ferrous ions, tungstate and selenite on the level of formate dehydrogenase in *Clostridium formicoaceticum* and formate synthesis from CO_2 during pyruvate fermentation. *Arch. Mikrobiol.* **96,** 103–118.

Anonymous (1965). "Mineral Facts and Problems" (1965 ed.). Bulletin 630. Bur. Mines. U.S. Dept. Interior, Washington, D.C.

Anonymous (1971). "Marine Chemistry. A Report of the Marine Chemistry Panel of the Committee on Oceanography." Nat. Acad. Sci., Washington, D.C.

Arceneaux, J. E. L., Davis, W. B., Downer, D. N., Haydon, A. H. and Byers, B. R. (1973). Fate of labeled hydroxamates during iron transport from hydroxamate-iron chelates. *J. Bacteriol.* **115,** 919–927.

Arcuri, E. J. and Ehrlich, H. L. (1977). Influence of hydrostatic pressure on the effects of the heavy metal cations of manganese, copper, cobalt and nickel on the growth of three deep-sea bacterial isolates. *Appl. Environ. Microbiol.* **33,** 282–288.

Aristovskaya, T. V. and Zavarzin, G. A. (1971). Biochemistry of iron in soil. *In* "Soil Biochemistry," Vol. 2 (Eds A. D. McLaren and J. Skujins), pp. 385–408. Marcel Dekker, Inc., New York.

Avakyan, Z. A. (1967). Comparative toxicity of heavy metals for certain micro-organisms. *Mikrobiologiya* **36,** 446–450.

Bartha, R. and Ordal, E. J. (1965). Nickel-dependent chemolithotrophic growth of two *Hydrogenomonas* strains. *J. Bacteriol.* **89,** 1015–1019.

Beck, J. V. (1967). The role of bacteria in copper mining operations. *Biotech. Bioeng.* **9,** 487–497.

Beck, J. V. and Brown, D. G. (1968). Direct sulfide oxidation in the solubilization of sulfide ores by *Thiobacillus ferrooxidans*. *J. Bacteriol.* **96,** 1433–1434.

Beckert, W. F., Moghissi, A. A., Au, F. H. F., Bretthauer, E. W. and McFarlane J. C. (1974). Formation of methyl mercury in a terrestrial environment. *Nature, Lond.* **249,** 674–675.

Bertilsson, L. and Neujahr, H. Y. (1971). Methylation of mercury compounds by methylcobalamin. *Biochemistry* **10,** 2805–2808.

Bertrand, D. (1974). Nickel, an active trace element for nitrogen-fixing micro-organisms. *C. R. Acad. Sci., Ser. D.* **278,** 2231–2235.

Bertrand, D. and De Wolff, A. (1973). Importance of nickel as trace element for *Rhizobium* of nodules of legumes. *C. R. Acad. Sci., Ser. D.* **276,** 1855–1858.

Bhattacharyya, P. (1970). Active transport of manganese in isolated membranes of *Escherichia coli*. *J. Bacteriol.* **104,** 1307–1311.

Bowen, H. J. M. (1966). "Trace Elements in Biochemistry." Academic Press, London and New York.

Bratt, J. (1967). Mercury compounds. *In* "Encyclopedia of Chemical Technology" (Vol. 13, 2nd revised ed.) (Eds H. F. Mark, J. J. McKetta and D. F. Othmer), pp. 235–248. Wiley-Interscience., New York, London and Sydney.

Brewer, P. G. and Spencer, D. W. (1969). A note on the chemical composition of the Red Sea Brines. *In* "Hot Brines and Recent Heavy Metal Deposits in the Red Sea" (Eds E. T. Degens and D. A. Ross), pp. 174–179. Springer Verlag, New York.

Brierley, C. L. and Brierley, J. A. (1973). A chemoautotrophic and thermophilic microorganism isolated from an acid hot spring. *Can. J. Microbiol.* **19,** 183–188.

Brierley, C. L. and Murr, L. E. (1973). Leaching: Use of a thermophilic and chemo-autotrophic microbe. *Science* **179,** 488–490.

Brock, T. D. (1974). "Biology of Microorganisms," 183 pp. Prentice-Hall, Inc. Englewood Cliffs, New Jersey.

Bromfield, S. M. (1956). Oxidation of manganese by soil microorganisms. *Aust. J. Biol. Sci.* **9,** 238–252.

Brunker, R. L. and Bott, T. L. (1974). Reduction of mercury to the elemental state by a yeast. *Appl. Microbiol.* **27,** 870–873.

Bryner, L. C. and Anderson, R. (1957). Microorganisms in leaching sulfide minerals. *Ind. Eng. Chem.* **49,** 1721–1724.

Bryner, L. C., Beck, J. V., Davis, D. B., and Wilson D. G. (1954). Microorganisms in leaching sulfide minerals. *Ind. Eng. Chem.* **46,** 2587–2592.

Bubela, B. (1970). Chemical and morphological changes in *Bacillus stearothermophilus* induced by copper. *Chem. Biol. Interact.* **2,** 107–116.

Cerbon, J. (1969). Arsenic-lipid complex formation during active transport of arsenate in yeast. *J. Bacteriol.* **97,** 658–662.

Cobet, A. B., Wirsen, C. and Jones, G. E. (1970). The effect of nickel on a marine bacterium, *Arthrobacter marinus* sp. nov. *J. Gen. Microbiol.* **62,** 159–169.

Cobet, A. B., Jones, G. E., Albright, J., Simon, H. and Wirsen, C. (1971). The effect of nickel on a marine bacterium: Fine structure of *Arthrobacter marinus. J. Gen. Microbiol.* **66,** 185–196.

Colmer, A. R. and Hinkle, M. E. (1947). The role of microorganisms in acid mine drainage: a preliminary report. *Science* **106,** 253–256.

Colmer, A. R., Temple, K. L. and Hinkle, M. E. (1950). An iron-oxidizing bacterium from the acid drainage of some bituminous coal mines. *J. Bacteriol.* **59,** 317–328.

Corbett, R. G. and Gorwitz, D. J. (1967). Composition of water discharged from bituminous coal mines in northern West Virginia. *Econ. Geol.* **62,** 848–851.

Corbin, J. L. and Bulen, W. A. (1969). The isolation and identification of 2,3-dihydroxybenzoic acid and 2-*N*, 6-*N*-*di*(2,3-dihydroxybenzoyl)-L-lysine formed by iron deficient *Azotobacter vinelandii. Biochemistry* **8,** 757–762.

Cox, G. B., Gibson, F., Luke, R. K. J., Newton, N. A., O'Brien, I. G. and Rosenberg, H. (1970). Mutations affecting iron transport in *Escherichia coli. J. Bacteriol.* **104,** 219–226.

Da Costa, E. W. B. (1972). Variation in the toxicity of arsenic compounds to microorganisms and the suppression of the inhibitory effects of phosphate. *Appl. Microbiol.* **23,** 46–53.

Davis, W. B. and Byers, B. R. (1971). Active transport of iron in *Bacillus megaterium*: Role of secondary hydroxamic acids. *J. Bacteriol.* **107,** 491–498.

De Castro, A. F. (1969). Bacterial reduction of iron oxides. Ph.D. thesis. Rensselaer Polytechnic Institute, Troy, New York.

De Castro, A. F. and Ehrlich, H. L. (1970). Reduction of iron oxide minerals by a marine Bacillus. *Antonie van Leeuwenhoek* **36,** 317–327.

Degens, E. T. and Ross, D. A., Eds (1969). "Hot Brines and Recent Heavy Metal Deposits in the Red Sea." Springer Verlag, New York.

De Grys, A. (1964). Copper distribution patterns in soils and drainage in Central Chile. *Econ. Geol.* **59,** 636–646.

Den Dooren De Jong, L. E. and Roman, W. B. (1971). Tolerance of *Azotobacter* for metallic and non-metallic ions. *Antonie van Leeuwenhoek* **37,** 119–124.

De Turk, W. E. and Bernheim, F. (1960). The inhibition of enzyme induction and ammonia assimilation in *Pseudomonas aeruginosa* by sulfhydryl compounds and by cobalt, and its reversal by iron. *Arch. Biochem. Biophys.* **90,** 218–223.

D'Itri, F. M. (1972). Sources of mercury in the environment. *In* "Environmental Mercury Contamination" (Eds R. Hartung and B. D. Dinman), pp. 5–25. Ann Arbor Science Publishers Inc., Ann Arbor, Michigan.

Duncan, D. W. and Walden, C. C. (1972). Microbiological leaching in the presence ferric iron. *Dev. Ind. Microbiol.* **13,** 66–75.

Duncan, D. W., Trussell, P. C. and Walden, C. C. (1964). Leaching chalcopyrite with *Thiobacillus ferrooxidans*: Effects of surfactants and shaking. *Appl. Microbiol.* **12,** 122–126.

Dyke, K. G. H., Parker, M. T. and Richmond, M. H. (1970). Penicillinase production and metal-ion resistance in *Staphylococcus aureus* cultures isolated from hospital patients. *J. Med. Microbiol.* **3,** 125–136.

Ehrlich, H. L. (1963a). Microorganisms in acid drainage from a copper mine. *J. Bacteriol.* **86,** 350–352.

Ehrlich, H. L. (1963b). Bacterial action on orpiment. *Econ. Geol.* **58,** 991–994.

Ehrlich, H. L. (1963c). Bacteriology of manganese nodules. I. *Bacterial action on* manganese in nodule enrichments. *Appl. Microbiol.* **11,** 15–19.

Ehrlich, H. L. (1964). Bacterial oxidation of arsenopyrite and enargite. *Econ. Geol.* **59,** 1306–1312.

Ehrlich, H. L. (1966). Reactions with manganese by bacteria from marine ferromanganese nodules. *Dev. Ind. Microbiol.* **7,** 279–286.

Ehrlich, H. L. (1968a). Bacteriology of manganese nodules. II Manganese oxidation by cell-free extract from a manganese nodule bacterium. *Appl. Microbiol.* **16,** 197–202.

Ehrlich, H. L. (1968b). Biogeochemistry of the minor elements. *In* "Soil Biochemistry," Vol. 2 (Eds A. D. McLaren and J. Skujins), pp. 361–384. Marcel Dekker Inc., New York.

Ehrlich, H. L. (1976) Manganese as an energy source for bacteria. *In* "Environmental Biogeochemistry," Vol. 2 "Metals Transfer and Ecological Mass Balances" (Ed. J. O. Nriagu), Vol. 2, pp. 633–644. Ann Arbor Science Publishers Inc. Ann Arbor, Michigan.

Ehrlich, H. L., Ghiorse, W. C. and Johnson, G. L. II (1972). Distribution of microbes in manganese nodules from the Atlantic and Pacific Oceans. *Dev. Ind. Microbiol.* **13,** 57–65.

Ehrlich, H. L., Yang, S. H. and Mainwaring, J. D. Jr. (1973). Bacteriology of manganese nodules. VI. Fate of copper, nickel, cobalt and iron during bacterial and chemical reduction of the manganese (IV). *Z. Allg. Mikrobiol.* **13,** 39–48.

Eisenstadt, E., Fisher, S., Der, Chi-Lui and Silver, S. (1973). Manganese transport in *Bacillus subtilis* W23 during growth and sporulation. *J. Bacteriol.* **113,** 1363–1372.

Emery, T. F. (1971). Role of ferrichrome as a ferric ionophore in *Ustilago sphaerogena. Biochemistry* **10,** 1483–1488.

Engel, W. B. and Owen, R. M. (1970). Metal-accumulating properties of fuel-utilizing bacteria. *Dev. Ind. Microbiol.* **11,** 196–209.

Enoch, H. G. and Lester, R. L. (1972). Effects of molybdate, tungstate, and selenium compounds on formate dehydrogenase and other enzyme systems in *Escherichia coli. J. Bacteriol.* **110,** 1032–1040.

Esposito, R. G. and Wilson, P. W. (1956). Trace metal requirement of *Azotobacter. Proc. Soc. Exp. Biol. Med.* **93,** 546–567.

Fagerstroem, T. and Jerneloev, A. (1971) Formation of methyl mercury from pure mercuric sulphide in aerobic organic sediment. *Water Res.* **5,** 121–122.

Fisher, S., Buxbaum, L., Toth, K., Eisenstadt, E. and Silver, S. (1973). Regulation of manganese accumulation and exchange in *Bacillus subtilis* W23. *J. Bacteriol.* **113,** 1373–1380.

Fox, S. I. (1967). Bacterial oxidation of simple copper sulfides. Ph.D. thesis. Rensselaer Polytechnic Institute, Troy, New York.

Friedman, B. A. and Dugan, P. R. (1968). Concentration and accumulation of metallic ions by the bacterium *Zooloea*. *Dev. Ind. Microbiol.* **9,** 381–388.

Furukawa, K. and Tonomura, K. (1971). Enzyme system involved in the decomposition of phenyl mercuric acetate by mercury-resistant *Pseudomonas*. *Agr. Biol. Chem.* **35,** 604–610.

Furukawa, K. and Tonomura, K. (1972a). Metallic mercury-releasing enzyme in mercury-resistant *Pseudomonas*. *Agric. Biol. Chem.* **36,** 217–226.

Furukawa, K. and Tonomura, K. (1972b). Induction of metallic mercury-releasing enzyme in mercury-resistant *Pseudomonas*. *Agric. Biol. Chem.* **36,** Suppl. 13, 2441–2448.

Furukawa, K., Suzuki, T. and Tonomura, K. (1969). Decomposition of organic mercurial compounds by mercury-resistant bacteria. *Rep. Ferment. Res. Inst.* (*Chiba*) **37,** 39–47.

Gleeson, C. F. and Coope, J. A. (1967). The distribution of metals in swamps in eastern Canada. *Can. Dep. Energy Mines. Res., Geol. Surv. Canada.* **66–54,** 145–166.

Gonye, E. R. Jr. and Jones, G. E. (1973). An ecological survey of open ocean and estuarine microbial populations. II. The oligodynamic effect of nickel on marine bacteria. *In* "Estuarine Microbial Ecology" (Eds L. H. Stevenson and R. R. Colwell), pp. 243–257. University of South Carolina Press, Columbia.

Hajj, H. and Makemson, J. (1976). Determination of growth of *Sphaerotilus discophorus* in the presence of manganese. *Appl. Environ. Microbiol.* **32,** 699–702.

Hanert, H. (1968). Investigations on isolation, physiology and morphology of *Gallionella ferruginea* Ehrenberg. *Arch. Mikrobiol.* **60,** 348–376.

Haydon, A. M., Davis, W. B., Arceneaux, J. E. L. and Byers, B. R. (1973). Hydroxamate recognition during iron transport from hydroxamate-iron chelates. *J. Bacteriol.* **115,** 912–918.

Hedges, R. W. and Baumberg, S. (1973). Resistance to arsenic compounds conferred by a plasmid transmissible between strains of *Escherichia coli*. *J. Bacteriol.* **115,** 459–460.

Imai, K., Sakaguchi, H., Sugio, T. and Tano, T. (1973). On the mechanism of chalcocite oxidation by *Thiobacillus ferrooxidans*. *J. Ferment. Technol.* **51,** 865–870.

Imura, N., Sukegawa, E., Pan, S.-K., Nagao, K., Kim, J.-Y., Kwan, T. and Ukita, T. (1971). Chemical methylation of inorganic mercury with methyl-cobalamin, a vitamin B_{12} analog. *Science* **172,** 1248–1249.

Ivanov, V. I. (1962). Effect of some factors on iron oxidation by cultures of *Thiobacillus ferrooxidans*. *Mikrobiologiya* **31,** 795–799.

Ivanov, V. I., Nagirnyak, F. I. and Stepanov, B. A. (1961). Bacterial oxidation of sulfide ores. I. Role of *Thiobacillus ferrooxidans* in the oxidation of chalcopyrite and sphalerite. *Mikrobiologiya* **30,** 688–692.

Jenne, E. A. (1968). Controls on Mn, Fe, Co, Ni, Cu and Zn concentrations in soils and water: the significant role of hydrous Mn and Fe oxides. *In* "Trace Inorganics in Water" (Ed. R. F. Gould), pp. 337–387. American Chemical Society, Washington, D.C.

Jensen, S. and Jerneloev, A. (1969). Biological methylation of mercury in aquatic organisms. *Nature, Lond.* **223,** 753–754.

Jerneloev, A. (1970). Release of methyl mercury from sediments with layers containing inorganic mercury at different depths. *Limnol. Oceanogr.* **15,** 958–960.

Jerneloev, A. and Martin, A.-L. (1975). Ecological implications of metal metabolism by microorganisms. *Ann. Rev. Microbiol.* **29,** 61–77.

Jones, G. E. (1973). An ecological survey of open ocean and estuarine microbial populations. I. The importance of trace metal ions to microorganisms in the sea. *In* "Estuarine Microbial Ecology" (Eds L. H. Stevenson and R. R. Colwell), pp. 233–241. University of South Carolina Press, Columbia.

Komura, I., Izaki, K. and Takahashi, H. (1970). Vaporization of inorganic mercury by cell-free extracts of drug-resistant *Escherichia coli. Agric. Biol. Chem.* **34,** 480–482.

Kee, N. S. and Bloomfield, C. (1961). The solution of some minor element oxides by decomposing plant materials. *Geochim. Cosmochim. Acta* **24,** 206–225.

Kendrick, W. B. (1962). Soil fungi of a copper swamp. *Can. J. Microbiol.* **8,** 639–647.

Kharkar, D. P., Turekian, K. K. and Bertine, K. K. (1968). Stream supply of dissolved silver, molybdenum, antimony, selenium chromium, cobalt, rubidium, and cesium to the oceans. *Geochim. Cosmochim. Acta* **32,** 285–298.

Khovrychev, M. P. (1973). Absorption of copper ions by cells of *Candida utilis. Mikrobiologiya* **42,** 839–844.

Klein, D. H. (1972). Some estimates of natural levels of mercury in the environment. *In* "Environmental Mercury Contamination" (Eds R. Hartung and B. D. Dinman), pp. 25–29. Ann Arbor Science Publishers Inc. Ann Arbor, Michigan.

Komura, I., Izaki, K. and Takahashi, H. (1970). Vaporization of inorganic mercury by cell-free extracts of drug resistant *Excherichia coli. Agric Biol. Chem.* **34,** 480–482.

Kossaya, T. A. (1967). Composition of manganese oxides in cultures of *Metallogenium* and *Leptothrix. Mikrobiologiya* **26,** 1024–1029.

Kondo, I., Ishikawa, T. and Nakahara, H. (1974). Mercury and cadmium resistances mediated by the penicillinase plasmid in *Staphylococcus aureus. J. Bacteriol.* **117,** 1–7.

Konrad, J. G. (1972). Mercury contents of bottom sediments from Wisconsin rivers and lakes. *In* "Environmental Mercury Contamination" (Eds R. Hartung and B. D. Dinman), pp. 52–58. Ann Arbor Science Publishers Inc., Ann Arbor, Michigan.

Kovalskii, V. V. and Letunova, S. V. (1974). Geochemical ecology of microorganisms. *Tr. Biogeokhim. Lab. Akad. Nauk (SSSR)* **13,** 3–56.

Krumbein, W. E. (1971). Manganese oxidizing fungi and bacteria in recent shelf sediments of the Bay of Biscay and the North Sea. *Naturwissenschaften* **58,** 56–57.

Kucera, S. and Wolfe, R. S. (1957). A selective enrichment method for *Gallionella ferruginea. J. Bacteriol.* **74,** 344–349.

Kuznetsov, S. I., Ivanov, M. V. and Lyalikova, N. N. (1963). *In* "Introduction to Geological Microbiology." McGraw-Hill Book Co. Inc., New York.

Laborey, F. and Lavollay, J. (1967). Toxicity of Zn^{+2} and Cd^{+2} during *Aspergillus niger* growth antagonisms of these ions, and Mg^{2+}, Zn^{2+} and Cd^{2+} interaction. *C. R. Acad. Sci., Ser. D.* **264,** 2937–2940.

Lackey, J. B. (1938). The flora and fauna of surface waters polluted by acid mine drainage. *Public Health Rep.* **53,** 1499–1507.

Landner, L. (1971). Biochemical model for the biological methylation of mercury suggested from methylation studies *in vivo* with *Neurospora crassa. Nature, Lond.* **230,** 452–454.

Leahy, J. J. (1969). Cobaltous ion effect on diauxie. *J. Bacteriol.* **100,** 1194–1197.

Leathen, W. W., Braley, S. A. Sr. and McIntyre, L. D. (1953). The role of bacteria in the formation of acid from certain sulfuritic constituents associated with bituminous coal. II. Ferrous iron oxidizing bacteria. *Appl. Microbiol.* **1,** 65–68.

Lewis, A. J. and Miller, J. D. A. (1977). Stannous and cuprous ion oxidation by *Thiobacillus ferrooxidans*. *Can. J. Microbiol.* **23,** 319–324.

Lieske, R. (1911). Beitraege zur Kenntniss der Physiologie von *Spiro-phyllum ferrugineum* Ellis, einem typischen Eisenbakterium. *Jahrb. Wiss. Bot.* **49,** 91–127.

Lieske, R. (1919). Zur Ernaehrungsphysiologie der Eisenbakterien. *Z. Bakt. Parasitk. Infektionskr. Hyg. Abt.* II, **49,** 413–425.

Loutit, J. S. (1970). Mating system of *Pseudomonas aeruginosa* strain I. VI. Mercury resistance associated with the sex factor (FP). *Genet. Res.* **16,** 179–184.

Luckey, M., Pollack, J. R., Wayne, R., Ames, B. N. and Neilands, J. B. (1972). Iron uptake in *Salmonella typhimurium*: Utilization of exogeneous siderochromes as iron carriers. *J. Bacteriol.* **111,** 731–738.

Lundgren, D. G., Vestal, J. R. and Tabita, F. R. (1972). The microbiology of mine drainage pollution. *In* "Water Pollution Microbiology" (Ed. R. Mitchell), pp. 69–88. Wiley-Interscience, New York.

Lyalikova, N. N. (1961). *Tr. In-ta Mikrobiol.* AN SSSR, Mo. 9.

Lyalikova, N. N. (1972). Oxidation of trivalent antimony to higher oxides as an energy source for the development of a new autotrophic organism *Stribiobacter* gen. n. *Dokl. Akad. Nauk SSSR, Ser. Biol.* **205,** 1228–1229.

Lyalikova, N. N. (1974). *Stibiobacter senarmontii*: a new microorganism oxidizing antimony. *Mikrobiologiya* **43,** 941–948.

Lyalikova, N. N., Shlain, L. B., Trofimov, V. G. (1974). Formation of minerals of antimony (V) under the effect of bacteria. *Izv. Akad. Nauk SSSR. Ser. Biol.* **3,** 440–444.

Lyalikova, N. N., Vedenina, I. Ya. and Romanova, A. K. (1976). Assimilation of carbon dioxide by a culture of *Stibiobacter senarmontii*. *Mikrobiologiya* **45,** 552–554.

Malanchuk, J. L. and Gruendling, G. K. (1973). Toxicity of lead nitrate to algae. *Water, Air, Soil Pollut.* **2,** 181–190.

Malouf, E. E. and Prater, J. D. (1961). Role of bacteria in the alteration of sulfide minerals. *J. Metals* **13,** 353–356.

Mann, P. J. G. and Quastel, J. H. (1946). Manganese metabolism in soils. *Nature, Lond.* **158,** 154–156.

Marchlewitz, B. and Schwartz, W. (1961). Untersuchungen ueber die Mikroben-Assoziation saurer Grubenwaesser. *Z. Allg. Mikrobiol.* **1,** 100–114.

Matsumara, F., Gotoh, Y. and Boush, G. M. (1971). Phenyl mercuric acetate: Metabolic conversion by microorganisms. *Science* **173,** 49–51.

McKenzie, R. M. (1967). The sorption of cobalt by manganese minerals in soils. *Aust. J. Soil. Res.* **5,** 235–246.

Moiseyev, A. N. (1971). A non-magmatic source for mercury ore deposits. *Econ. Geol.* **66,** 591–601.

Mosin, A. F., Petrova, K. M., Kharchuk, A. I. and Abaturov, Yu. D. (1974). The toxic effect of arsenate on yeast. *Mikrobiologiya* **43,** 94–98.

Nelson, J. D., Blair, W., Brinckman, F. E., Colwell, R. R. and Iverson, W. P. (1973). Biodegradation of phenylmercuric acetate by mercury-resistant bacteria. *Appl. Microbiol.* **26,** 321–326.

Nielsen, A. M. and Beck, J. V. (1972). Chalcocite oxidation and coupled carbon dioxide fixation by *Thiobacillus ferrooxidans*. *Science* **175,** 1124–1126.

Novick, R. P. and Roth, C. (1968). Plasmid-linked resistance to inorganic salts in *Staphylococcus aureus*. *J. Bacteriol.* **95,** 1335–1342.

Nowosielski, O., Knezek, B. D. and Ellis, B. G. (1971). Evaluation of toxicity of metals, NTA and EDTA by *Aspergillus niger* bioassay. *Am. Chem. Soc., Div. Water Air Waste Chem. Gen. Papers* **11,** 32–38.

O'Callaghan, R. J., Bundy, L., Bradley, R. and Paranchych, W. (1973). Unusual arsenate poisoning of the F pili of *Escherichia coli*. *J. Bacteriol.* **115,** 76–81.

Oster, K. A. and Golden, M. J. (1954). Fungistatic and fungicidal compounds. *In* "Antiseptics, Disinfectants, Fungicides, and Chemical and Physical Sterilization" (Ed. G. F. Reddish), pp. 548–565. Lea and Febiger, Philadelphia.

Ottow, J. C. G. (1968). Evaluation of iron-reducing bacteria in soil and the physiological mechanism of iron reduction in *Aerobacter aerogenes* (*Enterobacter aerogenes*). *Z. Allg. Mikrobiol.* **8,** 441–443.

Ottow, J. C. G. (1969). Influence of nitrate, chlorate sulfate, form of iron oxide and growth conditions of the extent of bacteriological reduction of iron. *Z. Pfl-Ernähr. Bodenk.* **124,** 238–253.

Ottow, J. C. G. (1970). Selection, characterization and iron-reducing capacity of nitrate reductaseless (nit$^-$) mutants of iron-reducing bacteria. *Z. Allg. Mikrobiol.* **10,** 55–62.

Ottow, J. C. G. (1971). Iron reduction and gley formation by nitrogen-fixing Clostridia. *Oecologia* **6,** 164–175.

Ou, J. T. and Anderson, T. F. (1972). Effect of Zn^{2+} on bacterial conjugation: Inhibition of mating pair formation. *J. Bacteriol.* **111,** 177–185.

Perfil'ev, B. V. and Gabe, D. R. (1965). The use of the microbial-landscape method to investigate bacteria which concentrate manganese and iron in bottom deposits. *In* "Applied Capillary Microscopy. The Role of Microorganisms in the Formation of Iron-Manganese Deposits" (Eds B. V. Perfil'ev, D. R. Gabe, A. M. Galperina, V. A. Rabinovich, A. A. Sapotnitskii, E. E. Sherman, E. P. Troshanov), pp. 9–54. Consultants Bureau, New York.

Praeve, P. (1957). Untersuchungen ueber die Stoffwechselphysiologie des Eisenbakteriums *Leptothrix ochrae* Kuetzing. *Arch. Mikrobiol.* **27,** 33–62.

Ramamoorthy, S. and Kushner, D. J. (1975a). Binding of mercuric and other heavy metal ions by microbial growth media. *Microb. Ecol.* **2,** 162–176.

Ramamoorthy, S. and Kushner, D. J. (1975b). Heavy metal binding components of river water. *J. Fish. Res. Board Canada* **32,** 1755–1766.

Razzell, W. E. and Trussell, P. C. (1963). Microbiological leaching of metallic sulfides. *Appl. Microbiol.* **11,** 105–110.

Renshaw, E. C., Rosenberg, B. and Van Camp, L. (1966). Platinum-induced growth in *Escherichia coli*. *Bacteriol. Proc.*, p. 74.

Romans, I. B. (1954). Oligodynamic metals. *In* "Antiseptics, Disinfectants, Fungicides, and Chemical and Physical Sterilization" (Ed. G. F. Reddish), pp. 388–428. Lea and Febiger, Philadelphia.

Ross, D. A. (1972). Red sea hot brine area: revisited. *Science* **175,** 1455–1457.

Sadler, W. R. and Trudinger, P. A. (1967). The inhibition of microorganisms by heavy metals. *Min. Dep.* **2,** 158–168.

Sartory, A. and Meyer, J. (1947). Contribution a l'etude du metabolisme hydrocarbone des bacteries ferrugenieuses. *C. R. Acad. Sci.* **225,** 541–542.

Sastry, K. S., Adiga, P. R., Venkatasubramanyam, V. and Sarma, P. S. (1962). Interrelation in trace-element metabolism in metal toxicities in *Neurospora crassa*. *Biochem. J.* **85,** 486–491.

Sauchelli, V. (1969). "Trace Elements in Agriculture." Van Nostrand Reinhold Co., New York, Cincinnati, Toronto, London and Melbourne.

Schrauzer, G. N., Weber, J. H., Beckham, T. M. and Ho, R. K. Y. (1971). Alkyl group transfer from cobalt to mercury: reaction of alkyl cobalamins, alkyl-cobaloximes, and of related compounds with mercuric acetate. *Tetrahedron Lett.* **3,** 275–277.

Schweisfurth, R. (1968). Untersuchungen ueber manganoxydierende und-reduzierende Mikroorganismen. *Mitt. Internat. Verein. Kimnol.* **14,** 179–186.

Schweisfurth, R. (1971). Manganoxydierende Pilze. I. Vorkommen, Isolierung und mikroskopische Untersuchungen. *Z. Allg. Mikrobiol.* **11,** 415–430.

Sherman, J. M. and Albus, W. R. (1923). Physiological youth in bacteria. *J. Bacteriol.* **8,** 127–139.

Silver, S. and Kralovic, M. L. (1969). Manganese accumulation by *Escherichia coli*: evidence for a specific transport system. *Biochem. Biophys. Res. Commun.* **34,** 640–645.

Silver, M. and Torma, A. E. (1974). Oxidation of metal sulfides by *Thiobacillus ferrooxidans* grown in different substrates. *Can. J. Microbiol.* **20,** 141–147.

Silver, S., Johnsenie, P. and King, K. (1970). Manganese active transport in *Escherichia coli. J. Bacteriol.* **104,** 1299–1306.

Silverman, M. P. (1967). Mechanism of bacterial pyrite oxidation. *J. Bacteriol.* **94,** 1046–1051.

Silverman, M. P. and Ehrlich, H. L. (1964). Microbial formation and degradation of minerals. *Adv. Appl. Microbiol.* **6,** 153–206.

Silverman, M. P. and Lundgren, D. G. (1959a). Studies on the chemoautotrophic iron bacterium *Ferrobacillus ferrooxidans.* I. An improved medium and a harvesting procedure for securing high cell yields. *J. Bacteriol.* **77,** 642–647.

Silverman, M. P. and Lundgren, D. G. (1959b). Studies on the chemoautotrophic iron bacterium *Ferrobacillus ferrooxidans. J. Bacteriol.* **78,** 326–331.

Silverman, M. P. and Munoz, E. F. (1970). Fungal attack on rock: Solubilization and altered infrared spectra. *Science* **169,** 985–987.

Singer, P. C. and Stumm, W. (1970). Acidic mine drainage: the rate-determining step. *Science* **167,** 1121–1123.

Smith, D. H. (1967). R factors mediate resistance to mercury, nickel, and cobalt. *Science* **156,** 1114–1116.

Sokolova-Dubinia, G. A. and Deryugina, Z. P. (1967). Process of iron–manganese concretion formation in Lake Punnus-Yarvi. *Mikrobiologiya* **36,** 1066–1076.

Sokolova-Dubinina, G. A. and Deryugina, Z. P. (1968). Influence of limnetic environmental conditions on microbial manganese ore formation. *Mikrobiologiya* **37,** 147–153.

Spangler, W. J., Spigarelli, J. L., Rose, J. M. and Miller, H. M. (1973a). Methyl mercury: bacterial degradation in lake sediments. *Science* **180,** 192–193.

Spangler, W. J., Spigarelli, J. L., Rose, J. M., Fillipin, R. S. and Miller, H. H. (1973b). Degradation of methylmercury by bacteria isolated from environmental samples. *Appl. Microbiol.* **25,** 488–493.

Starkey, R. L. (1973). Effect of pH on toxicity of copper to *Scytalidium* sp., a copper-tolerant fungus, and some other fungi. *J. Gen. Microbiol.* **78,** 217–225.

Storm, D. R. and Gunsalus, R. P. (1974). Methylmercury is a potent inhibitor of membrane adenyl cyclase. *Nature, Lond.* **250,** 778–779.

Sullivan, J. D. (1930). "Chemistry of Leaching Chalcocite." Technical Paper 473. U.S. Bureau of Mines.

Summers, A. O. and Lewis, E. (1973). Volatilization of mercuric chloride by mercury-resistant plasmid-bearing strains of *Escherichia coli*, *Staphylococcus aureus*, and *Pseudomonas aeruginosa*. *J. Bacteriol.* **113,** 1070–1072.

Summers, A. O. and Silver, S. (1972). Mercury resistance in a plasmid-bearing strain of *Escherichia coli*. *J. Bacteriol.* **112,** 1228–1236.

Tandon, S. P. and Mishra, M. M. (1969). Effect of sodium, manganese, and zinc on the activity of *Nitrobacter agilis*. *Zentrlbl. Bakteriol., Parasitenk., Infektionskr.,* Hyg. Abt 2. **123,** 399–402.

Temple, K. L. (1964). Syngenesis of sulfide ores: An evaluation of biochemical aspects. *Econ. Geol.* **59,** 1473–1491.

Temple, K. L. and Colmer, A. R. (1951). The autotrophic oxidation of iron by a new bacterium: *Thiobacillus ferrooxidans*. *J. Bacteriol.* **62,** 605–611.

Temple, K. L. and Delchamps, E. W. (1953). Autotrophic bacteria and the formation of acid in bituminous coal mines. *Appl. Microbiol.* **1,** 255–258.

Temple, K. L. and Le Roux, N. W. (1964). Syngenesis of sulfide ores: Desorption of adsorbed metal ions and their precipitation as sulfides. *Econ. Geol.* **59,** 647–655.

Thiel, G. A. (1925). Manganese precipitated by microorganisms. *Econ. Geol.* **20,** 301–310.

Tonomura, K., Makagami, T., Futai, F. and Maeda, D. (1968). Studies on the action of mercury-resistant microorganisms on mercurials. I. The isolation of a mercury-resistant bacterium and the binding of mercurials to the cells. *J. Ferment. Technol.* **46,** 506–512.

Torma, A. E. (1971). Microbiological oxidation of synthetic cobalt, nickel and zinc sulfides by *Thiobacillus ferrooxidans*. *Rev. Can. Biol.* **30,** 209–216.

Torma, A. E. and Subramanian, K. N. (1974). Selective bacterial leaching of lead sulfide concentrate. *Int. J. Min. Process.* **1,** 125–134.

Tortoriello, R. C. (1971). "Manganese Dioxide-Reduction by Microorganisms from Fresh Water Environments." Ph.D. thesis, Rensselaer Polytechnic Institute, Troy.

Trimble, R. B. and Ehrlich, H. L. (1968). Bacteriology of manganese nodules. III. Reduction of MnO_2 by two strains of nodule bacteria. *Appl. Microbiol.* **16,** 695–702.

Troshanov, E. P. (1968). Microorganisms reducing iron and manganese in ore-containing lakes of the Karelian Isthmus. *Mikrobiologiya* **37,** 934–940.

Troshanov, E. P. (1969). Conditions affecting the ability of bacteria to reduce iron and manganese in ore-bearing lakes of the Karelian Isthmas. *Mikrobiologiya* **38,** 634–643.

Tuovinen, O. H., Niemelae, S. I. and Gullenberg, H. G. (1971). Tolerance of *Thiobacillus ferrooxidans* to some metals. *Antonie van Leeuwenhoek* **37,** 489–496.

Tuttle, J. H., Randles, C. I. and Dugan, P. R. (1968). Activity of microorganisms in acid mine water. I. Influence of acid water on aerobic heterotrophs of a normal stream. *J. Bacteriol.* **95,** 1495–1503.

Tyler, P. A. (1970). Hyphomicrobia and the oxidation of manganese in aquatic ecosystems. *Antonie van Leeuwenhoek* **36,** 567–578.

Vonk, J. W. and Sijpesteijn, A. K. (1973). Methylation of mercuric chloride by pure cultures of bacteria and fungi. *Antonie van Leeuwenhoek* **39,** 505–513.

Walsh, F. and Mitchell, R. (1972). An acid-tolerant iron-oxidizing *Metallogenium*. *J. Gen. Microbiol.* **72,** 369–376.

Watson, S. W. and Waterbury, J. G. (1969). Sterile hot brines of the Red Sea. *In*

"Hot Brines and Recent Heavy Metal Deposits in the Red Sea" (Eds E. T. Degens and D. A. Ross), pp. 272–281. Springer Verlag, New York.
Wang, C. C. and Newton, A. (1969). Iron transport in *Escherichia coli*: Roles of energy-dependent uptake and 2,3-dihydroxybenzoylserine. *J. Bacteriol.* **98,** 1142–1150.
Weed, L. L. and Longfellow, D. (1954). Morphological and biochemical changes induced by copper in a population of *Escherichia coli. J. Bacteriol.* **67,** 27–33.
Weinberg, E. D. (1970). Biosynthesis of secondary metabolites: Roles of trace metals. *Adv. Microbial Physiol.* **4,** 1–44.
Winogradsky, S. (1888). Ueber Eisenbakterien. *Bot. Z.* **46,** 261–270.
Winslow, D.-E. A. and Hotchkiss, M. (1921/22). Studies on salt action. V. The influence of various salts upon bacterial growth. *Proc. Soc. Exp. Biol. Med.* **19,** 314–315.
Wood, J. M. (1974). Biological cycles for toxic elements in the environment. *Science* **183,** 1049–1052.
Wood, J. M., Kennedy, F. S. and Rosen, C. G. (1968). Synthesis of methyl-mercury compounds by extracts of a methanogenic bacterium. *Nature, Lond.* **220,** 173–174.
Yamada, M. and Tonomura, K. (1972a). Formation of methylmercury compounds from inorganic mercury by *Clostridium cochlearium. J. Ferment. Technol.* **50,** 159–166.
Yamada, M. and Tonomura, K. (1972b). Further study of formation of methylmercury from inorganic mercury by *Clostridium cochlearium* T-2. *J. Ferment. Technol.* **50,** 893–900.
Yamada, M. and Tonomura, K. (1972c). Microbial methylation of mercury in hydrogen sulfide-evolving environments. *J. Ferment. Technol.* **50,** 901–909.
Yamaguchi, N., Wada, O., Ono, T., Yazaki, K. and Toyakawa, K. (1973). Detection of heavy metal toxicity by *Tetrahymena pyriformis* culture method. *Ind. Health* **11,** 27–31.
Yang, S. H. (1974). Effect of manganese, nickel, copper and cobalt ions on some bacteria from the deep sea. Ph.D. thesis. Rensselaer Polytechnic Institute. Troy, New York.
Yang, S. H. and Ehrlich, H. L. (1976). Effect on four heavy metals (Mn, Ni, Cu and Co) on some bacteria from the deep sea. *In* "Proceedings of the Third International Biodegradation Symposium" (Eds J. M. Sharpley and A. M. Kaplan), pp. 867–873. Applied Science Publishers, London.
Zajic, J. E. (1969). "Microbial Biogeochemistry." Academic Press. New York and London.
Zajic, J. E. and Chiu, Y. S. (1972). Recovery of heavy metals by microbes. *Dev. Ind. Microbiol.* **13,** 91–100.
Zavarzin, G. A. (1961). Symbiotic culture of a new manganese oxidizing microorganism. Mikrobiologiya **30,** 393–395.
Zlochevskaya, I. V. and Rabotnova, I. L. (1966). Toxicity of lead to *Aspergillus niger. Mikrobiologiya* **35,** 1044–1052.

Chapter 11

Life Under Conditions of High Irradiation

A. NASIM and A. P. JAMES

National Research Council of Canada

I. Introduction 409
II. Radiation in the environment 410
A. Non-ionizing radiation 411
B. Ionizing radiation 411
III. Variation in radiation sensitivity 412
IV. Protection mechanisms 416
V. Nature of radiation damage 419
VI. DNA repair mechanisms 421
A. Dark repair systems 424
VII. *Micrococcus radiodurans* 426
VIII. Radiation-resistant strains 429
IX. Concluding remarks 430
Acknowledgements 431
References 431

I. Introduction

It is a well-documented fact that radiation in the environment is potentially destructive (Lea, 1955). However, except at high doses, the action of radiation is such that any cell has a finite probability of escaping damage. For this reason it might be supposed that the hazard to single celled organisms would be adequately circumvented by a high rate of reproduction. This is not the case, since these organisms have acquired additional means to protect themselves from the lethal or otherwise detrimental effects. These protective mechanisms are numerous, and many individual species utilize more than one device to cope with radiation injury.

Paradoxically, one of the consequences of radiation, mutation, is of selective advantage. It would seem then, that from an evolutionary point of view, a balance between radiation resistance and sensitivity is advantageous, and it is perhaps for this reason that protection is seldom if ever complete. This balance might be expected to differ for different organisms and, in fact, an extreme diversity does exist between species in the extent to which they are resistant to the lethal and mutagenic effects of radiation. The diversity provides considerable scope for investigating the phenomenon of radiation resistance and sensitivity.

Cellular mechanisms which confer radiation resistance can be divided into two broad categories. One utilizes devices which act to protect the cell from the induction of damage. The other involves mechanisms that repair radiation-induced lesions in DNA. Both aspects will be considered in this review, but the latter, i.e. repair, is under more extensive investigation at the present time and will, therefore, be dealt with in greater detail. This area of research has been widened by the exciting discovery that at least some of the repair pathways are relatively independent of essential metabolic activities of the cell. Because of this it has been possible to isolate strains which are defective in their ability to repair radiation-induced damage and yet remain viable. Such mutants are radiation sensitive, and as genetic tools they are proving to be remarkably useful in the elucidation of cellular repairs pathways. Research in this area has provided a much better understanding of radiation resistance.

The range of doses to which microorganisms can be exposed either intermittently or chronically has been vastly extended in recent years by the advent of man-made sources of radiation. Effectively, these sources have challenged microorganisms with an increased level of radiation in the environment. The response to this challenge is of practical as well as basic interest. For example, the use of high doses for food sterilization has raised questions concerning the induction or selection of radiation-resistant microorganisms, possibilities that could have unfortunate consequences.

This review is concerned with the kinds and levels of biologically destructive radiation to which microorganisms are exposed, with the nature of the damage, and with the methods by which protection is achieved. Other reviews that have considered this subject are those of de Serres (1961), Thornley (1963) and Kushner (1964). It should be noted that much of the impetus for research in this area has come from a desire to understand the hazards to human populations from exposure to both high and low levels of ionizing radiation. These studies have been successful in this regard, as has been demonstrated by recent advances in our understanding of ultraviolet-induced skin cancer (Epstein, 1970).

II. Radiation in the Environment

Radiation in the environment can be classified as non-ionizing and ionizing. Both are hazardous to microorganisms, but among natural radiations non-ionizing solar radiation presents the greatest potential for biological destruction.

A. Non-Ionizing Radiation

It has been estimated that only 0·00028% of solar radiation is important for lethality in bacteria (Pollard, 1974). Nevertheless, solar radiation provides an environment which is potentially highly destructive. The essential hazard of sunlight is evident in the behaviour of mutant organisms which are defective in their ability to utilize repair processes. As an example, more than 99·9% of the cells of a strain of *Escherichia coli* deficient in repair mechanisms are killed by exposure to sunlight for three minutes (Harm, 1969; Billen and Fletcher, 1974). A similar demonstration of the lethal effects of the sun's rays has been made with mutant strains of yeast (Resnick, 1970).

The lethal effects of solar radiation extend all the way up to 700 nm (Luckiesh, 1946) but the most harmful wavelengths fall within the ultraviolet region (below 400 nm). Although these rays are of low penetration, they give rise to photochemical reactions in single-celled organisms. The most important of these is the production of pyrimidine dimers in DNA, though other harmful photochemical reactions are also induced (Setlow, 1966).

The levels of ultraviolet light to which organisms are exposed have been comprehensively reviewed recently (Caldwell, 1971). Briefly, the lower limit at the earth's surface as far as wavelength is concerned, is approximately 290 nm (Koller, 1965). Intensity of global irradiation increases exponentially with increasing wavelength. For a wavelength of 307·5 nm, the annual average radiation at the equator has been calculated at approximately 56·4 Watt sec cm^{-2} (Schulze and Grafe, 1969). Intensity varies with solar angle and altitude, and is affected by such factors as the concentration of ozone, atmospheric turbidity, air pollution, and cloud cover. The concentration of ozone, which acts to filter solar ultraviolet, is a particularly critical factor. It has been postulated that a 1% decrease of ozone would lead to a 1 to 2·5% increase in "sunburn" ultraviolet (Robertson, 1972). The question of ozone concentration and solar ultraviolet radiation has been discussed in detail (Cutchis, 1974).

B. Ionizing Radiation

The various sources of ionizing radiation and estimates of the doses to which living organisms are now exposed or will be exposed in the future have been thoroughly reviewed by such committees as the United Nations Scientific Committee on the Effects of Atomic Radiation (UNSCEAR report, 1972) and the BEIR Committee report, 1972). Here, the information will be only briefly summarized.

Natural ionizing radiation is of two sources: extra-terrestrial and terrestrial.

The former originates as primary cosmic rays in outer space. These give rise to the secondary cosmic rays to which organisms are exposed. Dose rates at sea level are about 30 mrad year^{-1}. However, they vary somewhat with latitude and considerably with altitude, approximately doubling every 1500 m up to a few kilometres above sea level. Terrestrial radiation is emitted by radioactive nuclides in rocks, soil, the hydrosphere, and the atmosphere. Some of these radionuclides can be transferred to organisms and for this reason exposures may be of internal as well as of external origin. A recent estimate of the mean dose from external terrestrial radiation is 50 mrad year^{-1}. The estimate from internal radiation is 20 mrad year^{-1} (UNSCEAR, 1972).

1. Man-Made Radiations

Man-made ionizing radiations have their origin in nuclear weapon tests, power production from nuclear fission, medical diagnosis and therapy, and countless miscellaneous sources. In general, the sources are highly localized. Man and other higher organisms are only accidentally exposed to more acute doses, and the genetically significant dose is estimated to be only about 1% of that from natural sources (UNSCEAR, 1972). However, it is evident that naturally-occurring microorganisms, being unshielded, may be periodically or chronically exposed to very high levels under some circumstances. In view of this, the figures representing the levels of man-made radiation present in the environment are not strictly relevant to microorganisms.

III. Variation in Radiation Sensitivity

Species vary enormously in their natural resistance to radiation. This variation far exceeds that expected from any differences in the mean levels of radiation to which organisms are exposed. There is, in fact, little or no correlation between natural levels of exposure and species resistance. The reasons for this variation are obscure, but it seems likely that extreme resistance is a consequence of some other feature of a species.

Reliable comparisons of the radiation-resistance of different organisms from published information are somewhat difficult to make because estimates are frequently based on different doses, dose rates, or different kinds of radiation. Furthermore, such factors as cell stage and growth conditions affect survival, and these vary from publication to publication. With regard to the kind of radiation, the difficulty is somewhat alleviated by the fact that a general correlation exists between reaction to ionizing and non-ionizing radiation (Figs 1 and 2), a circumstance which will be considered later in more detail. The variation in radiation sensitivity among various organisms is illustrated in Figs 1 and 2, and the LD37 for different organisms is presented in Table 1.

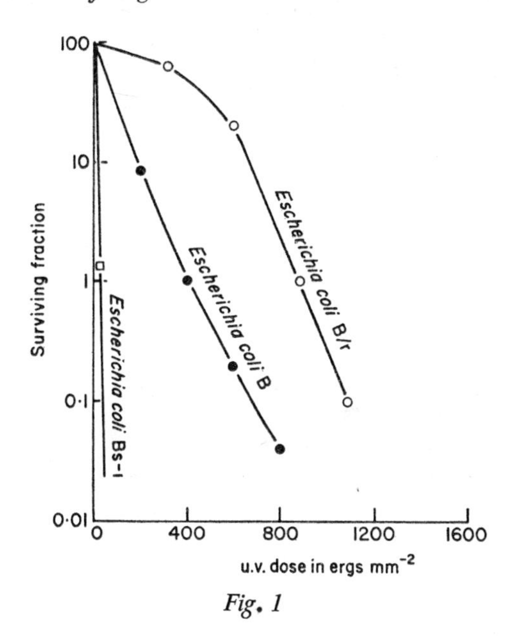

Fig. 1

Table 1

A comparison of the relative radiation sensitivity of different organisms

Organism	Radiation dose giving approximately 37% survival (LD37)		References
	Ultraviolet light (ergs mm^{-2})	Ionizing radiation (kR)	
Escherichia coli K12	500	2[a]	Setlow, R. B. (1967)
Micrococcus radiodurans	6000	150	Moseley and Laser (1965)
Amoeba	—	120	Bacq and Alexander (1961)
Paramaecium	—	350	Bacq and Alexander (1961)
Bodo marina	50,000	—	Lozina-Lozinskiy (1973)
Saccharomyces cerevisiae	800	3	Patrick *et al.* (1964)
Schizosaccharomyces pombe	1350	80	Nasim and Smith (1974)
Ustilago maydis	1300	—	Moore (1975)

[a] Gunther and Kohn (1956)

Measurements of resistance to ultraviolet light are usually made with a source whose emitted rays are largely a wavelength 254 nm. The fact that this wavelength is almost entirely depleted from solar radiation should be kept in mind in any consideration of resistance to solar radiation. Limited data have

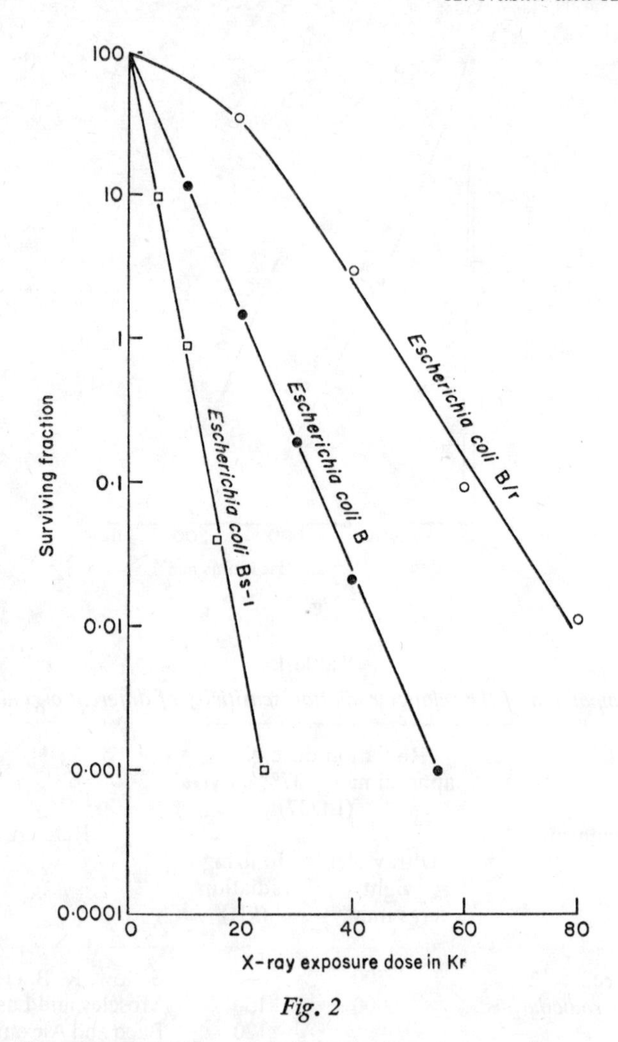

Fig. 2

suggested, however, that there is a general correlation between resistance to non-ionizing radiation from artificial and solar sources (Koller, 1965).

One of the most ultraviolet-resistant of microorganisms is the marine flagellate *Bodo marina.* According to Kamshilov, as quoted by Lozina-Lozinskiy (1973), this organism requires a dose of 112,000 ergs mm^{-2} to inactivate 90% of its cells. The same author has noted that Protozoa are in general more resistant than bacteria (5000 to 12,000 ergs mm^{-2} *v.* 4 to 250 ergs mm^{-2} for 90% inactivation). The resistances of different bacterial species have been more carefully compared. They cover a wide range, and examples are presented graphically in the dose-survival curves of Fig. 3.

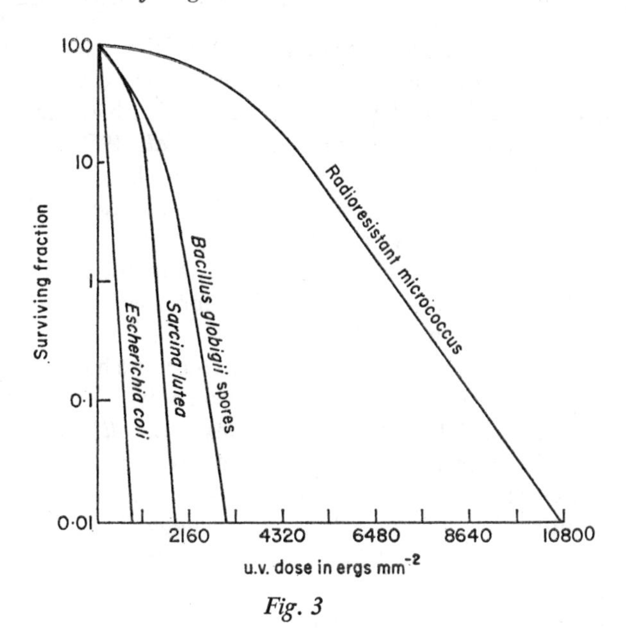

Fig. 3

For ionizing radiation the resistance pattern is comparable. Thus, the doses required to kill 50% of the cells in cultures of *E. coli*, yeast, amoeba, *B. mesentericus* and *Infusaria* rise from 5600 through 350,000 *R* (Bacq and Alexander, 1961).

Wide differences in naturally occurring radioresistance have also been noted within groups of species. For example, 100-fold differences in resistance to gamma rays have been detected among *Dematiacae* (Vasilevskaya and Zhdanova, 1972), and within one species of *Rodotorula* the sensitivity to ultraviolet light has been shown to vary by a factor of 30 (Solntseva, as quoted by Lozina-Lozinskiy, 1973). Strains of *Achromobacter-alcaligenes* have been shown to vary widely in resistance (Thornley, 1963) and a study of survival of spores of *Bacillus subtilis* after X-irradiation has revealed factor differences of 200 (Zamenhof *et al.*, 1965).

The absence of a correlation between species resistance and level of exposure in the natural environment does not always hold true within species. For instance, microorganisms isolated from radioactive mineral water sources have been shown to be 3 to 10 times more radio-resistant than similar organisms isolated from non-radioactive water (Kiselev *et al.*, 1961). Again a species of *Pseudomonas* has occurred as an "infection" of a research reactor where the average dosage was probably in excess of 10^6 Rep. (roentgen equivalent physical) (Fowler *et al.*, 1960).

Such instances of apparent adaptation have raised questions concerning the likelihood that procedures which utilize irradiation for purposes of sterilization will result in the appearance of radiation-resistant organisms. Indeed, the most radiation-resistant bacterium so far isolated. *Micrococcus radiodurans*, was first detected in canned meat which had been exposed to several mrad of gamma irradiation (Anderson *et al.*, 1956). This organism can withstand doses as high as 500 krad without showing any significant inactivation, and has been the subject of considerable research, much of which will be considered later in this review.

The possibility of "producing" radiation-resistant strains in the course of food sterilization has been one of the motives for a series of experiments in different laboratories. Some of these have been successful in producing resistant strains, others not. The experimental procedure has usually involved a series of progressively higher doses in the course of which stepwise increases in resistance are detected. In this manner, strains of *E. coli* have been obtained that are more resistant by factors ranging up to 10 (Gaden and Henley, 1953; Erdman *et al.*, 1961; Idziak and Thatcher, 1964; Wright and Hill, 1968). Radiation-resistant strains of *Salmonella* (Licciardello *et al.*, 1969; Davies *et al.*, 1973), Achromobacter group (Thornley, 1963) and of yeast (Moustacchi, 1965) have also been produced. The same procedure has been used to obtain ultraviolet-resistant strains of *E. coli* (Luckiesh and Knowles, 1948), *Bacillus subtilis* (Zamenhof *et al.*, 1965), and *Haemophilus* (Barnhart and Cox, 1968).

Irradiation can serve as an inducing agent or as a selective agent in such programmes, a fact which tends to blur information relating to whether a resistant strain was induced by the treatment or merely selected from spontaneous mutants already present in a culture. The question of origin might seem to be academic since a fraction of all spontaneous mutations are no doubt induced by natural radiations. Nevertheless, it is possible that radiations of high intensity are effective whereas low intensity natural radiations are ineffective. This would be the case if production of a resistant strain required the simultaneous occurrence of two or more mutational events in the same cell.

Changes in resistance due to the presence of single mutant genes are now under intensive investigation. A single gene conferring radiation resistance was first detected by Witkin (1946), who used radiation to isolate a strain of *E. coli* (B/r) which is resistant to both X-rays and ultraviolet light. The first radiation-sensitive mutant of *E. coli* (B_{s-1}) was isolated by Hill (1958). The dose-effect curves of these mutants are compared with a normal control in Figs 1 and 2. These and more recently obtained mutants have been used extensively to study the molecular basis of resistance.

IV. Protection Mechanisms

Early investigations of radiation-resistant organisms were concerned primarily with a search for intracellular substances which protect the organism from damage. Current investigations have concentrated more heavily on mechanisms that one way or another remove the lesions induced in DNA by irradiation. About the importance of these last-mentioned mechanisms there can be no doubt. Nevertheless, it seems likely that protective mechanisms are of subsidiary importance.

It has been noted that radio-resistant organisms are, in general, highly pigmented (Anderson *et al.*, 1956; Davis *et al.*, 1963), and the presence of such pigments has been considered a possible cause of resistance (Kilburn *et al.*, 1958; Krabbenhoft *et al.*, 1967). It was postulated that these substances act as energy sinks, thus preventing radiation or its products from reaching either DNA or any other critical target.

So far, experimental evidence has provided little support for this notion. Thus, using *M. radiodurans* or *Sarcina lutea,* the resistance of mutants deficient in caretenoid pigment has been compared with that of wild type. Results indicated that the pigment was ineffective as a protective agent against either ionizing radiation or ultraviolet light (Moseley, 1963; Mathews and Krinsky, 1965). Likewise, in a related investigation, an increase in concentration of pigment through an alteration in conditions of culture was found to have little effect on radio-resistance (Lewis *et al.*, 1973). Negative results have also been obtained with fungi (Mirchink *et al.*, 1971). It is noteworthy, however, that such pigments do provide protection against the damaging effects of visible light (Mathews and Sistrom, 1960; Mathews and Krinsky, 1965).

It has been suggested that resistance may involve the presence of specific metabolic products (Anderson *et al.*, 1959; Raj *et al.*, 1960; Duryee *et al.*, 1961). With this possibility in mind, an attempt has been made to isolate some intracellular radio-protective compound and to correlate its level with that of radio-resistance (Bruce, 1964). It was found that *E. coli* is indeed more radio-resistant when irradiated in the presence of extracts from *M. radiodurans* than when irradiated in buffer. However, extracts from *S. lutea* had the opposite effect. The radio-protective action of the extracts from *M. radiodurans* was considered to involve either oxygen removal by enzymatic action or other means (Stapleton *et al.*, 1953, 1955) or the modification of radical yields by a mechanism analogous to that of known chemical radio-protective agents. Bruce favoured the latter possibility because the peculiar sulphydryl metabolism of the sulphur amino acids of *M. radiodurans* suggested that these are comparable to sulphydryl-protective compounds. The extract produced a two-component response, sensitizing the test organism at low concentrations and protecting it at high concentrations. However, it was concluded that such a mechanism is insufficient to explain the extremely high resistance of *M. radiodurans*.

The idea that catalase level in the cell may be responsible for its radiation-resistance has been attractive. It has been shown that some of the more radiation-resistant bacteria possess higher levels of catalase. However the correlation may be coincidental rather than real; although there is some indication in a few instances that catalase level may contribute to radiation-resistance, the bulk of evidence does not support this idea and definitive experimental data are lacking (Clarke, 1952; Ogg *et al.*, 1956; Clayton *et al.*, 1958; Kushner, 1964).

Studies have also been made of changes in the physiological characteristics of the resistant strains of *E. coli* (Idziak and Thatcher, 1964). It has been noted that changes in resistance may be accompanied by marked variations in rates of cell division and in the mode and utilization of specific carbohydrates. Other features of these radiation resistant strains that have been investigated include nutritional requirements (Robern and Thatcher, 1968), enzyme synthesis and its repression (Robern and Thatcher, 1969), effect of *p*-fluorophenylalanine (Dickie *et al.*, 1968) and mesomes (Pontefract and Thatcher, 1970). None of these features seem to be of more than minor importance in determining the resistance of cells.

Cytological studies have been made of the resistant strains of *E. coli* obtained in the course of repeated exposures to radiation (Pontefract and Thatcher, 1964). It was observed that the length of cells increased with increasing radio-resistance; the most resistant strains contained a high percentage of cells which were 30 to 40 time longer than normal. The cells of these strains also exhibited a peculiar budding phenomenon. In fact, cell elongation has been a recognized consequence of radiation for many years (Gates, 1933). Usually a temporary phenomenon, it has been attributed to radiation-induced inhibition of cell division (Lea *et al.*, 1937). However in the resistant strains mentioned above, cell elongation was persistent, being a stable characteristic over a period of three years.

One important factor in the response of any cellular system to either physical or chemical agents is the composition of the cell wall. In the case of chemical mutagens, the cell wall structure may determine the permeability characteristics which affect sensitivity to agents. Although the structure of the cell wall does not influence penetration of ionizing radiation, it might nevertheless influence the radiation-resistance of a microbe. For example, a membrane-bound enzyme complex which is released or activated by radiation may play some role in the repair system(s) or in the cause of the ultimate inactivation. In a recent article (Gentner and Mitchel, 1975) it has been shown that exposure to ionizing radiation results in the release of a cell surface exonuclease in *M. radiodurans*. After a dose of 400 krad, which is sublethal in this organism, only 10% of the enzyme remains cell-associated. The extent of release is a function of the total dose.

An increase in the amount of DNA per cell has frequently been suggested as a factor in radio-resistance. Such an increase could occur either through an increase in frequency of nuclei per cell or through the acquisition of polyploidy. The filamentous nature of the resistant cells of *E. coli* suggest that the former is of importance in those cells. However, it was found that DNA content in these strains does not in fact differ from that in wild type (Idziak and Thatcher, 1964).

Studies of the GC content of eight different bacterial species demonstrated that there is an inverse relationship between GC content and resistance to X-rays. However, it was also demonstrated that the relationship between GC content and resistance to ultraviolet light is direct. Such correlations are not meaningful in the case of *M. radiodurans* which is resistant to both types of radiation; however, these may hold true to some extent in the absence of efficient repair systems. In fact, the base composition of *M. radiodurans* is the same as that of *Pseudomonas* which are extremely sensitive to ionizing radiation (Kaplan and Zavarine, 1962).

V. Nature of Radiation Damage

Identification of the critical target(s) for cell inactivation is essential for an understanding of resistance. Although there is evidence that nongenetic targets are of some importance at higher doses (Kimball, 1957; Gentner and Mitchell, 1975), most of the information now suggests that DNA is the critical target at lower doses. The impetus for research in this area has been provided by the concept of the target theory of radiation damage (Lea, 1955; Timofeeff-Ressovsky and Zimmer, 1947; Hitchinson and Pollard, 1961; Zimmer, 1961). According to this theory, energy will be dissipated in such a manner than a major fraction of the immediate damage will be concentrated in the vicinity of the original site, i.e. the target volume. The passage of a single particle or photon through this sensitive volume or target will lead to cellular inactivation. Although this theory has been modified in many ways, especially as it relates to the close association of lethality with the initial absorption events, it remains a strong stimulus to the formulation of ideas concerning the critical target for radiation damage.

For ultraviolet light, the evidence that damage to DNA is the prime cause of inactivation is very convincing. In particular, the wavelengths that are most effective in causing lethality and mutation are those which correspond to the absorption spectrum of nucleic acids (Hollaender and Emmons, 1941). Furthermore, the damage caused by ultraviolet irradiation has been shown to occur in DNA by experiments which demonstrated that the induction of pyrimidine dimers is one of the major causes of cell death (Setlow and Setlow, 1963).

For ionizing radiation, the evidence that DNA damage is involved is less direct. However, several observations indicate that DNA may also be the critical target for X-rays: (1) Studies of the sensitizing effect of incorporated base analogs such as 5-bromouracil (5-BU) in a number of biological systems have shown that the increase in X-ray sensitivity produced by 5-BU is closely paralleled by a corresponding increase in sensitivity of the transforming DNA extracted from the same culture (Szybalski and Lokiewicz, 1962). RNA was excluded as a possible target by showing that no sensitization is observed in *E. coli* grown in the presence of 5-fluorouracil which is incorporated into RNA and not into DNA (Kaplan *et al.*, 1962); (2) X-ray sensitivity of various organisms is correlated with their total DNA content (Kaplan and Moses, 1964); (3) Bacterial sensitivity to radiation is correlated with DNA base composition (Haynes, 1964; Kaplan and Zavarine, 1962) and (4) Ploidy of yeast influences its sensitivity to X-rays (Mortimer, 1961). The evidence for DNA being the critical target for cell inactivation has been presented in detail by Hutchinson (1966).

Following the demonstration that ultraviolet light reacts with frozen solutions of thymine to produce covalently linked dimers (Beukers and Berends, 1961) it became apparent that dimer formation is the most obvious consequence of ultraviolet irradiation of living cells. Dimers can be formed between adjacent thymine nucleotides, between cytosine nucleotides, or between cytosine and thymine nucleotides. The relative proportions of these depends on the ratio of thymine and cytosine in DNA and also upon the nearest neighbour ratios (Setlow and Carrier, 1966).

There is much evidence to support the conclusion that dimer formation in DNA is the major cause of killing and mutation in biological systems after exposure to ultraviolet irradiation (Setlow, 1966). For instance, mutants which are ultraviolet-sensitive are unable to remove dimers from their DNA (Setlow and Carrier, 1964; Boyce and Howard-Flanders, 1964). Again, many organisms contain an enzyme which monomerizes dimers in the presence of visible light; a large proportion of the cells of such organisms can be reactivated by post-ultraviolet treatment with visible light (Setlow *et al.*, 1965). The detrimental effect of dimer induction is considered to involve the inhibition of normal DNA synthesis (Setlow, 1964).

Dimer formation is not, however, the only cause of cell inactivation after ultraviolet irradiation. For instance, the photoproducts produced in bacterial spores do not include dimers (Donnelan and Setlow, 1965). Types of damage other than dimers include the hydroxylation of cytosine and uracil, cytosine–thymine adducts, crosslinking between DNA and protein, interstrand crosslinking of DNA, and chain breakage and denaturation of DNA. The importance of such damage is no doubt greater at higher fluences. Detailed reviews of the effects of ultraviolet light on DNA have recently been published (Rahn, 1972, 1973).

Solar irradiation at the earth's surface contains little ultraviolet light of the wavelength maximally absorbed by DNA. For this reason some studies have been concerned with identifying the damage responsible for inactivation by solar energy. Results with *E. coli* (Harm, 1966) have indicated that here, also, inactivation is largely due to the induction of pyrimidine dimers. However, the efficiency of induction is lower than with wavelengths of 254 nm. Experiments with yeast have shown that the damage caused by sunlight is photoreversible, a fact which implies the induction of pyrimidine dimers (Resnick, 1970). Recently it has been shown that pyrimidine dimers are formed in DNA exposed to 365 nm light (Tyrrell, 1973). These experiments were, however, carried out at extremely high doses in the range of 10^5 ergs mm^{-2} and are therefore open to other interpretations.

The damage to DNA produced by ionizing radiation can be classified into two categories, direct and indirect. Direct damage results in single or double strand breaks and is rather rare. Indirect damage is mediated by free radicals which also produce single and double strand breaks. This action can be quenched by histidine or SH-containing compounds (Strauss, 1968). Indirect damage is also mediated by a different class of free radicals; these produce chemical modification of pyrimidine bases (Cerutti, 1974).

There is evidence that base damage may be involved in the inactivation of *E. coli* (Paterson and Setlow, 1972) and in *M. radiodurans* (Hariharan and Cerutti, 1972). However, one of the major causes of inactivation of mature viruses by ionizing radiation is considered to be the production of double strand breaks, since most single stand breaks are repaired. Studies of the production of breaks in DNA of the coliphage T7, using a technique which permitted single and double strand breaks to be distinguished by their effects on the sedimentation constants of native and denatured DNA (Freifelder, 1966) showed that the number of inactivated phage particles was equal to the number which had received double strand breaks. In contrast, the number of single strand breaks, detected by the reduction in the molecular weight of alkali-denatured DNA, was not correlated with the number of lethal events.

VI. DNA Repair Mechanisms

A variety of cellular processes involved in the repair of damaged DNA are now considered to be primarily responsible for radiation resistance. The availability of defined mutant strains exhibiting extreme diversity in radiation sensitivity has been of great value in studying these processes.

Since the isolation of a radiation sensitive mutant of *E. coli* (Hill, 1958; Hill and Simson, 1961) large numbers of radiation-sensitive mutants have been obtained and characterized in bacteria, yeast, and other fungi, including *Neurospora crassa* (Chang and Tuveson, 1968), *Aspergillus nidulans* (Wright and Pateman, 1970), *Saccharomyces cerevisiae* (Cox and Parry, 1968; Snow, 1967), *Schizosaccharomyces pombe* (Haefner and Howerey, 1967; Schupbach, 1971; Nasim and Smith, 1975) and *Coprinus lagopus* (Rahman and Cowan, 1974). In addition, another class of mutants which are sensitive to ionizing radiation but normal to ultraviolet have been isolated in *E. coli* (Kato and Kondo, 1970) as well as in yeast (Game and Mortimer, 1974; Nasim and Smith, 1975).

By genetic crosses, double and triple mutants of yeast have been obtained which have lost the repair capacity altogether (Fabre, 1971; Khan *et al.*, 1970; Nasim and Smith, 1974). A comparison of the wild type *S. cerevisiae* with the supersensitive double and triple mutants showed that whereas the normal strain can tolerate nearly 16,000 dimers (37% survival), the double and triple mutants can tolerate only 50 dimers and 1 dimer respectively in DNA (Cox and Game, 1974). The reduced resistance of those double or triple mutants is good evidence for the existence of multiple pathways for repair of radiation damage.

Detailed discussions of cellular repair mechanisms in organisms ranging from viruses to human cell lines has been presented in recent review articles (Adler, 1966; Bridges and Munson, 1968; Cleaver, 1974; Grossman, 1974; Hanawalt, 1968; Hanawalt and Haynes, 1967; Haynes, 1966; Haynes *et al.*, 1968; Howard-Flanders, 1964, 1968a, 1968b; Howard-Flanders and Boyce, 1966; Rupert and Harm, 1966; Setlow, 1966; Setlow, 1966, 1967; Setlow and Setlow, 1972; Smith, 1971; Strauss, 1968). Here the different repair processes will be described only briefly.

Such mechanisms can be subdivided into photo repair and dark repair depending upon whether visible light is involved in the modification of damage to DNA. Photo repair refers specifically to the phenomenon of photoreactivation, and was first described by Kelner in *Actinomycetes* (Kelner, 1951). Photoreactivation apparently operates only upon pyrimidine dimers. The process involves an enzyme, first isolated and purified from Baker's yeast (Muhammed, 1966), which becomes associated with the dimers. The complex thus formed between enzyme and substrate is activated by visible light and the dimers are monomerized *in situ* state. Thus, the lethal effects of ultraviolet radiation are considerably reduced if irradiated cells are subsequently exposed to visible light of wavelengths ranging between 360 and 420 nm. The photoreactivating enzyme has now been isolated from a wide variety of organisms. The implications of this wide distribution have been discussed by Cook (1970). In microorganisms, no clear correlation has emerged regarding overall resistance and the ability or inability to photoreactivate. It has been suggested (Smith and Hanawalt, 1969) that organisms like *M. radiodurans* which have a very efficient dark repair system lack the photo repair system. However, this correlation may be an exception rather than a rule.

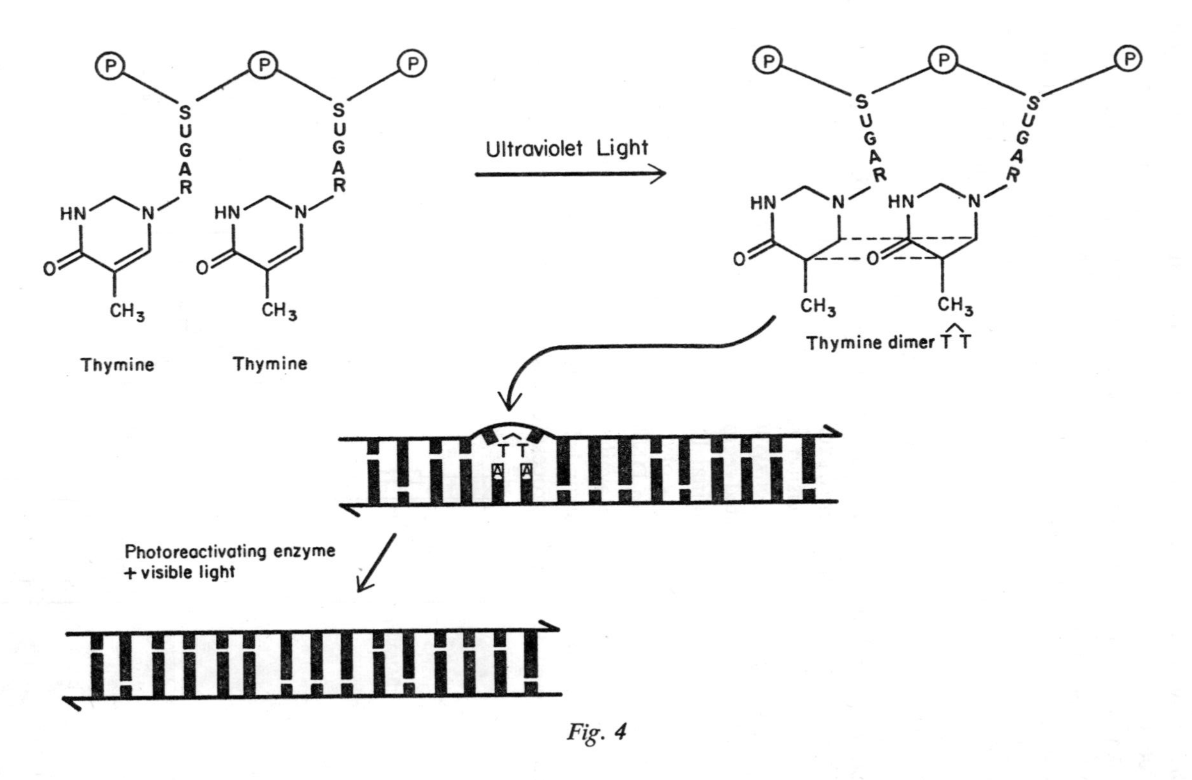
Ultraviolet Light
Thymine
Thymine
Thymine dimer T̂T
Photoreactivating enzyme
+ visible light

Fig. 4

Photoreactivation is schematically illustrated in Fig. 4. It has been a very powerful tool in studies of lethal and mutational damage, because the photoreversibility of any damage can be used to determine whether or not the production of pyrimidine dimers are involved in DNA damage.

Another kind of reactivation by visible light is photo protection. In this instance an increase in survival of cells results from exposure to visible light before ultraviolet irradiation. This phenomenon has been attributed by Jagger (1958) to the induction of division delay by visible light. This delay provides more time for the repair of damage caused by ultraviolet light.

A. Dark Repair Systems

"Dark repair" involves the repair of damage without the intervention of light. There are two known systems, excision repair and post replication recombinational repair; of these excision repair is the best understood. This repair requires the presence of enzymes which recognize distortions in DNA structure, remove the affected region, replace these with normal base sequence, and finally restore the original DNA structure by the process of resealing. Exposure of cells to a variety of inactivating agents can result in a number of different kinds of damage to DNA. A detailed knowledge of this system has been made possible by the availability of radiation-sensitive mutants which have assisted in the isolation and characterization of specific enzymes involved in the different steps of this process (Grossman, 1974). Sequential steps involved

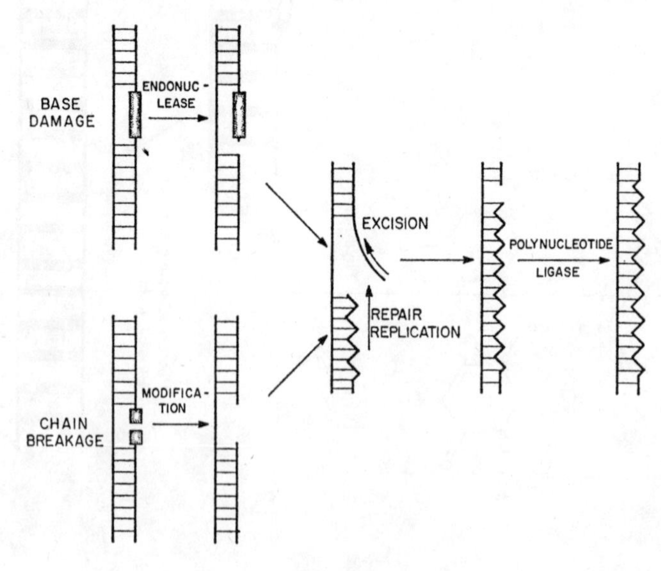

Fig. 5

in excision repair are shown in Fig. 5. In *E. coli* the process involves at least four different steps. The first step involves an incision close to the lesion by an endonuclease which recognizes distortions in DNA. Such ultraviolet-specific endonuclease has been isolated from *Micrococcus luteus* and *E. coli* (Grossman, 1974; Braun and Grossman, 1974). This incision is followed by the removal of pyrimidine dimers through exonuclease activity. The removal of dimers is accompanied by additional degradation of DNA, creating gaps which vary in size from 20 to 400 nucleotides. The gap is then filled by DNA polymerase, using the intact opposite strand as a template. The final step in this sequence is the sealing of the gap by ligase, thus restoring strand continuity. More detailed accounts of excision repair can be found in recent reviews (Cleaver, 1974; Setlow and Setlow, 1972).

The second type of dark repair, post replication recombinational repair, was first described by Howard-Flanders (1968a). As the term indicates, this repair system acts on DNA lesions after DNA replication has occurred. It has been reported (Howard-Flanders, 1968a) that the cell is capable of bypassing some of the dimers, leaving a gap in the strand opposite each of the pyrimidine dimers. It has been experimentally demonstrated that the number of gaps produced in this way roughly correspond to the number of dimers present in DNA (Rupp and Howard-Flanders, 1968). Two intact DNA strands are produced by a recombination-like event involving exchange between sister strands. In a more recent study (Ganesan, 1974) it has been shown that the gap-filling involves exchanges between irradiated parental DNA and unirradiated daughter strands. This study also shows that intact, dimer-free DNA molecules are not necessarily generated by gap-filling and instead dimers may be gradually diluted among successive generations of DNA molecules synthesized after irradiation.

Mutants have been isolated in both bacteria and yeast which are sensitive to killing by ionizing radiation but not to ultraviolet radiation. Their existence has been considered to indicate that some internal steps for the repair of damage caused by each of these radiations are independent. X-ray-sensitive bacteria mutants have been shown to lack the ability to repair single strand breaks (Kapp and Smith, 1970; McGrath and Williams, 1966). The mechanism(s) for repair of damage caused by ionizing radiation are less well understood than those involved in the repair of ultraviolet damage.

Comparison of repair systems in prokaryotes and eukaryotes suggest that these mechanisms are far more complex in eukaryotes. The increased complexity is apparent even in simple eukaryotes like yeast. In *S. cerevisiae* as many as 32 different loci are known to affect radiation sensitivity (Cox and Parry, 1968; Game and Cox, 1972). Similarly, in the fission yeast *Schizosaccharomyces pombe*, 22 independent loci have been reported to affect radiation sensitivity (Nasim and Smith, 1975).

There is little information about the factors which determine the efficiency of dark repair systems in different organisms. It seems likely that this is determined by the available number of molecules of the various repair enzymes within the cell. Other factors may include the efficiency of such enzymes, or the extent of degradation at the damaged points.

VII. *Micrococcus Radiodurans*

Our knowledge of DNA repair systems is based mainly on information derived from the use of radio-sensitive mutants in different organisms. This information has been applied to naturally-occurring resistant species of microorganisms. The experimental observations suggest that there is a positive correlation between radiation-resistance and efficiency of repair systems.

This review will concentrate its attention on *M. radiodurans* as an example of a highly resistant organism with efficient repair systems. The fluence of ultraviolet required to inactivate 90% of the cells of a ultraviolet-resistant strain of *E. coli* [K12 AB1157] is about 1000 ergs mm^{-2} (Boyce and Howard-Flanders, 1964), while 10 to 15,000 ergs mm^{-2} is needed to achieve the same effect with *M. radiodurans*. Similarly, for ionizing radiation no killing is observed at doses as high as 5×10^5 rad (Moseley and Laser, 1965).

At least three other radio-resistant species of *Micrococcus* have now been isolated. One of these was isolated from irradiated haddock fillets (Davis *et al.*, 1963). There are rough and smooth strains of this coccus as well as strains possessing appreciably less pigment. The smooth strain is more resistant to γ radiation than the rough strain and, at higher radiation doses, is comparable in resistance to *M. radiodurans*. Another species has been detected as an aerial contaminant (Lewis, 1971). An extremely radio-resistant orange-red pigmented *Micrococcus* has been isolated from irradiated Bombay duck. This isolate has been designated *Micrococcus radiophilus* (Lewis, 1973) and is more resistant to γ radiation than *M. radiodurans* (Lewis, 1971).

The extreme resistance of *M. radiophilus* is of particular interest; and a study of survival curves after γ radiation has revealed that the survival curve has a very large shoulder extending beyond 1·5 mrad. It is more resistant to γ radiation than any microbial species so far reported. It also exhibits a phenomenal resistance to ultraviolet radiation, a resistance higher than that reported so far for any bacterium. The ultraviolet survival curve is triphasic (Lewis and Kumta, 1972) being comprised of a very large shoulder extending up to 9000 ergs mm^{-2}, an exponential part, and than a pronounced tail beginning at 25,000 ergs mm^{-2} and extending well beyond 50,000 ergs mm^{-2}. The large shoulder has been considered to indicate the existence of an extremely efficient cellular repair system. Although this new isolate is of great potential interest, it has not so far been used in any biochemical studies of repair.

Detailed studies of the repair systems of *M. radiodurans* have been undertaken in the past decade. Various possible causes of resistance to ultraviolet have been considered (Setlow and Duggan, 1964). These include shielding of the DNA by other absorbing compounds, innate resistance of the DNA, effective repair mechanisms and high ploidy. To distinguish between such possibilities these authors have examined survival curves, thymine dimer formation as a function of ultraviolet dose, base ratio of the DNA, and the kinetics of DNA synthesis after irradiation. The experimental data showed that the DNA of *M. radiodurans* is about three times more resistant than that of *E. coli* to the induction of thymine dimers by ultraviolet radiation. Part of this difference in dimerization is accounted for by differences in base ratio, since *M. radiodurans* has a G+C/A+T ratio of 1·6 as compared with one for *E. coli* (Belozersky and Spirin, 1960). Differences in absorption per cell may also be of importance. The absorption per bacterium at 265 nm has been found to be somewhat lower for *M. radiodurans* than for *E. coli*. Whether the three times difference in dimerization can be completely attributed to differences in base ratios and absorption per cell is undetermined. At any rate the decrease in dimerization is entirely insufficient to explain the higher resistance of this organism.

A comparison of the effect of ultraviolet radiation on subsequent DNA synthesis has shown that the doses required to obtain the same delay are 20 times higher for *M. radiodurans* than *E. coli* (Setlow and Duggan, 1964). A very efficient mechanism for the removal of thymine dimers has been proposed to account for this difference. In a biochemical study of the repair efficiency of ultraviolet damage it was shown that the dimers are excised from the cellular DNA before resumption of DNA synthesis (Bolling and Setlow, 1966). It was concluded that the mechanism of dimer excision in this bacterium is so efficient that what death occurs is the result of other causes such as damage to deoxycytidine and proteins.

Two highly ultraviolet-sensitive mutants of this bacterium have been isolated (Moseley, 1967). These were found to be sensitive to ionizing radiation and the chemical mutagen nitrosoguanidine as well as to ultraviolet radiation. The mutants were found to be less efficient in both the excision of dimers and in the repair of single strand breaks (Moseley, 1969; Moseley and Mattingly, 1971; Bonura and Bruce, 1974).

Studies of the mutational response of this organism are limited. The increased resistance to killing is apparently accompanied by high resistance to mutation induction. For some of the sensitive mutants of *E. coli* it has been shown that increased sensitivity to lethal effects is accompanied by higher mutability (Witkin, 1966). In *M. radiodurans* such studies have been limited by the lack of well-defined mutational systems and by the absence of systems for genetic studies. However, mutation induction in *M. radiodurans* has been investigated by using a forward mutation system of the wild type to trimethoprim resistance as well as by reversion of a temperature-sensitive mutant (Sweet and Moseley, 1974). Both the wild type and the temperature-sensitive mutant failed to yield mutants after ultraviolet doses as high as 15,000 ergs. In contrast, in *E. coli* B/r mutations are induced by such low doses as 100 ergs (Witkin, 1969).

In another study, mutations to streptomycin resistance were investigated (Kerszman, 1975). No detectable levels of mutation were observed either after ultraviolet or ionizing radiation. Such observations suggest that the repair of radiation damage in *M. radiodurans* is quite accurate and error proof, a fact which needs to be emphasized and which may have evolutionary significance.

The repair of damage caused by ionizing radiation in *M. radiodurans* has also been considered in considerable detail. This organism shows 100% survival following exposure to 110 krad, a dose which has been shown to produce considerable physiochemical damage to the DNA of *E. coli* B/r (Alexander *et al.*, 1964).

Since most microorganisms are unable to repair double strand breaks (Kaplan, 1966), the possible repair of double strand scissions caused by γ rays in *M. radiodurans* would be quite adequate to account for its high resistance to ionizing radiation. Various authors using different experimental techniques have now obtained evidence which demonstrates the ability of this bacterium to repair DNA double strand breaks (Burrell *et al.*, 1971; Dean *et al.*, 1966; Kitayama and Matsuyama, 1968). Although there are some quantitative uncertainties in such data, since the methods used to examine the native molecular weight of DNA in X-irradiated cells exposes the DNA to shear, and hence some of the double strand breaks may arise from single strand breaks, the bulk of evidence very strongly support the repair of double strand breaks.

The enzymatic basis for the repair of radiation damage in this bacterium is beginning to be investigated. Evidence has been presented that the radiation-sensitive mutant UV 17 is defective in a DNA polymerase activity, possessing 1 to 2 % of that in wild type. Radiation-resistant strains obtained after nitrosoguanidine mutagenesis of UV-17 possessed a wild type level of the enzyme (Gentner, 1973 and personal communication). Exposure to ionizing radiation induced the release of a cell surface exonuclease (Gentner and Mitchel, 1975); this exonuclease was shown not to be the one involved in post-irradiation DNA degradation, but such an observation serves as a model for possible release of membrane-bound repair enzymes after irradiation. More detailed knowledge of the enzymology of DNA repair in this organism would be helpful in further understanding the molecular basis of radiation-resistance.

VIII. Radiation Resistant Strains

Other bacterial species have also been used in attempts to understand the basis of radiation-resistance. The data obtained from such studies are very similar to those from *M. radiodurans*. By using a strain of *Salmonella typhimurium* (Davies *et al.*, 1973) which is 20 times more resistant than the wild type to γ radiation, it was shown that on exposure to 20 or 50 krad the wild type strain degraded 30 to 50% of its DNA whereas the radioresistant strain degraded less than 15% after 4 h of post-irradiation incubation. Post-irradiation synthesis of DNA in the wild type strain was reduced after a dose of 20 krad and totally inhibited after exposure to 200 krad; with the radiation-resistant strain, DNA synthesis was delayed after a dose of 200 krad but not completely inhibited. The initial number of single strand breaks produced in the normal and resistant strains at equivalent doses of ionizing radiation was similar. However, the resistant strain, repaired strand breaks much more efficiently. These data are therefore consistent with the idea that radiation-resistance is largely determined by the repair efficiency of an organism. It should be noted, though, that a related study with *Salmonella thompson* (Allwood and Jordan, 1970) showed that the resistance of a mutant strain did not appear to be directly related to the ability to repair breaks. It is interesting that this strain, designated *S. thompson*/r, exhibited morphological abnormalities similar to those shown by γ radiation-resistant strains of *E. coli*. The cause of enhanced resistance of this isolate of *S. thompson* is not clear.

A great deal of work has been done with the mutants of *E. coli* already referred to in the section on variation in radiation sensitivity. In experiments designed to examine the repair efficiency of three such resistant strains, the rate of DNA synthesis after irradiation was determined. At low doses of 10 to 30 krad there was a permanent reduction of 70 to 90% in the rate of DNA synthesis in wild type *E. coli* (D37 = 4 krad) and 1γ (D37 = 20 krad). It was observed that there is a temporary inhibition in *E. coli* 12γ (D37 = 35 krad) and about 20% stimulation in the most radiation-resistant strain *E. coli* 6γ (D37 = 50 krad). In a related study, considerable differences were observed in the rates and extent of DNA degradation among these strains (Stavric *et al.*, 1968). It is evident that DNA synthesis and degradation are closely related to radiation-resistance. In *E. coli* it has been shown (Shaffer and McGrath, 1965) that less DNA is degraded in the resistant strain than in the radiation-sensitive strain.

The production and repair of strand breaks has been investigated in two strains of *E. coli* (Stavric *et al.*, 1971). A wild type strain and a radio-resistant mutant were exposed to increasing doses of γ-rays, and the progressive decrease in the sedimentation rate of DNA indicating single strand breaks was examined. Sedimentation of the DNA from irradiated cells of the radiation-resistant mutant was restored to the pre-irradiation rate after a period of post-irradiation growth. No such restoration was observed with irradiated cells of the parent strain, an indication that the strain was unable to repair most of the strand breaks.

Such studies are limited in usefulness by the fact that the resistant strains are not well-defined genetically. This limitation makes it difficult to interpret the data in terms of radiation-resistance. However, since the conclusions drawn from most of these experimentally-derived isolates are essentially similar to those known from the naturally occurring *M. radiodurans*, they provide further evidence of a positive correlation between repair efficiency and radiation-resistance.

IX. Concluding Remarks

The motive for studies of the response of microorganisms to radiation has been the concern for hazards which radiation poses for human populations. It was hoped that such studies would contribute to an understanding of basic mechanisms involved in cell inactivation and in the repair of radiation-induced damage. They have been highly successful.

Cellular mechanisms that guard against the harmful consequences of radiation are so widespread among microorganisms that the threat to them from this environmental agent is difficult to perceive. It is only when these mechanisms fail to act that the dimensions of the hazard, particularly from lower wave lengths of solar irradiation, become apparent. It is not surprising, then, that the elucidation of radiation resistance was largely initiated with the detection of radiation-sensitive mutants. Such mutants have not only exposed the hazard but have provided the tools for examining the mechanisms that control it.

To date, accumulated information suggests that, at lower doses, radiation-induced inactivation of cells results mainly from damage incurred in DNA. This conclusion is supported by the fact that, in prokaryotes, the majority of radiation-sensitive mutants so far investigated in detail have been defective in DNA-related functions. Furthermore, investigation of these mutants has demonstrated that resistance is commonly achieved, not by protecting DNA from the induction of lesions, but rather by mechanisms that involve the repair of DNA subsequent to the induction of these lesions. In fact, it now seems likely that purely protective mechanisms are of limited importance, as attempts to attribute radiation resistance to factors such as pigmentation, catalase level and other metabolic features of the cells have shown that the contribution of these factors is small.

Repair mechanisms involved in modifying radiation-induced damage have provided a focal point for investigations of the basis of radiation-resistance.

So far, three independent systems for repairing radiation-induced damage in DNA have been found. One involves the reversal of a photochemical reaction in the presence of visible light and a photoreactivating enzyme, a second one involves excision and replacement of the damaged region prior to DNA replication, and a third one involves repair subsequent to DNA replication. The first-mentioned is applicable only to pyrimidine dimers induced by non-ionizing radiation. Many organisms utilize all three systems to cope with radiation damage.

In lower eukaryotes the relationship between sensitivity and inability to repair DNA is less aparent than in prokaryotes; the number of sensitive mutants that have been detected is much larger, and few of these have been studied in detail. Nevertheless, most evidence suggests that repair of DNA is the main mechanism for resistance in there organisms as well. One may suspect that the greater number of different mutants in eukaryotes is a consequence of an increased complexity of the regulatory systems in these organisms. Certainly significant advances can be expected in this area of research.

Despite the universal occurrence of repair mechanisms in microorganisms, naturally-occurring differences in ability to withstand radiation do exist. A prime example of this variation is provided by the excessive resistance of *M. radiodurans* and its related species. The apparent absence of a relationship between level of resistance and the doses to which organisms are normally exposed suggest that repair pathways are important in cellular functions not directly related to radiation. This possibility does not, of course, reduce the usefulness of such organisms as *M. radiodurans* for investigating resistance. Indeed, investigations with this species have reinforced the evidence that resistance is correlated with repair capacity.

Of particular interest are mutants that are even more resistant to irradiation than the naturally-occurring strains from which they were derived. These mutants have not been studied as systematically as the sensitive mutants. A detailed analysis of such strains will no doubt provide further insight into the mechanism of resistance.

Acknowledgements

The authors are thankful to Ms. T. Brychcy, Mrs. E. Inhaber and Drs P. Duck, N. E. Gentner, M. A. Hannan and M. C. Paterson for many helpful suggestions and critical reading of the manuscript.

References

Alder, H. I. (1966). The Genetic Control of Radiation Sensitivity in Microorganisms. *In* "Advances in Radiation Biology" (Eds L. G. Augenstein, R. Mason and M. Zelle), Vol. 2, pp. 167–191. Academic Press, New York and London.

Alexander, P., Dean, C. J., Hamilton, L. D. G., Lett, J. T. and Parkins, G. (1964). Critical structures other than DNA as sites for primary lesions of cell death induced by ionizing radiations. *In* "Proceedings of the Symposium on Fundamental Cancer Research, Baltimore," p. 241. Williams and Wilkins, Baltimore.

Allwood, M. C. and Jordon, D. C. (1970). Apparent absence of relationship between repair of single strand breaks in DNA and gamma radiation resistance in *Salmonella thompson*. *Arch. Mikrobiol.* **70,** 161–166.

Anderson, A. W., Nordan, H. C., Cain, R. F., Parrish, G. and Duggan, D. (1956). Studies on a radio-resistant Micrococcus. I. Isolation morphology, cultural characteristics and resistance to gamma radiation. *Food Technol.* **10,** 575–578.

Anderson, A. W., Raj, H. D., Wagn, C. H., Duryee, F. L. and Elliker, P. R. (1959). Carbohydrate catabolism of a radiation-resistant bacterium. *Bacteriol. Proc.* **59,** 20.

Bacq, Z. M. and Alexander, P. (1961). "Fundamentals of Radiobiology," 389 pp. Pergamon Press, Oxford.

Barnhart, B. J. and Cox, S. H. (1968). Radiation-sensitive and radiation-resistant mutants of *Haemophilus influenzae*. *J. Bacteriol.* **96,** 280–282.

BEIR Report. (1972). The Effects on Populations of Exposure to Low Levels of Ionizing Radiation: Report of the Advisory Committee on the Biological Effects of Ionizing Radiations. National Academy of Sciences, National Research Council, Washington, D.C.

Belozersky, A. N. and Spirin, A. S. (1960). *In* "The Nucleic Acids" (Eds E. Chargoff and J. N. Davidson), Vol. 3, 147 pp. Academic Press, London and New York.

Beukers, R. and Berends, W. (1961). The effects of UV irradiation on nucleic acids and their components. *Biochim. Biophys. Acta* **49,** 181–189.

Billen, D. and Fletcher, M. M. (1974). Inactivation of dark-repair-deficient mutants of *Escherichia coli* by sunlight. *Int. J. Radiat. Biol.* **26,** 73–76.

Bolling, M. E. and Setlow, J. K. (1966). The resistance of *Micrococcus radiodurans* to ultraviolet radiation. III. A repair mechanism. *Biochim. Biophys. Acta.* **123,** 26–33.

Bonura, T. and Bruce, A. K. (1974). The repair of single strand breaks in a radiosensitive mutant of *Micrococcus radiodurans*. *Radiat. Res.* **57,** 260–275.

Boyce, R. P. and Howard-Flanders, P. (1964). Release of ultraviolet light-induced thymine dimers from DNA in *E. coli* K-12. *Proc. Natl. Acad. Sci. USA* **51,** 293–300.

Braun, A. and Grossman, L. (1974). An endonuclease from *Escherichia coli* that acts preferentially on UV-irradiated DNA and is absent from the *uvr A* and *uvr B* mutants. *Proc. Natl. Acad. Sci. USA* **71,** 1838–1842.

Bridges, B. A. and Munson, R. J. (1968). Genetic radiation damage and its repair in *Escherichia coli*. *In* "Radiation Research" (Eds M. Ebert and A. Howard), Vol. 4, pp. 95–188. Academic Press, London and New York.

Bruce, A. K. (1964). Extraction of a radio-resistant factor of *Micrococcus radiodurans*. *Radiat. Res.* **22,** 155–164.

Burrell, A. D., Feldschreiber, P. and Dean, C. J. (1971). DNA-membrane association and the repair of double breaks in X-irradiated *Micrococcus radiodurans*. *Biochim. Biophys. Acta* **247,** 28–53.

Caldwell, M. M. (1971). Solar UV irradiation and the growth and development of higher plants. *In* "Photophysiology" (Ed. A. C. Giese), Vol. 6, pp. 131–177. Academic Press, New York and London.

Cerutti, P. A. (1974). Excision repair of DNA base damage. *Life Sci.* **15,** 1567–1575.

Chang, L. and Tuveson, R. W. (1968). Ultraviolet-sensitive mutants in *Neurospora crassa*. *Genetics* **56,** 801–810.

Clarke, J. B. (1952). Catalase activity in *Escherichia coli*. *J. Bacteriol.* **64,** 527–530.

Clayton, R. K., Bryan, W. C. and Frederick, A. C. (1958). Some effects of ultraviolet on respiration in purple bacteria. *Arch. Mikrobiol.* **29,** 213–226.

Cleaver, J. E. (1974). Repair processes for photochemical damage in mammalian cells. *In* "Advances in Radiation Biology" (Eds J. T. Lett, H. Adler and M. Zelle), Vol. 4, pp. 1–75. Academic Press, New York and London.

Cook, J. S. (1970). Photoreactivation in animal cells. *In* "Photophysiology" (Ed. A. C. Giese), Vol. 5, pp. 191–233. Academic Press, New York and London.

Cox, B. S. and Game, J. (1974). Repair systems in Saccharomyces. *Mutat. Res.* **26,** 257–264.

Cox, B. S. and Parry, J. M. (1968). The isolation, genetics and survival characteristics of ultraviolet light-sensitive mutants in yeast. *Mutat. Res.* **6,** 37–55.

Cutchis, P. (1974). Stratospheric ozone depletion and solar ultraviolet radiation on earth. *Science* **184,** 13–19.

Davies, R., Sinskey, A. J. and Botstein, D. (1973). Deoxyribonucleic acid repair in a highly radiation-resistant strain of *Salmonella typhimurium*. *J. Bacteriol.* **114,** 357–366.

Davis, N. S., Silverman, G. J. and Masurovsky, E. B. (1963). Radiation resistant, pigmented coccus isolated from haddock tissue. *J. Bacteriol.* **86,** 294–298.

Dean, C. J., Feldschreiber, P. and Lett, J. T. (1966). Repair of X-ray damage to the deoxyribonucleic acid in *Micrococcus radiodurans*. *Nature, Lond.* **209,** 49–52.

de Serres, F. J. (1961). Some aspects of the influence of environment on the radio sensitivity of microorganisms. *Symp. Soc. Gen. Microbiol.* **11,** 196–216.

Dickie, N., Dennis, D. A. and Thatcher, F. S. (1968). Effect of *P. fluorophenylalanine* on radiation sensitivity in *Escherichia coli*. *Can. J. Microbiol.* **14,** 799–803.

Donnellan, J. E., Jr. and Setlow, R. B. (1965). Thymine photoproducts but not thymine dimers found in ultraviolet-irradiated bacterial spores. *Science* **149,** 308–310.

Duryee, F. L., Raj, H. D., Wang, C. H., Anderson, A. W. and Elliker, P. R. (1961). Carbohydrate metabolism in *Micrococcus radiodurans*. *Can. J. Microbiol.* **7,** 799–805.

Epstein, J. H. (1970). Ultraviolet carcinogenesis. *In* "Photophysiology" (Ed. A. C. Giese), Vol. 5, pp. 235–273. Academic Press, New York and London.

Erdman, I. E., Thatcher, F. S. and MacQueen, K. F. (1961). Studies on the irradiation of microorganisms in relation to food preservation. II. Irradiation resistant mutants. *Can. J. Microbiol.* **7,** 207–215.

Fabre, F. (1971). A UV supersensitive mutant in the yeast *Schizosaccharomyces pombe*: Evidence for two repair pathways. *Mol. Gen. Genet.* **110,** 134–143.

Fowler, E. B., Christenson, C. W., Jurwey, E. T. and Schafer, W. D. (1960). Bacterial "infection" of the omega west reactor. *Nucleonics* **18,** 102–105.

Freifelder, D. (1965). Mechanism of inactivation of coliphage T7 by X-rays. *Proc. Natl. Acad. Sci. USA* **54,** 128–134.

Freifelder, D. (1966). Lethal changes in bacteriophage DNA produced by X-rays. *Radiat. Res.* Suppl. **6,** 80–96.

Gaden, E. L. and Henley, E. J. (1953). Induced resistance to gamma irradiation in *E. coli*. *J. Bacteriol.* **65,** 727–732.

Game, J. C. and Cox, B. S. (1972). Epistatic interactions between four rad loci in yeast. *Mutat. Res.* **16,** 353–362.

Game, J. C. and Mortimer, R. K. (1974). A genetic study of X-ray sensitive mutants in yeast. *Mutat. Res.* **24,** 281–292.

Ganesan, A. K. (1974). Persistance of pyrimidine dimers during post-replication repair in ultraviolet light-irradiated *Escherichia coli* K12. *J. Mol. Biol.* **87,** 103–119.

Gates, F. L. (1933). The reaction of individual bacteria to irradiation with ultraviolet light. *Science* **77,** 350.

Gentner, N. E. (1973). DNA polymerase of *Micrococcus radiodurans* and its relation to repair of radiation damage. *Fed. Proc.* **32,** 452.

Gentner, N. E. and Mitchel, R. E. J. (1975). Ionizing radiation induced release of a cell surface nuclease from *Micrococcus radiodurans*. *Radiat. Res.* **61,** 204–215.

Grossman, L. (1974). Enzymes involved in the repair of DNA. *In* "Advances in Radiation Biology" (Eds J. T. Lett, H. Adler and M. Zelle), Vol. 4, pp. 77–129. Academic Press, New York and London.

Gunter, S. E. and Kohn, H. I. (1956). The effect of X-rays on the survival of bacteria and yeast. *J. Bacteriol.* **71,** 571–581.

Haefner, K. and Howerey, L. (1967). Gene controlled uv sensitivity in *Schizosaccharomyces pombe*. *Mutat. Res.* **4,** 219–221.

Hanawalt, P. C. (1968). Cellular recovery from photochemical damage. *In* "Photophysiology" (Ed. A. C. Giese), Vol. 4, pp. 203–251. Academic Press, London and New York.

Hanawalt, P. C. and Haynes, R. H. (1967). The repair of DNA. *Sci. Am.* **216,** 36–43.

Hariharan, P. V. and Cerutti, P. A. (1972). Formation and repair of X-ray-induced thymine damage in *Micrococcus radiodurans*. *J. Mol. Biol.* **66,** 65–81.

Harm, W. (1966). Repair effects in phage and bacteria exposed to sunlight. *Radiat. Res.* Suppl. **6,** 215.

Harm, W. (1969). Biological determination of the germicidal activity of sunlight. *Radiat. Res.* **40,** 63–69.

Haynes, R. H. (1964). Molecular localization of radiation damage relevant to bacterial inactivation. *In* "Physical Processes in Radiation Biology" (Eds L. Augenstein, R. Mason and B. Rosenberg), pp. 51–72. Academic Press, New York and London.

Haynes, R. H. (1966). The interpretation of microbial inactivation and recovery phenomena. *Radiat. Res.* Suppl. **6,** 1–29.

Haynes, R. H., Baker, R. M. and Jones, G. E. (1968). Genetic implications of DNA repair. *In* "Energetics and Mechanisms in Radiation Biology" (Ed. G. O. Philips), pp. 425–465. Academic Press, London and New York.

Hill, R. F. (1958). A radiation-sensitive mutant of *Escherichia coli*. *Biochim. Biophys. Acta.* **30,** 636–637.

Hill, R. F. and Simson, E. (1961). A study of radiosensitive and radioresistant mutants of *Escherichia coli* strain B. *J. Gen. Microbiol.* **24,** 1–14.

Hollaender, A. and Emmons, C. W. (1941). Wavelength dependence of mutation production in the ultraviolet with special emphasis on fungi. *Cold Spring Harbor Symp. Quant. Biol.* **9,** 179–186.

Howard-Flanders, P. (1964). Molecular mechanisms in the repair of irradiated DNA. *J. Genet.* **40,** 256–263.

Howard-Flanders, P. (1968a). DNA repair. *Ann. Rev. Biochem.* **37,** 175–200.

Howard-Flanders, P. (1968b). Genes that control DNA repair and genetic recombination in *Escherichia coli*. *Adv. Biol. Med. Phys.* **12,** 299–317.

Howard-Flanders, P. and Boyce, R. P. (1966). DNA repair and genetic recombination: Studies on mutants of *Escherichia coli* defective in these processes. *Radiat. Res.* Suppl. **6.** 156–184.
Hutchinson, F. (1966). The molecular basis for radiation effects on cells. *Cancer Res.* **26,** 2045–2052.
Hutchinson, F. and Pollard, F. C. (1961). Target theory and radiation effects on biological molecules. *In* "Mechanisms in Radiobiology" (Eds M. Errera and A. Forssberg), Vol. 1, pp. 71–92. Academic Press, New York and London.
Idziak, E. S. and Thatcher, F. S. (1964). Some physiological aspects of mutants of *Escherichia coli* resistant to gamma irradiation. *Can. J. Microbiol.* **10,** 683–697.
Jagger, J. (1958). Photoreactivation. *Bacteriol. Rev.* **22,** 99–142.
Kaplan, H. S. (1966). DNA strand scission and loss of viability after X-irradiation of normal and sensitized bacterial cells. *Proc. Natl. Acad. Sci. USA* **55,** 1442–1446.
Kaplan, H. S. and Moses, L. E. (1964). Biological complexity and radiosensitivity. *Science* **145,** 21–25.
Kaplan, H. S. and Zavarine, R. (1962). Correlation of bacterial radiosensitivity and DNA base composition. *Biochem. Biophys. Res. Commun.* **8,** 432–436.
Kaplan, H. S., Smith, K. C. and Tomlin, P. A. (1962). Effect of halogenated pyrimidines on radiosensitivity of *E. coli. Radiat. Res.* **16,** 98–113.
Kapp, D. S. and Smith, K. C. (1970). Repair of radiation-induced damage in *Escherichia coli. J. Bacteriol.* **103,** 49–54.
Kato, T. and Kondo, S. (1970). Genetic and molecular characteristics of X-ray-sensitive mutants of *Escherichia coli* defective in repair synthesis. *J. Bacteriol.* **104,** 871–881.
Kelner, A. (1951). Revival by light. *Sci. Am.* **181,** 22–25.
Kerszman, G. (1975). Induction of mutation to *Streptomycin* resistance in *Micrococcus radiodurans. Mutat. Res.* **28,** 9–14.
Khan, N. A., Brendel, M. and Haynes, R. H. (1970). Supersensitive double-mutants in yeast. *Mol. Gen. Genet.* **107,** 376–378.
Kilburn, R. E., Bellamy, W. D. and Terni, S. A. (1958). Studies on a radiation-resistant pigmented *Sarcina* sp. *Radiat. Res.* **9,** 207–215.
Kimball, R. F. (1957). Nongenetic effects of radiation on microorganisms. *Ann. Rev. Microbiol.* **11,** 199–220.
Kiselev, P. N., Kashkin, K. P., Boltaks, Yu. B. and Vitovskaya, G. A. (1961). Acquisition of radioresistance by microbe cells inhabiting media with increased levels of natural radiation. *Microbiology* (*USSR*) **30,** 194–198.
Kitayama, S. and Matsuyama, A. (1968). Possibility of the repair of double-strand scissions in *Micrococcus radiodurans* DNA caused by gamma-rays. *Biochem. Biophys. Res. Commun.* **33,** 418–422.
Koller, L. R. (1965). "Ultraviolet Radiation" (2nd ed.), 312 pp. Wiley, New York.
Krabbenhoft, K. L., Anderson, A. W. and Elliker, P. R. (1967). Influence of culture media on the radiation resistance of *Micrococcus radiodurans. Appl. Microbiol.* **15,** 178–185.
Kushner, D. J. (1964). Microbial resistance to harsh and destructive environmental conditions. *In* "Experimental Chemotherapy" (Eds R. J. Schnitzer and F. Hawking), Vol. 2, pp. 114–168. Academic Press, New York and London.
Lea, D. E. (1955). "Actions of Radiations on Living Cells" (2nd ed.), 416 pp. Cambridge University Press, London and New York.
Lea, D. E., Haines, R. B. and Coulson, C. A. (1937). The action of radiations on

bacteria. III. γ-rays on growing and non-proliferating bacteria. *Proc. R. Soc. London, Ser. B,* **123,** 1–21.

Lewis, N. F. (1971). Studies on a radio-resistant coccus isolated from Bombay duck (*Harpodon nehereus*). *J. Gen. Microbiol.* **66,** 29–35.

Lewis, N. F. (1973). Radio-resistant *Micrococcus radiophilus* sp. nov. isolated from irradiated Bombay duck (*Harpodon nehereus*). *Curr. Sci.* **42,** 504.

Lewis, N. F. and Kumta, U. S. (1972). Evidence for extreme UV resistance of *Micrococcus* sp. NCTC 10785. *Biochem. Biophys. Res. Commun.* **47,** 1100–1105.

Lewis, N. F., Madhavesh, D. A. and Kumta, U. S. (1973). Role of caretenoid pigments in radio-resistant micrococci. *Can. J. Microbiol.* **20,** 455–459.

Licciardello, J. J., Nickerson, J. T. K., Goldblith, S. A., Shannon, C. A. and Bishop, W. W. (1969). Development of radiation resistance in *Salmonella* cultures. *Appl. Microbiol.* **18,** 24–30.

Lozina-Lozinskiy, L. K. (1973). Resistance of unicellular organisms to ultraviolet radiation in relation to the problem of the existence of extraterrestrial life. *Nasa. Tech. Transl. TTF* **719,** 378–392.

Luckiesh, M. (1946). "Applications of Germicidal Erythmal and Infrared Energy." Vari Nostrand-Reinhold, Princeton, New Jersey.

Luckiesh, M. and Knowles, T. (1948). Radiosensitivity of *Escherichia coli* to ultraviolet energy (λ 2537) as affected by irradiation of preceding cultures. *J. Bacteriol.* **55,** 369.

Mathews, M. M. and Krinsky, N. I. (1965). The relationship between carotenoid pigments and resistance to radiation in non-photosynthetic bacteria. *Photochem. Photobiol.* **4,** 813–817.

Mathews, M. M. and Sistrom, W. R. (1960). The function of the caretenoid pigments in *Sarcina lutea. Arch. Microbiol.* **35,** 139–146.

McGrath, R. A. and Williams, R. W. (1966). Reconstruction *in vivo* of irradiated *Escherichia coli* deoxyribonucleic acid; the rejoining of broken pieces. *Nature, Lond.* **212,** 534–535.

Mirchink, T. G., Kashkina, G. B. and Abaturov, Yu. D. (1971). The resistance of fungi with various pigments to γ irradiation. *Mikrobiologiya* **41,** 83–86.

Moore, P. D. (1975). Radiation-sensitive pyrimidine auxotrophs of *Ustilago maydis.* I. Isolation and characterization of mutants. *Mutat. Res.* **28,** 355–366.

Mortimer, R. K. (1961). Factors controlling the radiosensitivity of yeast cells. *Brookhaven Symp. Biol.* **14,** 62–75.

Moseley, B. E. B. (1963). The variation in X-ray resistance of *Micrococcus radiodurans* and some of its pigmented mutants. *Int. J. Radiat. Biol.* **6,** 489.

Moseley, B. E. B. (1967). The isolation and some properties of radiation-sensitive mutants of *Micrococcus radiodurans. J. Gen. Microbiol.* **49,** 293–300.

Moseley, B. E. B. (1969). Repair of ultraviolet radiation damage in sensitive mutants of *Micrococcus radiodurans. J. Bacteriol.* **97,** 647–652.

Moseley, B. E. B. and Laser, H. (1965a). Repair of X-ray damage in *M. radiodurans. Proc. R. Soc. Ser. B.* **162,** 210–222.

Moseley, B. E. B. and Laser, H. (1965b). Similarity of repair of ionizing and ultraviolet radiation damage in *Micrococcus radiodurans. Nature, Lond.* **206,** 373–375.

Moseley, B. E. B. and Mattingly, A. (1971). Repair of irradiated transforming deoxyribonucleic acid in wild type and a radiation-sensitive mutant of *Micrococcus radiodurans. J. Bacteriol.* **105,** 976–983.

Moustacchi, E. (1965). Induction by physical and chemical agents of mutations for radioresistance in *Saccharomyces cerevisiae. Mutat. Res.* **2,** 403–412.

Muhammed, A. (1966). Studies on the yeast photoreactivating enzyme. I. A method for the large scale purification and some properties of the enzyme. *J. Biol. Chem.* **241,** 516–523.

Nasim, A. and Smith, B. P. (1974). Dark repair inhibitors and pathways for repair of radiation damage in *Schizosaccharomyces pombe. Mol. Gen. Genet.* **132,** 13–22.

Nasim, A. and Smith, B. P. (1975). Genetic control of radiation sensitivity in *Schizosaccharomyces pombe. Genetics* **79,** 573–582.

Ogg, J. E., Adler, H. I. and Zelle, M. R. (1956). Protection of *Escherichia coli* against UV irradiation by catalase and related enzymes. *J. Bacteriol.* **72,** 494–496.

Paterson, M. C. and Setlow, R. B. (1972). Endonucleolytic activity from *Micrococcus luteus* that acts on γ-ray-induced damage in plasmid DNA of *Escherichia coli* minicells. *Proc. Natl. Acad. Sci. USA* **69,** 2927–2931.

Patrick, M. H., Haynes, R. H. and Uretz, R. B. (1964). Dark recovery phenomena in yeast. I. Comparative effect with various inactivating agents. *Radiat. Res.* **21,** 144–164.

Pollard, E. C. (1974). Cellular and molecular effects of solar ultraviolet radiation. *Photochem. Photobiol.* **20,** 301–308.

Pontefract, R. D. and Thatcher, F. S. (1964). A cytological study of normal and radiation-resistant *Escherichia coli. Can. J. Microbiol.* **11,** 271–278.

Pontefract, R. D. and Thatcher, F. S. (1970). An electron microscope study of mesosomes in irradiation-resistant mutants of *Escherichia coli. J. Ultrastruct. Res.* **30,** 78–86.

Rahman, M. A. and Cowan, J. W. (1974). Ultraviolet light sensitive mutants of *Coprinus lagopus. Mutat. Res.* **23,** 29–40.

Rahn, R. O. (1972). Ultraviolet irradiation of DNA. *In* "Concepts in Radiation Cell Biology," pp. 1–56. Academic Press, New York and London.

Rahn, R. O. (1973). Denaturation in ultraviolet-irradiated DNA. *In* "Photophysiology" (Ed. A. C. Giese), Vol. 8, pp. 231–255. Academic Press, London and New York.

Raj, H. D., Duryee, F. L., Deeney, A. M., Wang, C. H., Anderson, A. W. and Elliker, P. R. (1960). Ultilization of carbohydrates and amino acids by *Micrococcus radiodurans. Can. J. Microbiol.* **6,** 289–298.

Resnick, M. A. (1970). Sunlight-induced killing in *Saccharomyces cerevisiae. Nature, Lond.* **226,** 377–378.

Robern, H. and Thatcher, F. S. (1968). Nutritional requirements of mutants of *Escherichia coli* resistant to gamma-irradiation. *Can. J. Microbiol.* **14,** 711–715.

Robern, H. and Thatcher, F. S. (1969). Gamma-irradiation resistant mutants of *Escherichia coli*: modified enzyme synthesis repressible by arginine and uracil. *Can. J. Microbiol.* **15,** 549–554.

Robertson, D. F. (1972). Ph.D. thesis, University of Queensland.

Rupert, C. S. and Harm, W. (1966). Reactivation after photobiological damage. *In* "Advances in Radiation Biology" (Eds L. G. Augenstein, R. Mason and M. Zelle), Vol. 2, pp. 1–81. Academic Press, New York and London.

Rupp, W. D. and Howard-Flanders, P. (1968). Discontinuities in the DNA synthesized in an excision-defective strain of *Escherichia coli* following ultraviolet irradiation. *J. Mol. Biol.* **31,** 291–304.

Schulze, R. and Gräfe, K. (1969). "The biologic effects of ultraviolet irradiation" (Ed. F. Urbach), pp. 359. Pergamon, New York.

Schüpbach, M. (1971). The isolation and genetic classification of UV-sensitive mutants of *Schizosaccharomyces pombe. Mutat. Res.* **11,** 361–371.

Setlow, J. K. (1966). The molecular basis of biological effects of ultraviolet radiation and photoreactivation. *In* "Current Topics in Radiation Research" (Eds M. Ebert and A. Howard), Vol. 2, pp. 195–248. North-Holland Publishing Company, Amsterdam.

Setlow, J. K. and Duggan, D. E. (1964). The resistance of *Micrococcus radiodurans* to ultraviolet radiation. I. Ultraviolet-induced lesions in the cell's DNA. *Biochim. Biophys. Acta* **87,** 664–668.

Setlow, J. K. and Setlow, R. B. (1963). Nature of the photoreactivable ultraviolet lesion in DNA. *Nature, Lond.* **197,** 560–562.

Setlow, J. K., Bolling, M. E. and Bollum, F. J. (1965). The chemical nature of photoreactivable lesions in DNA. *Proc. Natl. Acad. Sci. USA* **53,** 1430–1436.

Setlow, R. B. (1964). Molecular changes responsible for ultraviolet inactivation of the biological activity of DNA. *In* "Mammalian Cytogenetics and Related Problems in Radiobiology," pp. 291–307. Pergamon Press, Oxford.

Setlow, R. B. (1966). The repair of molecular damage to DNA. *In* "Radiation Research," pp. 525–537. North-Holland Publishing Company, Amsterdam.

Setlow, R. B. (1967). Repair of DNA. *In* "Regulation of Nucleic Acid and Protein Biosynthesis" (Eds V. V. Koningsberger and L. Bosch), pp. 51–62. Elsevier Publishing Company, Amsterdam.

Setlow, R. B. and Carrier, W. L. (1964). The disappearance of thymine dimers from DNA: An error-correcting mechanism. *Proc. Natl. Acad. Sci. USA* **51,** 226–231.

Setlow, R. B. and Carrier, W. L. (1966). Pyrimidine dimers in ultraviolet-irradiated DNA's. *J. Mol. Biol.* **17,** 237–254.

Setlow, R. B. and Setlow, J. K. (1972). Effects of radiation on polynucleotides. *Ann. Rev. Biophys. Bioeng.* **1,** 293–346.

Shaffer, C. R. and McGrath, R. A. (1965). The effects of X-irradiation on the DNA of *Escherichia coli*. *Exp. Cell Res.* **39,** 604–606.

Smith, K. C. (1971). The roles of genetic recombination and DNA polymerase in the repair of damaged DNA *In* "Photophysiology" (Ed. A. C. Giese), Vol. 6, pp. 209–278. Academic Press, New York and London.

Smith, K. C. and Hanawalt, P. C. (1969). "Molecular Photobiology Inactivation and Recovery." 230 pp. Academic Press, New York and London.

Snow, R. (1967). Mutants of yeast sensitive to ultraviolet light. *J. Bacteriol.* **94,** 571–575.

Stapleton, G. E., Billen, D. and Hollaender, A. (1953). Recovery of X-irradiated bacteria at suboptimal incubation temperatures. *J. Cell. Comp. Physiol.* **41,** 345–358.

Stapleton, G. E., Sbarra, A. J. and Hollaender, A. (1955). Some nutritional aspects of bacterial recovery from ionizing radiations. *J. Bacteriol.* **70,** 7–15.

Stavric, P., Dickie, N. and Thatcher, F. S. (1968). Effects of γ-irradiation on *Escherichia coli* wild type and its radiation-resistant mutants. II. Post-irradiation. degradation of DNA. *Int. J. Radiat. Biol.* **14,** 411–416.

Stavric, S., Dighton, M. and Dickie, N. (1971). DNA-strand breaks in two strains of *Escherichia coli* after exposure to γ-rays. *Int. J. Radiat. Biol.* **20,** 101–110.

Strauss, B. S. (1968). DNA repair mechanisms and their relation to mutation and recombination. *Curr. Top. Microbiol. Immunol.* **44,** 1–85.

Sweet, D. M. and Moseley, B. E. B. (1974). Accurate repair of ultraviolet-induced damage in *Micrococcus radiodurans*. *Mutat. Res.* **23,** 311–318.

Szybalski, W. and Lorkiewicz, Z. (1962). On the nature of the principal target of lethal and mutagenic radiation effects. *Abhandl. Deutsch. Akad. Wiss. Berlin, Kl. Med.* **1,** 63–71.

Thornley, M. J. (1963). Radiation resistance among bacteria. *J. Appl. Bact.* **26,** 334–345.

Timofeeff-Ressovsky, N. W. and Zimmer, K. G. (1947). "Das Trefferprinzip in der Biologie" Hirzel, Leipzig.

Tyrrell, R. M. (1973). Induction of pyrimidine dimers in bacterial DNA by 365 nm radiation. *Photochem. Photobiol.* **17,** 69–73.

United Nations Scientific Committee on the Effects of Atomic Radiation, Ionizing Radiation: Levels and Effects. (1972). Vol. 1, United Nations, New York.

Vasilevskaya, A. I. and Zhdanova, N. N. (1972). Resistance to γ-rays of some soil dark hyphomycetes. *Mikrobiol. Zhurnal* **34,** 444–447.

Witkin, E. M. (1946). Inherited differences in sensitivity to radiation in *Escherichia coli. Proc. Natl. Acad. Sci. USA* **32,** 59–68.

Witkin, E. M. (1966). Radiation-induced mutations and their repair. *Science* **152,** 1345–1353.

Witkin, E. M. (1969). The mutability toward ultraviolet light of recombination deficient strains of *Escherichia coli. Mutat. Res.* **8,** 9–14.

Wright, S. J. L. and Hill, F. C. (1968). The development of radiation-resistant cultures of *E. coli* by a process of "growth irradiation" cycles. *J. Gen. Microbiol.* **51,** 97–106.

Wright, P. J. and Pateman, J. A. (1970). Ultraviolet light sensitive mutants of *Aspergillus nidulans. Mutat. Res.* **9,** 579–587.

Zamenhof, S., Bursztyn, H., Reddy, T. K. R. and Zamenhof, P. J. (1965). Genetic factors in radiation resistance of *Bacillus subtilis. J. Bacteriol.* **90,** 108–115.

Zimmer, K. G. (1961). "Studies on Quantitative Radiation Biology" 124 pp. Oliver and Boyd, Edinburgh and London.

Subject Index*

A

Absidia corymbifera
temperature effect on growth, 168
thermophilic, 170, 179, 183
Acetobacter acidophilum
pH effect on growth, 280–281
Acholeplasma laidlawii
lipids as growth temperature determinants, 115
membrane proteins, 301
pH effect on lipid composition, 296
Achromobacter
psychrotrophic, in springs and wells, 31
Achromobacter-alcaligenes
radiation sensitivity, 415
Achromobacter parvalus
in Don Juan Pond, 27
Acid hot springs (*see also* Hot springs)
microorganisms in, 282–283
Acid mine-drainage
bacteriology of, 389–392
Acidophiles (*see also* pH, individual organisms)
algae, 284, 293
bacteria (*see also* pH, and individual organisms, 280–283, 288–290)
evolutionary importance, 305
fungi, including yeasts, 283–284
hydrogen ion pumps, 306
intracellular pH, 291–293
protozoa, 284
Acinetobacter
in lakes, 19
Aconitase
Bacillus steaothermophilus, 251
Acontium velatium
low pH and high Cu^{2+} effect on growth, 284
Acremonium alabamensis
temperature effect on growth, 170
Acrophialophora fusispora
temperature effect on growth, 170
Actinomyces levoris
pressure effect on bacteriophages of, 135–136.
Actinomyces olivaceus
pressure effect on bacteriophages of, 136
Actinomycetes
desiccation susceptibility in dried soil, 373–374
growth at high temperatures, 163
photoreactivation of irradiated DNA, 422–426
thermophilic, in self-heating of hay, 168
Actinoplanes
growth at high temperatures, 163
Actinopolyspora halophila
nutrition, 330
salt response, 323–324
Activation energy
microbial growth, 73–77
Active transport
halobacteria, effect of ions, 342
microorganisms living in extreme pH values, 294, 306
Adenosine triphosphate (ATP)
formation catalysed by purple membrane of extreme halophiles, 347
Adenosine triphosphatase (ATPase)
alkaline- requiring bacilli, 294
Bacillus stearothermophilus, 229
Streptococcus faecalis, effect of pressure, 118, 120, 132
Thiobacillus, 292

* Most microorganisms are listed individually, others may be found as types, e.g. thermophilic fungi, extreme halophiles, non-photosynthetic bacteria growing at high temperatures, etc. Individual heavy metals are usually not listed as such but under the heading: Heavy metals, arsenic and antimony

Adenyl cyclase
rat, effect of mercury, 396
Agaricus bisporus
growth on heated compost, 173
Agrobacterium radiobacter
pH effect on growth, 285
L-Alanine dehydrogenase
alkaline-requiring bacilli, 294
Alcaligenes
pressure effects, 139
Alcohol dehydrogenase
thermoadaptation in thermophiles, 240, 242
thermophile mutants, 251
Algae (*see also* individual organisms)
growing at high temperatures, 186, 187
water stress, effect on, 372, 374–376
Alkaline requiring bacilli
L-lactate dehydrogenase, 294
Alkalinophiles (*see also* pH, individual organisms)
algae, 286, 287
anion pumps, 306
bacteria, 285, 286, 288–290, 293, 294
fungi, including yeasts, 286, 287
hydrogen ion pumps, 306
intracellular pH, 291
protozoa, 286, 287
Alligator nests
thermophilic microorganisms, 183
Amino acid-activating enzymes
extreme halophiles, 331
Amino acyl tRNA synthetases
Escherichia coli, 334
Halobacterium cutirubrum, 334
psychrophiles and mesophiles, effect of temperature, 81–83
Aminopeptidases
Bacillus stearothermophilus, 229, 239
Escherichia coli, 239
thermostability, 239, 240
Ammonifying bacteria
pH effect on growth, 285
Amoeba
in hot springs, 188
pressure effects, 134
radiation sensitivity, 413–416
Amylase (usually α-amylase)
Bacillus caldolyticus, 229
Bacillus stearothermophilus, 228, 235, 261
hydrogen binding, 232
Micrococcus halobius, 349
sulphydryl content, 234
thermophilic bacilli, 229, 240, 261
Amylase, alkaline
alkaline-requiring bacilli, 294
Anabena
toxic levels of lead, 384
Anion pumps
alkalinophiles and acidophiles, 306
Antarctic (*see also* Chapter 2)
dry valleys, 2, 3
Antibiotics
effects on extreme halophiles, 335, 343, 355
produced by fungi, effects of water stress, 370, 371
production by thermophilic fungi, 181, 182
Aphanocapsa
thermophilic, 161
Aphanocapsa thermalis
temperature effect on growth, 161
Aphanothece halophytica,
salt response and physiology, 323, 357
Arrhenius equation
in kinetics of microbial growth, 73–77
Arrhenius plots
of transport by *Escherichia coli*, 96
Artemia salina
in saturated salt lakes, 1
Arthrobacter
dessication susceptibility in dried soil, 373, 374
from Antarctic air, 14
from caves, 17–18
temperature effect on growth, 11
Arthrobacter crystallopoietes
killing by pressure and temperature, 136
Arthrospira plantensis
pH effect on growth, 286
Aspartate transcarbamylase
extreme halophiles, 331, 332, 339
temperature effect on regulation, 94
Aspartokinase,
thermophiles, allosterism, 238

Aspergillus
pH effect on growth, 284
temperature effect on growth, 170
thermophilic, 170, 174, 179, 183
Aspergillus amstelodamii
a_w effect on growth, 321, 322
Aspergillus flavus
a_w effect on growth, 322
Aspergillus nidulans
radiation sensitive mutants, 421
Aspergillus niger
a_w effect on growth, 322
toxic levels of cobalt, lead, mercury, nickel, and zinc, 384
Atmosphere
microbial ecology, 13–16
Azotobacter
desiccation susceptibilities in dried soil, 373, 374
toxic levels of antimony, arsenic, cadmium, cobalt, copper, mercury, nickel, silver, uranium, and vanadium, 384
Azotobacter chroococcum
from polar soils, 40
Azotobacter indicus
from polar soils, 40
Azotobacter vinelandii
temperature effect on polysomes, 98

B

Bacilli (*see also* individual species)
action on natural metal-containing minerals, 29, 388
in deep sea sediments, 144–145
in Don Juan Pond, 27
in upper atmosphere, 15
membrane fluidity and rigidity, 221
pH effect on growth, 282, 283, 285, 286
pressure and oxygen effects, 139, 140
temperature effect on fatty acid composition, 97, 220, 221
temperature effect on growth, 32, 162, 165
Bacillus 21–1
moderately halophilic, 348
Bacillus acidocaldarius
intracellular pH, 291, 292
lipids, 297, 298
sulphonolipids, 298
temperature and pH effects on growth, 283, 295, 297
Bacillus alkalophilous
pH effect on growth, 285
Bacillus caldolyticus
amylase, 229
L-histidinol dehydrogenase, 242
lactate dehydrogenase, 237, 242
transformation studies, 242
Bacillus caldotenax
enzyme thermoadaptation, 241
temperature effect on regulation, 94
temperature effect on growth, 241
Bacillus cereus
negative pressure effect, 145
pH effect on growth, 285, 286
stability of glyceraldehyde-3-phosphate dehydrogenase, 236
thermostability of enzymes, 223
Bacillus circulans
pH effect on growth, 285
Bacillus coagulans
cell walls, 236
growth rate, 226
intracellular inorganic ion concentration, 236
maximum growth temperature, 219
protein and enzyme heat resistance, 237, 238
stability of glyceraldehyde-3-phosphate dehydrogenase, 231, 264
thermoadaptation, 242
temperature and pH effect on growth, 283
Bacillus insolitus
protein synthesis, 78, 80, 81
temperature-induced structural changes, 84
Bacillus licheniformis
pressure effect on penicillinase synthesis, 133
rod mutants, 325
temperature adaptation, 240
Bacillus megatherium
temperature effect on growth, 11
Bacillus mesentericus
radiation sensitivity, 415
Bacillus pasteurii
intracellular pH, 293

Bacillus pasteurii (*contd.*)
pH effect on growth, 285
urease, 293
Bacillus psychrophilus
protein synthesis, 78, 80, 81
temperature-induced structural changes, 84–87
Bacillus spp (*see also* Bacilli)
thermophilic, 162
Bacillus stearothermophilus, 218–267
accumulation of toxic metabolic products, 219
alkaline phosphatase, 221
aminopeptidases, 229, 239
amylase, 228, 235, 261
cardinal temperatures, 219
culture media requirements, 241, 242, 261
cysteine and thermostability of proteins, 234
deoxythymidine kinase, 229
DNA dependent RNA polymerase, 219, 229
enolase, 228
esterase, 229
enzymes, thermostability, 223, 241, 244, 261
feedback inhibition, 219
ferredoxin, 252, 254
fructose 1,6 diphosphate aldolase, 228
glucokinase, 228, 241
glucose-6-phosphate isomerase, 228
glutamine synthetase, 229–231, 233
glyceraldehyde-3-phosphate dehydrogenase, 228, 244–251, 266
growth rate, 226, 241
inorganic pyrophosphatase, 229
lipids, 220–222
low temperature effect on polysomes, 98
malate dehydrogenase, 229
membrane lipids, 222
membrane transport, 219
mutants, 251
NADP- isocitrate dehydrogenase, 228
osmotic rupture, 221
phosphatase, alkaline, 221
phosphofructokinase, 228
6-phosphogluconate dehydrogenase, 228
phosphoglycerate kinase, 228
pressure effect, 128
protein and nucleic acid synthesis and turnover, 225
protein thermostability, 225, 244, 261
protoplasts, 221
purification of glyceraldehyde-3-phosphate dehydrogenase, 224
repression of enzyme synthesis, 219
ribosomal proteins, 335
thermal death and membrane fluidity, 222
thermoadaptation, 240, 241
translational errors in, 219
Bacillus subtilis
cysteine, calcium and thermostability of proteins, 234
L-histidinol dehydrogenase, 242, 243
low temperature survival, 15
malate dehydrogenase, 223
radiation-resistant strains, 416
radiation sensitivity, 415
rod mutants, 325
temperature effect on growth, 11
transformation studies, 242
Bacillus thermoproteolyticus
thermolysin, 229, 258
Bacteria (*see also* individual species, microorganisms and microbial life)
baroduric (barotolerant), or pressure tolerant, 106, 109, 122, 123, 126–138
barophilic (*see also* Barophiles) 106, 107, 123
chemolithotrophic, growing at high temperatures, 166, 167
extremely halophilic (*see* Extreme halophiles)
in marine and deep sea animals, 4, 47, 48, 106
in soils (*see also* Soils)
microbial ecology of, 34
moderately halophilic (*see* Moderate halophiles)
nonphotosynthetic, upper temperature limits for growth, 165
psychrophilic, survival under pressure, 144
radiation sensitivity, 412, 416

sulphate reducing, from deep wells, 108
sulphate reducing, in Antarctic pools, 26
thermophilic (*see also* Thermophiles)
water relations, 373
Bacteria, marine
action on natural metal containing minerals, 388
as moderate halophiles, 324
in deep sea sediments, 45–47, 106, 122, 138, 144
psychrophilic, 42–49
pressure effects, 144
reactions with heavy metals, 393
sodium requirement for active transport, 342
sulphate reducing, effects of pressure, 139
Bacteria, methane
in permanently ice covered lakes, 26
methylation of mercury and methyl mercury, 396
Bacteriophage (*see also* Viruses)
lysozyme, cold sensitivity, 95
of extreme halophiles, 340
pressure killing, 124, 135, 136
Bacteriophage T4D
temperature sensitive mutants, 92
Bacteriorhodopsin
in purple membrane of extreme halophiles, 346, 347
Bacterium communis
toxic levels of cadmium, iron, manganese and mercury, 384
Bactoderma
pH effect on growth, 280
Bagasse
thermophilic microorganisms, 182
Barobiology, 105–147
Bifidobacterium thermophilum
temperature effect on growth, 163
Biodegradation
by fungi, water stress effects, 370, 371
by microorganisms in the guts of deep sea amphipods, 4, 106
in soils, temperature effects, 34, 39–40
of cellulose, in soils, 40
of leaves in rivers and streams, temperature effects, 30, 31
Biodeterioration
by thermophilic fungi, 172, 179, 184, 185
Birds' nests
thermophilic microorganisms, 182
Blue-green algae
growth at high temperatures, 161
water stress effects, 375, 376
Bodo marina
radiation sensitivity, 413–416
Brevibacterium
Antarctic air, 14
5-Bromouracil
radiation sensitizing effect, 420
Brucellaceae
pH effect on lipid composition, 296
Burgoa-Papulaspora sp
temperature effect on growth, 171

C

Cacao beans
thermophilic microorganisms, 182
Calcarisporium thermophile
temperature effect on growth, 170
Calothrix sp
temperature effect on growth, 161
thermophilic, 161
Candida
in Antarctic lakes, 26
Candida gelida
denaturation profiles of ribosomes, 79
protein synthesis, 80, 81
pyruvate decarboxylase, 88, 89
temperature induced changes in sedimentation profiles of ribosomes, 80
Candida nivalis
temperature-induced structural changes, 85, 86
Candida pseudotropicalis
a_w effect on growth, 322
Candida utilis,
denaturation profiles of ribosomes, 79
pyruvate decarboxylase, 88, 89
temperature induced changes in sedimentation profiles of ribosomes, 80
Carbamyl phosphate synthetase, *Salmonella typhimurium*, temperature effect on regulation, 94

Carbohydrate-derived membrane lipids, in thermophiles and thermoacidophiles, 304
Carboxypeptidase,
 comparison to thermolysin, 259
Carotinoid pigments
 in extreme halophiles, 320, 344, 351
Catalase
 in radiation resistance, 417, 418
Catalase, alkaline
 of alkaline-requiring bacilli, 294
Caulobacter
 pH effect on growth, 280
Caves
 microbial ecology, 16–18
Cell division,
 pressure effects, 130–134
 in radiation-resistant *Escherichia coli*, 418–419
Cell envelopes (*see also* Cell walls)
 external layer in *Spirillium* and *Acinetobacter*, 341
 of extreme halophiles, 340–347
  effects of ions, 341, 342
  external layers, 341, 343
  shape-determining layers, 341, 343, 344
  transport by vesicles, 342, 347
 of moderate halophiles, binding of ions, 328
Cell membranes and surfaces of acidophiles and alkalinophiles, 294–305
Cell structure
 of psychrophiles and mesophiles, effect of temperature, 83–87
Cell walls (*see also* Cell envelopes)
 extreme halophiles, 340, 341
 in radiation resistance, 418
 moderate halophiles, 350, 351
 Sulfolobus, 282, 299, 300
 Thiobacilli, 295
Cellulose wastes
 thermophilic bacteria, 175
Cephalosporium
 low pH and high Cu^{2+} effect on growth, 284
 temperature effect on growth, 170
Cephalosporium gramineum
 production of antibiotics, 371
Chaetomium sp
 temperature effect on growth, 169
 thermophilic, 169
Chlamydomonas
 intracellular pH, 293
 toxic levels of lead, 384
Chlamydomonas acidophila
 intracellular pH, 293
Chlorella
 pH effect on growth, 286
 temperature effect on growth, 186
Chloroflexus aurantiacus
 temperature effect on growth, 161
Chromatium sp
 temperature effect on growth, 161
Chromobacterium
 toxic levels of manganese, 384
Chrysosporium sp
 temperature effect on growth, 170
Chydorus
 pH effect on growth, 286
Chymotrypsin
 pressure denaturation, 114
Citrate synthase
 bacteria
 Halobacterium cutirubrum, 332
Clathrates
 hydrophobic interactions, 112
Clostridia
 mesophilic and psychrophilic, triosephosphate isomerase, 89
Clostridium sp
 protein synthesis, 81
 temperature effect on growth, 32, 162, 163, 165
 temperature-induced changes in sedimentation profiles of ribosomes, 80
 thermophilic, 162, 163
Clostridium 69
 ribosome thermostability, 225
Clostridium acidi-urici
 ferredoxin, 252, 254, 255
 formyltetrahydrofolate synthetase, 262
Clostridium cylindrosporum
 formyltetrahydrofolate synthetase, 262, 263
Clostridium formicoaceticum
 acetate synthesis, 224

formyltetrahydrofolate synthetase, 262, 263
Clostridium pasteurianum
cysteine and thermostability of proteins, 234
enzyme stability, 224
ferredoxin, 252–255
phosphofructokinase, 227
ribosome thermostability, 225
Clostridium tartarivorum
enzyme stability, 224
ribosome thermostability, 225
ferredoxin, 252–254, 257
Clostridium thermoaceticum
acetate synthesis, 224
cysteine and thermostability of proteins, 234
formyltetrahydrofolate synthetase, 234, 262, 263
Clostridium thermosaccharolyticum
enzyme stability, 224
ferredoxin, 252–254, 257
ribosomes, thermostability, 225
triose-phosphate isomerase, 228
Coal waste piles, 159
thermophilic microorganisms, 183
Cocchochloris elabens
salt response, 323
Cold environments
microbial ecology, 1–4, Chapter 2
Cold sensitive mutants
Escherichia coli, 4, 90–95
Cold sensitivity
enzymes of extreme halophiles, 339
hydrophobic interactions, 339
Compatible solutes, 329, 330, 351, 352
definition, 329, 330
yeasts and algae, 329, 330
Competition
thermophilic fungi, 181
Composts and compost piles
thermophilic microorganisms, 159, 173–177
Compressed gases
biological effects, 140–142
Cooling waters from industrial processes
thermophilic microorganisms, 160
Copper ore leaching piles
thermophilic microorganisms, 186
Coprinus sp
temperature effect on growth, 170
Coprinus lagopus
radiation sensitive mutants, 422
Corrosion
aluminum and steel,
pressure effects, 139
Corynebacterium
in Don Juan Pond, 27
toxic levels of manganese, 384
Corynebacterium sepedonicum
Antarctic air, 14
Cosmarium
toxic levels of lead, 384
Crenothrix
pH effect on growth, 282
Cryptococcus
in Antarctic lakes, 26
mercury transformation, 397
Ctenocladus circinnatus
water stress effects, 375
Cyanidium caldarium
temperature effects on growth, 186, 187
water stress effect, 374
intracellular pH, 293
Cyclopropane fatty acids
Bacillus acidocaldarius, 297
Thiobacillus thiooxidans and other bacteria, 296
Cytochrome oxidase
halobacteria, 339
Thiobacillus, 292
Cytophaga
from lake sediments, 20

D

Dactylaria gallopava
temperature effect on growth, 187
Dark repair
excision repair, 424
irradiated DNA, 422–426
post replication recombination repair, 424–425
radiation-sensitive mutants, 424
Dead Sea (*see* Salt lakes)
Dematiacae
radiation resistance, 415
Denaturation
thermal, 217–267

Denaturation of biopolymers
by pressure, 111–115, 124, 145
volume changes, 111–115
Denaturation of DNA
radiation induced, 420
Denaturation of ribosomes
of psychrophiles, mesophiles and thermophiles, by temperature, 79
Deoxyribonucleic acid (*see* DNA)
Deoxythymidine kinase
Bacillus stearothermophilis, 229
Desaturating enzyme
Bacillus megaterium, effect of growth temperature, 97
Deserts
algae, 375, 376
Desulfotomaculum nigrificans
temperature effect on growth, 163, 165
Desulfovibrio
Antarctic kettle holes, 26
Desulfovibrio thermophilus
temperature effect on growth, 164
Diatoms
temperature effect on growth, 186
Dilatometers, 110, 111
DNA
content in radiation resistance, 418–421
damage by ionizing radiation, 421
extreme halophiles, 340, 356
polymerase from *Micrococcus luteus*, 425
pressure effects, 117
repair by photoreactivation in halobacteria, 320, 340
repair mechanisms, 421–431
DNA-dependent RNA polymerase
Bacillus stearothermophilus, 219, 229
halobacteria, 331, 336, 339
Thermus aquaticus, 229
Thermus thermophilus, 229
DNA/DNA hybridization
between extreme halophiles, 340
DNA/RNA hybridization
between extreme halophiles, 340
Don Juan Pond (Lake) (*see* Salt lakes)
Dunaliella
glucose-6-phosphate dehydrogenase, effects of glycerol and NaCl, 329
glycerol, 329, 377
mutants with altered salt range for growth, 325
salt response, 329
water stress effects, 375–377
Dunaliella tertiolecta (*see Dunaliella*)
Dunaliella viridis (*see Dunaliella*)
Dung
thermophilic microorganisms, 183

E

Ectothiorhodospira halophila,
salt response and physiology, 323–324, 357
Electron spin resonance (*see* Membrane fluidity and rigidity)
Electrostatic interactions
envelopes of extreme halophiles, 302, 342, 343, 345
proteins of extreme halophiles, 337, 338
pressure effect, 113, 127, 128
temperature effect, 128
Thermoplasma, 302
Emeriella nidulans
temperature effect on growth, 169
Enolase
Bacillus stearothermophilus, 228
hydrogen bonding, 232
secondary structure, 233
stability in *Thermus aquaticus*, 232
thermostability, 261
thermostability in *Thermus* X-1, 261
Thermus X-1, 228
Thermus aquaticus, 228
Enteric bacteria
pH effect on growth, 286
Enterococcus
pH effect on growth, 286
Entner-Doudoroff pathway
halobacteria, 331
Environmental limits
definition, 165
Enzymes (*see also* individual enzymes)
extreme halophiles, 331–334
psychrophiles, temperature inactivation, 87–90
thermophiles, 217–267, especially, 223–235

thermophilic fungi, 176
Enzyme kinetics
pressure effects on, 119–121
temperature effects on, 250–251
Escherichia coli
aminopeptidases, 239
a_w effect on growth, 322
base damage by radiation, 421
cold sensitive for histidine biosynthesis, 93
cold-sensitive mutants, 95
phosphoribosy-ATP-pyrophosphorylase, 93
deficient in radiation repair mechanisms, 411
DNA polymerase, 425
β- galactosidase, thermostability, 265
glyceraldehyde-3-phosphate dehydrogenase, 247
growth temperature effect on fatty acid composition, 95–97
intracellular inorganic ion concentrations, 237
killing by temperature and pressure, 136
lipid composition of membranes, effect on active transport, 96, 97
low temperature survival, 15, 98
membrane properties and regulation, 223
nucleic acid synthesis, 225
oxygen effects, 140, 146
pH effects, 288
phosphofructokinase, 227
pressure and oxygen effects, 140
properties of proteins, 337
pressure, temperature and pH effects on growth, 107, 130, 131
protein synthesis, 225
radiation resistant strains, 416, 418
radiation sensitivity, 413, 416, 421, 422
ribosomal proteins, 335
ribosomes,
effect of NaCl, 354
effect of pressure, 117
temperature effects on growth, 74, 76, 77
toxic levels of cobalt, mercury, nickel, and silver, 384
tryptophane synthetases
pressure effects, 118
ultraviolet-specific endonuclease, 425
use of RNA's by amino acyl tRNA synthetase of *Halobacterium cutirubrum*, 334
Esterase
Bacillus stearothermophilus, 229
Estuaries
microorganisms, 356
Eubacteriales
pH effect on lipid composition, 296
Euglena gracilis
pH effect on growth, 286
Eukaryotes
growth at high temperatures, 168–193
Eutrepia
pH effect on growth, 284
Evolutionary adaptations
to extreme environments, 165, 166
Exobiology, 1, 3
Exonuclease
Micrococcus radiodurans, 418
Extreme halophiles (*see also* Halobacteria and Halococci, 6, Chapter 8, especially) 330–347
envelopes,
charge shielding by ions (*see also* Electrostatic interactions) 302, 342, 343, 345
glycoproteins, 343, 344
hydrophobic interactions, 343
enzyme activity, effects of ions, 331–334, 336–340
enzyme purification, 355
enzymes, regulatory properties, 332–334, 352
flagella, 355
isocitrate dehydrogenase, 331
lipids, 343–346
effects of ions, 345, 346
nutrition, 330, 355
proteins, hydrophobic interactions, 338–340
proteins, properties, 336–338
charge shielding by ions (*see also* Electrostatic interactions), 337, 338
protein synthesis, 327, 334–336, 352

Extreme halophiles (*contd.*)
  similarities with thermophilic acidophiles, 303, 304
  taxonomic state, 335, 340, 356

F

Farmer's lung disease
  associations with thermophiles, 179
Fatty acid synthetase
  Halobacterium cutirubrum, 331
Feedback inhibition (*see* Regulatory functions)
Ferredoxin
  mesophiles and thermophiles, 252–257
Fish
  pressure effects on enzymes, 120, 121
Flagella
  extreme halophiles, 355
  Thiobacillus, 295, 296
  Vibrio alginolyticus, 355
  volume changes in dissociation, 112, 118
Flavobacterium
  caves, 17
  lakes, 19, 20
  low temperature survival, 15
  pH effect on growth, 286
  springs and wells, 31
Flavobacterium thermophilum (*see Thermus thermophilus*)
Food
  chilled and frozen, microorganisms in, 12
  preservation
    by high solute concentrations and low water activity, 319, 373
    by pressure, 145
  radiation-resistant microorganisms, 410, 416, 426
  radiation sterlization, 410, 416
  salted, microorganisms in, 320, 348
  spoilage by halophilic bacteria, 319, 324
  sterilization by pressure, 137, 138
Formic hydrogenase
  formation in psychrophiles, 78, 79
Formic hydrogenlyase
  formation in psychrophiles, 78, 79
Formyltetrahydrofolate synthetase, 234, 262, 263
  clostridia, 228, 262, 263
  Micrococcus aerogenes, 262
  physicochemical properties, 262
  pigeon, 262
  thermostability, 262
Fresh water
  microbial ecology, 18–31
Fructose-1,6 diphosphatase
  fish, effects of pressure, 121
  temperature effect on allosterism, 238
Fructose 1,6 diphosphate aldolase
  Bacillus stearothermophilus, 228
Fumarase
  Bacillus stearothermophilus, 251
Fumaroles
  microorganisms in, 159
Fungal diseases of plants
  influence of water stress, 370, 371
Fungi (*see also* individual organisms)
  water stress effects, 370, 371
  in acid coal wastes and streams, 284
  growing at high temperatures, 168–186, 189–193
  pH effect on growth, 283, 284
  pigments in radiation resistance, 417
  thermophilic, 168–186, 189–193
Fusarium sp
  pH effect on growth, 284

G

β-Galactosidase
  Escherichia coli,
    effect of primary structure on thermostability, 265
    pressure effect on synthesis, 118
Gallionella ferruginea
  ferrous iron oxidation, 394
Garbage dumps
  thermophilic microorganisms in, 185
Gas vacuoles
  halobacteria, 107, 336, 337
  pressure effect, 107
Gases, anaesthetic
  biochemical effects, 141, 142
  effects on animals, 142
  narcotic effects on cells, 141, 142, 146

Geothermal power plants
  thermophilic microorganisms in, 185
Geothermal soils
  thermophilic microorganisms in, 182, 187
Gloeothece linaris
  pH effect on growth, 286
Glucokinase, 241
  Bacillus stearothermophilus, 228, 241
  thermostability, 228, 241
β-1,3-gluconase
  alkaline-requiring bacilli, 294
Glucose oxidation
  psychrophilic and mesophilic *Bacillus*, 88
Glucose-6-phosphate isomerase
  Bacillus stearothermophilus, 228
β-Glucosidase
  psychrophilic basidiomycete, 88
Glutamic oxaloacetic transaminase
  Bacillus cereus, 223
  Bacillus stearothermophilus, 223
Glutamine synthetase,
  Bacillus stearothermophilus, 229, 231, 233
Glyceraldehyde-3-phosphate dehydrogenase (GPDH)
  Bacillus coagulans,
    salt effect, 236, 237
    stability, 231, 264
  Bacillus stearothermophilus, 228, 244–251, 266
  lobster muscle, 248–250
  mesophiles and thermophiles, 223, 224, 228, 230, 231, 234, 236–238, 240, 242–251
  Thermus aquaticus, 228, 243–250
  Thermus thermophilus, 228, 224, 250, 251
  thermoadaptation, 241
Glycolytic enzymes
  thermophiles and mesophiles, 224
Glycoproteins
  extreme halophiles, 343, 344
Grain
  stored, self-heating thermophilic microorganisms, 178, 184
Gram negative aerobes
  growth at high temperatures, 164
Guanine and cytosine content
  extreme halophiles, 340
  radiation resistance, 419

H

Habitats
  thermophilic microorganisms, 172–186
Haemophilus
  radiation-resistant strains, 416
Halobacteria (*see also* Extreme halophiles)
  carbohydrate metabolism, 330, 331
  intracellular ions, 326, 328
  menadione reducatase, 339
  nutrition, 330, 355
  pressure effects on growth, 127
  protein synthesis, 127, 327, 334–336, 352
  ribosomal proteins, 335, 336, 356
  RNA-dependent DNA polymerase, 336, 339
  salt response, 323–357
  threonine deaminase, 333, 339
Halobacterium cutirubrum (*see also* Halobacteria)
  isocitric dehydrogenase, 332
  malic enzyme, 332, 333
Halobacterium halobium (see Halobacteria)
Halobacterium marismortuii (*see* Halobacteria)
Halobacterium saccharovorum (*see* Halobacteria)
Halococci (*see also* Extreme halophiles)
  cell walls, 341
  salt response, 323
Hansenula
  toxic levels of mercury, nickel and silver, 384
Hansenula polymorpha
  temperature effect on growth, 169
  thermotolerant, as a source of single-cell protein, 169, 184
Hay
  stored, self-heating thermophilic microorganisms, 177
  thermophilic microorganisms in, 184

Heavy metal
resistance
genetics and mechanisms, 386
toxicity, 6, Chapter 10
Heavy metals, arsenic and antimony, Chapter 10
effect on microbial growth, 381–398
effect of pressure, 144
environmental sources, 381–385
mechanisms of toxicity, 387
High salt and solute concentrations, Chapter 8
effects on internal solutes, 325–330
High temperature adaptation and resistance, 5, and Chapter 6
High temperature habitats, 159–193
Histidine biosynthesis
cold-sensitive mutants, 93
L-Histidinol dehydrogenase
Bacillus caldolyticus 242–243
Bacillus subtilis, 242–243
Homoserine dehydrogenase
thermophiles, allosterism, 238
Hot acid soils
chemistry and microorganisms, 282
Hot springs, 159
thermophilic microorganisms, 166, 167, 182, 186–188
Hot water taps
thermophilic microorganisms, 185
Human parasites
thermophilic fungi, 186
Humicola sp
temperature effect on growth, 170
thermophilic, 170, 179, 182
Hydrocarbons
degradation in deep seas, 138
Hydrogen bonds
effects of pressure, 113
Hydrogen ion (*see also* pH)
effect on heavy metal toxicity, 288
effect on cell surface charge, 288
effect on nutrient and energy requirements, 288
in external environment, 287–289
intracellular, 289–294
measurement, 290, 291
Hydrogen ion pumps
acidophiles, 306
alkalinophiles, 306
Hydrogenomonas thermophilus
temperature effect on growth, 164–165
Hydrolases
mesophiles and thermophiles, cysteine, calcium and thermostability, 234
Hydrophobic interactions
assembly processes, 339, 112, 116–118
cold-sensitive mutants, 94
envelopes of extreme halophiles, 343
membranes of *Thermoplasma acidophilum*, 303
pressure effects, 111–114, 127
proteins, 111–112
proteins of extreme halophiles, 338–340
temperature effects, 128
volume changes, 111–114

I

Incubator birds
thermophilic microorganisms in nests, 182
Industrial fermentations
thermophilic microorganisms in, 184
Inorganic pyrophosphatase
Bacillus stearothermophilus, 229
Thiobacillus, 292
Insects
in grain, allowing growth of thermophiles, 178, 179
pH effect on growth, 284
Insolation
providing heat for growth of thermophilic fungi, 181, 182
Ions (*see also* Chapter 8)
binding to cell structures, 327, 328, 345, 346
effect on hydrophobic interactions, 338–340
in cytoplasm, 327, 351–354
Iron oxidase
Thiobacillus, 292
Iron uptake
bacteria and fungi, 386
Isocitrate dehydrogenase
Bacillus stearothermophilus, 228
extreme halophiles, 331, 332
Saccharomyces cerevisiae, 330

Saccharomyces rouxii, 330
thermoadaptation, 241
thermophiles, 228, 240
Thermus flavus, 228

J

Jet aircraft fuels
thermophilic microorganisms in, 185

K

Kakabekia umbellata,
pH effect on growth, 286
Kinetics of microbial growth at low temperature, 73–77

L

Lac operon of *Escherichia coli*
pressure effects, 118
L-Lactate dehydrogenase
alkaline-requiring bacilli, 294
Bacillus caldolyticus, 237, 242
thermophiles, allosterism, 238
Lactic acid bacteria
growth at high temperatures, 43
Lactobacillus
a_w effect on growth, 322
temperature effect on growth, 163, 165
thermophilic, 163
Lagoons, 319, 356
Lake Assal (*see* Salt lakes)
Lakes
alkaline, 285
ice covered, microbial ecology, 21–27
Leptospira biflexa var. *thermophila*
temperature effect on growth, 164, 165
Leptothrix ochracea
ferrous iron oxidation, 394
Lichens
growth at high temperatures, 187
water stress effects, 371–373, 376
Lignin
thermophilic bacteria, 175
Lignite
pressure-sensitive and resistant bacteria, 138, 139
Lipids (*see also* Membranes)
acidophiles and thermoacidophiles, 295–305
Bacillus acidocaldarius, 297
composition affecting membrane function at low temperatures, 95–98
cyclization in acidophiles, 304–305
extreme halophiles, 115, 343–346, 351
moderate halophiles, 350, 351
phase changes at different temperatures and pressures, 115, 116
psychrophiles and mesophiles, 95–98
Thiobacillus, 295, 296
Lipids
ether containing,
extreme halophiles, 344, 351
Sulfolobus, 299, 300, 302
isoprenoid
extreme halophiles, 344, 346, 347, 351
Sulfolobus and *Thermoplasma*, 299–302
Lipopolysaccharide
Thiobacillus and *Salmonella*, 295
Lysis and leakage
temperature induced in psychrophiles, 85–87
Lysozyme
bacteriophage, cold sensitivity, 95
pressure effects, 121

M

Malate dehydrogenase
alkaline-requiring bacilli, 294
thermophilic and mesophilic bacteria, 223, 229
thermoplasma, 291
Vibrio marinus, 87
Malate synthetase
salt linkages and thermostability, 233
Malbranchea pulchella var. *sulfurea*
temperature effect on growth, 170
Malic enzyme
Halobacterium cutirubrum, 332, 333
Manganese nodules (ferromanganese nodules)
effects of pressure on bacteria, 139
Manure
thermophilic microorganisms, 183

Marasmius foetidus
pH effect on growth, 284
Marine pollution
pressure effects, 138–139
Marine pseudomonad B-16
internal ionic concentrations, 326
Mars
environment, 2
Mastigocladus laminosus
temperature effect on growth, 161
Membranes (*see also* Cell envelopes, Cell membranes, Cell surfaces and Lipids)
fluidity and rigidity
acidophiles and thermophiles, 5, 221–223, 298, 300, 301, 303, 305
extreme halophiles, 301, 303, 345
thermophilic bacilli, 221, 222
permeability at low temperatures, effect of lipid composition, 95–98
Membranes, purple (*see* Purple membrane)
Membrane function
temperature adaptation, 4, 220–223
Membrane lipids
thiobacilli, 295, 296
Menadione reductase
halobacteria, 339
Merulius lacrymans
temperature induced structural changes, 86
Metal ions, toxic (*see* Heavy metals, arsenic and antimony)
Metalloendopeptidases (*see* Thermolysin)
Metallogenium
in acid-mine drainage, 389
pH effect on growth, 282
Methane oxidizing bacteria
growth at high temperature, 164
Methane producing bacteria
growth at high temperatures, 163
Methanobacterium thermoautotrophicum
temperature effect on growth, 163, 165
Methylene tetrahydrofolate dehydrogenase
Clostridium formicoaceticum, 224
Methylmercury
formation by biological and abiological means, 396
Methylococcus capsulatus
temperature effect on growth, 164
Metmyoglobin
pressure denaturation, 114
Microbial interactions
with manganese and iron compounds, 392–395
with mercury, 395–397
Microbial life, in
acid hot springs, 1, Chapters 6 and 7
the cold, mechanisms, 4, Chapter 3
depth of the sea, 1, 4, Chapter 4
dry rock surfaces, 1, Chapter 9
high irradiation, Chapter 11
high temperatures, Chapters 5 and 6
hot springs, 1, Chapters 5 and 7
mine effluents, 1, Chapters 7 and 10
salt lakes, 1, Chapter 8
toxic heavy metals, 1, Chapter 10
under very low nutrient supply, 7
Microbispora sp
temperature effect on growth, 163, 165
thermophilic, 163
Micrococcus
Antarctic air, 14
Don Juan Pond, 27
pH effect on growth, 282
Micrococcus aerogenes
ferredoxins, 233, 254, 257
formyltetrahydrofolate synthetase, 262
Micrococcus cryophilus
mesophilic mutants, 74
protein synthesis, 78, 81–83
temperature effect on growth, 74
Micrococcus halobius
salt concentration effect on growth, 348
Micrococcus halodenitrificans (*see Paracoccus halodenitrificans*)
Micrococcus luteus, 425
ultraviolet-specific endonuclease, 425
Micrococcus radiodurans
base damage by radiation, 421
exonuclease, 418
pigments in radiation resistance, 412
photoreactivation of irradiated DNA, 422
radiation sensitivity, 413–416

Micrococcus varians
salt concentration effect on growth, 348
Microorganisms
from radioactive minerals, 415
growing at low temperature, 73–99
growing in extreme pH values (*see also* Acidophiles, Alkalinophiles), 279–306
growing under high pressure (*see also* Barobiology, Bacteria, baroduric and barophilic, Pressure), 106–147
historical aspects, 122–126
growing under water stress in nature, 369–377
in acid hot springs, 2, Chapter 4, 5 and 7
in acid mine waters, 3, 389–392
salt and solute tolerant (*see also* Bacteria, extremely and moderately halophilic), 320–325
Micropolyspora
temperature effect on growth, 165
Microsporium
pH effect on growth, 282
Microtubules
pressure effects, 116, 117, 125, 133, 134
volume changes in dissociation, 112
Mines and mine effluents
bacteria in, 281, 282, 389–392
Minimum growth temperatures
biochemical factors affecting, 73–99
Mitosis
pressure effect, 116, 117
Moderate halophiles (*see also Vibrio costicola*), 317–357, especially 347–351
enzyme activity, effects of ions, 349, 350
internal ionic concentrations, 326–329, 357
polyamines in protein-synthesizing systems, 354
ribosomal proteins, 336
salt concentration effects on growth, 321–323, 348, 356
Moraxella
in lakes, 19
Mortierella truficola
temperature effects on growth, 168
Mosses
effects of water stress, 376
Mucopeptide (*see also* Peptidoglycan, Cell walls)
lacking in mycoplasma, *Sulfolobus* and extreme halophiles, 341
Mucor sp
temperature effects on growth, 168
thermophilic, 168, 176, 179
Municipal composts
thermophilic bacteria, 175
Mushroom compost
thermophilic microorganisms, 173, 174, 176, 184
Mutations, radiation-induced
resistance, 409–431
Mutants
cold-sensitive, growth temperature, 90–95
Escherichia coli with altered lipid composition, 96, 97
frequency of cold- and heat-sensitive, 92, 93
obligately osmophilic *Saccharomyces rouxii*, 325
psychrophilic, 90, 91
radiation-sensitive, 410, 420–428
rod mutants of bacilli, effects of salt, 325
temperature-sensitive, effects of solutes, 324–325
with altered solute response, 325
Mycoplasma
growth at high temperatures, 164
pH effects on membranes, 222
Mycorrhizal association
water stress effect, 370, 371
Mycoses
caused by *Mucor pusillus*, 183
Mycrocyclus
pH effect on growth, 280
Mycrocystis aeruginosa
pH effect on growth, 286
Myokinase
thermostability, 226
Myriococcum albomyces
temperature effect on growth, 169

Myxobacteria
in lakes, 19

N

NADP oxidase
alkaline-requiring bacilli, 294
Bacillus stearothermophilus, 221
Naegleria fowleri
temperature effect on growth, 188
Navicula
toxic levels of lead, 384
Neurospora crassa
anaesthetic gasses, effects, 141
methylation of methyl-mercury, 396
radiation-sensitive mutants, 421
Nitrate-reducing bacteria
pH effect on growth, 285
Nitrobacter
pH effect on growth, 285
Nitrosomonas
pH effect on growth, 285
Non-photosynthetic bacteria
growth at high temperatures, 167, 168
upper temperature limits, 165
Nuclear production reactors (*see also* Radiation, man-made)
as source of thermophiles, 190

O

Oil palm kernels
thermophilic microorganisms in stacks, 182
Oscillatoria sp
temperature effects on growth, 161
thermophilic, 161
Osmophilic yeasts, 322, 325
obligate osmophiles, 325
Oxidation and fermentation
psychrophilic and mesophilic, 88
Oxygen
hyperbaric, toxicity, 140, 141, 146
production by thermophilic fungi, 177
Ozone
effect on radiation intensity, 411

P

Paecilomyces sp
temperature effect on growth, 170
thermophilic, 170
Papulaspora thermophila
temperature effect on growth, 171
Paracoccus halodenitrificans, 357
a_w effect on growth, 321, 322, 348
internal ionic concentrations, 326–328
Paramecium
radiation sensitivity, 413–416
Pasteurella tularensis
low temperature survival, 15
Peat
heavy metals, 381, 382
thermophilic microorganisms, 182
Pectinase
alkaline-requiring bacilli, 294
Penicillium
a_w effect on growth, 322
pH effect on growth, 284
resistance to heavy metals, 387
temperature effect on growth, 170
thermophilic, 170
Penicillium variable
pH effect on growth, 286
Peptostreptococcus elsdenii
ferredoxins, 254
pH (*see also* Hydrogen ion)
effect on bacterial growth at different pressures, 107, 114, 115
effect on growth of *Escherichia coli* at different temperatures and pressures, 107
effect on protein denaturation by pressure, 114
high, microbial life at, 285–287 (*see also* Alkalinophiles)
low, microbial life at, 280–284 (*see also* Acidophiles)
Phagocytosis
pressure effect, 116
Phanerochaete chrysosporium
temperature effect on growth, 170
thermotolerant, 170, 184
Phormidium sp
temperature effect on growth, 161
thermophilic, 161
Phosphatase
alkaline, *Bacillus stearothermophilus*, 221
Phosphofructokinase
Bacillus stearothermophilus, 228

Clostridium pasteurianum, 227
Escherichia coli, 227
thermophiles,
allosterism, 238
cysteine content, 234
Thermus aguaticus, 228
Thermus X-1, 228
6-Phosphogluconate dehydrogenase
Bacillus stearothermophilus, 228
thermophilic clostridia, 224
Phosphoglycerate kinase
Bacillus stearothermophilus, 228
hydrogen binding, 232
cysteine content, 234
Phosphoribosy-ATP pyrophosphorylase
cold sensitive mutants of *Escherichia coli*, 93
Photophosphorylation
purple membranes, 346, 347
Photobacterium fischeri
protein synthesis, 78
Photo repair, 422
irradiated DNA, 422–424
Photoreactivation
Actinomycetes, 422
extremely halophilic bacteria, 320, 340
Photosynthetic microorganisms, growing at
high temperatures, 161
low pH, 187
Phycomyces blakesleeanus
pH effect on growth, 284
Pigments
in radiation resistance, 417, 427
Planococcus halophilus
temperature effect on salt range for growth, 324
salt concentration effect on growth, 348
Plantomyces
pH effect on growth, 280
Plasmids
heavy metal resistance, 386
Plasticizers
thermophilic bacteria, 175
Plastics
in composting, 175
Plectonema nostocorum
pH effect on growth, 286
Pleurocapsa sp
temperature effect on growth, 161
thermophilic, 161
Polyamines
in protein synthesizing systems of moderate halophiles, 354
Polymer aggregation
volume changes and pressure effects, 116–118, 121, 125, 133, 134
Polyols
as "compatible solutes" in yeasts, 330
Polysome formation
effect of low temperature, 98, 99
Polytomella caeca
pH effect on growth, 284
Potassium
in extreme halophiles, 326, 327, 331
pressure effect on *Streptococcus faecalis*, 133
Power plants
thermophilic microorganisms in cooling pipes and effluents, 185
Pressure
deep wells, 107
industrial equipment, 108
lakes, 107, 130
negative,
effect on *Bacillus cereus*, 145
plants, 128
relations with temperature and volume, 110–121
seas, 105, 138, 139
sterilizing, 106, 108, 135–138
effects of salts, 136, 137
sterilizing of food, 137, 138
technology, 123, 124, 144
underground deposits, 108, 138
Pressure effects (*see also* Volume changes, Temperature effects)
activated complexes, 119, 120
active transport, 132–134
aluminium corrosion, 139
amoebae, 137
bacilli and spores, 137
bacteria in lignite, 138, 139
bacteria in manganese nodules, 139
biological equilibria, 110–121
biological reactions, 126
cell division, 130–134

Pressure effects (*contd.*)
cell membranes, 133, 146
denaturation of biopolymers, 111–115, 121
endospores of *Bacillus*, 146
enzymes, 119–121 (*see also* individual enzymes)
flagella, 118
heavy metals on bacterial growth, 144
hydrogen bond, 113
hydrophobic interactions, 111–114, 127
industrial enzymatic processes, 109
industrial processes, 125
lysogenic state, 135
marine pollution, 138, 139
microbial growth, 105–147, especially 106, 107, 116, 126–138
influence of environment, 127, 128, 132, 146
microtubules, 116, 117, 125, 133, 134
mutation reversion in *Salmonella*, 146
penicillinase synthesis, 133, 146
polymer aggregation, 116–118, 121, 125, 133, 134
potassium in *Streptococcus faecalis*, 133
rates of biochemical reactions, 119–122
reaction rates, 125, 126
regulatory functions, 120, 121, 134
repressor binding to operator, 118
ribosomes, 117
RNA synthesis, 117
steel corrosion, 139
synthesis of DNA, RNA, and protein, 117, 118, 131–133
Pressure-temperature interrelations in
chemical reactions, 125
microbial growth, 127, 128, Chapter 4
Prokaryotes
at high temperatures, 161–168
Protease
alkaline-requiring bacilli, 294
formation in a psychrophilic *Pseudomonas*, 79
Protein synthesis
Bacillus insolitus, 78, 80, 81
Bacillus psychrophilus, 78, 80, 81
Bacillus stearothermophilus, 225
bacteria in manganese nodules, effect of pressure, 139
Candida gelida, 80, 81
extreme halophiles, 127, 327, 334–336, 352
mesophiles, 81–83
Micrococcus cryophilus, 78, 81–83
moss under dry conditions, 376
Photobacterium fischeri, 78
pressure effects, 117, 118, 131, 132, 133, 144
psychrophiles *in vitro*, 79–83
psychrophilic and mesophilic bacteria, 81–83
radiation resistant *Escherichia coli*, 418
Thermus thermophilus, 225
Thiobacillus thiooxidans, 292, 293
Vibrio costicola, 351, 354
Protozoa
adaptation to high temperatures, 188, 189
growth at high temperatures, 187–189
pH effect on growth, 284
radiation sensitivity, 413–416
Pseudomonads
in springs and wells, 31
Pseudomonas
desiccation susceptibility in dried soil, 373, 374
growth at high temperatures, 164
in caves, 17
in lakes, 19
mercury transformation, 397
protein synthesis, 78, 79, 83
psychrophilic, protease, 79
radiation sensitivity, 415
temperature effect on growth, 32, 74, 76, 77
toxic levels of cadmium, cobalt, silver, mercury and nickel, 384
Pseudomonas 101
internal ionic concentrations, 326
salt concentration effect on growth, 348
Pseudomonas aeruginosa
a_w effect on growth, 322
psychrophilic mutants, 91
Pseudomonas Ba_1
salt concentration effect on growth, 348

Pseudomonas bathycetes
  pressure effects, 106, 132
  salt and pressure effects on ribosomes, 144
Pseudomonas C-1
  toxic levels of copper, 384
Pseudomonas fluorescens
  pressure effects, 132
  transduction of psychrophily to *Pseudomonas aeruginosa*, 91
Pseudomonas pyocyanea
  toxic levels of lead, 384
Pseudomonas syringae
  role in the production of ice nuclei particles, 14
Pseudonocardia thermophila
  temperature effect on growth, 163, 165
Psychrophiles (*see also* Obligate psychrophiles, Psychrophilic, Psychrotrophes, Chapters 2 and 3)
  definition, 10–13
  determinants of low maximum growth temperatures, 77–90
  protein synthesis, 81–83
  temperature-induced changes in sedimentation profiles of ribosomes, 80
Psychrophilic environments
  definitions, 10–13
Pschrotrophs
  definitions, 10–13
Pulpwood
  thermophilic microorganisms, 179
Purple membrane
  halobacteria, 336, 337, 343, 346, 347, 355, 356
Pyrimidine dimers
  DNA, 411, 419, 420, 425, 427
  photorepair, 422
  radiation induced, 411
  repair, 420, 421–426
Pyruvate decarboxylase
  Candida gelida, 88, 89
  Candida utilis, 88, 89
Pyruvate kinase
  mesophiles and thermophiles, 223, 224

R

Radiation
  damage caused, 419–421
  in environment, 410–412
  ionizing, 411–412
  non-ionizing, 411
  protective mechanisms, 416–419
  repair systems, 425, 430, 431 (*see also* Radiaton, Dark repair, Photoreactivation in prokayotes and eukaryotes)
  resistance, 416
  sensitivity, 6
  solar effects on microorganisms, 411
  variation between microbial species, 412
Regulatory functions
  low temperature effects, 493, 494
  pressure effects, 120, 121, 134
Relative humidity
  definition, 370
Rhizopus sp
  temperature effect on growth, 168
  thermophilic, 168
Rhizopus nigricans
  a_w effect on growth, 322
Rhodanase
  Thiobacillus, 292
Rhodotorula
  from copper mine wastes, 283
  radiation sensitivity, 415
  toxic levels of nickel, mercury and silver, 384
Rhizobium
  pH effect on growth, 285
Ribonuclease
  temperature effect, 266, 267
Ribonucleic acid (*see* RNA)
Ribonucleotide reductase
  thermophile allosterism, 238
Ribosomal function (*see* Protein synthesis)
Ribosomal proteins (*see also* Ribosomes)
  halobacteria, 335, 336, 356
  moderate halophiles, 336
  Vibrio costicola, 336, 350
Ribosomes (*see also* Ribosomal proteins)
  Candida gelida and *Candida utilis*, temperature effects, 79, 80
  changes in composition with growth temperature, 79

Ribosomes (*contd.*)
Clostridium sp, temperature effects, 80, 225
cold-sensitive mutations, 93, 94
Escherichia coli, 117, 335, 354
halobacteria, 334–336, 352
low temperature effects, 4, 98
pressure effects, 117
Pseudomonas bathycetes, 144
psychrophiles, mesophiles and thermophiles, temperature stability, 79, 80, 225
temperature-induced changes in mesophiles, 80
thermal stability, in psychrophilic, mesophilic and thermophilic *Clostridium* sp, 225
Vibrio costicola, 350
volume changes in dissociation, 112
Rivers and streams
microbial ecology, 27–31
RNA-dependent RNA polymerase
halobacteria, 336, 339
RNA synthesis
pressure effects, 117

S

Saccharomyces cerevisiae
arsenate transport, 386
a_w effect on growth, 322
effects of glycerol and NaCl on isocitrate dehydrogenase, 330
radiation-sensitive mutants, 422
radiation sensitivity, 413–416, 425
temperature effect on regulation, 94
Saccharomyces rouxii
a_w effect on growth, 322
effects of glycerol and NaCl on isocitrate dehydrogenase, 330
glycerol, 330
salt response, 322, 330
Saccharomyces sp
pH effect on growth, 283
Salmonella
a_w effect on growth, 321
radiation resistant strains, 416
Salmonella oranienburg
a_w effect on growth, 321
Salmonella typhimurium
temperature effect on regulation, 94
tryptophan synthetases, pressure effects, 118
Salt lakes, 319, 320, 356, 375, 376
composition, 319, 320
microbial ecology, 27, 320, 356
Salterns
microbial ecology, 319, 356
Sarcina
pH effect on growth, 282
Sarcina lutea
pigments in radiation resistance, 417
Saunas
thermophilic microorganisms, 185
Sawdust piles
thermophilic microorganisms, 185
Schizosaccharomyces pombe
radiation sensitivity, 413, 425
radiation sensitive mutants, 422
Sclerotinia borealis
temperature-induced structural changes, 84
Scolecobasidium sp
temperature effect on growth, 170
thermophilic, 170
Scytalidium
copper tolerance, 386
Seaweed
thermophilic microorganisms in, 182
Seaweed piles
as high temperature habitat, 159
Secondary metabolism
thermophilic fungi during insolation, 182
Sediments
mercury transformation, 397
thermophilic microorganisms in, 184
Self-heating
microbial, 172, 176, 179, 184
Serratia
toxic levels of silver, 384
Serratia marinorubra
killing by pressure and temperature, 136
Soils
acid, 374–375, Chapter 4
alkaline, 285, 287, Chapter 4
mercury transformation, 397
microbial ecology, 31-42, Chapter 4
temperatures, 31–36, 39–41, Chapter 4
thermophilic microorganisms, 180, 186, Chapter 4

water availabilities, 370, 373
water potential, 371
solar heating, 181
Solutes, intracellular, (*see* Ions, Compatible solutes)
intracellular, Chapter 8
non ionic effect on enzymes, 330
Sphaerospora saccata
temperature effect on growth, 169
Sphaerotilus discophorus
toxic levels of manganese, 384
Spirillum
temperature effect on growth, 77
manganese oxidation by, 393
Spirochetes
growth at high temperatures, 164
Spirulina species
temperature effect on growth, 161
thermophilic, 161
Spores
thermophilic fungi, 182
Spore-formers
growth at high temperatures, 162
Sporendonema sebii
a_w effect on growth, 322
Sporotrichum thermophile
temperature effect on growth, 170
Springs
thermophilic bacteria, 166
Staphylococcus aureus
anaesthetic gases effect, 141, 146
a_w effect on growth, 321
pH effect on lipid composition, 296
Staphylococcus epidermidis
salt response, 323
Steam line discharge sites
thermophilic microorganisms, 185
Steam lines
growth of non-photosynthetic bacteria, 160
Stilbella thermophila
temperature effect on growth, 170
Stibiobacter
antimony transformation, 385
Storage fungi
thermophilic fungi, 179
Streptococcus faecalis
anaesthetic gasses effect, 142
intracellular pH, 290
pressure effect on growth, 127–129, 132
pH effect on growth, 286
Streptococcus lactis
intracellular pH, 290
Streptococcus thermophilus
temperature effect on growth, 163
Streptomyces sp
temperature effect on growth, 163, 165
thermophilic, 163
water stress effect, 374
Streptosporangium album var *thermophilum*
temperature effect on growth, 163, 165
Successional change
blue green algae, 166
Successions
thermophilic fungi, 174, 177–179
Sugar cane
thermophilic microorganisms, 182
Succinate dehydrogenase
mesophiles and thermophiles, 240
Sulfolobus
acid mine-drainage, 389
action on natural metal-containing minerals, 388
cell wall lipids, 304
cell walls, 303, 341
glycerol diether lipids, 299, 300, 302, 303
growth at low pH and high temperature, 282, 283
iron oxidation, 391
Sulfolobus acidocaldarius
cell walls, 299
growth at low pH and high temperatures, 295
intracellular pH, 291
lipids, 299
pili, 299
temperature effect on growth, 164, 165
Sulphate-reducing bacteria
growth at high temperatures, 164
pH effect on growth, 285
Sulphonolipids
Bacillus acidocaldarius, 298, 304, 305
Sulphur-oxidizing
bacteria, growth at high temperatures, 163, 166, 167
enzyme system
of *Thiobacilli*, 292

Sun-heated
habitats, 188
litter, 159
mud, 180
rock, 159
soils, 159, 186
Superoxide dismutase
and oxygen toxicity, 140
effects of pressure, 140
Surfaces
microbial growth, 354, 355
Survival
microbial under pressure (*see* Chapter 4)
Symploca sp
temperature effect on growth, 161
thermophilic, 161
Synechococcus sp
temperature effect on growth, 161
thermophilic, 161
water stress effect, 375
Synechocystis sp
temperature effect on growth, 161
thermophilic, 161

T

Talaromyces sp
temperature effect on growth, 169
thermophilic, 169
Targets
of radiation damage, 419
Temperature
adaptation in regulatory properties of enzymes, 93, 94, 352
limits for growth, 160
shift, effect on growth of psychrophiles and mesophiles, 75, 76
Temperature effects on (*see also* Pressure effects, Volume changes)
cell division in psychrophiles, 84
lipid composition of membranes, 95–98, 220–223
microbial life, Chapters 2–6
regulatory functions, 93, 94, 231, 352
Tetrahymena
anaesthetic gases effect, 142
pressure effects, 131, 133, 143
toxic levels of cadmium, 384
Tetrathionase
Thiobacillus, 292
Thermal pollution
thermophilic microorganisms, 188
Thermoactinomycetes sp
temperature effect on growth, 163, 165
thermophilic, 163
Thermoadaptation
of enzymes in thermophiles, 238–243
Thermolysin, 229, 234, 258–262
Bacillus thermoproteolyticus, 229, 258
physicochemical properties, 258
sulphydryl content, 234
thermostability, 258, 261
Thermomicrobium roseum
temperature effect on growth, 164, 165
Thermomyces ibadanemsis
temperature effect on growth, 171
Thermonospora
temperature effect on growth, 165
Thermophiles
carbohydrate-derived membrane lipids, 304
enzymes, 5, Chapter 6
mechanisms and molecular aspects, 217–267
membrane fluidity and rigidity, 298
Thermophilic acidophiles
similarities with extreme halophiles, 303, 304
Thermophilic fungi 189–193 (*see also* Chapter 4)
air, 184
cellulolytic, 175
definition, 171
pathogens of humans, 183
Thermoplasma
cell wall lipids, 304
electrostatic interactions, 302
glycerol diether lipids, 302, 303
histone-like nucleoproteins, 303, 305
lipids, 302
malate dehydrogenase, 291
nutrition, 302
Thermoplasma acidophilum
cell wall, 301
from coal refuse piles, 283
growth at low pH and high temperature, 283, 295
intracellular pH, 291

membrane proteins, 301
pH effect on membranes, 222
temperature effect on growth, 164, 165
walls and membranes, 222, 299, 301
Thermotolerant
fungi, definition, 171
yeast grown on hydrocarbons, 184
Thermus
as extreme thermophiles, 219
temperature effect on growth, 164, 165, 167, 168
thermophilic, 164
Thermus aquaticus
cysteine and thermostability of proteins, 234
DNA-dependent RNA polymerase, 229
enolase stability, 232
glyceraldehyde-3-phosphate dehydrogenase, 228, 243, 250
growth temperature effect on the fatty acid composition, 221
hot springs, 218
lipids, 304
phosphofructokinase, 228
Thermus flavus
NADP-isocitrate dehydrogenase, 228
Thermus thermophilus
DNA-dependent RNA polymerase, 229
glyceraldehyde-3-phosphate dehydrogenase, 228, 244, 250, 251
lipids, 304
polypeptide synthesis, 225
Thermus X-1
cysteine and thermostability of proteins, 234
enolase, thermostability, 261
phosphofructokinase, 228
Thiobacilli
cell walls, 295
interaction with minerals, 281
membrane lipids, 295
pH optima of enzyme systems, 292
Thielavia sp
temperature effect on growth, 169
thermophilic, 169
Thiobacillus
iron oxidase, 292
rhodanase, 292
temperature effect on growth, 164, 165
tetrathionase, 292
thermophilic, 164
Thiobacillus ferrooxidans
acid mine-drainage, 389
acid production, 281
action on natural metal-containing minerals, 388
arsenopyrite oxidation, 386
cell wall structure, 295
intracellular pH, 292
oxidation of metals, 390–392
pH effect on growth, 281, 282
Thiobacillus organoparus
pH effect on growth, 282
Thiobacillus thiooxidans
acid mine-drainage, 389
acid production, 281
cell wall fine structure, 295
flagella, 295
intracellular pH, 292
oxidation of metals, 391, 392
pH effect on growth, 281, 282
temperature effect on growth, 163
Thiovobium
pH effect on growth, 280
Threonine deaminase
halobacteria, 333, 339
thermophiles,
allosterism, 238
thermostability, 231
Vibrio costicola, 349, 350
Thymine dimers (*see* Pyrimidine dimers)
Tobacco products
thermophilic microorganisms, 182
Torula
toxic levels of cobalt, mercury and silver, 384
Torula thermophila
temperature effect on growth, 171
Torulopsis candida
temperature effect on growth, 171
Trebouxia
water stress effect, 376
Triose-phosphate isomerase
from psychrophilic, mesophilic and thermophilic clostridia, 89
thermophilic, from *Clostridium thermosaccharolyticum*, 228

Tritirachium sp
temperature effects on growth, 171
Tryptophane synthetases
Escherichia coli and *Salmonella typhimurium*
pressure effects, 118
Typhula idahoensis
temperature effects on growth and metabolism, 88

U

Ultraviolet-specific endonuclease
Micrococcus luteus, 425
Upper temperature limits for
life, 166
photosynthesis, 165
Urease
Bacillus pasteurii, 293
Uridine kinase
thermophile allosterism, 238
Ustilago maydis
radiation sensitivity, 413

V

Vibrio alginolyticus
flagella, 355
salt concentration effect on growth, 348
Vibrio Ant. 303
survival, 145
Vibrio costicola (*see also* Moderate halophiles)
a_w effect on growth, 321, 322, 348, 349
enzymes, regulatory properties, 349, 350
internal ionic concentrations, 326–329, 349
nutrition, 348
temperature effect on salt range for growth, 324
protein synthesis, 351–354
ribosomal proteins, 336
ribosomes, 350
Vibrio marinus
malic dehydrogenase, 87
pressure effect on growth, 127, 128
temperature effect on growth, 75
temperature induced structural changes, 86, 87
Vibrio metchnikovii
a_w effect on growth, 322
temperature effect on growth, 75
Vibrio parahaemolyticus
temperature effect on growth, 12
Vibrio percolas
pressure effects, 139
Vibrio psychroerythrus
temperature-induced structural changes, 84, 85
Vibrio sp
in lakes, 19
Viruses (*see also* Bacteriophages)
inactivation by ionizing radiation, 421
in deep sea sediments, 136
killing by pressure, 135, 136
Volcanoes
microorganisms, 159–160
Volume changes in (*see also* Pressure effects, Temperature effects)
chemical and biochemical reactions, 111–121
hydrophobic interactions, 111–114

W

Water
effects of anions on structure, 339
effects of cold, 339
in cytoplasm, 318, 327
Water activity (a_w) (*see also* Water availability, Water potential)
definition, 320
effecting microbial growth, 6, 320–322, 354
intracellular as affected by external a_w, 325, 355
Water availability (*see* Water activity, Water potential, Relative humidity)
Water potential
definition, 370
matric,
definition, 370
effects, 369–377
osmotic
definition, 370
effects, 369–377
Water relations
lichens, 371–373

Water stress, 61, 354, 369–377
  effect on the nitrogen cycle, 374
  in Antarctic soils, 36
  mechanisms of damage, 376, 377
Water structure
  and hydrophobic interactions, 339, 355
Wood chip piles
  thermophilic microorganisms in, 179

X

Xeromyces bisporus
  a_w effect on growth, 322

Y

Yams
  thermophilic microorganisms in, 182
Yeasts (*see also* individual organisms)
  from copper mine wastes, 283
  internal solutes, 329, 330, 349
  intracellular pH, 290
  osmophilic, 322, 325
  psychrophilic and mesophilic, 75
  temperature effect on growth, 75
  toxic levels of arsenic, 384
Yersinia pestis
  temperature effect on growth, 11, 32